Barron's Review Course Series

Let's Review:

Geometry

Lawrence S. Leff, M.S.
Former Assistant Principal, Mathematics Supervision
Franklin D. Roosevelt High School
Brooklyn, New York

Dedication

To Rhona . . .
For the understanding,
for the sacrifices,
for the love,
. . . and with love.

All inquiries should be addressed to:
Barron's Educational Series, Inc.
250 Wireless Boulevard
Hauppauge, New York 11788
www.barronseduc.com

Library of Congress Catalog Card No. 2008010704

ISBN-13: 978-0-7641-4069-3
ISBN-10: 0-7641-4069-8

Library of Congress Cataloging-in-Publication Data

Leff, Lawrence S.
Let's review. Geometry / Lawrence S. Leff.
p. cm.—(Barron's review course series)
Includes index.
ISBN-13: 978-0-7641-4069-3
ISBN-10: 0-7641-4069-8
1. Geometry—Problems, exercises, etc. 2. Geometry—Outlines, syllabi, etc. 3. Geometry—Examinations—Study guides. 4. High schools—New York (State)—Examinations—Study guides. I. Title.

QA459.L39 2008
516.0076—dc22

2008010704

PRINTED IN CANADA
9 8 7 6 5 4 3 2 1

TABLE OF CONTENTS

PREFACE

This book organizes all of the various topics, concepts, and skills required by the *new* NYS Geometry Regents Examination in a way that is easy for students to understand and convenient for teachers to use when planning their daily lessons.

What Special Features Does The Book Have?

- *Reflects New Core Curriculum*
  This book is tailored to meet the needs of students enrolled in the new Regents geometry course. Topics new to the geometry curriculum such as midpoint and concurrency theorems, similarity and proportionality theorems, logical connectives, and aspects of solid geometry are included. Required geometric constructions are collected and placed in the back of the book for easy reference. Although emphasis is placed on reasoning and proof, algebra-based exercises are included to help reinforce geometric concepts as well as to maintain algebraic skills.
- *A Compact Format Designed for Self-Study and Rapid Learning*
  The clear writing style quickly identifies essential ideas while avoiding unnecessary details. Helpful diagrams, convenient "Math Fact" summaries, and numerous step-by-step demonstration examples will be appreciated by students who need an easy-to-read book that provides complete and systematic preparation for both classroom and the Regents geometry tests.
- *Develops and Then Builds on a Strong Foundation*
  Success in studying geometry depends on having a strong foundation. In addition to building a working knowledge of basic geometric terms and concepts, the first chapter features a thorough yet easy to understand introduction to the nature and format of a geometric proof. Different types of mathematical reasoning as well as the logical relationships between conditional statements are also discussed.
- *Practice Exercises with Answers*
  Modern geometry textbooks typically include few, if any, multiple-choice questions and contain a limited number of proof exercises. Exercise sections in *Let's Review: Geometry* feature a large sampling of "Regents-type" multiple-choice questions as well as a generous selection of original proof exercises that will help students prepare for both classroom tests and the Geometry Regents Examination. Answers to the practice exercises are intended to give students valuable feedback that will lead to greater understanding and higher test grades.

- *In-Depth Coverage of Coordinate and Transformation Geometry*
  Coordinate formulas and general coordinate proofs are integrated into the chapter that focuses on quadrilaterals so that both coordinate methods and the properties of special quadrilaterals can be further reinforced. Transformation concepts and techniques are carefully developed in a separate chapter.

Who Will Benefit from Using This Book?

- *Students Who Want To Raise Their Grades*
  Students enrolled in Regents-level mathematics classes that lead to the new Geometry Regents Examination will find this book helpful when they need either additional explanation and practice on a troublesome topic currently being studied in class or want to review specific topics before a classroom test or the Geometry Regents Examination.
- *Teachers Who Wish an Additional Resource When Lesson Planning*
  Because teachers will find this book to be a valuable lesson-planning aid as well as an indispensable source of classroom exercises, homework problems, and test questions, they will want to include *Let's Review: Geometry* in their personal, departmental, and school libraries.
- *School Districts and Mathematics Departments*
  School districts and mathematics departments who want to align their mathematics curricula with the set of prescribed topics and performance indicators of the New York State geometry course will find this book particularly helpful. The table of contents shows at a glance how the various topics, skills, and concepts required by the new geometry course can be organized into a manageable and cohesive set of lessons.

LAWRENCE S. LEFF

CHAPTER 1
THE LANGUAGE OF GEOMETRY

1.1 BUILDING A GEOMETRY VOCABULARY

KEY IDEAS

Cement and bricks are used to give a house a strong foundation. The building blocks of geometry take the form of *undefined terms*, *defined terms*, *postulates*, and *theorems*.

Undefined Terms

Some terms in geometry can be described but cannot be defined using simpler terms. *Point*, *line*, and *plane* are undefined terms.

Description of Term	Figure	Notation
A **point** indicates location. Although it has no size, a point is represented by a dot that you can see.	A, B, C	Named by a capital letter: points A, B, and C.
A **line** is a set of continuous points that extends endlessly in opposite directions. Although a line has no width, a drawing of it does.	m, B, A	Named by a lower case letter or by two points on the line written in either order: line m, $\overleftrightarrow{AB}$, or $\overleftrightarrow{BA}$.
A **plane** is a flat surface with no thickness. Unlike its picture representation, a plane extends indefinitely in all directions.	P	Named by a capital letter placed in a corner: plane P.

Defined Terms

Definitions are expressed using undefined terms, previously defined terms, and English words used in everyday conversation.

Defined Term	Diagram	Notation
A **line segment** is a part of a line consisting of two points, called **endpoints**, and the set of all points on the line between them.	A B	Named by its two endpoints written in either order: $\overline{AB}$ or $\overline{BA}$.
A **ray** is a part of a line consisting of an endpoint and the set of all points on one side of that endpoint.	B K	Named by its endpoint, written first, and another point on the ray: $\overrightarrow{KB}$. The arrow on top always points to the right.
Opposite rays are rays that have the same endpoint and lie on the same line.	W K X	$\overrightarrow{KX}$ and $\overrightarrow{KW}$.
Collinear points are points that all lie on the same line.	A B C D	Points *A*, *B*, and *C* are *collinear*. Points *B*, *C*, and *D* are *non*-collinear.
An **angle** is the union of two rays having the same endpoint, called the **vertex** of the angle. The two rays are the **sides** of the angle.	A Side Interior of angle B Side C Vertex	Named using three letters with the vertex letter in the middle: $\angle A\underline{B}C$ or $\angle C\underline{B}A$. The name of this angle can be shortened to $\angle B$.

Naming Angles with the Same Vertex

In Figure 1.1, three different angles share vertex D: $\angle ADB$, $\angle CDB$, and $\angle ADC$. Referring to any of these angles simply as $\angle D$ would create confusion. Instead, you can name a particular angle at vertex D by marking that angle with a number. Thus, $\angle 1$ refers to $\angle ADB$, and $\angle 2$ is another name for $\angle CDB$.

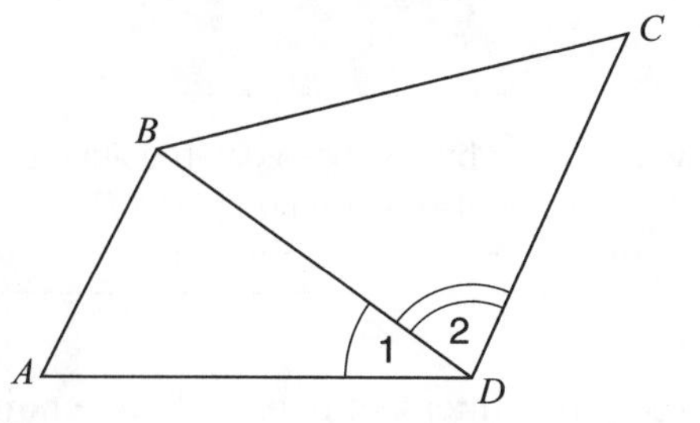

Figure 1.1 Naming an angle at vertex *D*.

Theorems and Postulates

Facts in geometry take the form of postulates and theorems.

- A **theorem** is a statement that can be proved using valid methods of reasoning. The familiar fact that "The sum of the measures of the three angles of a triangle is 180" is a theorem. Much of your study in geometry will involve investigating geometric relationships that lead to theorems, proving those theorems, and then using those theorems to help prove new theorems.
- Not everything can be proved. Some basic mathematical truths, called *postulates*, are needed as a beginning. A **postulate** or **axiom** is a statement that is assumed true without proof.

Beginning Postulates

In mathematics, *determine* means exactly one. Here are some beginning postulates.

Postulate 1
Two points *determine* a line.

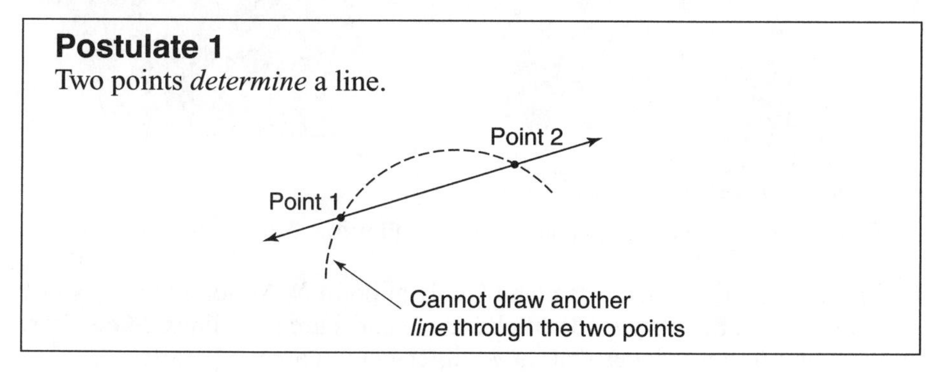

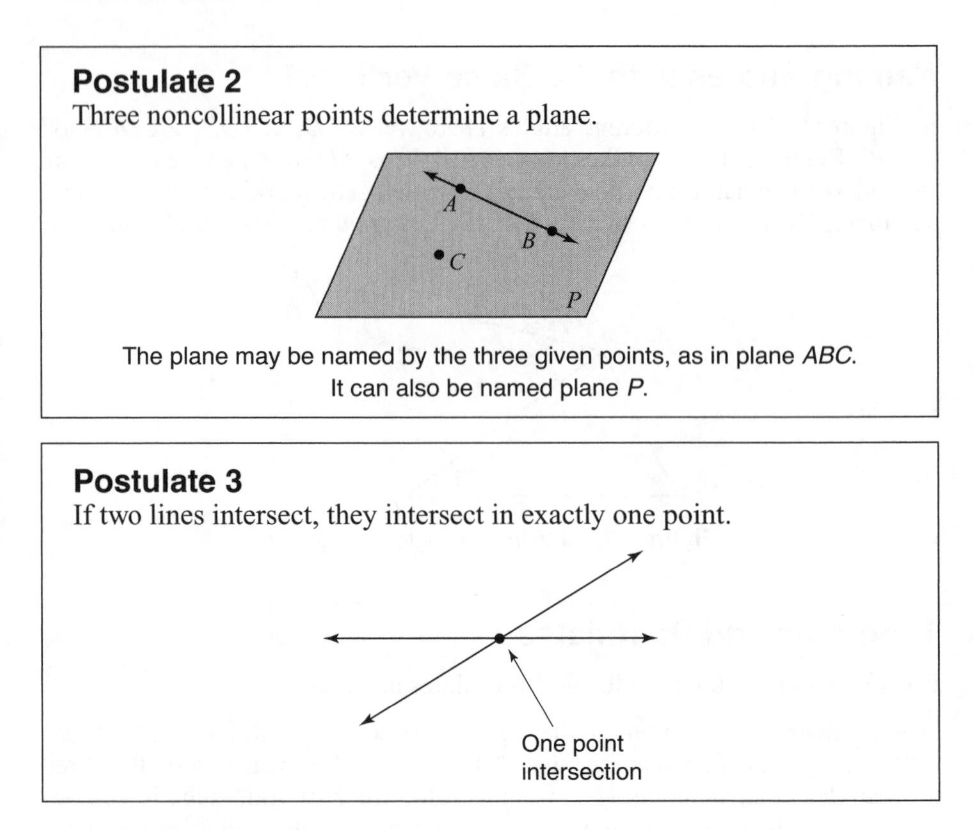

Postulate 2

Three noncollinear points determine a plane.

The plane may be named by the three given points, as in plane *ABC*. It can also be named plane *P*.

Postulate 3

If two lines intersect, they intersect in exactly one point.

Parallel and Skew Lines

A set of points that lie in the same plane are **coplanar**. Lines such as lines m, q, and r in Figure 1.2 lie in the same plane and, as a result, are **coplanar**. If lines m and q do not intersect, then line m *is parallel to* line q, which may be abbreviated as $m \parallel q$. **Parallel lines** are coplanar lines that do *not* intersect.

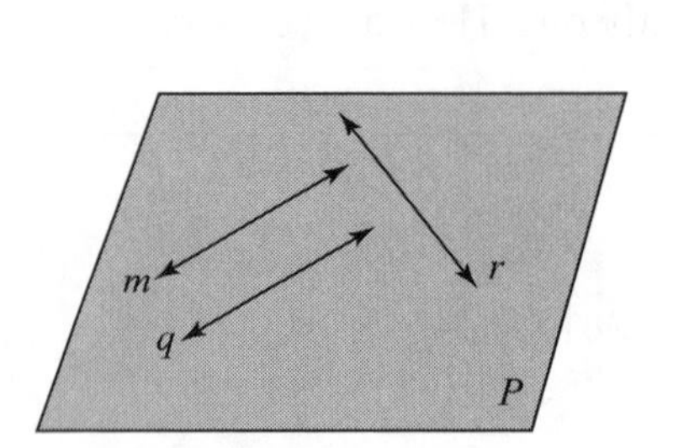

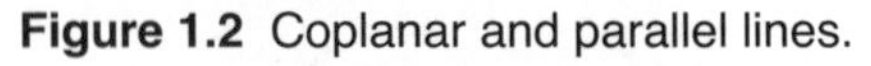

Figure 1.2 Coplanar and parallel lines.

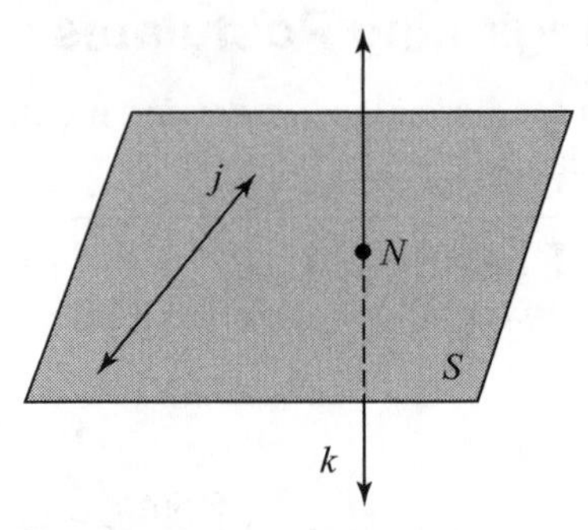

Figure 1.3 Noncoplanar lines.

In Figure 1.3, line k intersects plane S only at point N. Although lines j and k do not intersect, they are *not* parallel. Lines j and k are *skew* lines. **Skew lines** are lines in different planes that do *not* intersect but are not parallel.

Parallel and Intersecting Planes

Each pair of sides of the cube in Figure 1.4 are either in parallel planes or in planes that intersect. Planes *ABCD* and *FGHE* are parallel planes. Planes *ABCD* and *DEHC* intersect in $\overleftrightarrow{CD}$. Planes *BGHC* and *EFGH* intersect in $\overleftrightarrow{GH}$. This example suggests the following postulate.

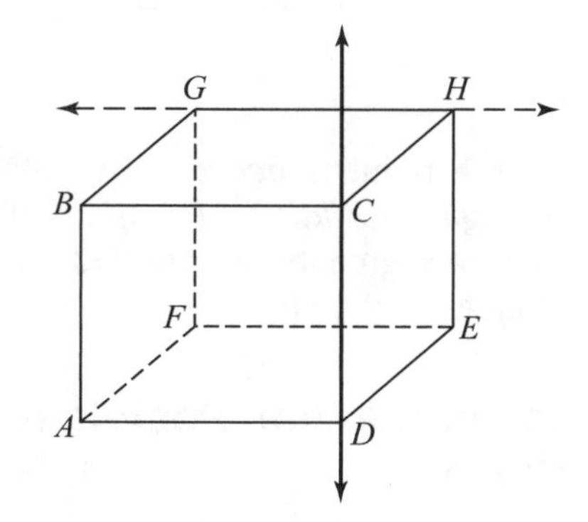

Figure 1.4 Parallel and intersecting planes.

> **Postulate 4**
> If two planes intersect, they intersect in a line.

Measuring a Line Segment: The Ruler Postulate

The *Ruler Postulate* relates the sets of points that form a line to the set of real numbers. The real number that corresponds to a particular point on a line is its **coordinate**. In Figure 1.5, the length or measure of $\overline{AB}$ is the absolute value of the difference in the coordinates of its endpoints. Thus, $AB = |5 - 3| = 2$, where *AB*, without an overbar, represents the length of $\overline{AB}$.

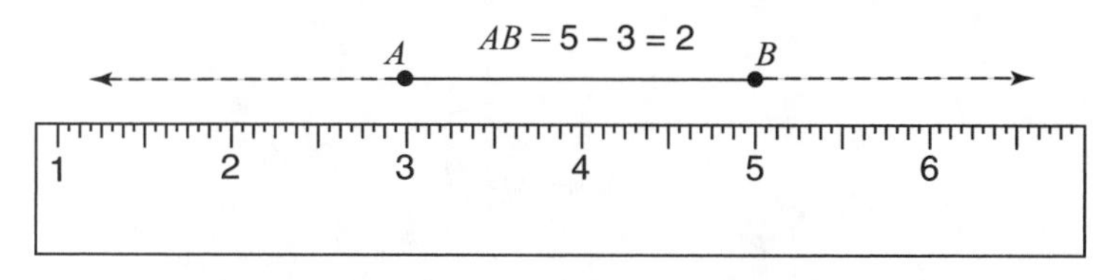

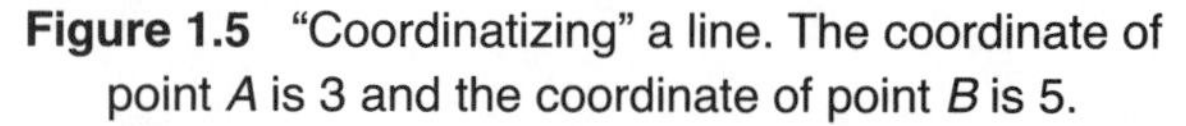

Figure 1.5 "Coordinatizing" a line. The coordinate of point *A* is 3 and the coordinate of point *B* is 5.

Postulate 5: Ruler Postulate

The set of points on a line can be paired in one-to-one fashion with the set of real numbers such that there is exactly one real number that represents the **length** or **measure** of a line segment. This number is the absolute value of the difference between the coordinates of the endpoints of the line segment.

Segment Addition

It is understood that when a point is *between* two other points, the three points are collinear. The *Segment Addition Postulate* allows you to add the measures of the two shorter segments determined by the three points to obtain the length of the longest segment.

Postulate 6: Segment Addition Postulate

If point X is *between* points A and B, then

$AX + XB = AB$

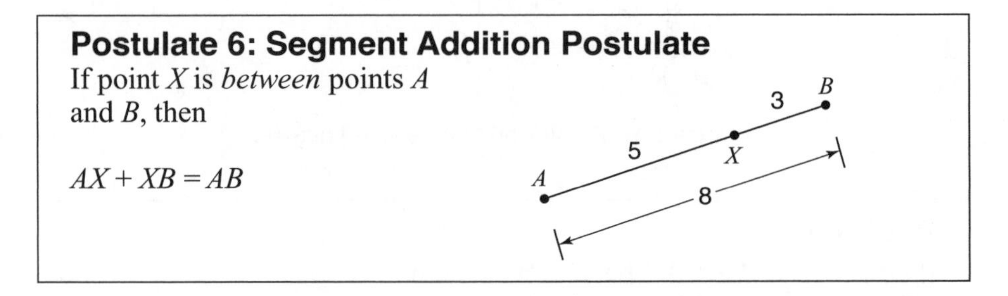

Measuring an Angle: The Protractor Postulate

One degree (1°) is $\frac{1}{360}$th of a full circular rotation. When a **protractor** is used to measure angles, as in Figure 1.6, only angles that range from 0° to 180°, or one-half of a complete rotation, are considered. The measure of $\angle AOP$ is 50°, which is written as $m\angle AOP = 50$. When measure notation ($m\angle$) is used, the degree symbol is omitted.

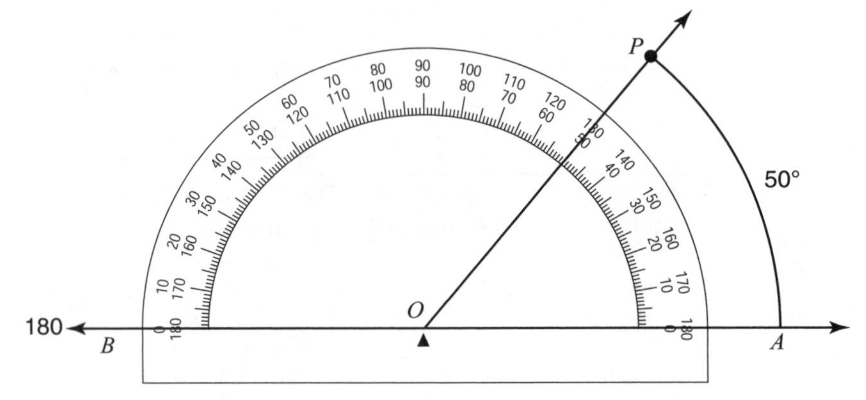

Figure 1.6 Finding the degree measure of an angle.

The *Protractor Postulate* relates the set of all rays in a semicircle rotation about a point to the set of real numbers from 0 to 180.

> **Postulate 7: Protractor Postulate**
> The set of all rays that can be drawn on one side of $\overleftrightarrow{AOB}$ with O as their endpoint can be paired in one-to-one fashion with the set of real numbers from 0 to 180. The **measure of an angle** is the unique real number obtained by taking the absolute value of the difference between the real numbers paired with the rays that form its sides.

In Figure 1.6, $\overrightarrow{OA}$ is paired with 0 and $\overrightarrow{OB}$ is paired with 180. Since $\overrightarrow{OP}$ is paired with 50, $m\angle AOP = |50 - 0| = 50$. Because the *measures* of angles and the *lengths* of line segments represent real numbers, they can be used in arithmetic operations.

Classifying Angles

Angles can be classified according to how their measures compare to 90 or 180 as shown in Figure 1.7.

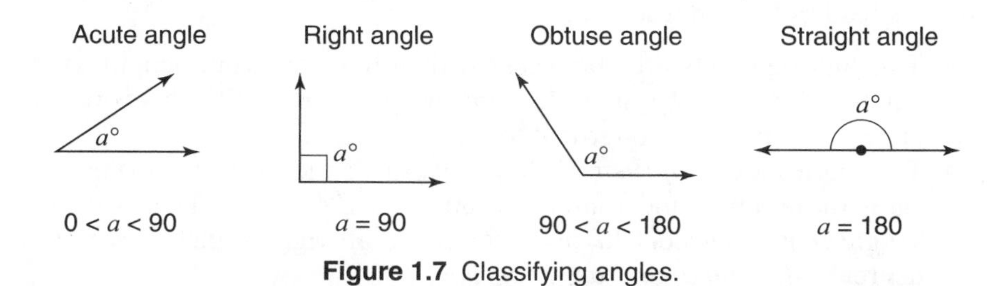

Figure 1.7 Classifying angles.

Adjacent Angles

Adjacent angles, as in Figure 1.8, are pairs of angles that have the same vertex, share one side, but do not have any interior points in common. The two non-common sides are the **exterior sides** of the adjacent angles. If the exterior sides of a pair of adjacent angles lie in opposite rays, the two angles form a **linear pair**, as in Figure 1.9. Since the measure of a straight angle is 180, the measures of a linear pair of angles must add up to 180.

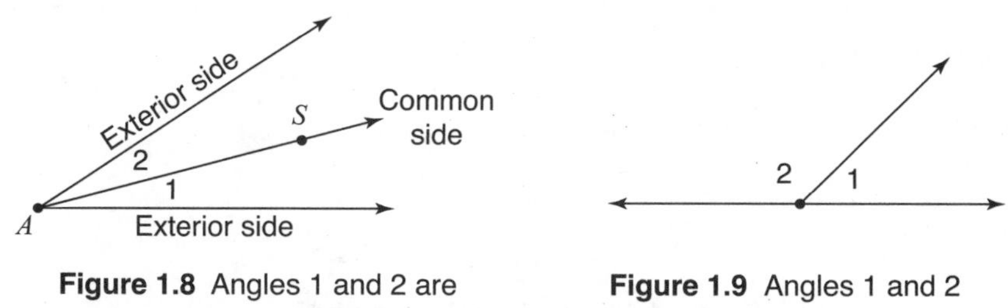

Figure 1.8 Angles 1 and 2 are adjacent with common side $\overrightarrow{AS}$.

Figure 1.9 Angles 1 and 2 form a linear pair.

Angle Addition Postulate

A ray is *between* the sides of an angle if it lies in the interior of the angle and its endpoint is the vertex of the angle. The *Angle Addition Postulate* allows you to add the measures of two adjacent angles to obtain the measure of the largest angle.

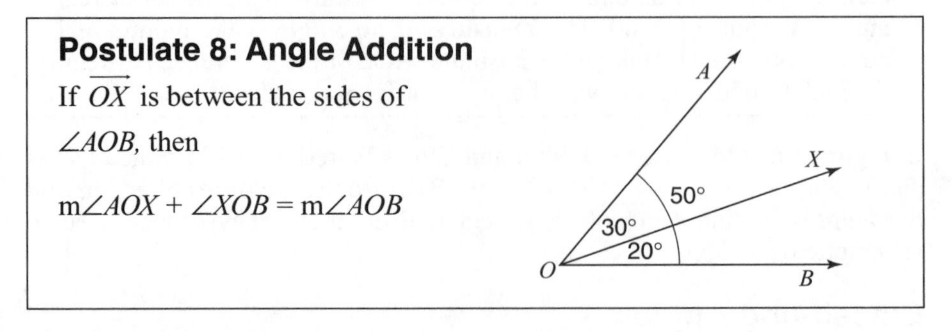

Congruent Segments and Angles

Figures that have the same size and the same shape are **congruent**. The symbol $\cong$ is read as "is congruent to."

- Two line segments are *congruent* if they have the same length. If the lengths of $\overline{AB}$ and $\overline{XY}$ are each 2 inches, then $\overline{AB} \cong \overline{XY}$. This is read as "line segment *AB* is congruent to line segment *XY*."
- Two angles are *congruent* if they have the same measure. If the measures of angles *A* and *B* are each 60, then $\angle A \cong \angle B$. This is read as "angle *A* is congruent to angle *B*." Since all right angles measure 90 degrees, *all right angles are congruent.*

Midpoint and Bisector

The **midpoint** of a line segment is the point that divides the segment into two congruent segments. In Figure 1.10, if *M* is the midpoint of $\overline{AB}$, then:

- $AM = MB$ or $\overline{AM} \cong \overline{MB}$;
- $AM = \frac{1}{2}AB$; and
- $BM = \frac{1}{2}AB$.

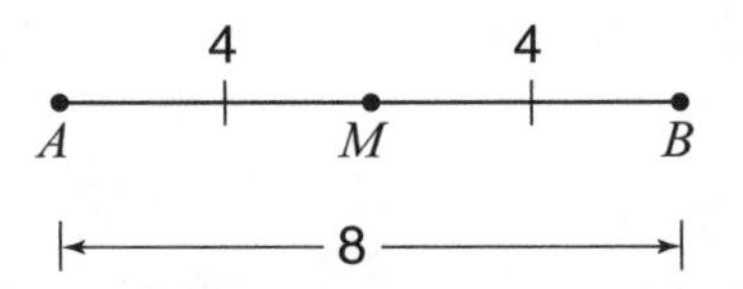

Figure 1.10 *M* is the midpoint of $\overline{AB}$.

The matching vertical bars on either side of M through $\overline{AB}$, called **tick marks**, are drawn to indicate that line segments AM and MB are congruent. A **bisector** of $\overline{AB}$ is another segment, line, or ray that passes through its midpoint, as in Figure 1.11. Only line *segments*, not lines, have midpoints. A line segment has exactly one midpoint, but can have more than one bisector.

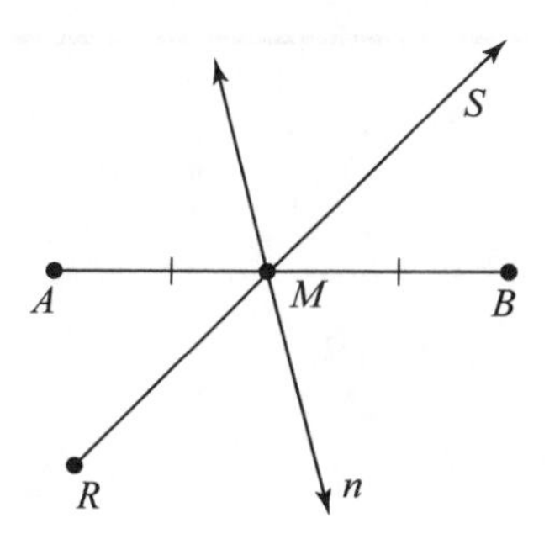

Figure 1.11 $\overrightarrow{RS}$ and line n are bisectors of $\overline{AB}$.

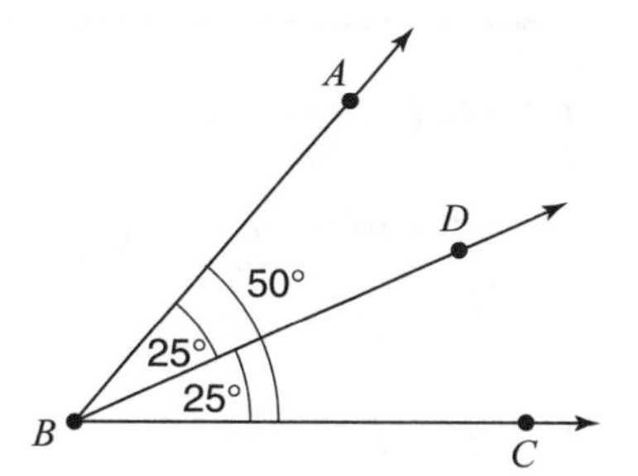

Figure 1.12 $\overrightarrow{BD}$ bisects $\angle ABC$.

Angle Bisector

An **angle bisector** is a ray, line segment, or line that divides an angle into two congruent adjacent angles, as in Figure 1.12. An angle has exactly one bisector. If $\overrightarrow{BD}$ is the bisector of $\angle ABC$, then

- $m\angle ABD = m\angle CBD$ or $\angle ABD \cong \angle CBD$;
- $m\angle ABD = \frac{1}{2}m\angle ABC$; and
- $m\angle CBD = \frac{1}{2}m\angle ABC$.

Reversing a Definition

It is sometimes necessary to draw a conclusion by *reversing* a definition. When a definition is reversed, its distinguishing characteristic is written first. The reverse of the midpoint definition is: "A point that divides a segment into two congruent segments is the midpoint of that segment."

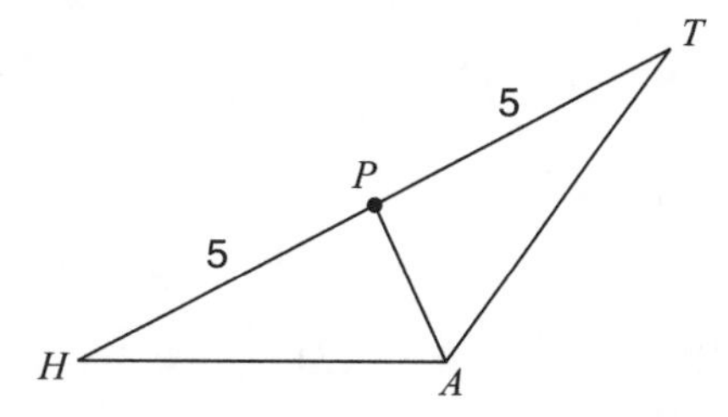

Figure 1.13 Concluding P is a midpoint.

In Figure 1.13, $HP = TP = 5$. By using the *reverse* of the definition of midpoint, you can conclude that P is the midpoint of $\overline{HT}$.

Check Your Understanding of Section 1.1

A. *Multiple Choice.*

1. If M is the midpoint of $\overline{AB}$, which statement is *false*?

(1) $\frac{AB}{2} = MB$ (3) $AM + AB = MB$

(2) $AB - MB = AM$ (4) $AM = MB$

2. If C is the midpoint of $\overline{AB}$ and D is the midpoint of $\overline{AC}$, which statement is *true*?

(1) $AC > BC$ (3) $DB = AC$

(2) $AD < CD$ (4) $DB = 3CD$

3. Point X is the midpoint of $\overline{AB}$, Y is the midpoint of $\overline{BX}$, and Z is the midpoint of $\overline{BY}$. If the length of $\overline{YZ}$ is 2, what is the length of $\overline{AB}$?

(1) 8 (2) 16 (3) 32 (4) 4

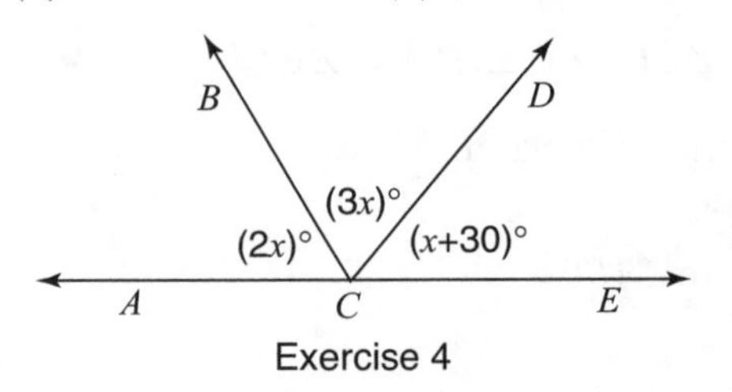

Exercise 4

4. In the accompanying figure, $\overleftrightarrow{ACE}$ is a straight line. What is $m\angle BCE$?

(1) 50 (2) 55 (3) 125 (4) 130

5. In which diagram does $AB + BC - AC = 0$?

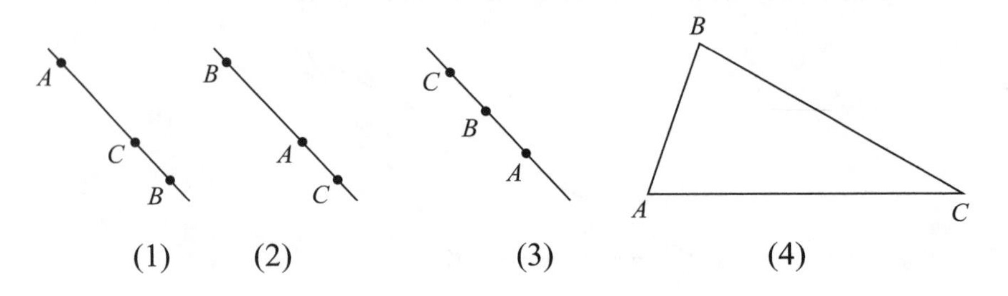

6. Line segments AB and CD intersect at E. If E is the midpoint of $\overline{AB}$ and $AE > CE$, then

(1) $CE = DE$
(2) $CE > DE$
(3) $AE < DE$
(4) $AE > DE$

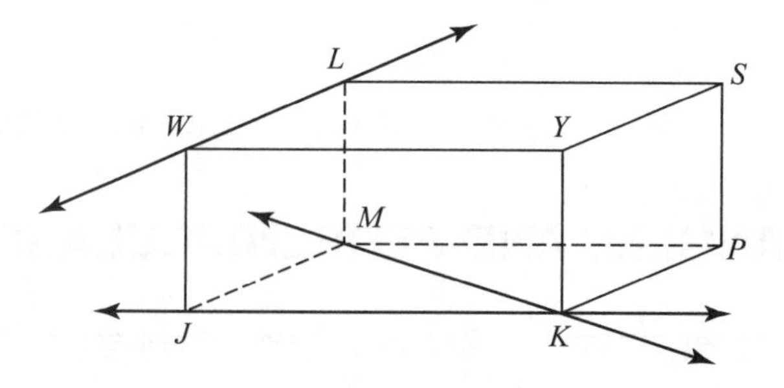

Exercises 7 and 8

7. In the accompanying figure, planes SYK and WJY intersect in

(1) $\overline{KY}$ (2) $\overleftrightarrow{JK}$ (3) $\overleftrightarrow{LW}$ (4) $\overleftrightarrow{KM}$

8. In the accompanying figure, which are a pair of skew lines?

(1) $\overleftrightarrow{WY}$ and $\overleftrightarrow{JK}$
(2) $\overleftrightarrow{YK}$ and $\overleftrightarrow{LS}$
(3) $\overleftrightarrow{LW}$ and $\overleftrightarrow{KM}$
(4) $\overleftrightarrow{SY}$ and $\overleftrightarrow{LS}$

B.*Show or explain how you arrived at your answer.*

9. Angle NYC is a right angle. If $m\angle NYC = 11x + 13$, find the value of x.

10. Line segments AB and XY are congruent. If $AB = 5n - 8$ and $XY = 2n + 19$, what is the length of $\overline{AB}$?

11. Angles RPL and LPH are adjacent angles. If $m\angle RPL = x - 5$, $m\angle LPH = 2x + 18$, and $m\angle RPH = 58$, what is the measure of $\angle LPH$?

12. Points P, N, and Z are collinear, and $NZ = 8$, $PN = 14$, and $PZ = 6$. If point M is the midpoint of $\overline{PN}$ and point W is the midpoint of $\overline{NZ}$, what is the length of $\overline{MW}$?

13. Ray PQ bisects $\angle HPJ$. If $m\angle QPJ = 2x - 9$ and $m\angle QPH = x + 29$, what is the measure of $\angle HPJ$?

14. Point R is the midpoint of $\overline{XY}$. If $XR = 3n + 1$ and $YR = 16 - 2n$, what is the length of $\overline{XY}$?

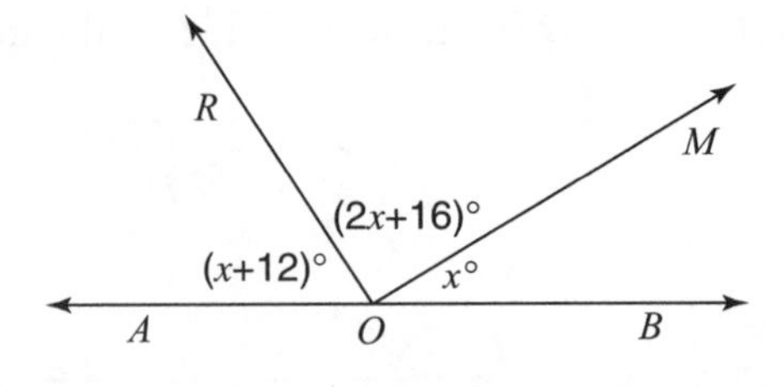

15. In the accompanying diagram, $\overleftrightarrow{AOB}$ is a straight line. Find m$\angle AOM$.

1.2 TRIANGLES AND PERPENDICULARS

KEY IDEAS

The *triangle* is one of the basic types of geometric figures. Many complicated figures can be studied by separating them into triangles. A **triangle** is a closed figure bounded by three line segments whose endpoints are three non-collinear points. Each of the three points is a **vertex** of the triangle. A triangle is named by its three vertices (plural of vertex). In the accompanying figure, the vertices of $\triangle NYS$ are points N, Y, and S. Line p is *perpendicular* to $\overline{YS}$ as it forms a right angle with that side as indicated by the square box drawn at their point of intersection.

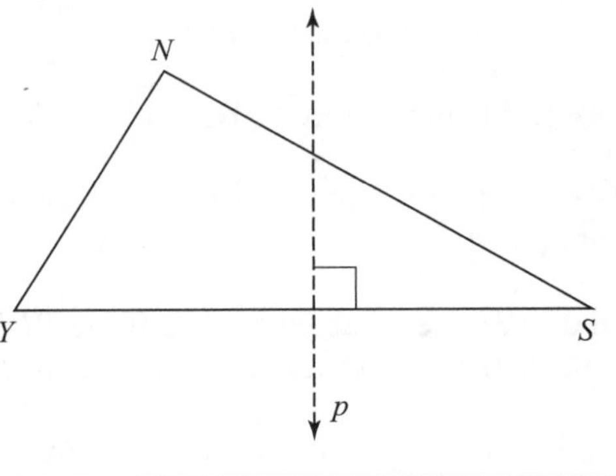

The Six Parts of a Triangle

A triangle has six *parts*: three sides and three angles. When referring to an angle of a triangle, a single letter may be used provided there is no possibility of confusion. It is sometimes convenient to label a side of a triangle using the lowercase form of the lettered vertex of the angle facing that side, as illustrated in the accompanying table. The **interior** of a triangle is the region where the interiors of all three angles of the triangle overlap.

Triangle	Three Sides	Three Angles
	a or $\overline{BC}$	$\angle A$ or $\angle B\underline{A}C$
	b or $\overline{AC}$	$\angle B$ or $\angle A\underline{B}C$
	c or $\overline{AB}$	$\angle C$ or $\angle A\underline{C}B$

Classifying Triangles

A triangle may be classified according to the number of sides that have the same length, as in Figure 1.14.

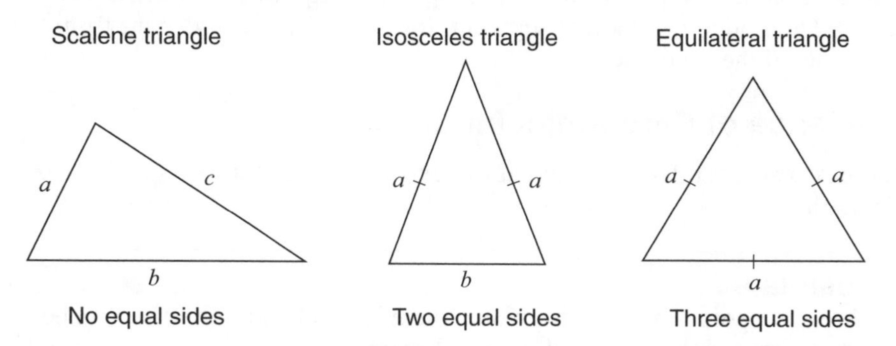

Figure 1.14 Triangles classified by number of congruent sides.

A triangle may also be classified by the measure of its greatest angle, as in Figure 1.15.

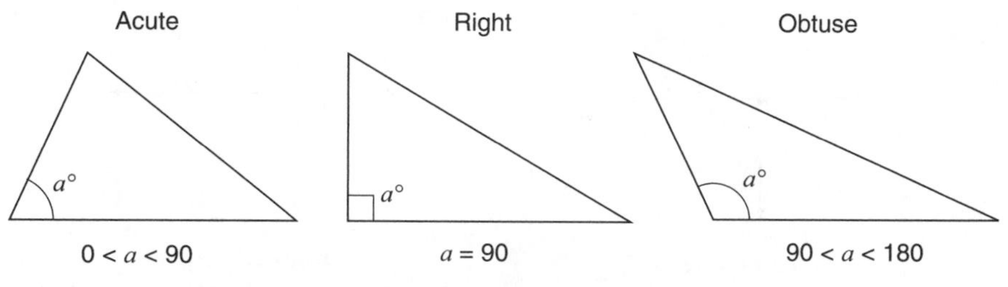

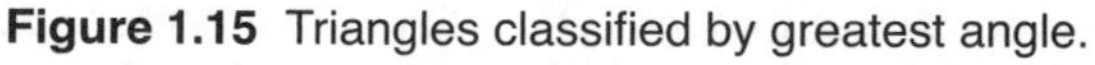

Figure 1.15 Triangles classified by greatest angle.

Perpendicular Lines and Distance

The notation $j \perp k$ indicates that line j is perpendicular to line k, as shown in Figure 1.16 where the symbol $\perp$ is read as "is perpendicular to."

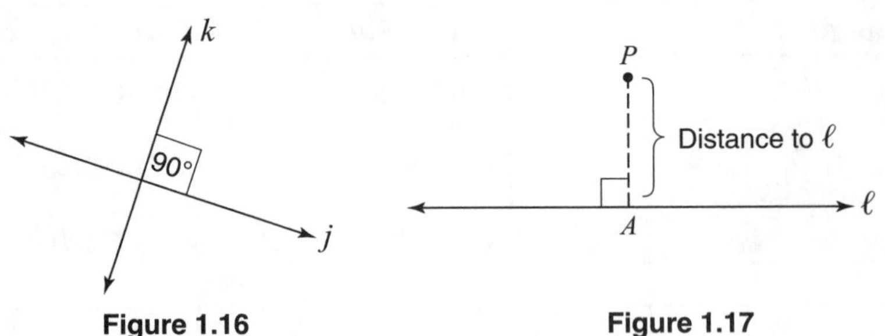

Figure 1.16
Perpendicular lines.

Figure 1.17
Distance from a point to a line.

The **distance from a point to a line** is the length of the perpendicular segment drawn from the point to the line, as illustrated in Figure 1.17. A point that is the same distance from two other points is **equidistant** from those two points. The midpoint of a line segment, for example, is equidistant from the endpoints of the segment.

Existence of Perpendicular Lines

The next two postulates guarantee the existence and uniqueness of perpendicular lines.

> **Postulate 9**
> Through a point on a line in a plane, there is exactly one line that can be drawn perpendicular to the line. See Figure 1.18.

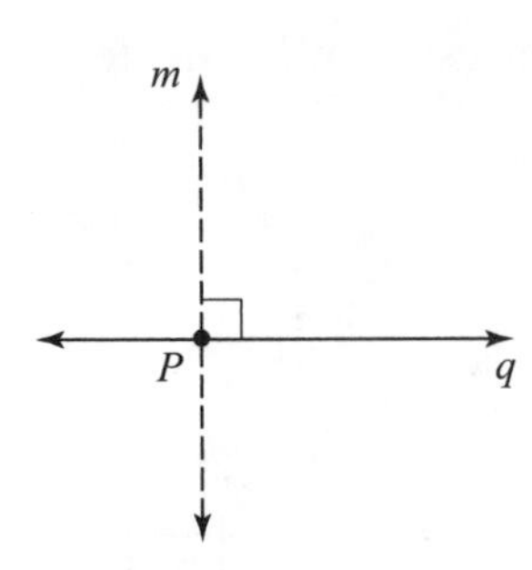

Figure 1.18
Drawing a perpendicular through point P when P is on line q

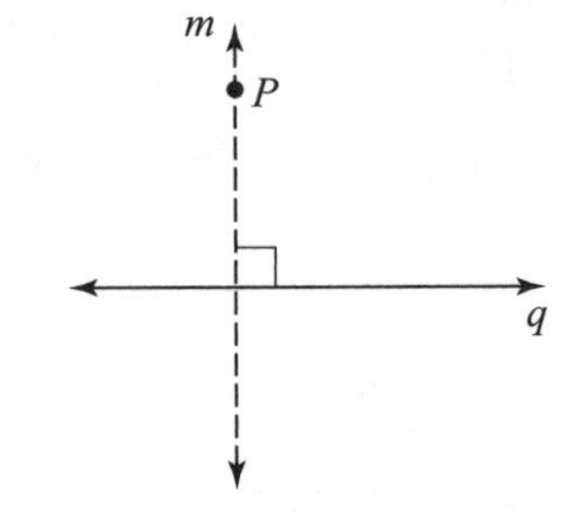

Figure 1.19
Drawing a perpendicular through point P when P is *not* on line q.

> **Postulate 10**
> Through a point *not* on a line in a plane, there is exactly one line that can be drawn perpendicular to the line. See Figure 1.19.

Median and Altitude

From each vertex of a triangle, a *median* and an *altitude* can be drawn to the opposite side.

- A **median** of a triangle is a line segment whose endpoints are a vertex of the triangle and the midpoint of the opposite side. In Figure 1.20, $\overline{CM}$ is the median to side $\overline{AB}$ since point M is the midpoint of $\overline{AB}$.

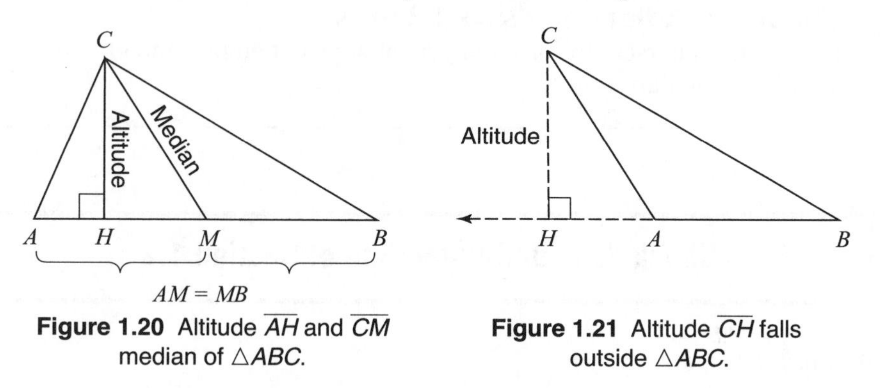

Figure 1.20 Altitude $\overline{AH}$ and $\overline{CM}$ median of $\triangle ABC$.

Figure 1.21 Altitude $\overline{CH}$ falls outside $\triangle ABC$.

- An **altitude** of a triangle is a segment drawn from a vertex of the triangle perpendicular to the line containing the opposite side, as in Figure 1.20. An altitude may fall outside a triangle, as in Figure 1.21.

Every triangle has three medians and three altitudes. A median and an altitude of a triangle may coincide.

Perpendicular Bisector

If line k is perpendicular to $\overline{AB}$ and also happens to pass through the midpoint of $\overline{AB}$, as in Figure 1.22, then line k is the **perpendicular bisector** of $\overline{AB}$.

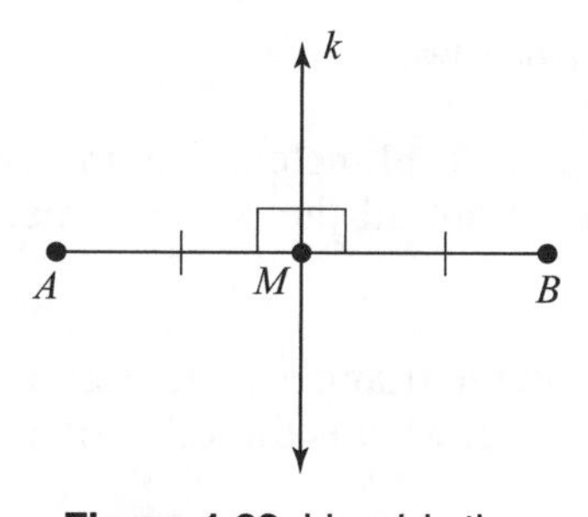

Figure 1.22 Line k is the perpendicular bisector of $\overline{AB}$.

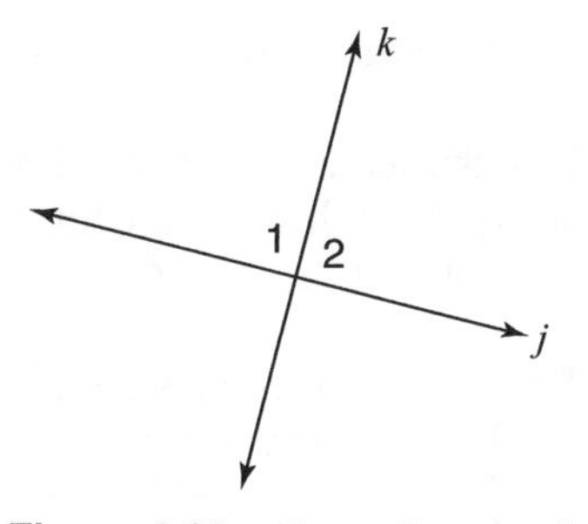

Figure 1.23 $\angle 1 \cong \angle 2 \Rightarrow j \perp k$.

When Lines Are Perpendicular

If $m\angle 1 = 85$ and $m\angle 2 = 95$ in Figure 1.23, then lines j and k are *not* perpendicular. If adjacent angles 1 and 2 are congruent, however, then $m\angle 1 = m\angle 2 = 90$. Because angles 1 and 2 are right angles, $j \perp k$. In the label that accompanies Figure 1.23, the symbol $\Rightarrow$ means "implies."

Theorem: Adjacent $\angle s \Rightarrow \perp$ Lines
If two lines intersect to form congruent adjacent angles, then the lines are perpendicular.

Check Your Understanding of Section 1.2

***A.**Multiple Choice.*

1. If an altitude of a triangle lies in the exterior of the triangle, the triangle must be
(1) acute (2) obtuse (3) isosceles (4) right

2. The lengths of the three sides of a triangle are represented by $4x - 5$, $3x$, and $2x + 7$. If the perimeter of the triangle is 65, the triangle is
(1) scalene (2) isosceles (3) equilateral (4) right

3. If the perimeter of an equilateral triangle is represented by $6x - 9$, the length of a side of this triangle can be expressed as
(1) $x - 9$ (2) $2x - 3$ (3) $3x - 3$ (4) x

***B.**Show or explain how you arrived at your answer.*

4. The perimeter of an isosceles triangle is 73 centimeters. The measure of one of the sides is 21 centimeters. What are all the possible measures of the other two sides?

5. The length of each side of an equilateral triangle is represented by $3x - 2$. If the perimeter of the triangle is 42, what is the value of x?

6. The perimeter of equilateral triangle ABC is 54. Median $\overline{AM}$ is drawn to side $\overline{BC}$. If $BM = 4x - 22$, find x.

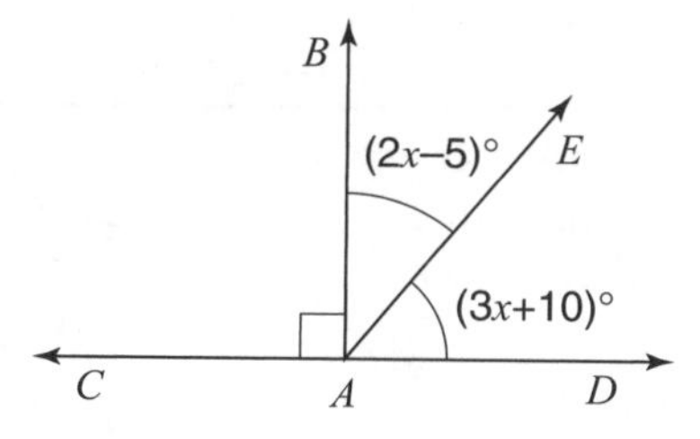

7. In the accompanying figure, $\overrightarrow{AB} \perp \overleftrightarrow{CAD}$. What is the measure of $\angle CAE$?

1.3 SPECIAL PAIRS OF ANGLES

KEY IDEAS

Complementary angles are two angles whose measures add up to 90. Each angle is the complement of the other. **Supplementary angles** are two angles whose measures add up to 180. Each angle is the supplement of the other. When two lines intersect, the opposite pairs of angles, called **vertical angles**, are congruent.

Complementary and Supplementary Angles

Two angles that form a complementary or supplementary pair may or may not be adjacent.

Theorems	Diagram
• If the exterior sides of two adjacent angles are perpendicular, then the two angles are complementary.	B, 60°, 2, 30°, 1, A, O If $\overrightarrow{OA} \perp \overrightarrow{OB}$, then adjacent angles 1 and 2 are complementary.
• If the exterior sides of two adjacent angles are opposite rays, then the angles are supplementary.	130°, 2, 1, 50°, A, O, B If angles 1 and 2 form a linear pair, they are supplementary.

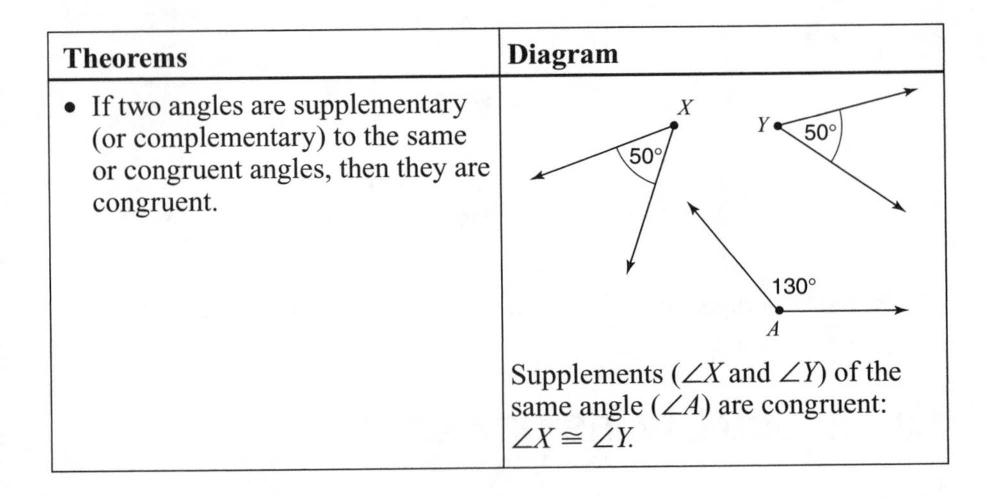

Theorems	Diagram
• If two angles are supplementary (or complementary) to the same or congruent angles, then they are congruent.	Supplements ($\angle X$ and $\angle Y$) of the same angle ($\angle A$) are congruent: $\angle X \cong \angle Y$.

Vertical Angles

When two lines intersect, pairs of angles whose sides are opposite rays are called **vertical angles**. In Figure 1.24, because vertical angles $\angle 1$ and $\angle 3$ are both supplementary to angle $\angle 2$, they are congruent. Similarly, vertical angles 2 and 4 are both supplementary to $\angle 1$ so they too are congruent.

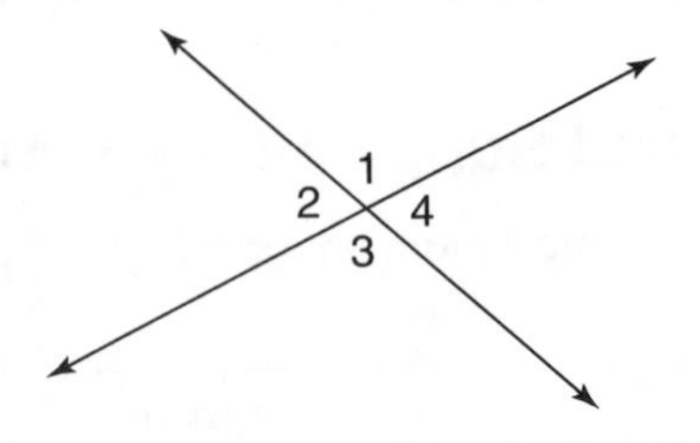

Figure 1.24 $\angle 1 \cong \angle 3$ and $\angle 2 \cong \angle 4$.

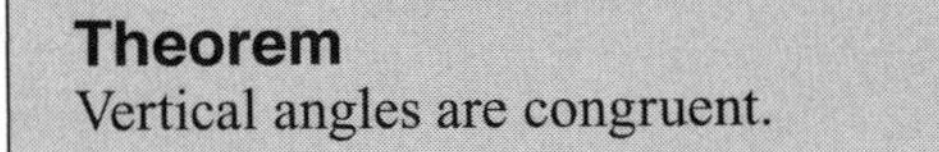

Theorem
Vertical angles are congruent.

Example 1

If the measure of an angle exceeds twice the measure of its supplement by 30, what is the measure of the angle?

Solution: If x = the measure of the angle, then $180 - x$ = the measure of the supplement of the angle.

$$x = 2(180 - x) + 30$$
$$x = 360 - 2x + 30$$
$$x + 2x = 390$$
$$3x = 390$$
$$\frac{3x}{3} = \frac{390}{3}$$
$$x = \mathbf{130}$$

Example 2

In the accompanying diagram, what is the value of x?

Solution: First find the measures of the vertical angles represented in terms of y. Since vertical angles are equal in measure:

$$3y - 18 = 2y + 5$$
$$3y - 2y = 18 + 5$$
$$y = 23$$

Thus, $3y - 18 = 3(23) - 18 = 69 - 18 = 51$. Since $\angle x$ is supplementary to this angle, $x = 180 - 51 = \mathbf{129}$.

Check Your Understanding of Section 1.3

A. *Multiple Choice*

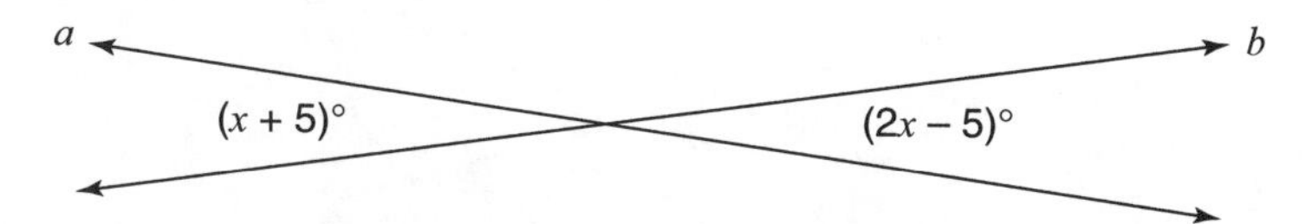

1. In the accompanying diagram, line a intersects line b. What is the value of x?
(1) -10 (2) 5 (3) 10 (4) 90

2. The measures of two complementary angles are represented by $3x + 15$ and $2x - 10$. What is the value of x?
(1) 17 (2) 19 (3) 35 (4) 37

3. What is the supplement of an angle whose measure is represented by $3x$?
(1) $90 - 3x$ (2) $3x - 90$ (3) $180 - 3x$ (4) $3x - 180$

4. In two supplementary angles, the measure of one angle is 6 more than twice the measure of the other. The measures of these two angles are
(1) 28° and 62° (3) 58° and 122°
(2) 32° and 58° (4) 62° and 118°

B. *Show or explain how you arrived at your answer.*

5. $\overleftrightarrow{AB}$ and $\overleftrightarrow{CD}$ intersect at E. If m$\angle AEC = 5x - 20$ and m$\angle BED = x + 52$, find m$\angle CEB$.

6. If the measure of an angle exceeds four times the measure of its complement by 25, what is the measure of the angle?

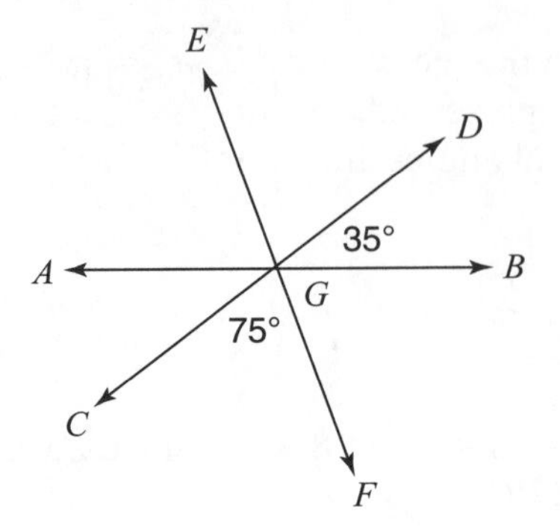

7. In the accompanying diagram, $\overleftrightarrow{AB}$, $\overleftrightarrow{CD}$, and $\overleftrightarrow{EF}$ intersect at G. Find m$\angle AGE$.

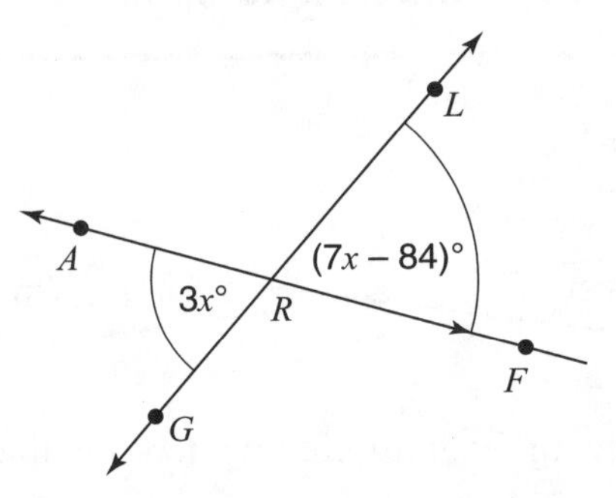

8. In the accompanying diagram, $\overleftrightarrow{AF}$ and $\overleftrightarrow{LG}$ intersect at R. Find m$\angle ARG$.

9. Two vertical angles are complementary. Find the measure in each angle.

10. The measure of the greater of two supplementary angles is five times the measure of the smaller angle. What is the measure of the larger angle?

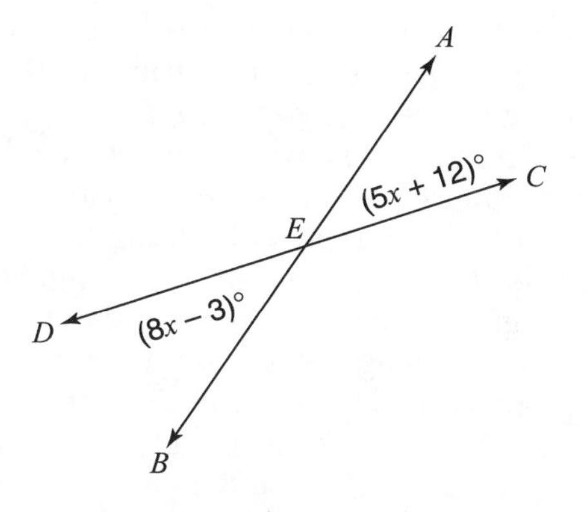

11. In the accompanying diagram, $\overleftrightarrow{AB}$ and $\overleftrightarrow{CD}$ intersect at E, $m\angle AEC = 5x + 12$, and $m\angle DEB = 8x - 3$. Find $m\angle AED$.

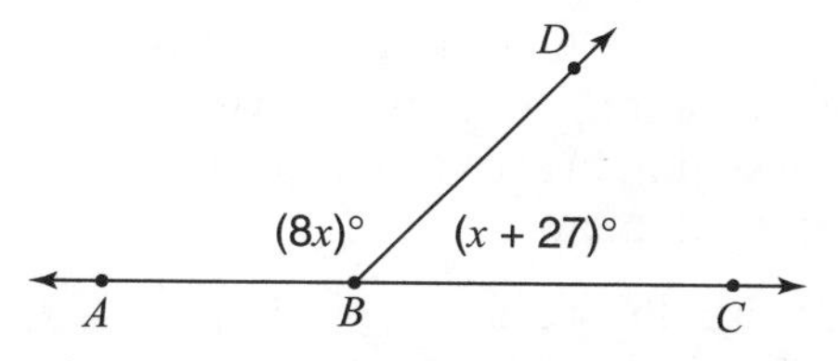

12. In the accompanying diagram, angles ABD and CBD form a linear pair. If $m\angle ABD = 8x$, and $m\angle CBD = x + 27$, find x.

1.4 REASONING AND PROOF

KEY IDEAS

Proving a statement is the process of using valid methods of reasoning to show that it is true. The finished product is called a **proof**. Although there is no standard format of a proof, the *two-column* proof is commonly used in beginning geometry courses. The properties of equality used in solving equations may also be applied to the measures of segments and angles.

Inductive Reasoning

Inductive reasoning is based on making a limited number of observations, discovering a pattern, and then drawing a general conclusion based on that pattern. Amy transfers to a new high school. Because the first three students Amy meets in her homeroom class are not friendly, she concludes that all the students in her homeroom class are unfriendly. Amy reached her conclusion, known as a **conjecture**, by using inductive reasoning. Conjectures based on inductive reasoning may prove to be either true or false. In this example, Amy's conjecture cannot be assumed to be true.

Although inductive reasoning is *not* a valid method of proof, it often serves a useful purpose in helping to discover mathematical relationships. For example, after drawing several isosceles triangles and carefully measuring the angles opposite the congruent sides, you might conclude that the angles opposite the congruent sides of *every* isosceles triangle are congruent. By no means does this reasoning process represent a valid proof. It does, however, suggest a *possible* mathematical truth worthy of further investigation. If this conjecture about isosceles triangles can eventually be proved using valid principles of reasoning, then the conjecture becomes a theorem or proven fact.

Deductive Reasoning

Unlike inductive reasoning, which moves from observing specific cases to making a broad generalization, **deductive reasoning** uses accepted facts and logical methods of reasoning to arrive at a specific conclusion. For example, consider the following argument:

1. Every triangle has three sides.
2. Carlos drew a triangle.

$\therefore$3. The figure Carlos drew has three sides.

The three-dot notation $\therefore$ is an abbreviation for "therefore." The first two statements represent accepted facts, and the third statement is the conclusion reached through *deductive reasoning*. Deductive reasoning is the central type of reasoning used in geometry.

Indirect Reasoning

Indirect reasoning uses "the process of elimination" to reach a conclusion.

- Suppose a police detective finds that there are exactly three people, *X*, *Y*, and *Z*, who could have committed a particular crime. Upon further investigation, the detective finds that both *X* and *Z* were observed to be in a different town when the crime was committed. The police detective reasons *indirectly* that since it is an impossibility for either *X* or *Z* to be guilty, *Y* must be guilty as that person is the only remaining suspect. This is a valid method of reasoning provided that all of the possibilities or "suspects" have been correctly identified and each can be ruled out except one.
- An indirect proof of a mathematical statement typically begins by assuming its *opposite*. After showing that this assumption leads to some sort of contradiction, you know that the original statement that you needed to prove must be true, as this is the only other possibility.

Indirect proofs based on obtaining a contradiction are sometimes needed to prove geometric relationships that are difficult or impossible to prove directly using deductive reasoning.

Circular Reasoning

A reasoning process may be valid or invalid. *Circular reasoning* occurs when one argues that X is true because Y is true, and Y is true because X is true. This is never a valid type of reasoning.

Proof by Counterexample

Sometimes the easiest way of *disproving* a statement is to find a single instance in which the statement is *not* true. For example, Ben claims that all prime numbers are odd whole numbers. Sue disproves Ben's statement simply by presenting 2, an even prime number, as a counterexample.

Congruence and Equality Properties

You are already familiar with the properties of equality, although you might not remember their names. For example, when you write $x = 3$ instead of $3 = x$, you have used the *symmetric property* of equality. The *substitution property* is applied when a root of an equation is checked by replacing the variable with that root in the original equation. The properties of the congruence and equality relations that you need to know are summarized in the accompanying table.

Properties of Congruence and Equality

Property	Example
Reflexive Property: A quantity is congruent (or equal) to itself.	$\overline{AB} \cong \overline{AB}$, and $m\angle 1 = m\angle 1$.
Symmetric property: The quantities on either side of a congruence (equal) sign can be interchanged.	If $\angle 1 \cong \angle 2$, then $\angle 2 \cong \angle 1$. If $AB = CD$, then $CD = AB$.
Transitive property: If two quantities are congruent (equal) to the same quantity, they are congruent (equal) to each other.	If $\angle 1 \cong \angle 2$ and $\angle 2 \cong \angle 3$, then $\angle 1 \cong \angle 3$.
Substitution property: A quantity may be substituted for its equal.	If $m\angle 1 + m\angle 2 = 90$ and $\angle 2 \cong \angle 3$, then $m\angle 1 + m\angle 3 = 90$.

Any relation that satisfies the reflexive, symmetric, and transitive properties is called an **equivalence relation**.

Writing a Two-Column Proof

A proof in geometry typically includes these four elements:

① GIVEN: { The set of facts that you know.

② PROVE: { What you need to show.

③ **Labeled Figure**
The figure may be provided or you may need to draw it yourself.

④ ARGUMENT: { The step-by-step reasoning that shows what you need to "Prove" must be true.

A mathematical argument can be presented in different ways. When writing a proof, the ancient geometers explained their reasoning in a series of paragraphs. Most beginning geometry students, however, find it helpful to organize and record their thinking using a two-column table format. In a two-column proof, the left column shows your reasoning as a set of numbered statements. The right column lists the supporting reason for each statement.

Example 1

Write a two-column proof.

Given: $\overline{BD}$ bisects $\angle ABC$,
$\angle 3$ is complementary to $\angle 1$,
$\angle 4$ is complementary to $\angle 2$.
Prove: $\angle 3 \cong \angle 5$.

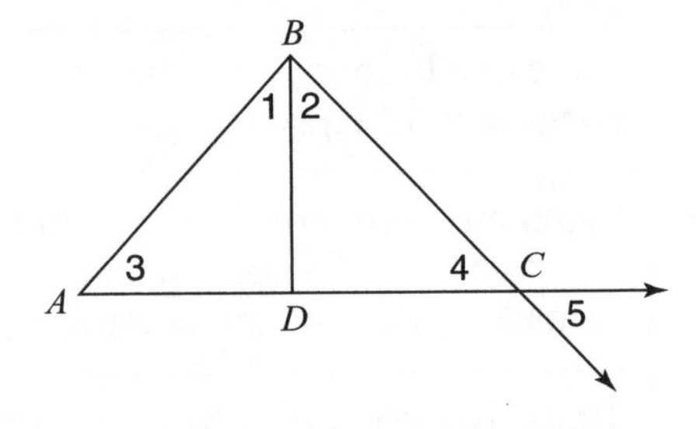

Solution: Before you attempt to write a proof, you need to have a plan. The key element of a plan is a figure with all the known facts marked off.

- Mark off the figure with parts that you know are congruent from the "Given." Also, mark off any other pair of parts that are congruent such as vertical angles.

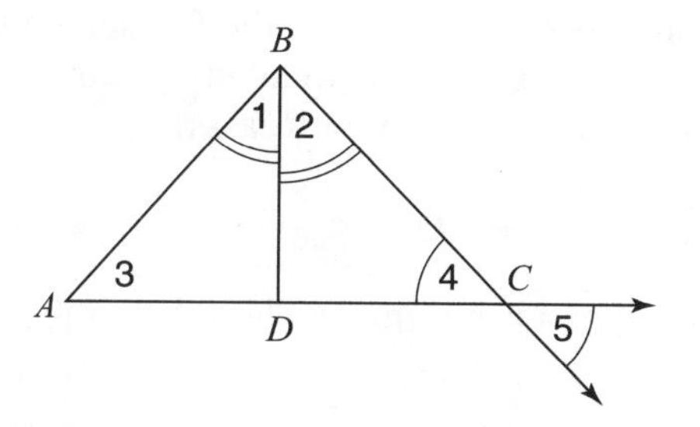

- Based on the figure, think of a logical chain of statements that ends in what you need to prove:

  $\angle 3 \cong \angle 4$ (complements of congruent angles)
  $\angle 4 \cong \angle 5$ (vertical angles)
  $\therefore \angle 3 \cong \angle 5$ (transitive property)

- Once you have a plan, draw a two-column table. Label the left column "Statement" and the right column "Reason" :

Statement	**Reason**
1. $\overline{BD}$ bisects $\angle ABC$.	1. Given.
2.	2.

- Using your plan as a guide, write the proof. The first statement in the proof is usually, but not always, a fact from the Given. The last statement in the proof is always the statement in the "Prove." Make sure each statement and its corresponding reason have matching numbers. Here is the completed proof.

Proof

Statement	**Reason**
1. $\overline{BD}$ bisects $\angle ABC$.	1. Given.
2. $\angle 1 \cong \angle 2$.	2. Definition of angle bisector.
3. $\angle 3$ is complementary to $\angle 1$, $\angle 4$ is complementary to $\angle 2$.	3. Given.
4. $\angle 3 \cong \angle 4$.	4. If two angles are complementary to congruent angles, then they are congruent.
5. $\angle 4 \cong \angle 5$.	5. Vertical angles are congruent.
6. $\angle 3 \cong \angle 5$.	6. Transitive property.

You can tell from this proof that facts from the Given are introduced in the proof only when they are needed, as in Statements 1 and 3. Furthermore, only certain types of statements may be used in the Reason column: facts from the Given (Reasons 1and 3); a definition (Reason 2); earlier theorems (Reasons 4 and 5); and a property of congruence (Reason 6).

MATH FACTS

The Reason column of a proof may include only these types of statements:

- A fact included in the Given.
- A property of equality or congruence.
- A definition, postulate, or theorem already proved.

Drawing Conclusions from Diagrams

Do not be misled by a diagram. Special properties such as midpoint, bisectors, and congruence may *not* be assumed from the diagram alone. You may *not* assume lines are perpendicular or parallel simply because they look that way in a figure. Here are some of the things you may and may not assume from the accompanying figure.

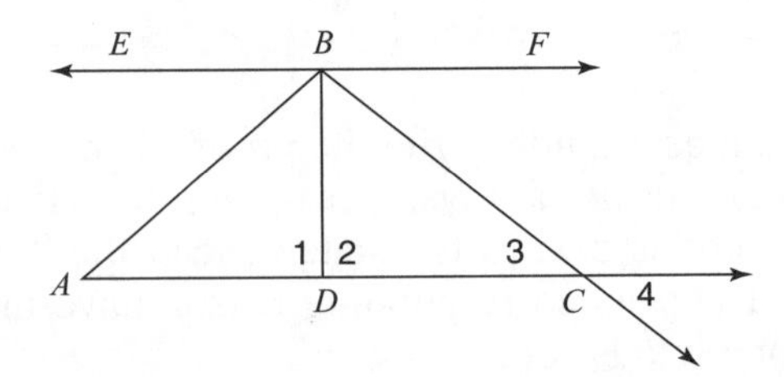

Some Things You May Assume	Some Things You May *Not* Assume
1. Figure ABC is a triangle; $\overrightarrow{BC}$ and $\overrightarrow{ADC}$ are rays; $\overleftrightarrow{EBF}$ is a line.	1. $\angle 1 \cong \angle 2$ or $\overline{AB} \cong \overline{BC}$.
2. $\overline{BD}$ intersects $\overline{AC}$ at point D.	2. $\overline{BD}$ bisects $\angle ABC$.
3. $\angle 1$ and $\angle 2$ are adjacent angles.	3. D is the midpoint of $\overline{AC}$.
4. $\angle 3$ and $\angle 4$ are vertical angles.	4. $\overline{BD} \perp \overline{AC}$.
5. Point D is between points A and C; point B is between points E and F.	5. $\overleftrightarrow{EBF} \parallel \overleftrightarrow{AC}$
6. $AD + DC = AC$.	6. Side length relationships such as $AC > AB$.
	7. Angle measure relationships such as $m\angle ABC > m\angle A$.

Proofs Involving Arithmetic Operations

Sometimes a proof depends on using an arithmetic property of equality. Because arithmetic operations can only be performed on the *measures* of segments and angles, it may be necessary to convert from congruence to measure and then back again to congruence. For convenience, any necessary conversions between congruence and equality will be accomplished "mentally" as illustrated in the next example.

Example 2

Write a two-column proof.

Given: $\overline{AB} \cong \overline{CD}$.
Prove: $\overline{AC} \cong \overline{BD}$.

Solution: Adding BC to the measures of each of the given segments leads to the desired relationship.

Proof

Statement	Reason
1. $\overline{AB} \cong \overline{CD}$.	1. Given.
2. $\overline{BC} \cong \overline{BC}$.	2. Reflexive property.
3. $\underbrace{AB + BC}_{AC} = \underbrace{CD + BC}_{BD}$.	3. Addition property of equality.
4. $AC = AB + BC$, $BD = CD + BC$.	4. Segment addition postulate.
5. $\overline{AC} \cong \overline{BD}$.	5. Substitution property.

Steps 1–3: convert from $\cong$ to $=$
Steps 4–5: convert from $=$ to $\cong$

The pattern of addition illustrated in this proof arises so frequently that it will be convenient to agree on the consolidation of certain obvious steps. In subsequent proofs, the use of the Segment Addition Postulate (see Statement 4) or its counterpart, the Angle Addition Postulate, will be understood and not necessarily shown as a separate step. Any required conversions between congruence and measure will continue to be performed "mentally" and not shown as individual statements.

Example 3

Write a two-column proof.

Given: $ABCD$ is a square, $\overline{AM} \cong \overline{CP}$.
Prove: $\triangle MBP$ is isosceles.

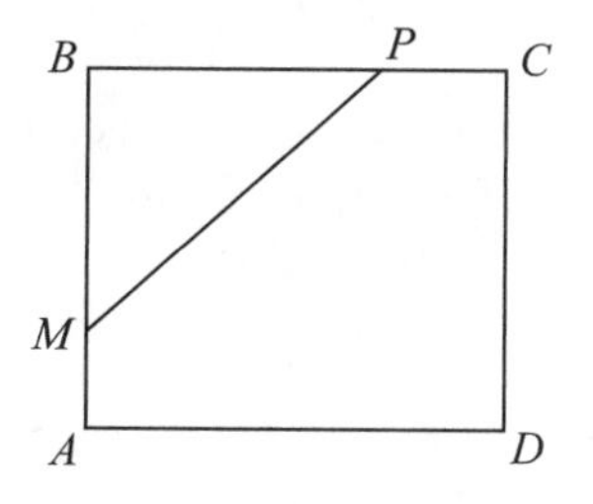

Solution: Use subtraction to show that $\overline{MB} \cong \overline{PB}$, which makes $\triangle MBP$ isosceles.

Proof

Statement	Reason
1. $ABCD$ is a square.	1. Given.
2. $\overline{AB} \cong \overline{BC}$.	2. The four sides of a square are congruent.
3. $\overline{AM} \cong \overline{CP}$.	3. Given.
4. $\underbrace{AB - AM}_{MB} = \underbrace{BC - CP}_{PB}$.	4. Subtraction property of equality.
5. $\overline{MB} \cong \overline{PB}$.	5. Substitution property.
6. $\triangle MBP$ is isosceles.	6. If a triangle has two congruent sides, then it is an isosceles triangle.

MATH FACTS

To decide whether addition or subtraction should be used when combining the measures of segments or angles, use these guidelines:

- When the pair of segments or angles that need to be proved congruent are *greater than* those that are given, use addition.
- When the pair of segments or angles that need to be proved congruent are *smaller than* those that are given, use subtraction.

Multiplication Property

Multiplying equal quantities by the same nonzero number produces an equivalent equation. The expression "halves of equals are equal" describes the special case in which the multiplying factor is $\frac{1}{2}$.

In Figure 1.25, $AB = BC$. As points E and F are midpoints of $\overline{AB}$ and $\overline{BC}$, respectively,

$AE = \frac{1}{2}AB$ and $CF = \frac{1}{2}BC$.

Because halves of equals are equal, $AE = CF$ or, equivalently, $\overline{AE} \cong \overline{CF}$. Similarly, $\overline{BE} \cong \overline{BF}$.

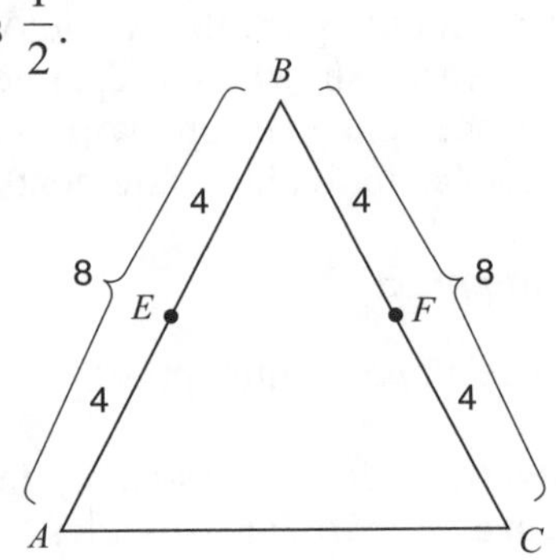

Figure 1.25 Multiplication property.

Example 4

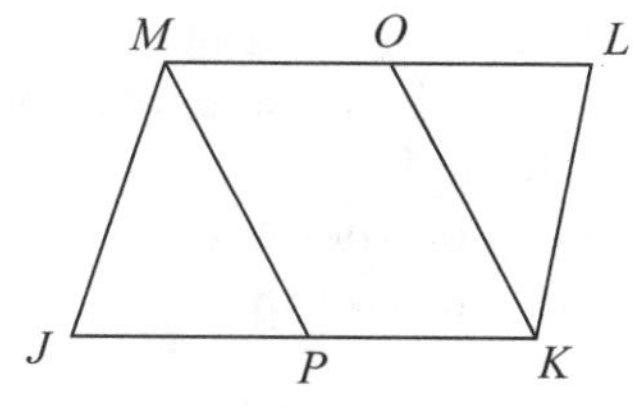

Write a two-column proof.
Given: $\overline{JK} \cong \overline{LM}$,
P is the midpoint of $\overline{JK}$,
O is the midpoint of $\overline{LM}$.
Prove: $\overline{JP} \cong \overline{LO}$.

Solution: Use the property that halves of equals (congruent segments) are equal (congruent).

Proof

Statement	**Reason**
1. P is the midpoint of $\overline{JK}$, O is the midpoint of $\overline{LM}$.	1. Given.
2. $JP = \frac{1}{2}JK$ and $LO = \frac{1}{2}LM$.	2. Definition of midpoint.
3. $\overline{JK} \cong \overline{LM}$.	3. Given.
4. $\overline{JP} \cong \overline{LO}$.	4. Halves of equals are equal and, as a result, congruent.

Check Your Understanding of Section 1.4

A. *Multiple Choice*

1. After Sara studies the number sequence

$$1, 1, 2, 3, 5, 8,13, \ldots$$

she concludes that 21 is the next number in the sequence. Which type of reasoning did Sara use to arrive at her conclusion?
(1) circular
(2) inductive
(3) deductive
(4) indirect

2. On $\overline{ABC}$ and $\overline{DEF}$, $\overline{AB} \cong \overline{DE}$ and $\overline{BC} \cong \overline{EF}$. It follows that
(1) $AB + DE = BC + EF$
(2) $AB \times EF = BC \times DE$
(3) $AB - BC = EF - DE$
(4) $\overline{AD} \cong \overline{CF}$

3. Given two real numbers, x and y. Syed argues that if x is *not* less than y and also x is *not* greater than y, then it must be the case that x equals y. This argument is an example of
(1) indirect reasoning
(2) inductive reasoning
(3) circular reasoning
(4) deductive reasoning

4. Given the statement: "A triangle cannot have two right angles." In order to prove this statement by the indirect method, it should be assumed that a triangle
   (1) does not have a right angle
   (2) has two right angles
   (3) has one right angle
   (4) does not have two right angles

B. *Write a two-column proof.*

5. Given: $\angle 1 \cong \angle 2$.
   Prove: $\angle TOM \cong \angle BOW$.

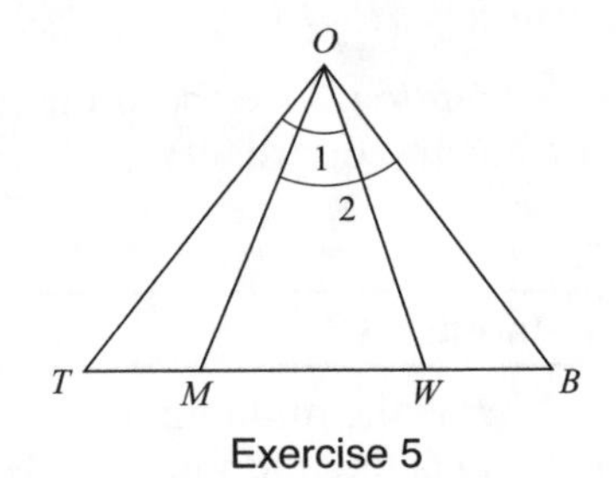

Exercise 5

6. Given: $\overline{AC}$ and $\overline{BD}$ bisect each other at E, $\overline{AC} \cong \overline{BD}$.
   Prove: $\triangle AED$ is isosceles.

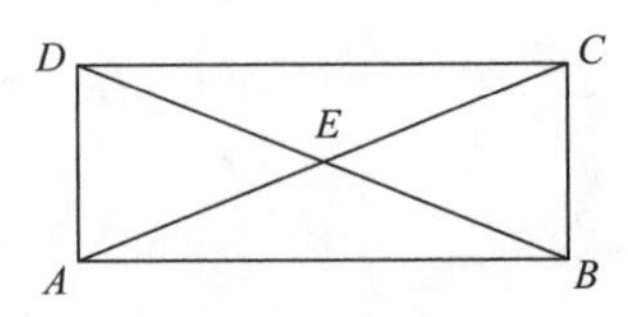

Exercise 6

7. Given: $\overline{AB} \cong \overline{CD}$, $\overline{DE} \cong \overline{BF}$, $\triangle ADF$ is equilateral.
   Prove: $\overline{CE} \cong \overline{AD}$.

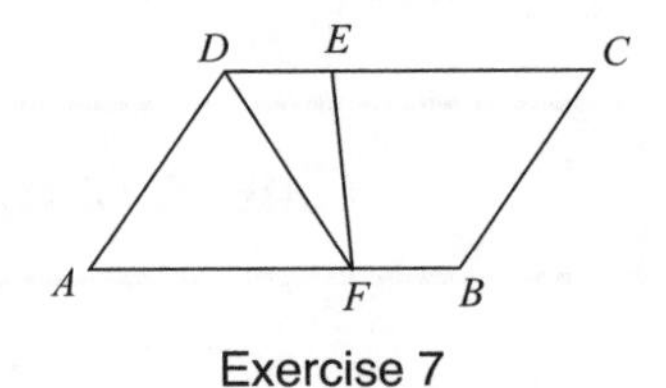

Exercise 7

8. Given: $\angle 1 \cong \angle 2$, $\overline{AD}$ bisects $\angle BAC$, $\overline{CD}$ bisects $\angle BCA$.
   Prove: $\angle 3 \cong \angle 4$.

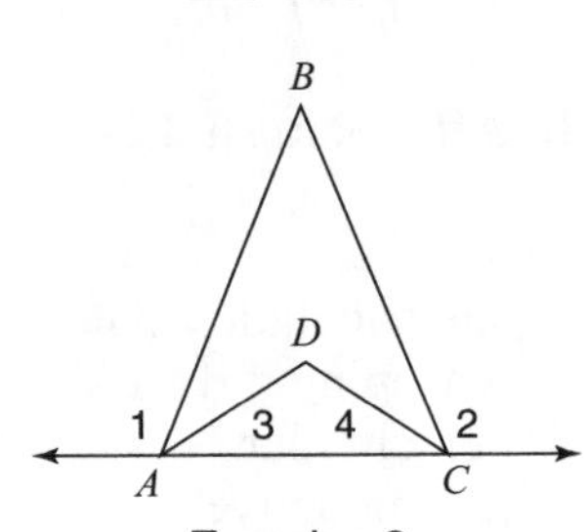

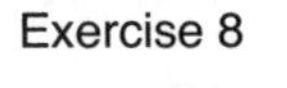
Exercise 8

9. Given: $\overline{SJ} \cong \overline{SK}$, $\overline{JR} \cong \overline{MR}$, $\overline{KT} \cong \overline{MT}$, M is the midpoint of $\overline{RT}$.
   Prove: $\overline{SR} \cong \overline{ST}$.

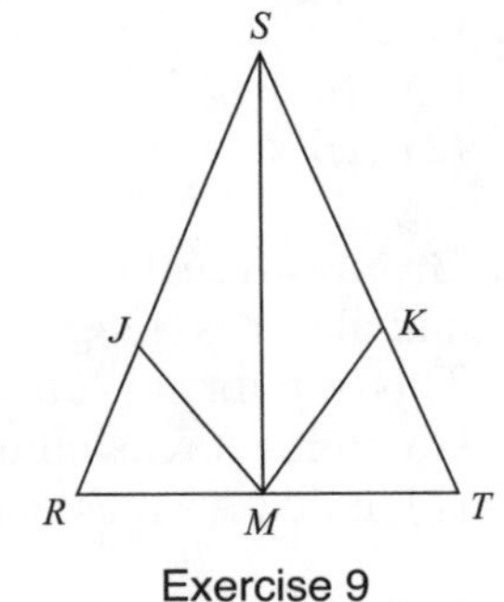

Exercise 9

1.5 LOGICAL STATEMENTS

KEY IDEAS

A **statement** is a sentence that can be judged as either true or false, but not both. The truth (T) or falsity (F) of a statement is its **truth value**. Statements such as "I like milk" and "I do *not* like milk" are negations of each other and have opposite truth values. Two statements can be joined by *AND* or *OR* to form a **compound** statement.

Statements and Their Negations

To form the **negation** of a statement, insert the word NOT so that the original statement and its negation have opposite truth values.

STATEMENT: The capital of New York State is Buffalo. **[FALSE]**

NEGATION: The capital of New York State is not Buffalo. **[TRUE]**

or

It is not true that the capital of New York State is Buffalo.

Example 1

Write the negation of each statement.

a. Parallel lines do not intersect.
b. $m\angle A > 35$

Solution: a. Parallel lines intersect.
b. $m\angle A \not> 35$ or $m\angle A \leq 35$

Symbolic Notation for Negation

Sometimes it is convenient to use a shorthand notation in which letters such as p and q serve as placeholders for actual statements. The negation of p is denoted by $\sim p$ and is read as "not p." If p is the statement, "Monday is the day after Sunday," then $\sim p$ represents, "Monday is *not* the day after Sunday." Because a statement and its negation always have opposite truth values, it is not possible for p and $\sim p$ to have the same truth value.

Connecting Statements with *AND* or *OR*

Two different statements may be combined using the word AND or OR to form a new statement.

- A **conjunction** uses *AND* to connect two statements. The conjunction "It is spring *and* the birds are chirping" is true only when the two statements joined by the word AND are both true. Each of the statements that make up a conjunction is called a **conjunct**. The symbol for conjunction is $\wedge$. The symbolic notation $p \wedge q$ represents the conjunction of statements p and q.
- A **disjunction** uses *OR* to connect two statements. The disjunction "I will study *or* I will go to the movies" is true only when at least one of the statements joined by the word *or* is true. Each of the statements that make up a disjunction is called a **disjunct**. The symbol for disjunction is $\vee$. The symbolic notation $p \vee q$ represents the disjunction of statements p and q.

Truth Tables

A **truth table** summarizes the truth values a compound statement takes on for all possible combinations of truth values of the simple statements it comprises. In each of the accompanying truth tables, the first two columns give the possible combinations of truth values for statements p and q. On each row of the truth table, the last column shows the truth value for the compound statement using the truth values of p and q on that same row.

p	q	$p \wedge q$
T	T	T
T	F	F
F	T	F
F	F	F

p	q	$p \vee q$
T	T	T
T	F	F
F	T	T
F	F	F

Example 2

Let p represent "It is cold in January" and q represent "It snows in August." Express each of the following statements in symbolic form.

a. It is cold in January or it does not snow in August.
b. It is not cold in January and it does not snow in August.
c. It is not true that it is cold in January and it snows in August.

Solution: a. Write the disjunction of p and the negation of q:

$$p \vee \sim q$$

b. Write the conjunction of the negation of p and the negation of q:

$$\sim p \wedge \sim q$$

c. Write the negation of the conjunction of p and q:

$$\sim (p \wedge q)$$

Example 3

Let p represent "All right angles are congruent" and q represent "Vertical angles are always supplementary." Determine the truth value of each compound statement:

a. $\sim p \vee q$ b. $p \wedge \sim q$ c. $\sim p \vee \sim q$ d. $\sim (p \wedge q)$

Solution: Statement p is true and statement q is false since vertical angles are always congruent. Substitute these truth values into each logic statement:

a. $p \vee \sim q$	b. $p \wedge \sim q$	c. $\sim p \vee \sim q$	d. $\sim (p \wedge q)$
$\sim$(T) $\vee$ F	T $\wedge$ $\sim$(F)	$\sim$(T) $\vee$ $\sim$(F)	$\sim$(T $\wedge$ F)
F $\vee$ F	T $\wedge$ T	F $\vee$ T	$\sim$ (F)
False	**True**	**True**	**True**

Drawing Inferences

Based on the truth value of a compound statement, you may be able to draw a conclusion about the truth value of a related statement.

Example 4

Given: "Ben has a driver's license or Ben is not 18 years old " is false. What is the truth value of "Ben is 18 years old"?

Solution:

- Because the given disjunction is false, both disjuncts must be false.
- Hence, the truth value of the statement "Ben is not 18 years old" is false.
- Since a statement and its negation have opposite truth values, the truth value of the statement "Ben is 18 years old" is **true.**

Example 5

Given: "I will buy a new suit or I will not go to the dance" is true and "I will buy a new suit" is false. What is the truth value of the statement, "I will go to the dance"?

Solution:

- Because the given disjunction is true, at least one of the two disjuncts must also be true.
- It is given that the disjunct "I will buy a new suit" is false. This means that the disjunct "I will not go to the dance" is true.
- Since a statement and its negation have opposite truth values, the truth value of "I will go to the dance" is **false**.

Check Your Understanding of Section 1.5

A. *Multiple Choice.*

1. Let p represent "It is cold" and let q represent "It is snowing." Which expression can be used to represent "It is cold and it is not snowing"?
(1) $\sim p \wedge q$ (2) $p \wedge \sim q$ (3) $p \vee \sim q$ (4) $\sim p \vee q$

2. Let p represent "The figure is a triangle" and q represent "The figure does not contain two obtuse angles." Which expression can be used to represent "The figure is not a triangle and the figure contains two obtuse angles."
(1) $\sim p \wedge q$ (2) $p \wedge \sim q$ (3) $\sim p \wedge \sim q$ (4) $\sim(p \wedge q)$

3. If p represents "Math is fun" and q represents "Math is difficult," which expression can be used to represent "It is not true that Math is not fun or math is difficult."
(1) $\sim(p \vee q)$ (2) $\sim(p \wedge q)$ (3) $\sim p \vee q$ (4) $\sim(\sim p \vee q)$

4. Which statement is always false?
(1) $p \vee \sim q$ (2) $\sim p \wedge q$ (3) $\sim p \vee p$ (4) $q \wedge \sim q$

5. Given the true statements: "Jason goes shopping or he goes to the movies" and "Jason does not go to the movies." Which statement must also be true?
(1) Jason stays home.
(2) Jason goes shopping.
(3) Jason does not go shopping.
(4) Jason does not go shopping and he does not go to the movies.

6. If $p \wedge \sim q$ is true, then which is true?
(1) p and q are both true.
(2) p is false and q is true.
(3) p is true and q is false.
(4) p and q are both false.

B. *Show or explain how you arrived at your answer.*

7. Let x represent "Mr. Ladd teaches mathematics" and let y represent "Mr. Ladd is the football coach." Write in symbolic form: "Mr. Ladd does not teach mathematics and Mr. Ladd is the football coach."

8. Let p represent the statement "Perpendicular lines intersect at right angles" and q represents the statement "Parallel lines do not intersect." Determine the truth value of each compound statement:
a. $p \wedge \sim q$
b. $\sim p \vee q$
c. $\sim p \wedge \sim q$
d. $\sim(p \vee q)$
e. $\sim(\sim p \wedge \sim q)$

9. Each part that follows consists of a set of three sentences. The truth values of the first two sentences are given. Determine the truth value of the third sentence.

a. It rains or it is cold. It is cold. It rains.	TRUE FALSE ?
b. The month is June and it is *not* warm. The month is June. It is warm.	FALSE TRUE ?
c. I will study or I will *not* pass the test. I will not study. I will pass the test.	TRUE ? TRUE
d. I will *not* work at camp this summer or I will attend summer school. I will work at camp this summer. I will not attend summer school.	FALSE TRUE ?
e. The month is *not* January and it is *not* snowing. The month is January. It is not snowing.	? TRUE TRUE

10. Let p represent "I go to the beach" and q represent "I get a sunburn."

a. Using p and q, write the following two statements in symbolic form:

(1) It is not the case that I went to the beach and I got a sunburn.

(2) I did not go to the beach or I did not get a sunburn.

b. In the last two columns of the accompanying truth table, enter the symbolic forms of statements (1) and (2) determined in part a. Then complete the truth table.

		Symbolic Form of Statement (1)	Symbolic Form of Statement (2)
p	q		
T	T		
T	F		
F	T		
F	F		

c. Determine if statements (1) and (2), written in part a, are logically equivalent. Justify your answer.

1.6 CONDITIONAL STATEMENTS

KEY IDEAS

An If–Then statement is called a **conditional statement**. Associated with each conditional statement are three other conditional statements: the *converse*, the *inverse*, and the *contrapositive*. Certain pairs of these conditionals always agree in their truth values and, as a result, are **logically equivalent**.

Truth Value of a Conditional

A **conditional statement** is a statement that has the form, "If p, then q," as in

"If I live in Albany (hypothesis), then I am a New Yorker (conclusion)."

The "If" part of a conditional statement is the **hypothesis** and the part that follows "then" is the **conclusion**. A conditional statement is false when there is at least one situation for which the hypothesis is true, but the conclusion is false. Such a situation is called a **counterexample**. Consider the conditional statement

"If the month has 31 days, then it is winter."

To prove that the conditional statement is false, it is only necessary to find a single counterexample. The month of July, which has 31 days, serves as a counterexample.

MATH FACTS

A conditional statement is true *except* in the single instance when the hypothesis is true and the conclusion is false. A specific situation for which this occurs is called a **counterexample**.

Forming Related Conditional Statements

By interchanging or negating both parts of a conditional statement, or by doing both, three related conditional statements can be formed.

Type of Statement	Forming a Conditional from "If p, then q"
Converse	Interchange p and q: "If q, then p"
Inverse	Negate both p and q: "If $\sim p$, then $\sim q$"
Contrapositive	Interchange and negate both p and q: "If $\sim q$, then $\sim p$"

Here is an example:

ORIGINAL:	If I live in Albany, then I am a New Yorker.	**(TRUE)**
CONVERSE:	If I am a New Yorker, then I live in Albany.	**(FALSE)**
INVERSE:	If I do not live in Albany, then I am not a New Yorker.	**(FALSE)**
CONTRAPOSITIVE:	If I am *not* a New Yorker, then I do not live in Albany.	**(TRUE)**

This example illustrates the truth value relationships between pairs of related conditional statements:

- A conditional statement and its converse may have the same or may have opposite truth values. In the previous example, the original statement was true, but its converse was false.
- A conditional statement and its contrapositive always agree in their truth values.
- The converse and inverse always have the same truth values.

MATH FACTS

Statements that always agree in their truth values are **logically equivalent**.

- A conditional statement and its contrapositive are logically equivalent, as are the converse and inverse.
- Starting with a true conditional statement, you can form another conditional that must be true by writing its contrapositive.

Example 1

Write the converse of each statement and indicate the truth values.

a. If two angles are right angles, then the angles are congruent.

b. If two lines are parallel, then the two lines do not intersect

Solution: The converse of a true statement may be true or may be false.

a.	ORIGINAL:	If two angles are right angles, then the angles are congruent."	**(TRUE)**
	CONVERSE:	If two angles are congruent, then the angles are right angles.	**(FALSE)**
b.	ORIGINAL:	If two lines are parallel, then the two lines do not intersect.	**(TRUE)**
	CONVERSE:	If two lines do not intersect, then the two lines are parallel.	**(TRUE)**

Example 2

Given the true statement "If I study, then I pass the test." Which statement must also be true?

(1) I study if I pass the test.
(2) If I do not study, then I do not pass the test.
(3) If I do not pass the test, then I did not study.
(4) If I pass the test, then I study.

Solution: Because it is given that the original statement is true, its contrapositive must also be true. Form the contrapositive by negating and then interchanging the If and Then statements:

ORIGINAL: If I study, then I pass the test.

CONTRAPOSITIVE: If I do *not* pass the test, then I did *not* study.

Look for the contrapositive among the answer choices.
The correct choice is **(3)**.

Role of Conditional Statements

Theorems are often expressed in the form "If p, then q" where statement p is the *hypothesis* or what is given and statement q is the *conclusion* or what needs to be proved. A theorem that will eventually be proved is,

If two sides of a triangle are congruent (hypothesis or Given), **then** the angles opposite them are congruent (conclusion or Prove).

From this If–Then statement, you can determine the Given, Prove, and diagram:

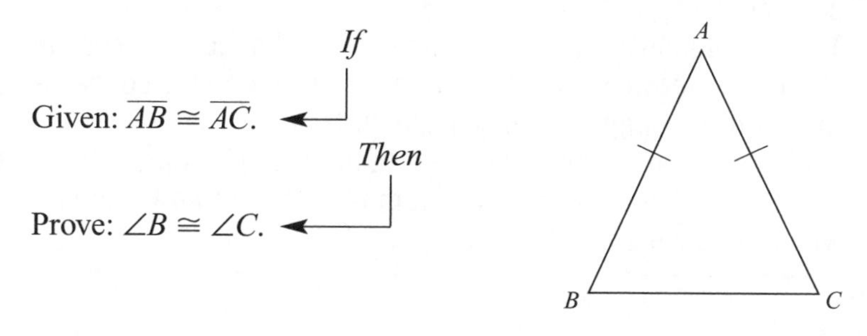

After a proposed theorem is proved, forming related conditionals may suggest additional relationships that can be investigated. For instance, once the above theorem is proved, a likely question to ask is: If two *angles* of a triangle are congruent, are the sides opposite them congruent?

Biconditional Statements

The definition of a right angle can be written as the conditional,

ORIGINAL: If an angle is a right angle, then the angle measures 90°.

Since a definition is reversible, the definition could also be written as

CONVERSE: If an angle measures 90°, then the angle is a right angle.

As both forms of the definition are true, their conjunction is true:

CONJUNCTION {
If an angle is a right angle, then the angle measures 90°
AND
If an angle measures 90°, then the angle is a right angle.

Combining a conditional statement and its converse in this way forms a *biconditional statement*. A **biconditional statement** is the conjuction of a conditional statement and its converse. A biconditional can be shortened by connecting the hypothesis and the conclusion of the original conditional statement with the phrase "if and only if" :

An angle is a right angle (Term being defined) *if and only if* the angle measures 90° (Distinguishing characteristic).

MATH FACTS

A **biconditional** statement has the general form,

$$\underbrace{\text{[Fact 1]}}_{\text{hypothesis}} \textbf{ if and only if } \underbrace{\text{[Fact 2]}}_{\text{conclusion}}$$

- When a definition is written as a biconditional, Fact 1 corresponds to the term that is being defined and Fact 2 is its distinguishing characteristic.
- The biconditional form of a definition emphasizes its reversibility. Most often, however, the two parts of a definition are connected by the word "is" simply because it is shorter.
- A biconditional is true only when both of its parts are either both true or both false. A proved theorem is expressed as a biconditional when its converse is also true.

Representing Conditional Statements Symbolically

A conditional statement can be represented symbolically as $p \rightarrow q$, read as "p implies q." The biconditional of p and q is denoted by $p \leftrightarrow q$ and read as "p if and only if q."

- The conditional $p \rightarrow q$ is always true except in the single instance when p is true and q is false.
- The biconditional $p \leftrightarrow q$ is true only when p and q are either both true or both false.

The accompanying table summarizes all the possible forms of a conditional statement.

Conditional	Sentence Form	Symbolic Form
Original	If p, then q.	$p \rightarrow q$
Converse	If q, then p.	$q \rightarrow p$
Inverse	If not p, then not q.	$\sim p \rightarrow \sim q$
Contrapositive	If not q, then not p.	$\sim q \rightarrow \sim p$
Biconditional	p if and only if q.	$p \leftrightarrow q$

(Original and Contrapositive: logically equivalent; Converse and Inverse: logically equivalent)

Check Your Understanding of Section 1.6

A. *Multiple Choice.*

1. What is the converse of the statement "If Alicia goes to Albany, then Ben goes to Buffalo"?
(1) If Alicia does not go to Albany, then Ben does not go to Buffalo.
(2) Alicia goes to Albany if and only if Ben goes to Buffalo.
(3) If Ben goes to Buffalo, then Alicia goes to Albany.
(4) If Ben does not go to Buffalo, then Alicia does not go to Albany.

2. Which statement is the inverse of "If the waves are small, I do not go surfing"?
(1) If the waves are not small, I do not go surfing.
(2) If I do not go surfing, the waves are small.
(3) If I go surfing, the waves are not small.
(4) If the waves are not small, I go surfing.

3. Which statement is logically equivalent to the statement "If you are an elephant, then you do not forget"?
(1) If you do not forget, then you are an elephant.
(2) If you do not forget, then you are not an elephant.
(3) If you are an elephant, then you forget.
(4) If you forget, then you are not an elephant.

4. Which statement is expressed as a biconditional?
(1) Two angles are congruent if they have the same measure.
(2) If two angles are both right angles, then they are congruent.
(3) Two angles are congruent if and only if they have the same measure.
(4) If two angles are congruent, then they are both right angles.

5. Let p represent "I like cake" and let q represent "I like ice cream." Which expression represents "If I do not like cake, then I do not like ice cream"?
(1) $\sim p \vee \sim q$ (2) $\sim p \wedge \sim q$ (3) $\sim p \rightarrow \sim q$ (4) $\sim(p \rightarrow q)$

6. Which statement is logically equivalent to "If I did not eat, then I am hungry"?
(1) If I am not hungry, then I did not eat.
(2) If I did not eat, then I am not hungry.
(3) If I am not hungry, then I did eat.
(4) If I am hungry, then I did eat.

7. Let p represent "$x > 5$" and let q represent "x is a multiple of 3." If $x = 12$, which statement is false?
(1) $p \leftrightarrow q$ (2) $p \wedge q$ (3) $\sim p \vee q$ (4) $\sim p \vee \sim q$

8. Let p represent "x is an even number greater than 10" and let q represent "x is *not* evenly divisible by 4." For which value of x is $p \wedge \sim q$ a true statement ?
(1) 8 (2) 14 (3) 18 (4) 20

9. Which statement is true when p is false and q is true?
(1) $p \leftrightarrow q$ (2) $p \wedge q$ (3) $\sim p \vee q$ (4) $\sim(p \vee q)$

10. Which statement is always true?
(1) $p \wedge \sim p$ (2) $p \vee \sim p$ (3) $p \rightarrow \sim p$ (4) $p \leftrightarrow \sim p$

11. Which statement is logically equivalent to "If it is Saturday, then I am not in school"?
(1) If I am not in school, then it is Saturday.
(2) If it is not Saturday, then I am in school.
(3) If I am in school, then it is not Saturday.
(4) If it is Saturday, then I am in school.

12. Which statement is the contrapositive of "If a triangle is a right triangle, then it has two complementary angles"?
(1) If a triangle is a right triangle, then it does not have two complementary angles.
(2) If a triangle does have two complementary angles, then it is not a right triangle.
(3) If a triangle is not a right triangle, then it has two complementary angles.
(4) If a triangle does not have two complementary angles, then it is a right triangle.

13. Which statement is the converse of "If the sum of two angles is 180°, then the angles are supplementary"?
(1) If two angles are supplementary, then their sum is 180°.
(2) If the sum of two angles is not 180°, then the angles are not supplementary.
(3) If two angles are not supplementary, then their sum is not 180°.
(4) If the sum of two angles is not 180°, then the angles are supplementary.

14. Which statement is expressed as a biconditional?
(1) Two angles are congruent if they have the same measure.
(2) If two angles are both right angles, then they are congruent.
(3) Two angles are congruent if and only if they have the same measure.
(4) If two angles are congruent, then they are both right angles.

15. What is the inverse of the statement "If Bob gets hurt, then the team loses the game"?
(1) If the team loses the game, then Bob gets hurt.
(2) Bob gets hurt if the team loses the game.
(3) If the team does not lose the game, then Bob does not get hurt.
(4) If Bob does not get hurt, then the team does not lose the game.

16. Given the true statement: "If a person is eligible to vote, then that person is a citizen." Which statement must also be true?
(1) Kayla is not a citizen; therefore, she is not eligible to vote.
(2) Juan is a citizen; therefore, he is eligible to vote.
(3) Marie is not eligible to vote; therefore, she is not a citizen.
(4) Morgan has never voted; therefore, he is not a citizen.

17. What is the converse of the statement "If the Sun rises in the east, then it sets in the west"?
(1) If the Sun does not set in the west, then it does not rise in the east.
(2) If the Sun does not rise in the east, then it does not set in the west.
(3) If the Sun sets in the west, then it rises in the east.
(4) If the Sun rises in the west, then it sets in the east.

18. Which statement cannot be written as a true biconditional?
(1) If the two lines are congruent, then they have the same measure.
(2) If two angles have the same measure, then they are congruent.
(3) If two angles are vertical angles, then they are congruent.
(4) If two lines are perpendicular, the lines intersect to form congruent adjacent angles.

B. *Show or explain how you arrived at your answer.*

19. Let r represent "You may vote in the general election" and let s represent "You are *at least* 18 years old." Using r and s, write in symbolic form: "You may *not* vote in the general election if and only if you are *less than* 18 years old."

20. Let p represent "The triangle is equilateral" and q represent "The triangle is a right triangle."

a. Write in symbolic form, If the triangle is a right triangle, then it is not equilateral.

b. Using p and q, write in symbolic form a statement that is logically equivalent to the statement written in part a.

21. Let p represent "It is raining" and let q represent "I am going swimming." Write each of these sentence in symbolic form:

a. I am going swimming if and only if it is not raining.

b. If it is not raining, then I am going swimming.

c. It is not true that if it is raining, then I am not going swimming.

22. a. Write the inverse of $\sim p \rightarrow q$ in symbolic form.

b. What is the inverse of the statement "If I do not buy a ticket, then I do not go to the concert"?

(1) If I buy a ticket, then I do not go to the concert.
(2) If I buy a ticket, then I go to the concert.
(3) If I go to the concert, then I buy a ticket.
(4) If I do not go to the concert, then I do not buy a ticket.

23. Given the definition, "An isosceles triangle is a triangle that has two congruent sides."

a. Rewrite the definition as a conditional statement.

b. Write a statement that is logically equivalent to the statement written in part a.

c. Rewrite the given definition as a biconditional statement.

24. Let p represent "The water temperature is 100°C" and q represent "The water boils."

a. Using p and q, write this statement in symbolic form: "If the water temperature is not 100°C, then the water does not boil."

b. Write in symbolic form the inverse of the statement in part a.

c. Write in symbolic form a statement that is logically equivalent to the statement in part a.

CHAPTER 2
PARALLEL LINES AND POLYGONS

2.1 ANGLES FORMED BY PARALLEL LINES

KEY IDEAS

A **transversal** is a line that intersects two or more other lines at different points, as in Figure 2.1. If the lines intersected by a transversal happen to be parallel, then any two of the eight angles formed are either congruent or supplementary.

Naming Pairs of Angles

In Figure 2.1, the four angles between lines ℓ and m are **interior angles** ($\angle$s 3, 4, 5, and 6) and the remaining four angles ($\angle$s 1, 2, 7, and 8) are **exterior angles**. Pairs of angles are given special names according to their position relative to the transversal.

Type of Angle	Diagram
Alternate interior angles: $\angle 3$ and $\angle 6$; $\angle 4$ and $\angle 5$. Corresponding angles: $\angle 1$ and $\angle 5$, $\angle 2$ and $\angle 6$ $\angle 3$ and $\angle 7$, $\angle 4$ and $\angle 8$. Alternate exterior angles: $\angle 1$ and $\angle 8$, $\angle 2$ and $\angle 7$.	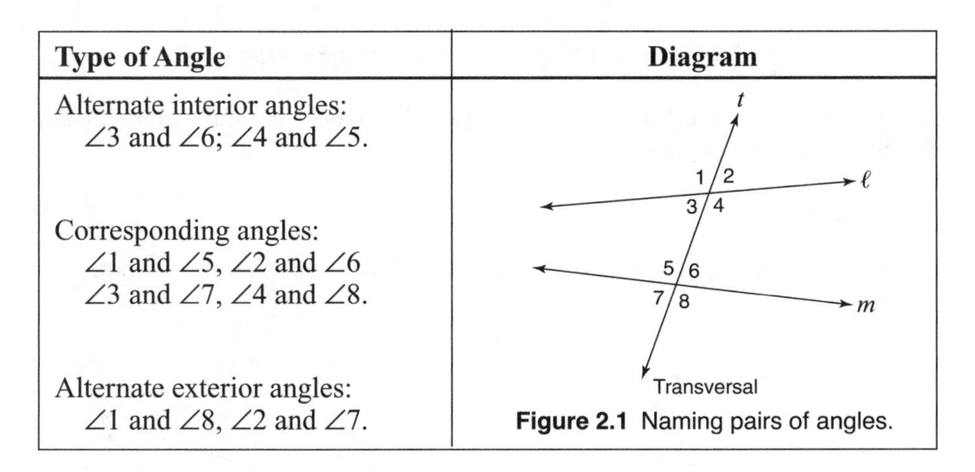 **Figure 2.1** Naming pairs of angles.

From the accompanying table, you can tell that

- **Alternate interior angles** are pairs of interior angles on opposite sides of the transversal that have different vertices.
- **Corresponding angles** are pairs of angles on the same side of the transversal consisting of one interior angle and one exterior angle that have different vertices.

- **Alternate exterior angles** are pairs of exterior angles on opposite sides of the transversal that have different vertices.

Recognizing Special Angle Pairs

Pairs of alternate interior angles trace out ***Z***–shapes as illustrated in Figure 2.2.

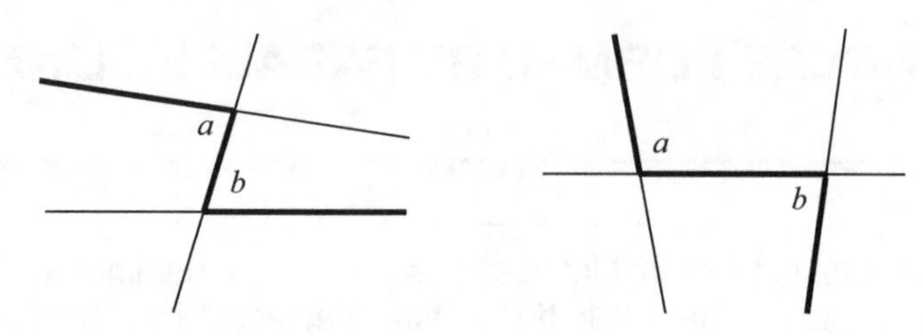

Figure 2.2 Alternate interior angles forming a *Z*-shaped letter.

Pairs of corresponding angles form ***F***–shapes, as shown in Figure 2.3.

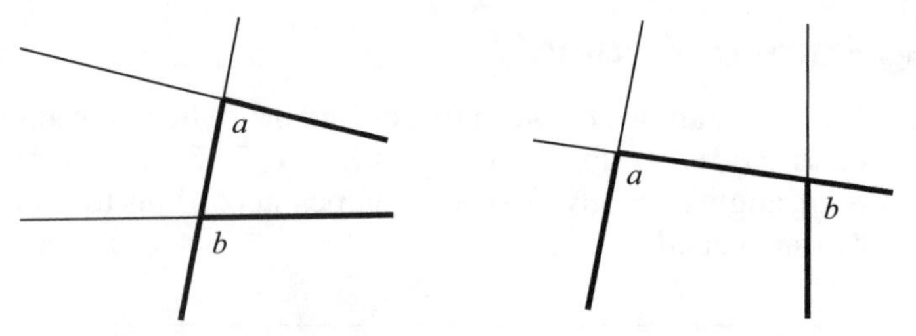

Figure 2.3 Corresponding angles forming an ***F***–shaped Letter.

In different diagrams these ***Z***–shapes and ***F***–shapes may appear reversed, flipped sideways, or rotated.

Example 1

Classify angles 1 and 2 as either alternate interior angles or corresponding angles.

a.

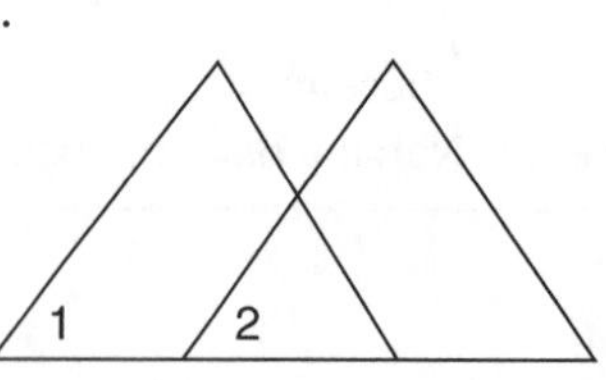

b.

Solution:

a. The sides of angles 1 and 2 trace out an "***F***-shape" so they are *corresponding angles:*

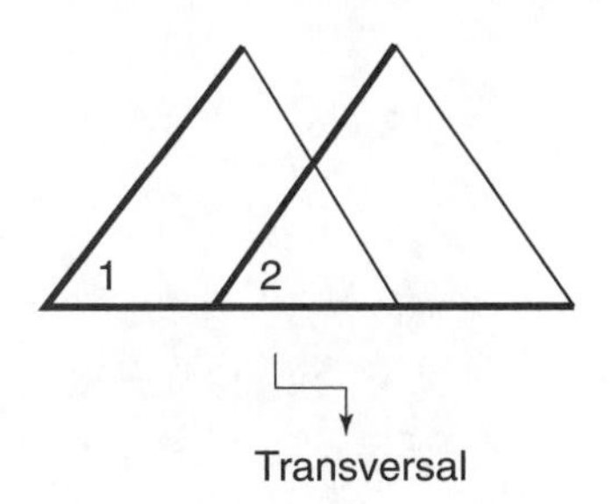

b. The sides of angles 1 and 2 trace out a "***Z***-shape" so they are *alternate interior angles:*

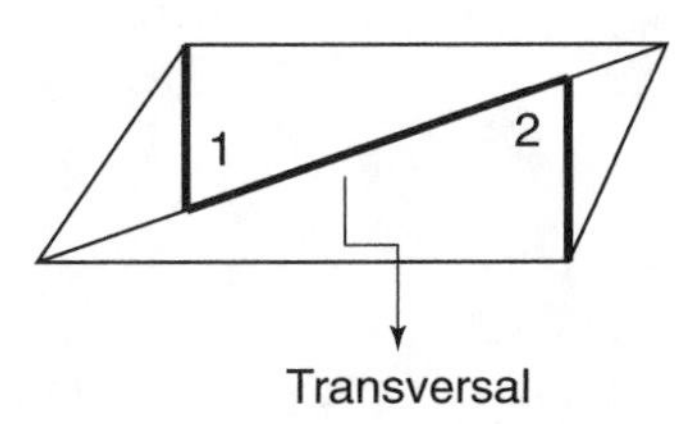

Parallel Lines in a Plane

As a starting point, we assume that *if* lines are parallel, then corresponding angles are congruent.

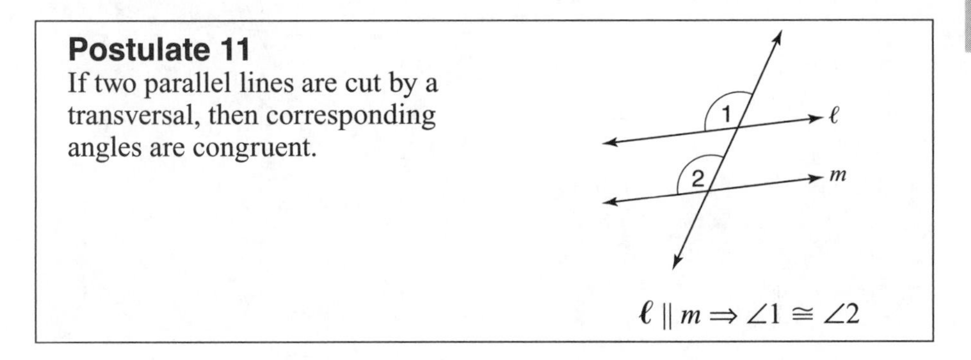

Postulate 11
If two parallel lines are cut by a transversal, then corresponding angles are congruent.

$\ell \parallel m \Rightarrow \angle 1 \cong \angle 2$

Using this postulate, other angle properties of parallel lines can be derived.

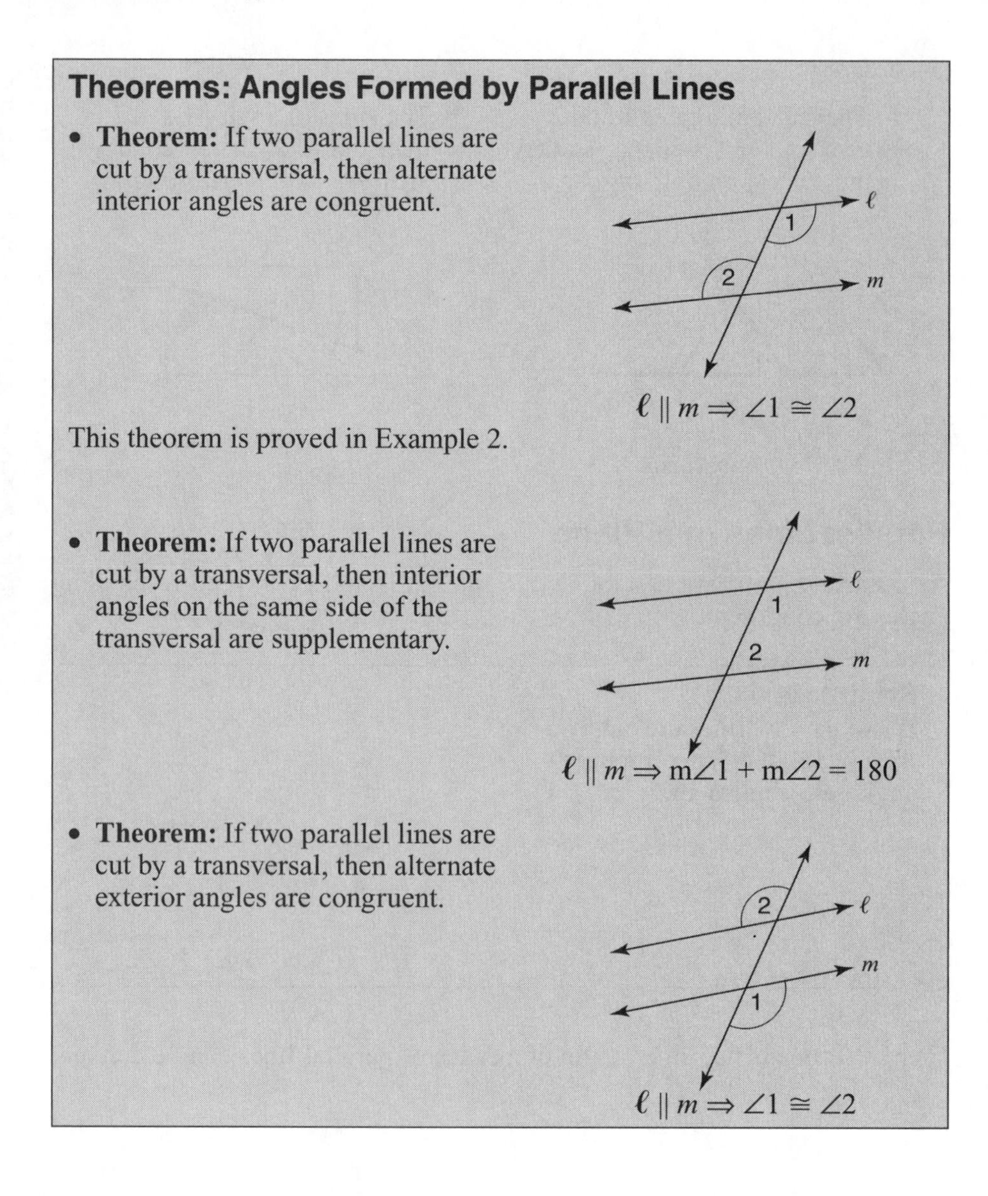

Theorems: Angles Formed by Parallel Lines

- **Theorem:** If two parallel lines are cut by a transversal, then alternate interior angles are congruent.

$\ell \parallel m \Rightarrow \angle 1 \cong \angle 2$

This theorem is proved in Example 2.

- **Theorem:** If two parallel lines are cut by a transversal, then interior angles on the same side of the transversal are supplementary.

$\ell \parallel m \Rightarrow m\angle 1 + m\angle 2 = 180$

- **Theorem:** If two parallel lines are cut by a transversal, then alternate exterior angles are congruent.

$\ell \parallel m \Rightarrow \angle 1 \cong \angle 2$

Example 2

Prove: If two parallel lines are cut by a transversal, then alternate interior angles are congruent.

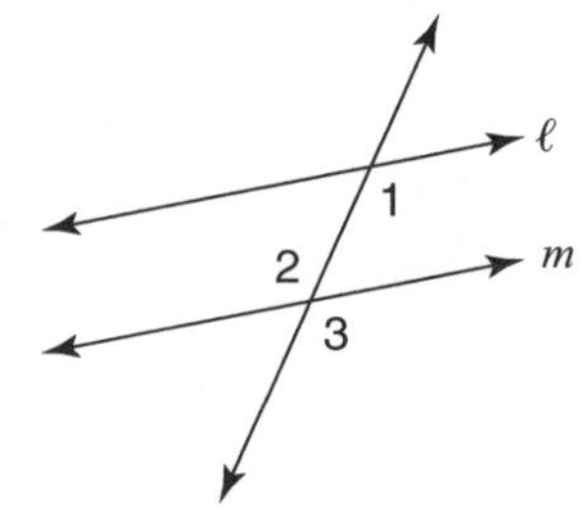

Solution: Set up the proof:

Given: $\ell \parallel m$.
Prove: $\angle 1 \cong \angle 2$.

Proof

Statement	Reason
1. $\ell \parallel m$.	1. Given.
2. $\angle 1 \cong \angle 3$.	2. If two lines are parallel, corresponding angles are congruent.
3. $\angle 3 \cong \angle 2$.	3. Vertical angles are congruent.
4. $\angle 1 \cong \angle 2$.	4. Transitive property.

Example 3

In the accompanying diagram, $\ell \parallel m$. What is the value of x?

Solution: Since corresponding angles formed by parallel lines are congruent or, equivalently, equal in measure:

$$
\begin{aligned}
3x - 40 &= 2x - 10 \\
3x &= 2x - 10 + 40 \\
3x &= 2x + 30 \\
3x - 2x &= 30 \\
\mathbf{x} &= \mathbf{30}
\end{aligned}
$$

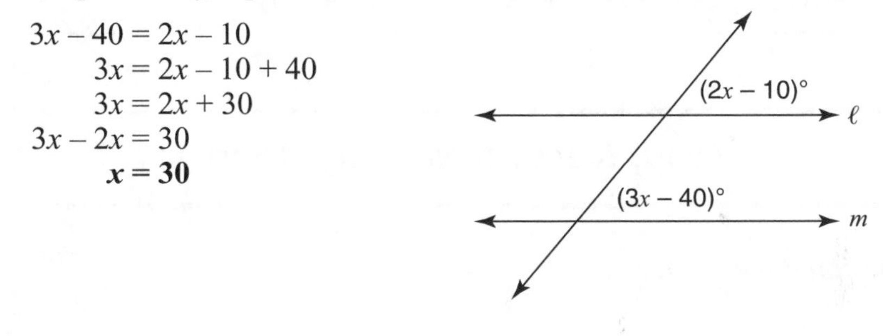

Example 4

In the accompanying diagram, $\overleftrightarrow{ALB} \parallel \overleftrightarrow{CJD}$ and $\overleftrightarrow{LJ}$ is a transversal. If $m\angle BLJ = 6x - 7$ and $m\angle LJD = 7x + 5$, what is the measure of $\angle JLA$?

Solution: First find $m\angle LJD$. Then use the fact that $m\angle JLA = m\angle LJD$.

- Because $\overleftrightarrow{ALB} \parallel \overleftrightarrow{CJD}$, same side interior angles are supplementary:

$$
\begin{aligned}
m\angle BLJ + m\angle LJD &= 180 \\
(6x - 7) + (7x + 5) &= 180 \\
13x - 2 &= 180 \\
\frac{13x}{13} &= \frac{182}{13} \\
x &= 14
\end{aligned}
$$

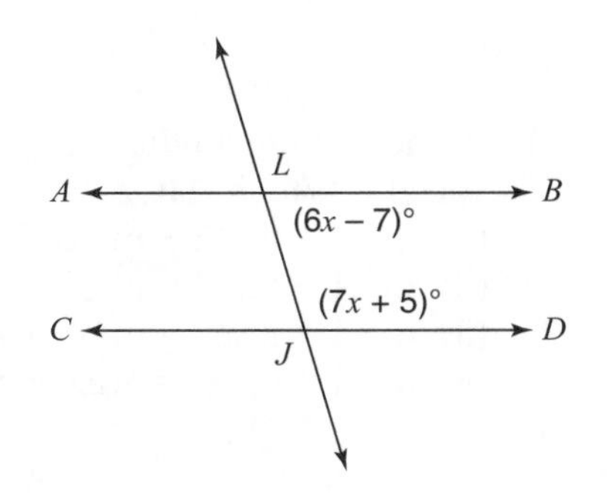

- Find m$\angle LJD$ by substituting 14 for x:

$$\begin{aligned} m\angle LJD &= 7x + 5 \\ &= 7(14) + 5 \\ &= 103 \end{aligned}$$

- Since $\angle JLA$ and $\angle LJD$ are alternate interior angles formed by parallel lines:

$$m\angle JLA = m\angle LJD = \mathbf{103}$$

MATH FACTS

Properties of Parallel Lines

When two *parallel* lines are cut by a transversal, each pair of angles formed are either congruent or supplementary:

- Corresponding angles are congruent
- Alternate interior angles are congruent
- Alternate exterior angles are congruent
- Same side interior angles are supplementary
- Same side exterior angles are supplementary

Check Your Understanding of Section 2.1

A. *Multiple Choice.*

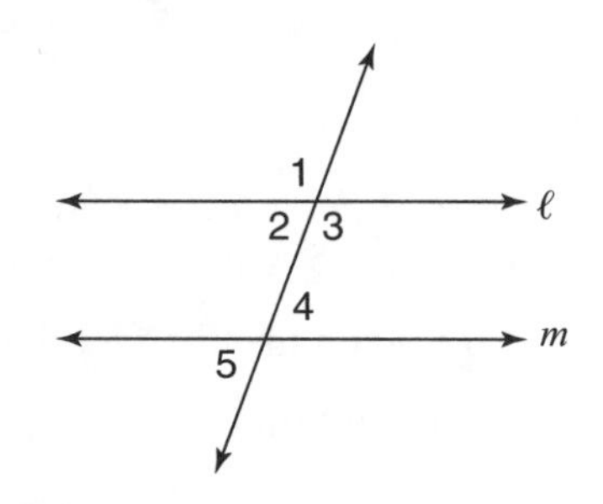

1. In the accompanying diagram, parallel lines ℓ and m are cut by transversal t. Which statement is true?
   (1) $m\angle 1 + m\angle 2 + m\angle 5 = 360$
   (2) $m\angle 1 + m\angle 2 + m\angle 3 = 180$
   (3) $m\angle 1 + m\angle 2 = m\angle 3 + m\angle 4$
   (4) $m\angle 1 + m\angle 3 = m\angle 4 + m\angle 5$

2. In the accompanying diagram, parallel lines $\overleftrightarrow{AB}$ and $\overleftrightarrow{CD}$ are cut by transversal $\overleftrightarrow{EF}$ at P and Q, respectively. Which statement must *always* be true?
(1) $m\angle APE = m\angle CQF$
(2) $m\angle APE + m\angle CQF = 90$
(3) $m\angle APE < m\angle CQF$
(4) $m\angle APE + m\angle CQF = 180$

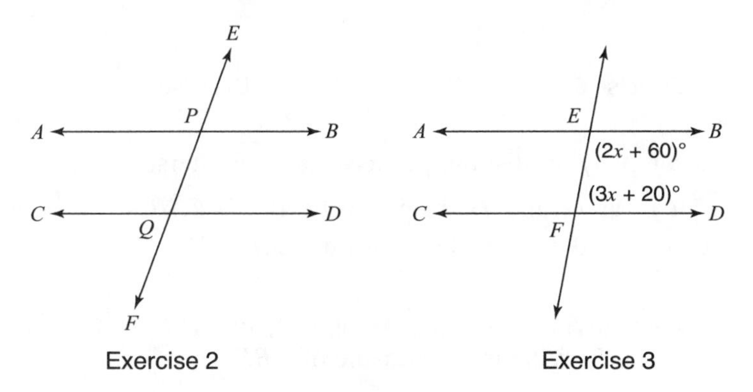

Exercise 2 Exercise 3

3. In the accompanying diagram, parallel lines $\overleftrightarrow{AB}$ and $\overleftrightarrow{CD}$ are cut by transversal $\overleftrightarrow{EF}$. If $m\angle BEF = 2x + 60$ and $m\angle EFD = 3x + 20$, what is $m\angle BEF$?
(1) 100 (2) 20 (3) 140 (4) 40

4. In the accompanying diagram, $\overrightarrow{AD} \parallel \overrightarrow{BC}$ and $\overrightarrow{AC}$ bisects $\angle BAD$. If $m\angle ABC = x$, what is the measure of $\angle 1$ in terms of x?
(1) $90 - x$ (2) $\frac{90-x}{2}$ (3) $90 - \frac{x}{2}$ (4) $\frac{90+x}{2}$

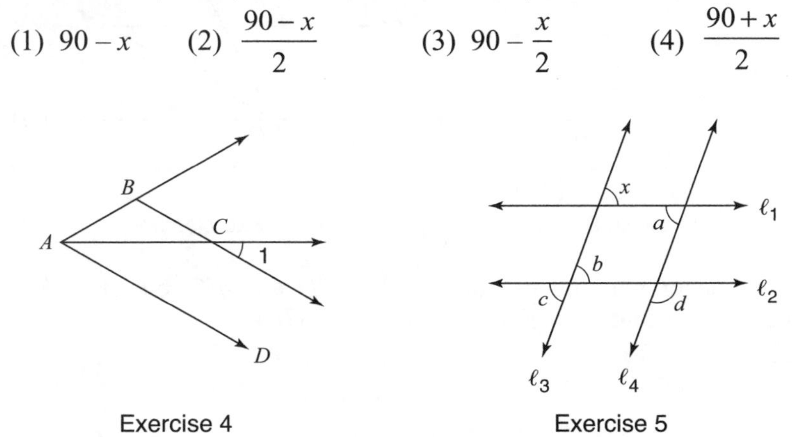

Exercise 4 Exercise 5

5. If, in the accompanying diagram, $\ell_1 \parallel \ell_2$ and $\ell_3 \parallel \ell_4$, then $\angle x$ is *not* always congruent to which angle?
(1) a (2) b (3) c (4) d

B. *Show or explain how you arrived at your answer.*

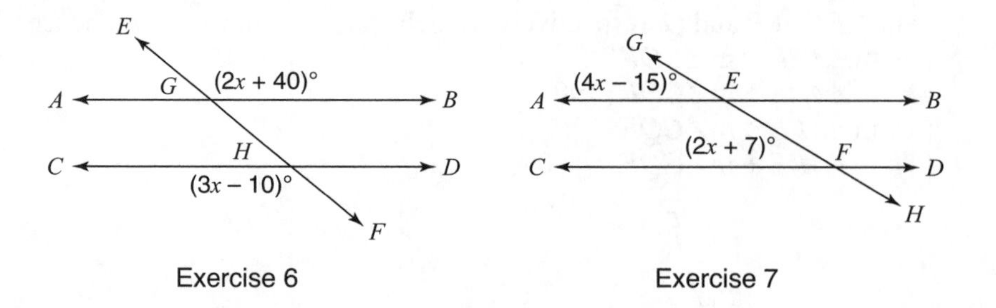

Exercise 6 Exercise 7

6. In the accompanying diagram, transversal $\overleftrightarrow{EF}$ intersects parallel lines $\overleftrightarrow{AB}$ and $\overleftrightarrow{CD}$ at G and H, respectively. If $m\angle EGB = 2x + 40$ and $m\angle FHC = 3x - 10$, what is the measure of $\angle DHE$?

7. In the accompanying diagram, $\overleftrightarrow{AB} \parallel \overleftrightarrow{CD}$, $m\angle AEG = 4x - 15$, and $m\angle CFE = 2x + 7$. What is the measure of $\angle BEF$?

8. In the accompanying diagram, $\overleftrightarrow{WX} \parallel \overleftrightarrow{YZ}$; $\overleftrightarrow{AB}$ and $\overleftrightarrow{CD}$ intersect $\overleftrightarrow{WX}$ at E and $\overleftrightarrow{YZ}$ at F and G, respectively. If $m\angle CEW = m\angle BEX = 50$, find $m\angle EGF$.

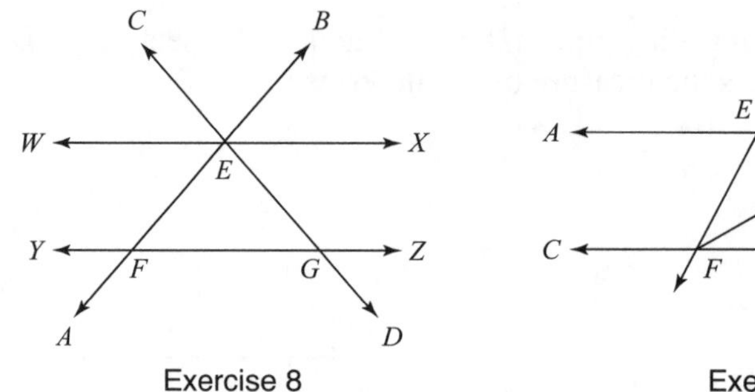

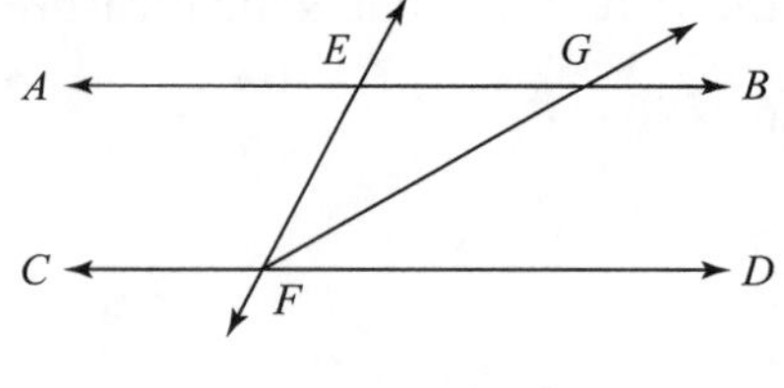

Exercise 8 Exercise 9

9. In the accompanying diagram, $\overleftrightarrow{AB} \parallel \overleftrightarrow{CD}$ and $\overleftrightarrow{FG}$ bisects $\angle EFD$. If $m\angle EFG = x$ and $m\angle FEG = 4x$, find $m\angle EGF$.

C. *Write a proof.*

10. a. Given: $p \parallel q$, $m \perp p$.
Prove: $m \perp q$.
b. Express the result of what you proved in part a as a theorem.

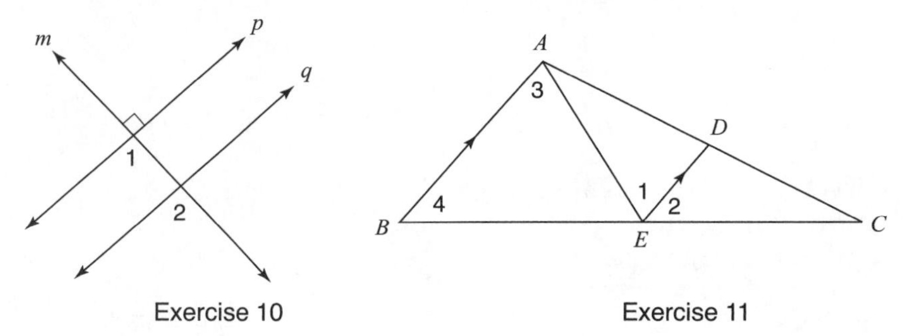

Exercise 10 Exercise 11

11. Given: $\overline{AB} \parallel \overline{DE}$, $\overline{DE}$ bisects $\angle AEC$.
Prove: $\angle 3 \cong \angle 4$.

2.2 PROVING LINES PARALLEL

KEY IDEAS

Methods for proving lines are parallel are based on the converses of the properties of parallel lines presented in the previous lesson.

Determining When Lines Are Parallel

As a starting point, we assume that when corresponding angles are congruent, lines are parallel.

Postulate 12
If two lines are cut by a transversal such that a pair of corresponding angles are congruent, then the lines are parallel.

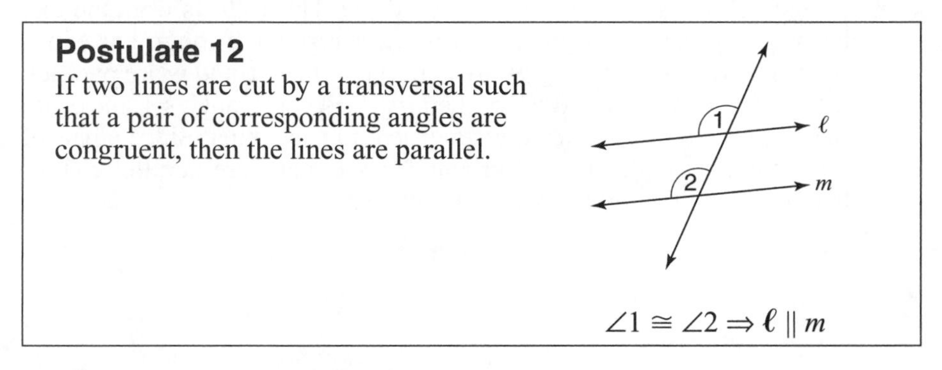

$\angle 1 \cong \angle 2 \Rightarrow \ell \parallel m$

Using this postulate, additional ways of proving that lines are parallel can be derived.

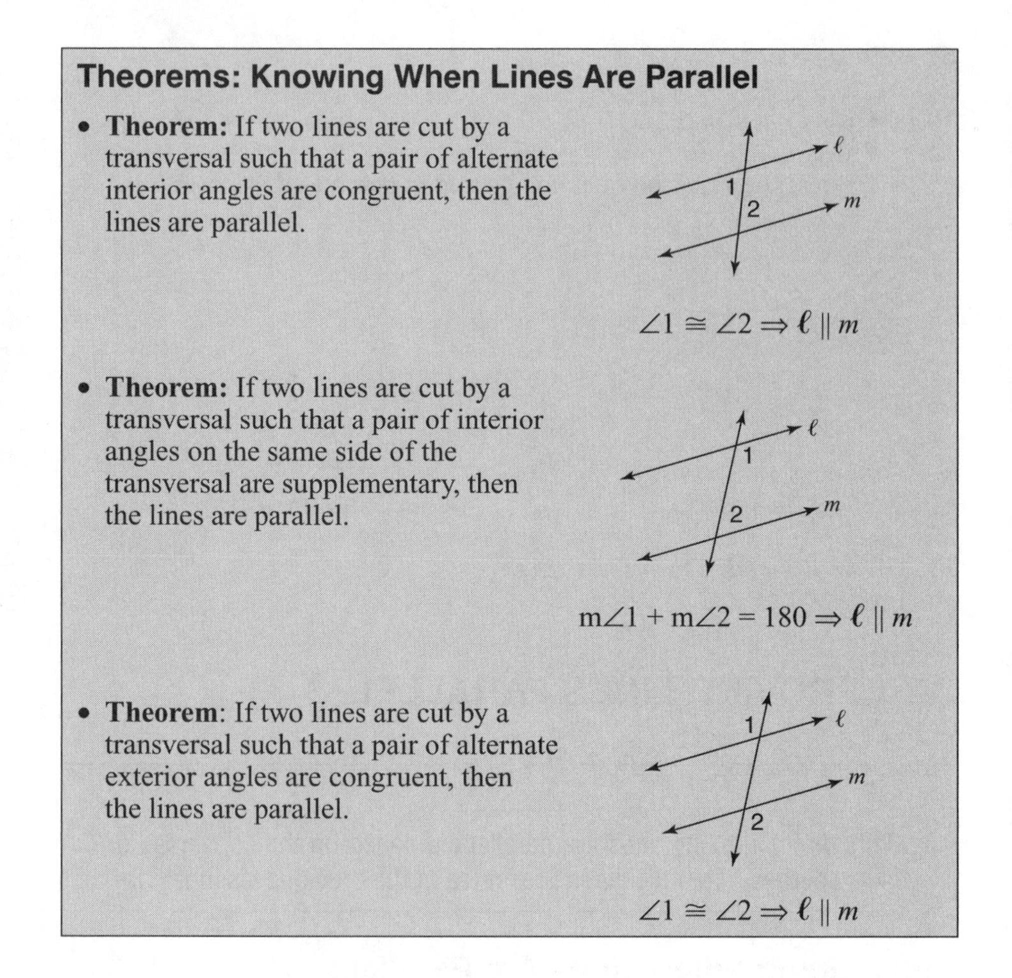

Theorems: Knowing When Lines Are Parallel

- **Theorem:** If two lines are cut by a transversal such that a pair of alternate interior angles are congruent, then the lines are parallel.

$$\angle 1 \cong \angle 2 \Rightarrow \ell \parallel m$$

- **Theorem:** If two lines are cut by a transversal such that a pair of interior angles on the same side of the transversal are supplementary, then the lines are parallel.

$$m\angle 1 + m\angle 2 = 180 \Rightarrow \ell \parallel m$$

- **Theorem**: If two lines are cut by a transversal such that a pair of alternate exterior angles are congruent, then the lines are parallel.

$$\angle 1 \cong \angle 2 \Rightarrow \ell \parallel m$$

It can also be shown that if two lines that are perpendicular or parallel to the same line, the lines are parallel.

- In Figure 2.6, if $\ell \perp t$ and $m \perp t$, then $\ell \parallel m$. This follows from the fact that angles 1 and 2 are right angles and, as a result, are congruent. Since a pair of corresponding angles are congruent, lines ℓ and m are parallel.
- In Figure 2.7, if $p \parallel q$ and $r \parallel q$, then $p \parallel r$. Because angles 1 and 3 are each congruent to $\angle 2$, they are congruent to each other. Since a pair of corresponding angles are congruent, lines p and r are parallel. Thus, there is a transitive property of parallelism.

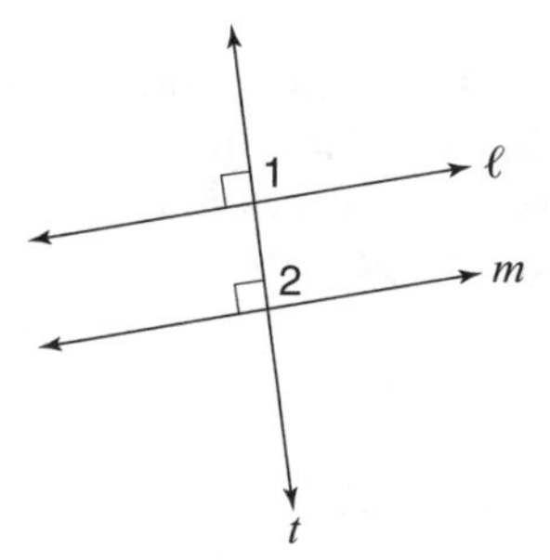

Figure 2.6 Lines perpendicular to the same line.

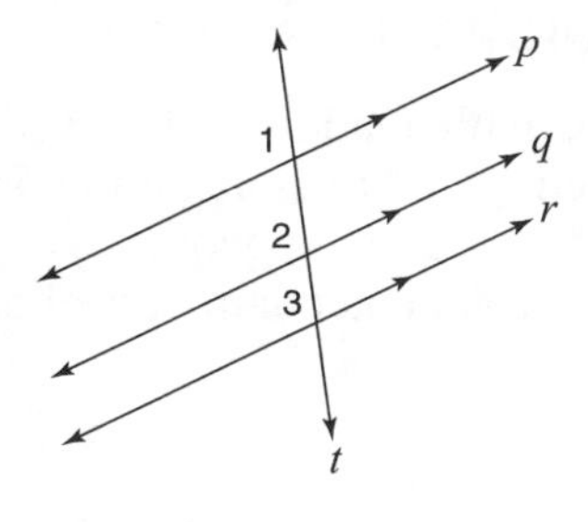

Figure 2.7 Lines parallel to the same line.

MATH FACTS

Ways of Proving Lines Parallel

To prove that two lines are parallel, show that any one of the following statements is true:

- A pair of corresponding angles are congruent.
- A pair of alternate interior or alternate exterior angles are congruent.
- A pair of same side interior angles or same side exterior angles are supplementary.
- The two lines are perpendicular to the same line.
- The two lines are parallel to the same line.

Example 1

Given: $\overline{BC} \parallel \overline{AD}$, $\angle 2 \cong \angle 3$.

Prove: $\overrightarrow{AB} \parallel \overline{CD}$.

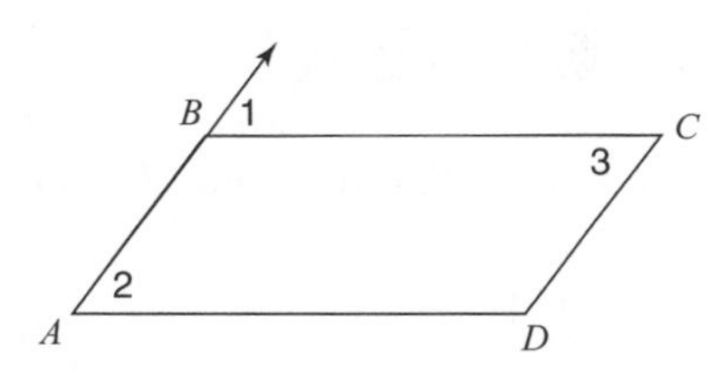

Solution: To prove $\overrightarrow{AB} \parallel \overline{CD}$, show that $\angle 1 \cong \angle 3$.

Proof

Statement	**Reason**
1. $\overline{BC} \parallel \overline{AD}$.	1. Given.
2. $\angle 1 \cong \angle 2$.	2. If two lines are parallel, then corresponding angles are congruent.
3. $\angle 2 \cong \angle 3$.	3. Given.
4. $\angle 1 \cong \angle 3$.	4. Transitive property.
5. $\overrightarrow{AB} \parallel \overline{CD}$.	5. If two lines are cut by a transversal such that a pair of alternate interior angles are congruent, then the lines are parallel.

Euclid's Parallel Postulate

Euclid, a Greek mathematician who lived in approximately 300 B.C., is credited with collecting and organizing the postulates, definitions, and theorems that are studied in beginning geometry courses. The *parallel postulate* is one of the most important and controversial of Euclid's postulates.

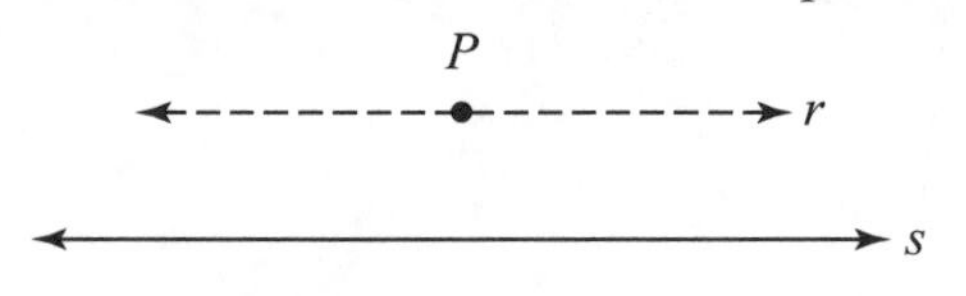

> **Postulate 13: Parallel Postulate**
> Through a point not on a line, there is exactly one line parallel to the given line.

The statement of the parallel postulate offered here is actually a simpler version of the one given by Euclid. It was formulated by John Playfair in 1795. Through the centuries, mathematicians have debated the validity of the parallel postulate. Geometries have been created that do not accept the parallel postulate, assuming instead that through a point not on a line there are *no* parallel lines or there is *more than one* parallel line. These *non–Euclidean* geometries are studied in more advanced geometry courses.

Check Your Understanding of Section 2.2

A. *Multiple Choice*

1. Which figure contains a pair of parallel lines?

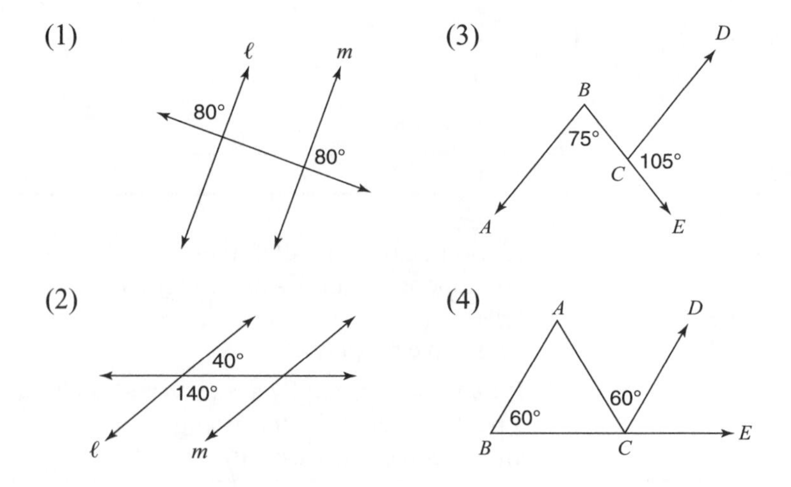

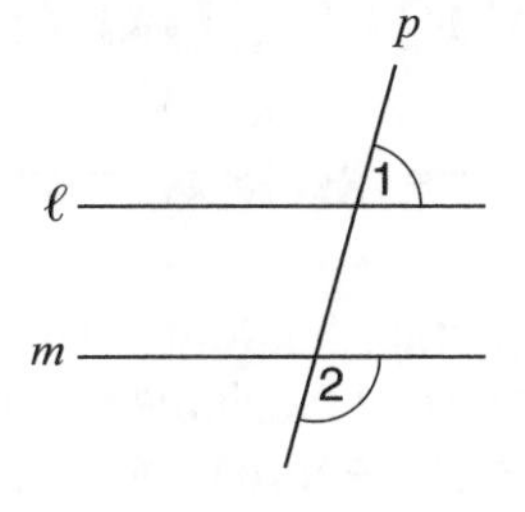

2. In the accompanying diagram, ∠1 and ∠2 are supplementary. Which is always true?
(1) $\ell \perp p$ (2) $\ell \perp m$ (3) $p \parallel \ell$ (4) $p \parallel m$

B. *Show or explain how you arrived at your answer.*

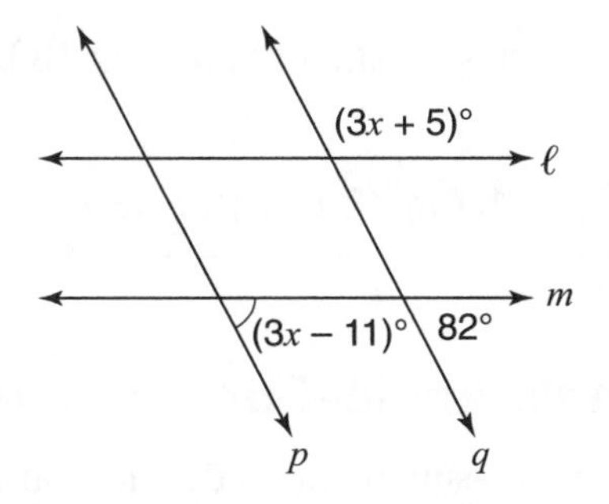

3. In the accompanying diagram, lines p and q are parallel. Determine whether lines ℓ and m are parallel. Give a reason for your answer.

C. *Write a proof.*

4. Given: $\overrightarrow{BD}$ bisects $\angle ABC$,
$\angle 1 \cong \angle 2$.
Prove: $\overleftrightarrow{AD} \parallel \overleftrightarrow{BC}$.

5. Given: $\overline{TC} \perp \overline{AC}$, $\overline{HK} \perp \overline{AC}$,
$\overline{KH}$ bisects $\angle AHC$.
Prove: $\angle 1 \cong \angle 4$.

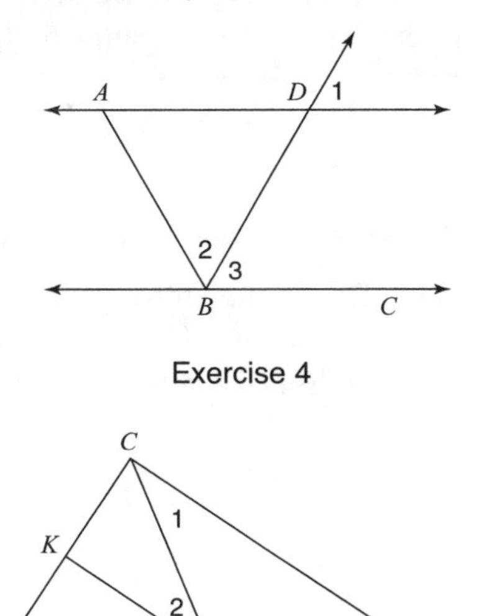

Exercise 4

Exercise 5

2.3 TRIANGLE ANGLE SUM THEOREM

KEY IDEAS

The accompanying diagrams illustrate that, after ∠1 and ∠3 have been "torn off," their sides can be aligned with one of the sides of ∠2 so that the exterior sides of the three angles form a straight line.

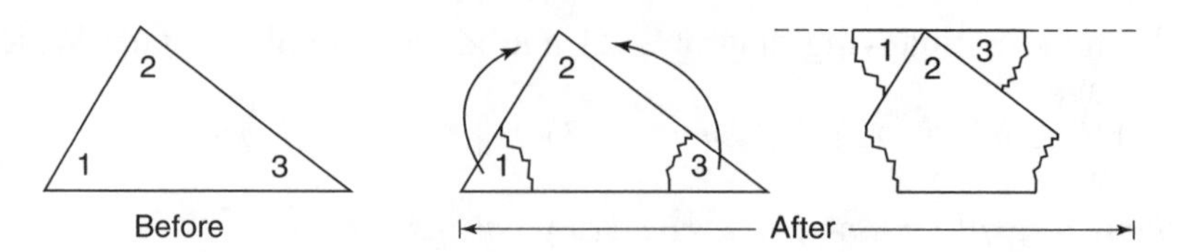

Since the degree measure of a straight angle is 180, this experiment suggests that

$$m\angle 1 + m\angle 2 + m\angle 3 = 180$$

Proof of the Triangle Angle–Sum Theorem

The proof of the triangle angle–sum theorem depends on the Parallel Postulate.

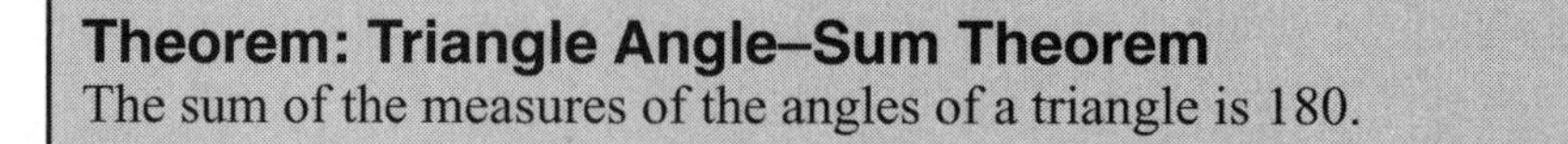

Theorem: Triangle Angle–Sum Theorem
The sum of the measures of the angles of a triangle is 180.

Given: $\triangle ABC$

Prove: $m\angle 1 + m\angle 2 + m\angle 3 = 180$.

Plan: Draw a line through B and parallel to $\overline{AC}$. Then make the appropriate angle substitutions.

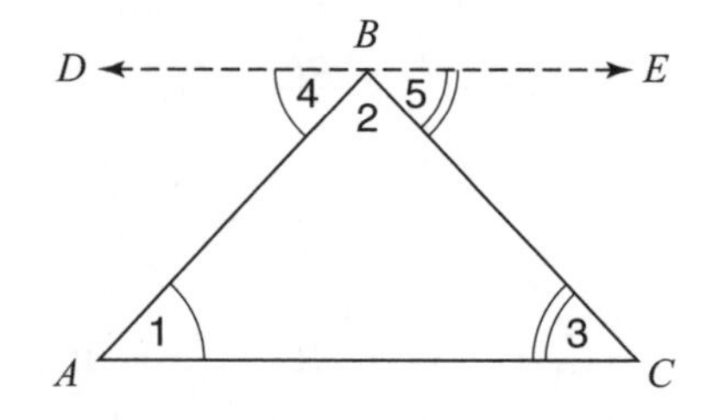

Proof

Statement	Reason
1. Draw $\overleftrightarrow{DE}$ through B and parallel to $\overline{AC}$.	1. Parallel postulate.
2. $m\angle DBC + m\angle 5 = 180$.	2. If two angles form a linear pair, they are supplementary.
3. $m\angle DBC = m\angle 4 + m\angle 2$.	3. Angle addition postulate.
4. $m\angle 4 + m\angle 2 + m\angle 5 = 180$.	4. Substitution property.
5. $m\angle 4 = m\angle 1$ and $m\angle 5 = m\angle 3$.	5. If two lines are parallel, then alternate interior angles are equal in measure.
6. $m\angle 1 + m\angle 2 + m\angle 3 = 180$.	6. Substitution property.

Sometimes a theorem has some immediate consequences worth noting. A **corollary** is a theorem that is a byproduct of another theorem from which it can easily be proved.

Corollaries To Triangle–Sum Theorem

- Corollary 1: A triangle can have, *at most*, one right or obtuse angle.
- Corollary 2: The acute angles of a right triangle are complementary.
- Corollary 3: If two angles of a triangle are congruent to two angles of another triangle, then the third pair of angles are also congruent.

Example 1

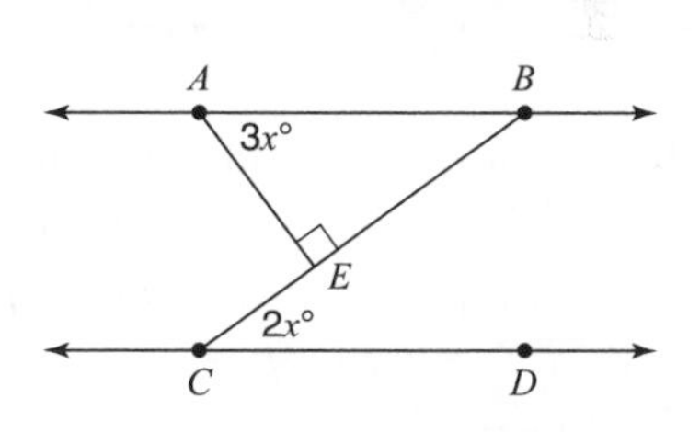

In the accompanying figure, $\overleftrightarrow{AB} \parallel \overleftrightarrow{CD}$, is parallel to $m\angle BCD = 2x$, $m\angle BAE = 3x$. Find the value of x.

Solution: Since alternate interior angles formed by parallel lines have the same degree measure, $m\angle ABE = m\angle BCD = 2x$. In right triangle AEB, angles BAE and ABE are the acute angles and, as a result, are complementary. Hence:

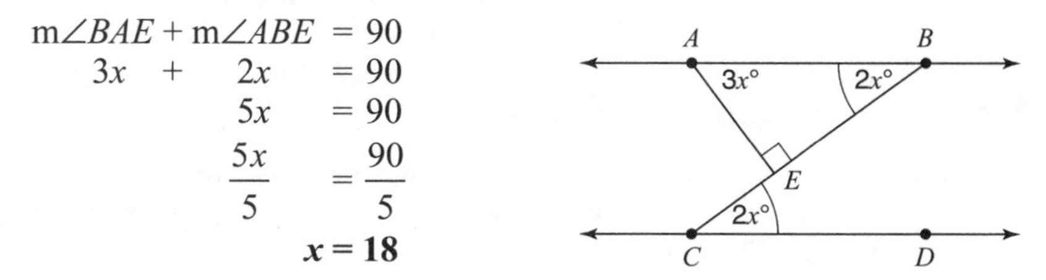

$$\begin{aligned} m\angle BAE + m\angle ABE &= 90 \\ 3x + 2x &= 90 \\ 5x &= 90 \\ \frac{5x}{5} &= \frac{90}{5} \\ \mathbf{x} &= \mathbf{18} \end{aligned}$$

***Example* 2**

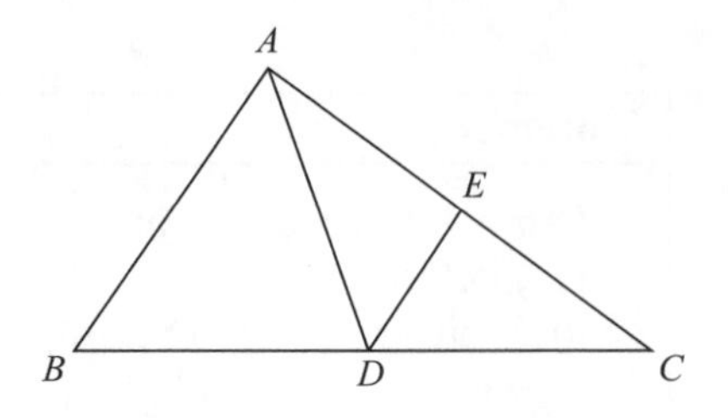

In the accompanying diagram of $\triangle ABC$, $\overline{DE} \perp \overline{AEC}$. If $m\angle ADB = 80$ and $m\angle CDE = 60$, what is the measure of $\angle DAE$?

Solution: Since angles *ADB*, *ADE*, and *CDE* form a straight angle, the sum of their degree measures is 180. Hence,

$$\begin{aligned} 80 + m\angle ADE + 60 &= 180 \\ m\angle ADE &= 180 - 140 \\ &= 40 \end{aligned}$$

In $\triangle ADE$,

$$\begin{aligned} m\angle DAE + m\angle ADE + m\angle AED &= 180 \\ m\angle DAE + \quad 40 \quad + \quad 90 \quad &= 180 \\ m\angle DAE &= 180 - 130 \\ &= \mathbf{50} \end{aligned}$$

***Example* 3**

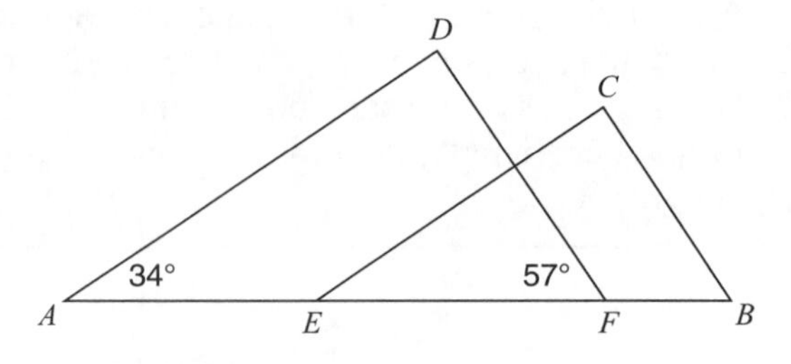

In the accompanying diagram, $\overline{AD} \parallel \overline{EC}$, $\overline{DF} \parallel \overline{CB}$, $m\angle DAE = 34$, and $m\angle DFE = 57$. Find $m\angle ECB$.

Solution: Because $\overline{AD} \parallel \overline{EC}$, corresponding angles are congruent so $m\angle CEB = m\angle DAE = 32$. Similarly, since $\overline{DF} \parallel \overline{CB}$, $m\angle CBE = m\angle DFE = 57$. In $\triangle CEB$,

$$\begin{aligned} m\angle ECB + m\angle CEB + m\angle CBE &= 180 \\ m\angle ECB + \quad 34 \quad + \quad 57 \quad &= 180 \\ m\angle ECB &= 180 - 91 \\ &= \mathbf{89} \end{aligned}$$

Exterior Angles of a Triangle

At each vertex of a triangle an *exterior* angle may be formed by extending one side of the triangle. In Figure 2.8, an exterior angle is formed at vertex *C* by extending $\overline{AC}$. Thus,

$$m\angle 1 + c = 180$$
and
$$(a + b) + c = 180$$

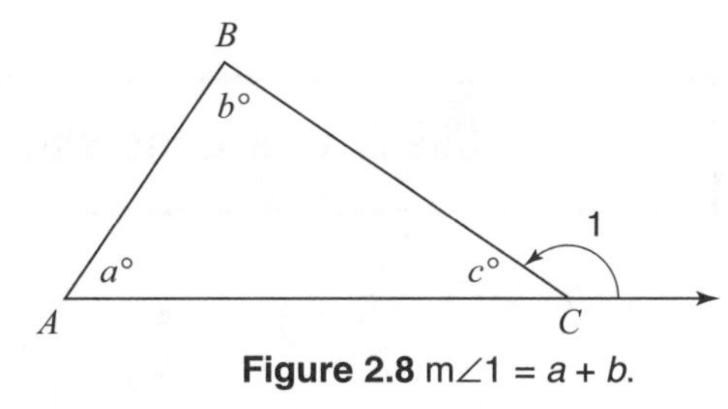

Figure 2.8 m∠1 = *a* + *b*.

By comparing corresponding sides of the two equations, it must be the case that $m\angle 1 = a + b$. The angles whose measures are represented by a and b are the two interior angles of the triangle that are the most distant from ∠1. With respect to ∠1, these angles are non–adjacent or "remote" interior angles.

> **Theorem: Exterior Angle of a Triangle**
> The measure of an exterior angle of a triangle is equal to the sum of the measures of the two remote (non–adjacent) interior angles of the triangle.

Example 4

Find the value of x.

a.

b.

c.

Solution:

a.
$$2x + 30 + 60 = 110$$
$$2x = 110 - 90$$
$$2x = 20$$
$$\frac{2x}{2} = \frac{20}{2}$$
$$\boldsymbol{x = 10}$$

b.
$$3x - 10 = (x + 15) + 45$$
$$3x - 10 = x + 60$$
$$3x = x + 70$$
$$2x = 70$$
$$\frac{2x}{2} = \frac{70}{2}$$
$$\boldsymbol{x = 35}$$

c.

$x + 50 = 110$, so $\boldsymbol{x = 60}$.

Check Your Understanding of Section 2.3

A. *Multiple Choice*

1. In the accompanying diagram, $\overline{RT}$ is extended to W, $\overrightarrow{RQ}$ and $\overrightarrow{TP}$ intersect at S to form $\triangle RST$, $m\angle PSQ = 40$, and the measure of exterior angle WRQ is 135. What is the measure of $\angle STR$?
(1) 85 (2) 95 (3) 105 (4) 175

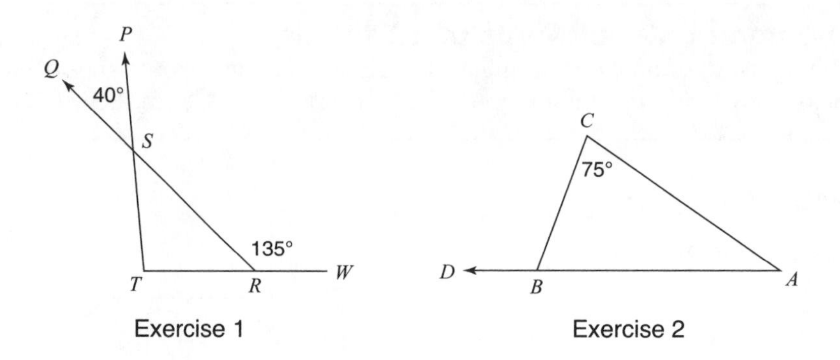

Exercise 1 Exercise 2

2. In the accompanying diagram of $\triangle ABC$, $\overline{AB}$ is extended to D, $m\angle ACB = 75$, and the measure of exterior angle CBD exceeds three times $m\angle CAB$ by 5. What is $m\angle CAB$?
(1) 25 (2) 35 (3) 45 (4) 60

3. In the accompanying diagram of $\triangle ABC$, $\angle BCD$ is an exterior angle formed by extending $\overline{AC}$ to D, $m\angle A = x + 30$, $m\angle B = 2x$, and $m\angle BCD = 120$. What is the value of x?
(1) 20 (2) 30 (3) 60 (4) 90

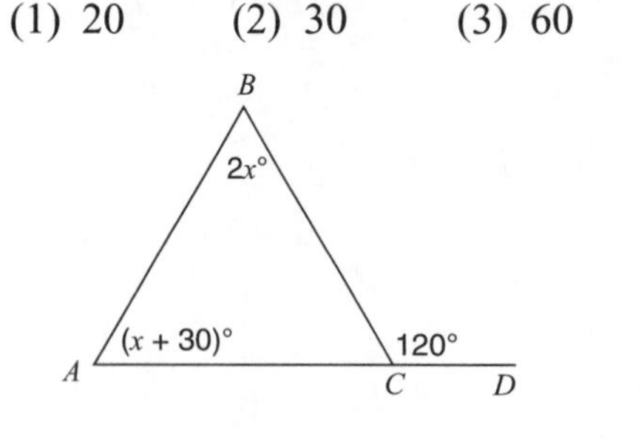

Exercise 3

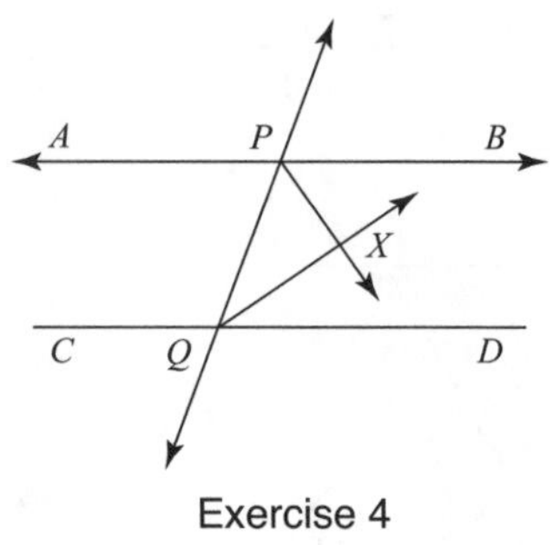

Exercise 4

4. In the accompanying diagram, parallel lines $\overleftrightarrow{AB}$ and $\overleftrightarrow{CD}$ are cut by transversal $\overleftrightarrow{PQ}$ at points P and Q, $\overleftrightarrow{PX}$ bisects $\angle QPB$, and $\overleftrightarrow{QX}$ bisects $\angle PQD$. Which statement is *always* true?
(1) $\angle PXQ$ is a right angle.
(2) $m\angle PXQ = m\angle APQ$.
(3) $m\angle PXQ = m\angle XPB$.
(4) $m\angle PXQ + m\angle DQX = 180$.

5. If the measures of the angles of a triangle are represented by $5x - 7$, $7x + 6$, and $4x - 11$, the triangle is
(1) acute (2) right (3) obtuse (4) equiangular

6. Two angles of a triangle measure 72 and 46. Which could be the measure of an exterior angle of this triangle?
(1) 46 (2) 62 (3) 108 (4) 144

7. The measure of one acute angle of an obtuse triangle is 20 more than the measure of the other acute angle. What is a possible measure of the *larger* acute angle?
(1) 50 (2) 55 (3) 60 (4) 65

8. In the accompanying diagram of $\triangle ABC$, $\overline{BC}$ is extended to D, $m\angle B = 2y$, $m\angle BCA = 6y$, and $m\angle ACD = 3y$. What is the measure of $\angle A$?
(1) 15 (2) 17 (3) 20 (4) 24

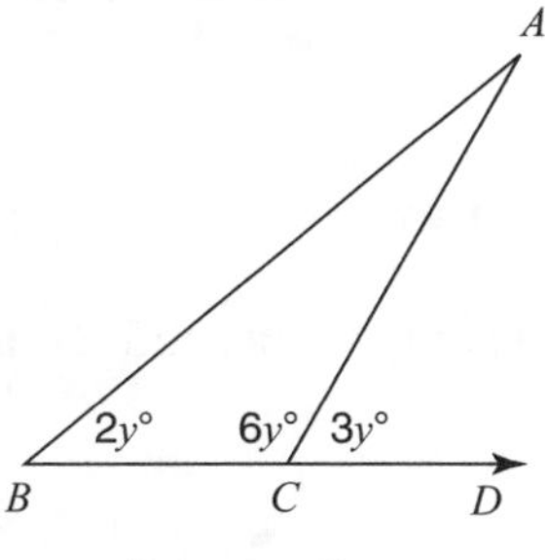

Exercise 8

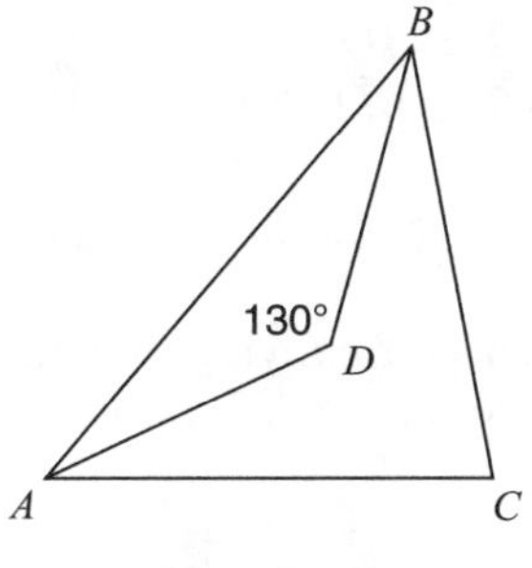

Exercise 9

9. In the accompanying diagram, the bisectors of $\angle A$ and $\angle B$ in acute triangle ABC meet at D, and $m\angle ADB = 130$. What is the measure of $\angle C$?
(1) 50 (2) 60 (3) 70 (4) 80

10. In the accompanying diagram of $\triangle ABC$, $\overline{AC}$ is extended to D, $\overline{DEF}$ is drawn, $m\angle B = 50$, $m\angle BFE = 105$, and $m\angle ACB = 65$. What is the measure of $\angle D$?

(1) 40 (2) 45 (3) 50 (4) 55

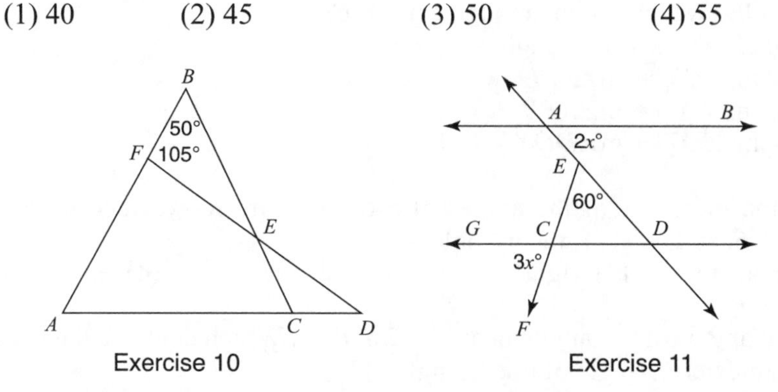

Exercise 10 Exercise 11

11. In the accompanying diagram, $\overleftrightarrow{AB} \parallel \overleftrightarrow{GCD}$, $\overleftrightarrow{AED}$ is a transversal, and $\overline{EC}$ is extended to F. If $m\angle CED = 60$, $m\angle DAB = 2x$, and $m\angle FCG = 3x$, what is the $m\angle GCE$?

(1) 36 (2) 72 (3) 108 (4) 144

12. In the accompanying diagram of $\triangle ABD$, C is a point on $\overline{AD}$, and $\overline{BC}$ is drawn.
Which statement must be true?

(1) $\overline{BC} \perp \overline{AD}$
(2) $\overline{AC} \cong \overline{CD}$
(3) $\overline{AB} \cong \overline{BD}$
(4) $\overline{AB} \perp \overline{BD}$

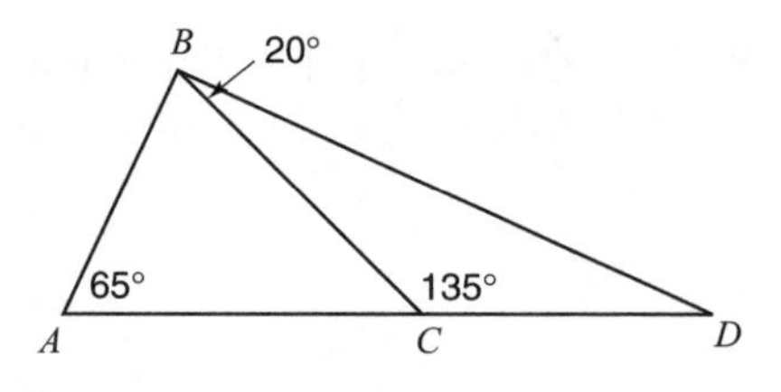

B. *Show or explain how you arrived at your answer.*

13. In the accompanying diagram, $\overleftrightarrow{ABCD}$ is a straight line, and $\angle E$ in $\triangle BEC$ is a right angle. What does $a + d$ equal?

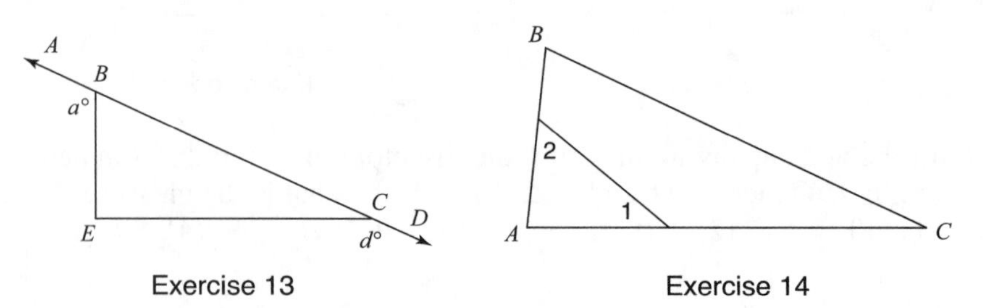

Exercise 13 Exercise 14

14. In the accompanying diagram of $\triangle ABC$, $m\angle 1 = 40$, $m\angle 2 = 55$ and $m\angle B = 70$. Find $m\angle C$.

15. In the accompanying diagram, line a is parallel to line b, and line t is a transversal. If $m\angle 1 = 97$ and $m\angle 2 = 44$, find $m\angle 3$.

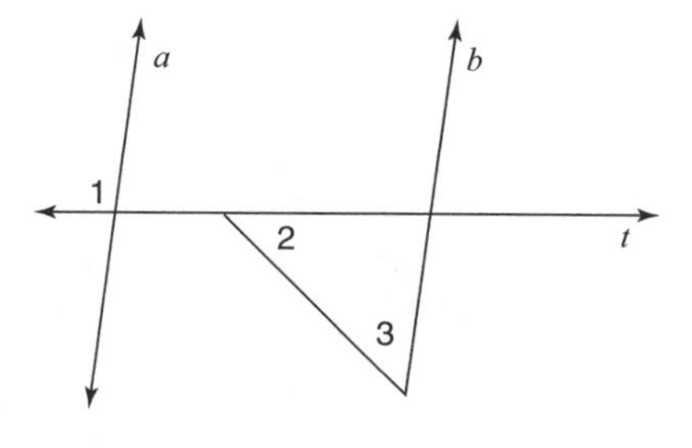

16. In the accompanying diagram, $\overleftrightarrow{ABC} \parallel \overline{DE}$, $\overline{BC}$ bisects $\angle DBC$, $m\angle FDE = 25$, and $m\angle DFB = 105$. What is $m\angle ABD$?

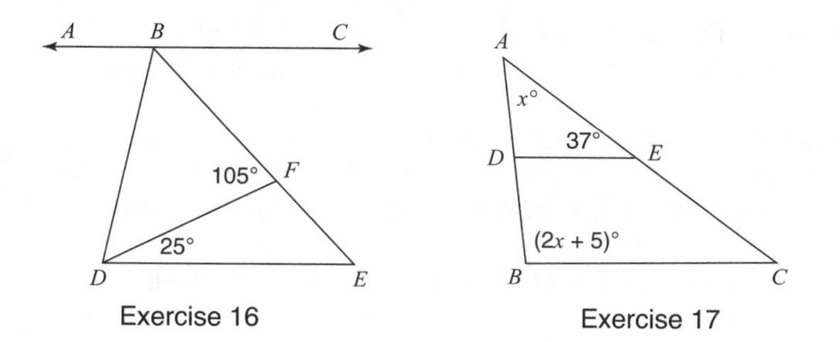

17. In the accompanying diagram, $\overline{DE} \parallel \overline{BC}$, $m\angle AED = 37$, $m\angle A = x$, and $m\angle B = 2x + 5$. What is the measure of $\angle ADE$?

2.4 POLYGONS AND THEIR ANGLES

KEY IDEAS

A triangle is the simplest type of *polygon*. It has three sides and three angles. A polygon with *n* sides has *n* interior angles. A simple formula can be used to find the sum of the measures of the interior angles of a polygon with any number of sides.

Definition of a Polygon

The term *polygon* means "many angles." A **polygon** is a closed figure whose sides are line segments that intersect at their endpoints. The polygon in Figure 2.9 has five sides. Each "corner" point where two sides intersect is a **vertex** of the polygon. The vertices of this polygon are A, B, C, D, and E. Line segment BE is a *diagonal*. A **diagonal** of a polygon is a line segment joining two nonconsecutive vertices.

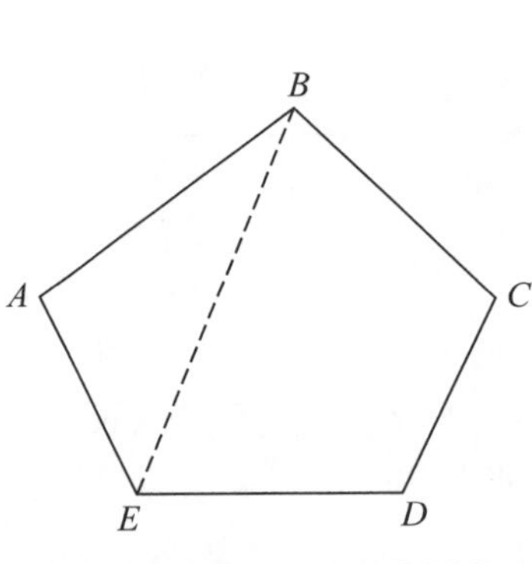

Figure 2.9 Polygon *ABCDE* with diagonal $\overline{BE}$.

Sides	Name
4	Quadrilateral
5	Pentagon
6	Hexagon
8	Octagon
9	Nonagon
10	Decagon
12	Dodecagon
n	n–gon

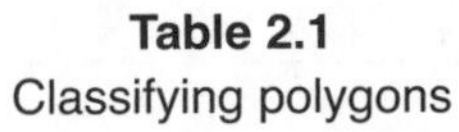

Table 2.1
Classifying polygons

Beginning with any of its lettered vertices, a polygon is named by listing consecutive vertices in either clockwise or counterclockwise order. The polygon in Figure 2.9 can be named in many different ways including *ABCDE* and *EDCBA*.

A **triangle** is a polygon with three sides. Table 2.1 lists other polygons and their names.

Convex Polygons

Polygon *ABCDEF* in Figure 2.10 is **convex** since a line drawn through any two interior points intersects the polygon in exactly two points. The measure of each interior angle of a convex polygon is always less than 180.

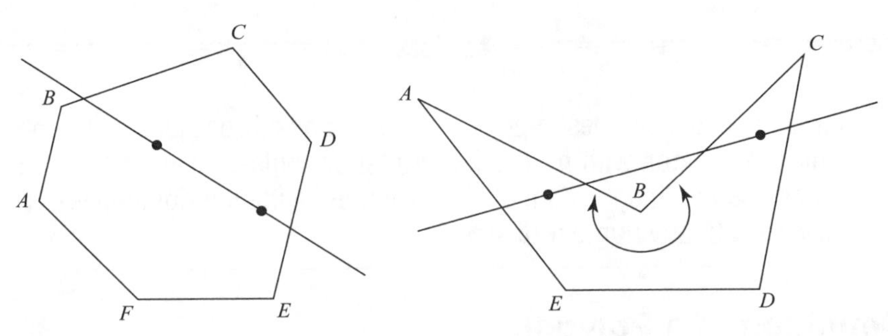

Figure 2.10 A convex polygon. **Figure 2.11** A concave polygon.

A polygon is nonconvex or **concave** if there are two points in the interior of the polygon such that the line through them (extended, if necessary) intersects the polygon in more than two points, as in Figure 2.11. The measure of the interior angle at vertex *B* is *greater than* 180, which also indicates that the polygon is concave. Only convex polygons are studied in this course.

Regular Polygons

In a polygon, the measures of the sides, the angles, or both the angles and sides may be equal. A **regular polygon** is a polygon in which each interior angle has the same measure (**equiangular**) *and* each side has the same length (**equilateral**). A square is a regular polygon because its four sides have the same length and its four right angles have the same measure.

Angle–Sum Formula for Polygons

A diagonal of a quadrilateral divides the quadrilateral into two triangles. The sum of the measures of the angles of these two triangles must be the same as the sum of the measures of the four angles of the quadrilateral. That sum is 2×180 or 360. Figure 2.12 shows that the number of triangles into which a polygon can be separated is always 2 less than the number of sides of the polygon.

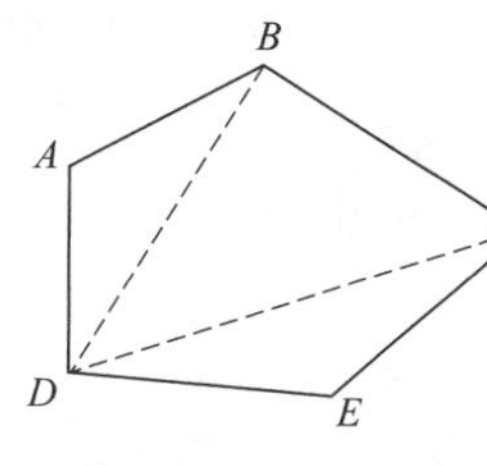

5 sides ⇒ **3** △s
Angle sum = $\underline{3} \times 180$

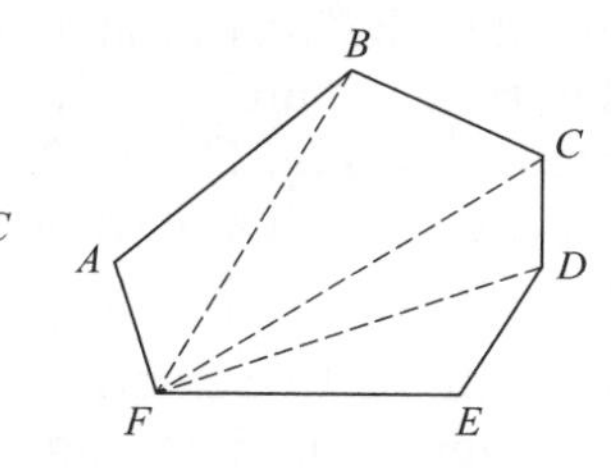

6 sides ⇒ **4** △s
Angle sum = $\underline{4} \times 180$

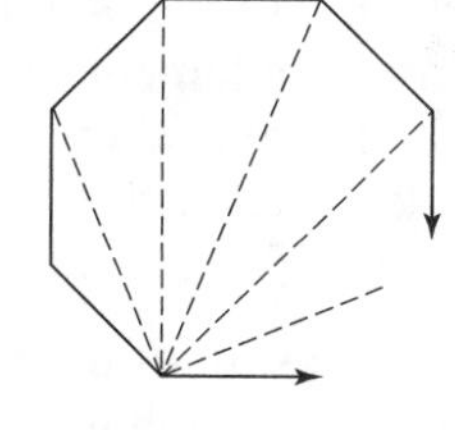

n sides ⇒ $\mathbf{(n-2)}$ △s
Angle sum = $\underline{(n-2)} \times 180$

Figure 2.12 Dividing a polygon with n sides into $n-2$ triangles.

> **Theorem: Polygon Angle–Sum Formula**
> The sum, S, of the measures of the interior angles of a convex polygon with n sides is given by the formula
>
> $$S = (n-2) \times 180.$$

Example 1

What is the sum of the measures of the interior angles of a stop sign, which is in the shape of an octagon?

Solution: An octagon has eight sides. Evaluate the formula $S = (n-2) \times 180$ for $n = 8$:

$$\begin{aligned} S &= (8-2) \times 180 \\ &= 6 \times 180 \\ &= \mathbf{1080} \end{aligned}$$

Example 2

Determine the number of sides of a polygon in which the sum of the measures of the angles is 900.

Solution: If n represents the number of sides of the polygon, then

$$900 = (n - 2) \times 180$$
$$900 = 180n - 360$$
$$1260 = 180n$$
$$\frac{180n}{180} = \frac{1260}{180}$$
$$\mathbf{n = 7 \text{ sides}}$$

Exterior Angles of a Polygon

At each vertex of a polygon, an *exterior* angle may be formed by extending one side of the polygon such that the interior and exterior angles at that vertex are supplementary. In Figure 2.13, angles 1, 2, 3, and 4 are exterior angles and the sum of their degree measures is 360. If the polygon is equiangular or regular, then each exterior angle, as well as each interior angle, has the same measure. For a regular pentagon,

- The sum of the measures of the five exterior angles is 360.
- As the polygon is equiangular, the measure of each exterior angle is $\frac{360}{5} = 72$.
- Since an exterior angle is supplementary to an interior angle at each of the vertices, the measure of each interior angle is $180 - 72 = 108$.

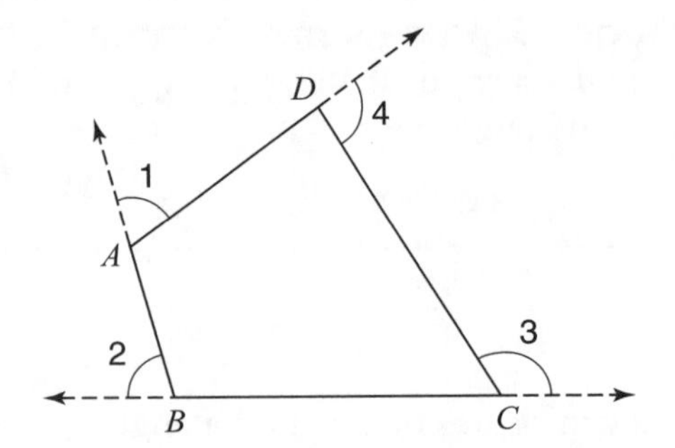

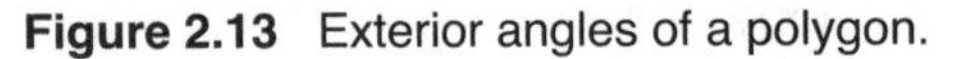
Figure 2.13 Exterior angles of a polygon.

Exterior Angle Theorems for Polygons

- **Theorem: Sum of Exterior Angles**
  The sum of the measures of the exterior angles of a convex polygon with any number of sides, one exterior angle at each vertex, is 360.
- **Theorem: Interior and Exterior Angle Relationships**
  If a *regular* polygon has n sides, then the measure of each exterior angle is $\frac{360}{n}$ and the measure of each interior angle is $180 - \left(\frac{360}{n}\right)$.

Example 3

Find the measure of each interior angle and each exterior angle of a regular decagon.

Solution: A decagon has ten sides.

- The measure of each exterior angle is $\frac{360}{10} = \mathbf{36}$.
- The measure of each interior angle is $180 - 36 = \mathbf{144}$.

Example 4

If the measure of each interior angle of a regular polygon is 150, how many sides does the polygon have?

Solution: Let n represent the number of sides of the regular polygon.

- Since each interior angle measures 150, each exterior angle measures

$$180 - 150 = 30.$$

- As $\frac{360}{n} = 30$, $30n = 360$ so $n = \frac{360}{30} =$ **12 sides**.

Check Your Understanding of Section 2.4

A. Multiple Choice

1. The number of sides of a regular polygon for which the measure of an interior angle is equal to the measure of an exterior angle is
(1) 8 (2) 6 (3) 3 (4) 4

2. If the measure of an interior angle of a regular polygon is 108, what is the total number of sides in the polygon?
(1) 5 (2) 6 (3) 7 (4) 4

3. If the measures of four interior angles of a pentagon are 116, 138, 94, and 88, what is the measure of the remaining interior angle?
(1) 76 (2) 104 (3) 120 (4) 144

4. A stop sign in the shape of a regular octagon is resting on a brick wall, as shown in the accompanying diagram. What is the measure of angle x?
(1) 45 (2) 60 (3) 120 (4) 135

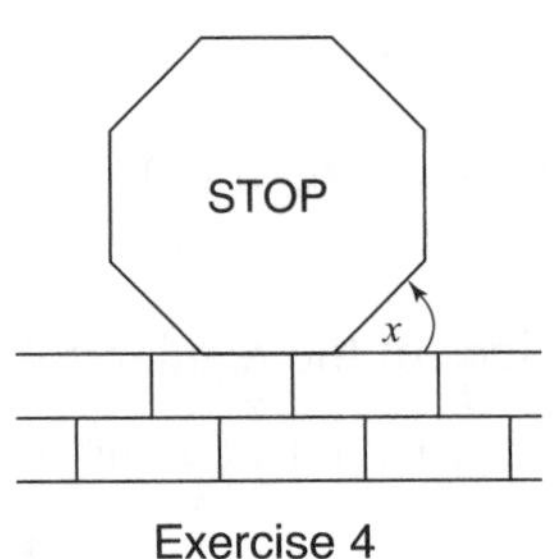

Exercise 4

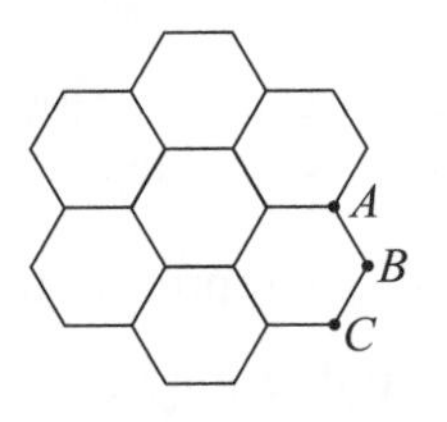

Exercise 5

5. The accompanying figure represents a section of bathroom floor tiles shaped like regular hexagons. What is the measure of ABC?
(1) 60 (2) 90 (3) 120 (4) 150

6. If the sum of the measures of the interior angles of a polygon is 1620, how many sides does the polygon have?
(1) 9 (2) 10 (3) 11 (4) 12

7. Which of the following could *not* represent the measure of an exterior angle of a regular polygon?
(1) 72 (2) 15 (3) 27 (4) 45

8. If each interior angle of a regular polygon measures 170, what is the total number of sides in the polygon?
(1) 10 (2) 17 (3) 18 (4) 36

B. *Show or explain how you arrived at your answer.*

9. What is the sum of the measures of the interior angles of a polygon that has 13 sides?

10. Find the number of sides of a regular polygon in which the measure of an interior angle is three times the measure of an exterior angle.

11. Find the number of sides in a regular polygon in which the measure of an interior angle exceeds six times the measure of an exterior angle by 12.

CHAPTER 3
CONGRUENT TRIANGLES AND INEQUALITIES

3.1 PROVING TRIANGLES CONGRUENT

KEY IDEAS

Two triangles are congruent when the three sides and the three angles of one triangle are congruent to the corresponding parts of the other triangle. Fewer than six measurements may determine both the shape and size of a triangle. There is exactly one triangle, for example, that can be constructed with side lengths of 3, 4, and 5 inches. The measures of the three angles of a triangle become determined once its three sides are fixed in length. If a second triangle is constructed with sides that also measure 3, 4, and 5 inches, then the two triangles must be congruent as their corresponding angles must agree.

Corresponding Parts of Congruent Triangles

Corresponding parts of two triangles are the pairs of sides or angles that are in the same relative positions in the two figures. Although congruent triangles have the same size and shape, they may be positioned differently (turned, flipped, or shifted) as in Figure 3.1. The two triangles can be made to coincide by pairing their vertices such that

$$A \leftrightarrow R, B \leftrightarrow S, \text{ and } C \leftrightarrow T$$

Pairs of matching vertices are referred to as **corresponding vertices**.

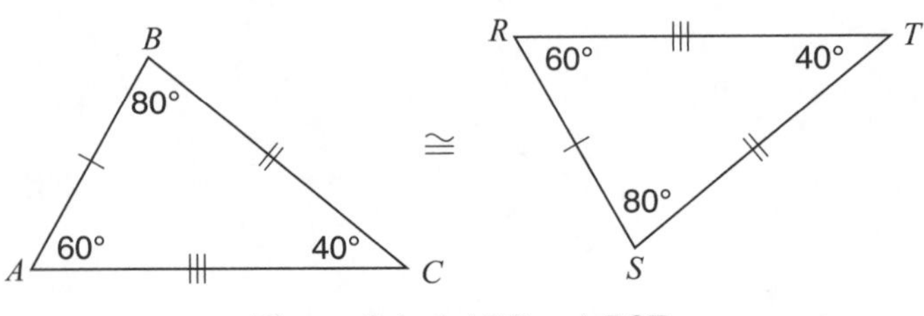

Figure 3.1 $\triangle ABC \cong \triangle RST$.

Pairs of corresponding vertices determine pairs of **corresponding angles**.

Corresponding sides are opposite corresponding angles.

Corresponding Angles	Corresponding Sides
$\angle A$ and $\angle R$	$\overline{BC}$ and $\overline{ST}$
$\angle B$ and $\angle S$	$\overline{AC}$ and $\overline{RT}$
$\angle C$ and $\angle T$	$\overline{AB}$ and $\overline{RS}$

When writing a congruence relation, the order in which the vertices are written matters. Corresponding vertices of the two triangles are written in the same relative position, as in

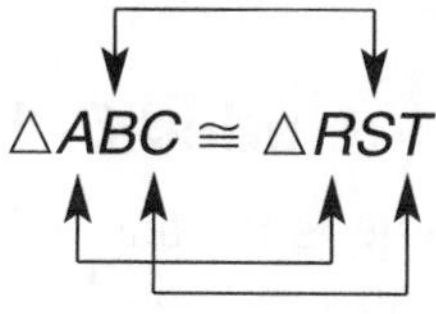

Triangle "Rigidity"

Certain combinations of three side–angle measurements "rigidly determine" a triangle such that it can have exactly one shape and size. When two triangles share such a set of three measurements, the two triangles must be congruent as the remaining three measurements of each of the triangles become determined. This "rigidity property" of triangles provides short-cut methods for proving triangles congruent.

Side–Side–Side Postulate: SSS ≅ SSS

A triangle is rigidly determined when a given set of Side-Side-Side measurements, which is abbreviated as S-S-S, is used to construct it.

Postulate 14: SSS Postulate
If three sides of one triangle are congruent to the corresponding sides of another triangle, then the two triangles are congruent.

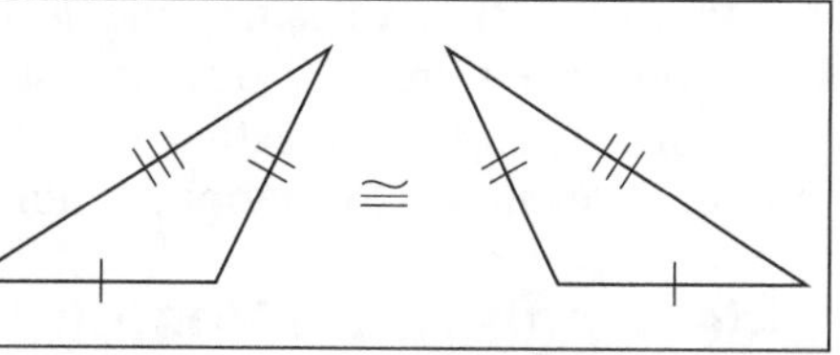

Example 1

Given: $\overline{AB} \cong \overline{DE}$, $\overline{BC} \cong \overline{EF}$, $\overline{AF} \cong \overline{DC}$.
Prove: $\triangle ABC \cong \triangle DEF$.

Solution: From the Given you know that two pairs of corresponding sides are congruent. Adding FC to both AF and DC gives the third pair of congruent sides, $\overline{AC} \cong \overline{DF}$. The two triangles are congruent by SSS ≅ SSS.

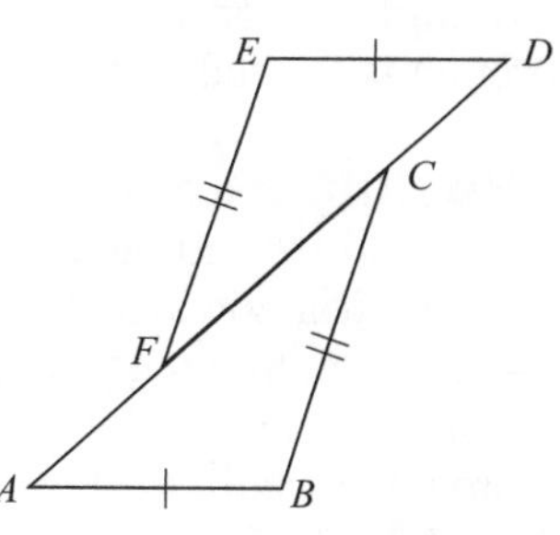

Proof

Statement		Reason
1. $\overline{AB} \cong \overline{DE}$.	Side	1. Given.
2. $\overline{BC} \cong \overline{EF}$.	Side	2. Given.
3. $AF = DC$.		3. Given.
4. $FC = FC$.		4. Reflexive property.
5. $\underbrace{AF + FC}_{AC} = \underbrace{DC + FC}_{DF}$		5. Addition property of equality.
6. $\overline{AC} \cong \overline{DF}$.	Side	6. Substitution property.
7. $\triangle ABC \cong \triangle DEF$.		7. SSS $\cong$ SSS.

Included Angles and Included Sides

An **included angle** of a triangle is the angle formed by two of its sides. In Figure 3.2, $\angle B$ is included by sides $\overline{AB}$ and $\overline{BC}$. A set of three parts of a triangle consisting of two sides and the included angle are in a ***S**ide–**A**ngle–**S**ide* formation, which is abbreviated as SAS.

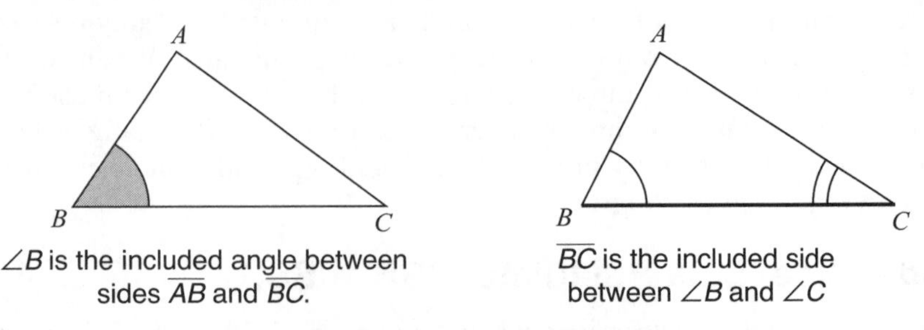

$\angle B$ is the included angle between sides $\overline{AB}$ and $\overline{BC}$.

$\overline{BC}$ is the included side between $\angle B$ and $\angle C$

Figure 3.2 Included angles and sides.

An **included side** of a triangle is the side that two of its angles have in common. In Figure 3.2, side $\overline{BC}$ is *included by* $\angle B$ and $\angle C$. A set of three parts of a triangle consisting of two angles and the included side is in an ***A**ngle–**S**ide–**A**ngle* formation, which is abbreviated as ASA.

Side–Angle–Side Postulate: SAS $\cong$ SAS

A triangle becomes determined when a given set of side–angle–side measurements is used to construct it.

Postulate 15: SAS Postulate

If two sides and the included angle of one triangle are congruent to the corresponding parts of another triangle, then the two triangles are congruent.

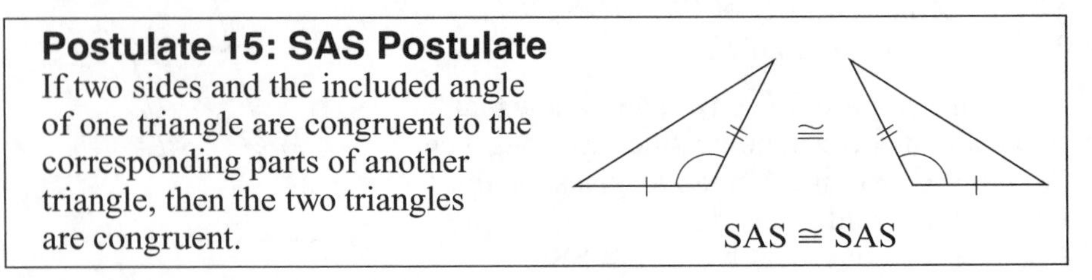

Example 2

Given: $\overline{AB} \cong \overline{BD}, \overline{BC}$ bisects $\angle ABD$.
Prove: $\triangle ABC \cong \triangle DBC$.

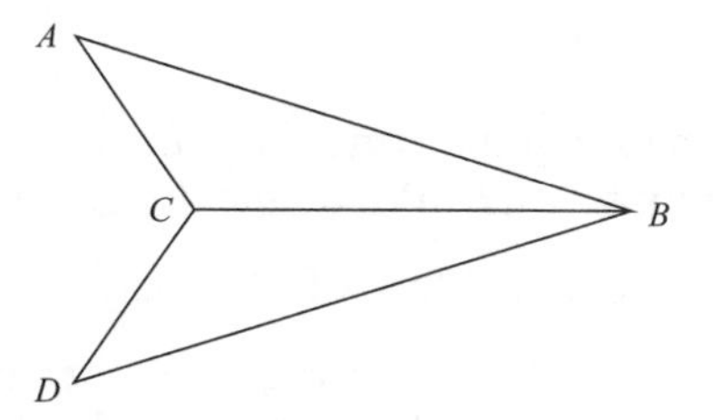

Solution: Mark off the diagram with the Given. For convenience, label the congruent angles formed by the angle bisector with numbers.

- Indicate that $\overline{BC}$ is a side of both triangles by drawing a cross mark, **X**, through it.
- Use SAS ≅ SAS.
- Write the proof.

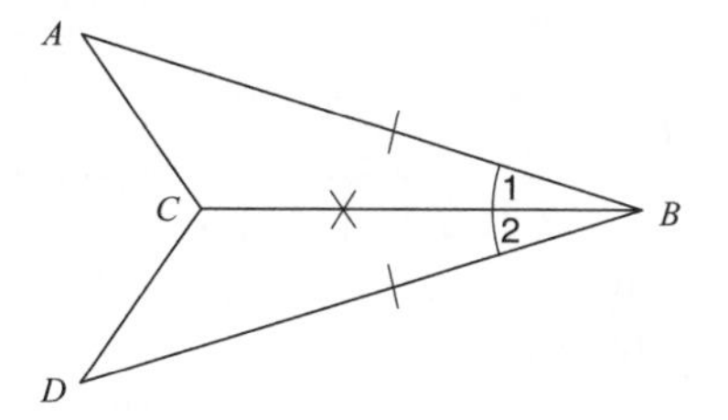

Proof

Statement		**Reason**
1. $\overline{AB} \cong \overline{BD}$.	Side	1. Given.
2. $\overline{BC}$ bisects $\angle ABD$.		2. Given.
3. $\angle 1 \cong \angle 2$.	Angle	3. An angle bisector divides an angle into two congruent angles.
4. $\overline{BC} \cong \overline{BC}$.	Side	4. Reflexive property.
5. $\triangle ABC \cong \triangle DBC$.		5. SAS ≅ SAS.

Angle–Side–Angle Postulate: ASA ≅ ASA

A triangle becomes determined when a given set of angle–side–angle measurements are used to construct it.

Postulate 16: ASA Postulate
If two angles and the included side of one triangle are congruent to the corresponding parts of another triangle, then the two triangles are congruent.

ASA ≅ ASA

Example 3

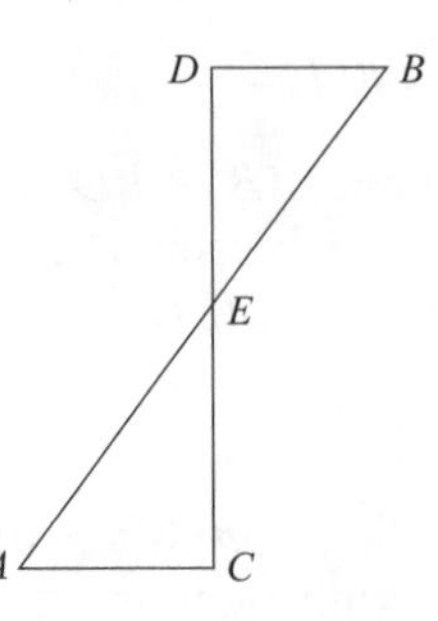

In the accompanying diagram, $\overline{AC} \perp \overline{CD}$, $\overline{BD} \perp \overline{CD}$, $\overline{AB}$ bisects $\overline{CD}$. Prove that $\triangle ACE \cong \triangle BDE$.

Solution: Mark off the diagram with the Given.

- Mark off the congruent vertical angles. For convenience, label these angles with numbers.
- Use ASA ≅ ASA.
- Write the proof.

Proof

Statement		Reason
1. $\overline{AC} \perp \overline{CD}$, $\overline{BD} \perp \overline{CD}$.		1. Given.
2. Angles C and D are right angles.		2. Perpendicular lines intersect to form right angles.
3. $\angle C \cong \angle D$.	Angle	3. All right angles are congruent.
4. $\overline{AB}$ bisects $\overline{CD}$.		4. Given.
5. $\overline{CE} \cong \overline{DE}$.	Side	5. A bisector divides a line into two congruent segments.
6. $\angle 1 \cong \angle 2$.	Angle	6. Vertical angles are congruent.
7. $\triangle ACE \cong \triangle BDE$.		7. ASA ≅ ASA.

Angle–Angle–Side Theorem: AAS ≅ AAS

In Figure 3.3, two angles and a nonincluded side of $\triangle ABC$ are congruent to the corresponding parts of $\triangle PQR$. Are the two triangles congruent?

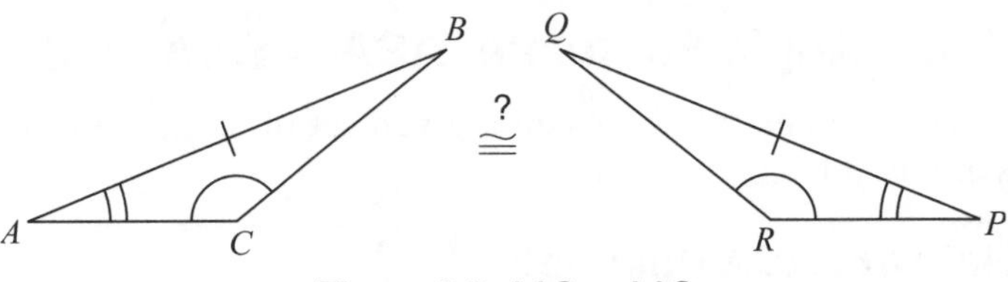

Figure 3.3 AAS ≅ AAS.

Because of the Triangle–Angle Sum Theorem, the remaining pair of angles of the triangles, $\angle B$ and $\angle Q$, are congruent. This makes $\triangle ABC \cong \triangle PQR$ by the ASA Postulate. Because the two triangles are congruent, AAS ≅ AAS can be used as a method for proving two triangles are congruent.

> **Theorem: Angle–Angle–Side Theorem**
> If two angles and the nonincluded side of one triangle are congruent to the corresponding parts of another triangle, then the two triangles are congruent [AAS ≅ AAS].

***Example* 4**

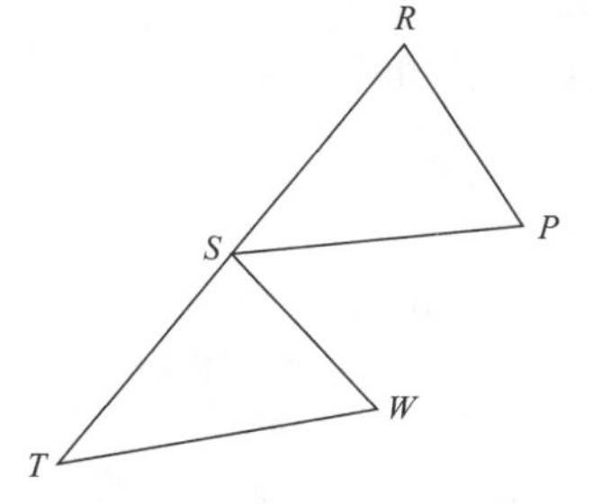

Given: $\overline{RP} \parallel \overline{SW}$, $\overline{SP} \parallel \overline{TW}$, and $\overline{SP} \cong \overline{TW}$.
Prove: $\triangle SRP \cong \triangle TSW$.

Solution:

- Mark the pairs of congruent corresponding angles formed by the two pairs of parallel lines. Insert numbers to make these angles easy to refer to in the proof.
- Since the congruent sides are *not* included by the congruent pairs of angles, use AAS ≅ AAS.
- Write the proof:

Proof

Statement		Reason
1. $\overline{RP} \parallel \overline{SW}$.		1. Given.
2. $\angle 1 \cong \angle 2$	Angle	2. If two lines are parallel, then corresponding angles are congruent.
3. $\overline{SP} \parallel \overline{TW}$.		3. Given.
4. $\angle 3 \cong \angle 4$.	Angle	4. Same as 2.
5. $\overline{SP} \cong \overline{TW}$.	Side	5. Given.
6. $\triangle SRP \cong \triangle TSW$.		6. AAS ≅ AAS.

Hypotenuse–Leg Theorem: HL ≅ HL

A *right* triangle becomes determined when a given set of hypotenuse–leg measurements are used to construct it.

Theorem: Hypotenuse–Leg Theorem
If the hypotenuse and a leg of one right triangle are congruent to the corresponding parts of another right triangle, then the two right triangles are congruent [HL ≅ HL].

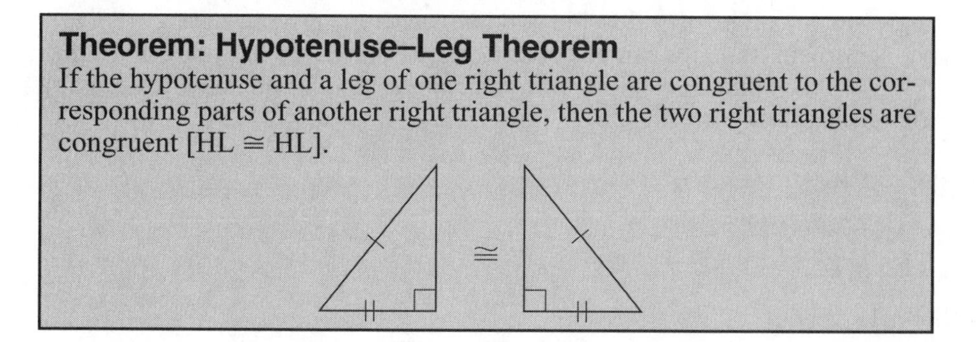

The next example makes use of the familiar fact that a square has four congruent sides and four right angles.

Example 5

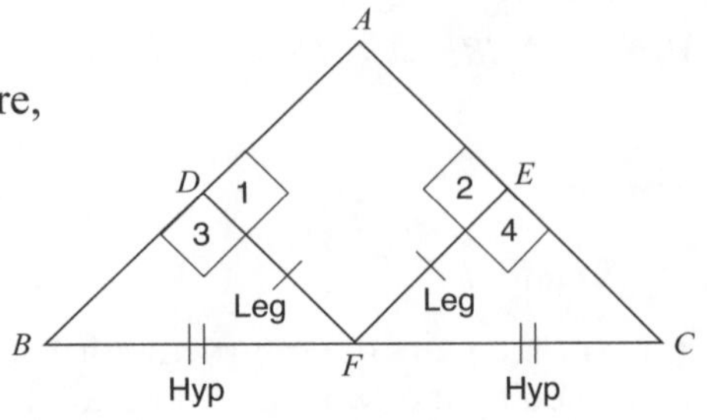

Given: Quadrilateral $AEFD$ is a square, F is the midpoint of $\overline{BC}$.
Prove: $\triangle BDF \cong \triangle CEF$.

Solution: Since $\overline{BF} \cong \overline{CF}$ and $\overline{DF} \cong \overline{EF}$, prove right $\triangle BDF \cong$ right $\triangle CEF$ by HL $\cong$ HL.

Proof

Statement		Reason
1. $AEFD$ is a square.		1. Given.
2. Angles 1 and 2 are right angles.		2. A square contains four right angles.
3. Angles 3 and 4 are right angles.		3. Supplements of right angles are right angles.
4. $\triangle BDF$ and $\triangle CEF$ are right triangles.		4. A triangle that contains a right angle is a right triangle.
5. F is the midpoint of $\overline{BC}$.		5. Given.
6. $\overline{BF} \cong \overline{CF}$.	Hyp	6. A midpoint of a line segment divides the segment into two congruent segments.
7. $\overline{DF} \cong \overline{EF}$.	Leg	7. The sides of a square are congruent.
8. $\triangle BDF \cong \triangle CEF$.		8. HL $\cong$ HL.

Methods Insufficient for Congruence

Not all sets of three measurements determine exactly one triangle. If angle–angle–angle or side–side–angle measurements are given, *different* noncongruent triangles can be constructed, as illustrated in Figure 3.4. Thus, AAA $\cong$ AAA and SSA $\cong$ SSA do *not* represent valid methods for proving triangles are congruent.

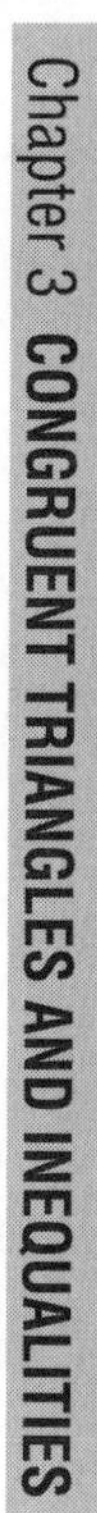

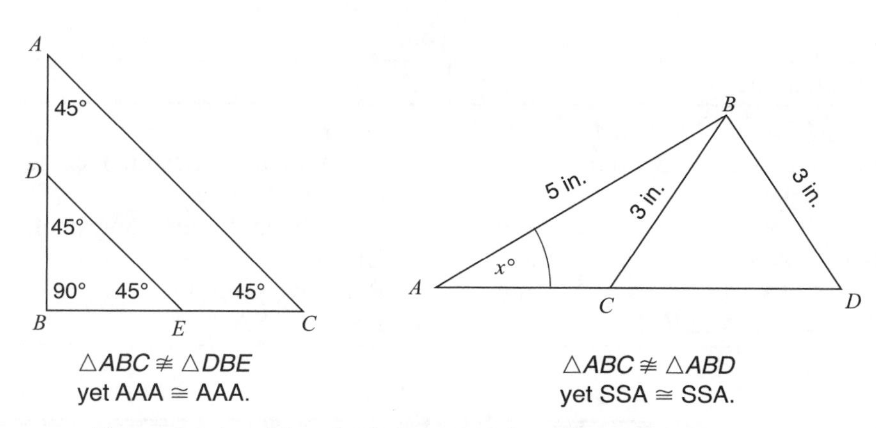

Figure 3.4 Invalid methods for proving triangles are congruent.

Proving Overlapping Triangles Congruent

When trying to prove overlapping triangles congruent, look for a side or an angle that is shared by the two triangles. After copying the diagram in your notebook, "separate" the two triangles by outlining their sides using two pencils of different colors. If you don't have colored pencils handy, outline the sides of one of the triangles with a thick, heavy line.

Example 6

Given: $\overline{MK} \perp \overline{JL}$, $\overline{LP} \perp \overline{JM}$, $\overline{JK} \cong \overline{JP}$.

Prove: $\triangle JKM \cong \triangle JLP$.

Solution: Outline the sides of $\triangle JKM$.

- Mark off the diagram with the Given.
- Mark off $\angle J$, the part shared by the overlapping triangles.
- Use ASA ≅ ASA.

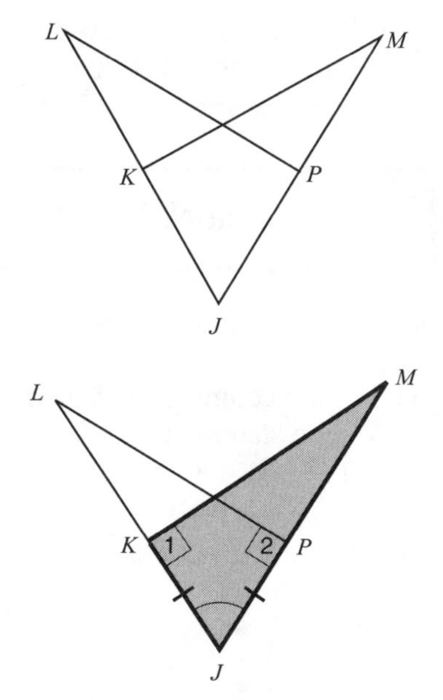

Proof

Statement	Reason
1. $\overline{MK} \perp \overline{JL}$, $\overline{LP} \perp \overline{JM}$.	1. Given.
2. Angles 1 and 2 are right angles.	2. Perpendicular lines intersect to form right angles.
3. $\angle 1 \cong \angle 2$. Angle	3. All right angles are congruent.
4. $\overline{JK} \cong \overline{JP}$. Side	4. Given.
5. $\angle J \cong \angle J$. Angle	5. Reflexive property.
6. $\triangle JKM \cong \triangle JLP$.	6. ASA $\cong$ ASA.

MATH FACTS

Before you write a proof involving congruent triangles, make a plan.

- Mark off the diagram with the Given. Then mark off any additional pairs of congruent parts such as common angles or sides, vertical angles, supplements (or complements) of the same angle, and so forth.
- Decide which congruence method you will use:

  ALL TRIANGLES: SAS $\cong$ SAS, ASA $\cong$ ASA, SSS $\cong$ SSS, or AAS $\cong$ AAS

  RIGHT TRIANGLES: HL $\cong$ HL

- Do ***not*** use AAA $\cong$ AAA or SSA $\cong$ SSA. These are *not* valid methods for proving triangles congruent.

Check Your Understanding of Section 3.1

A. *Multiple Choice*

1. In the accompanying diagram, diagonal $\overline{AC}$ bisects $\angle BAD$ and $\angle BCD$. Which statement can be used to prove $\triangle ABC \cong \triangle ADC$?

(1) AAS $\cong$ AAS (3) SAS $\cong$ SAS

(2) SSS $\cong$ SSS (4) ASA $\cong$ ASA

B A C D

Exercise 1

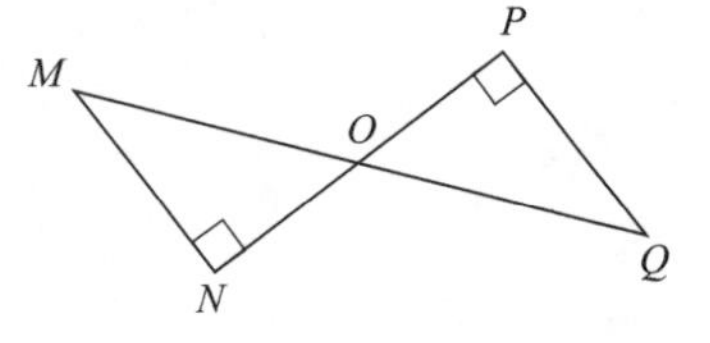

Exercise 2

2. In the accompanying diagram, $\overline{MN} \perp \overline{CD}$, $\overline{QP} \perp \overline{NP}$, O is the midpoint of $\overline{NP}$, and $\overline{MN} \cong \overline{QP}$. Which reason would be *least* likely to be used to prove $\triangle MNO \cong \triangle QPO$?

(1) Hy-Leg $\cong$ Hy-Leg (3) SAS $\cong$ SAS
(2) SSS $\cong$ SSS (4) ASA $\cong$ ASA

3. In the accompanying diagram, $\overline{RS} \cong \overline{ST}$, $\overline{SR} \perp \overline{QR}$, and $\overline{ST} \perp \overline{QT}$. Which method can be used to prove $\triangle QRS \cong \triangle QTS$?

(1) HL $\cong$ HL (3) SAS $\cong$ SAS
(2) SSS $\cong$ SSS (4) ASA $\cong$ ASA

Exercise 3

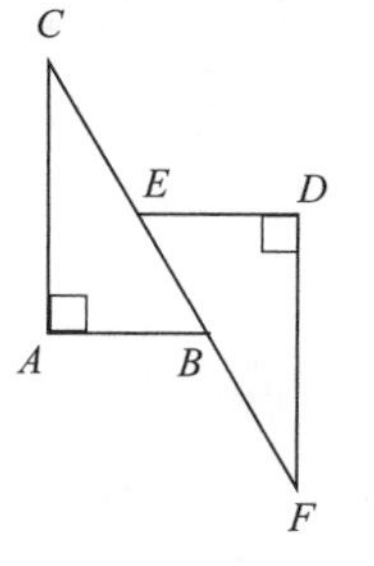

Exercise 4

4. In the accompanying diagram, $\overline{CA} \perp \overline{AB}$, $\overline{ED} \perp \overline{DF}$, $\overline{ED} \parallel \overline{AB}$, $\overline{CE} \cong \overline{BF}$, $\overline{AB} \cong \overline{ED}$. Which statement would *not* be used to prove $\triangle ABC \cong \triangle DEF$?

(1) HL $\cong$ HL (3) AAS $\cong$ AAS
(2) SSS $\cong$ SSS (4) SAS $\cong$ SAS

B. *Write a proof.*

5. Given: $\overline{BD}$ bisects $\angle ABC$,
$\angle BAD \cong \angle BCD$.
Prove: $\triangle ABD \cong \triangle CBD$.

6. Given: $\overline{AB} \perp \overline{BC}$, $\overline{DE} \perp \overline{EF}$
$\overline{BC} \parallel \overline{FE}$, $\overline{AF} \cong \overline{CD}$.
Prove: $\triangle ABC \cong \triangle DEF$.

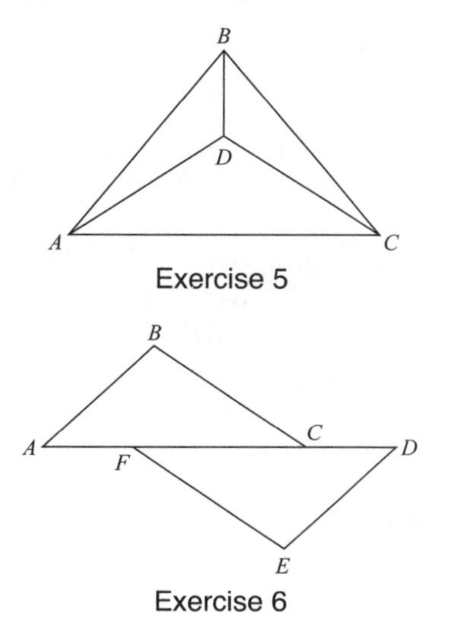

Exercise 5

Exercise 6

7. In the accompanying diagram of quadrilateral $MATH$, $\overline{AM} \parallel \overline{TH}$, $\overline{RA} \parallel \overline{SH}$ and $\overline{RA} \cong \overline{SH}$.
Prove: $\triangle ARM \cong \triangle HST$.

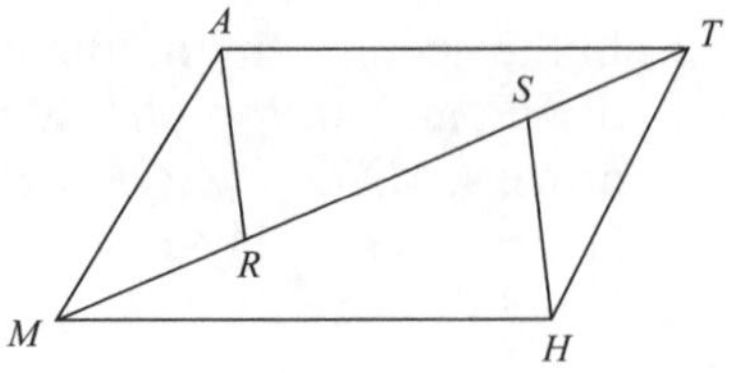

Exercise 7

8. In the accompanying diagram, $\overline{AD} \perp \overline{BC}$, $\angle 1 \cong \angle 2$, $\overline{AD}$ bisects $\angle BAC$.
Prove: $\triangle AED \cong \triangle AFD$.

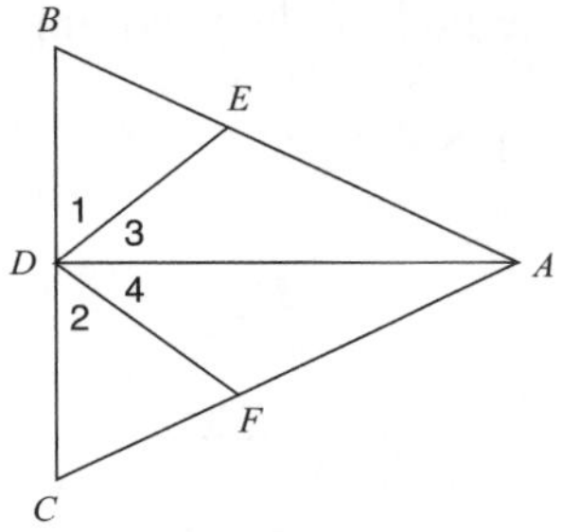

Exercise 8

9. Given: $\overline{BD} \perp \overline{AC}$, $\overline{AE} \perp \overline{BC}$, $\overline{BD} \cong \overline{AE}$.
Prove: $\triangle AEB \cong \triangle BDA$.

10. Given: $\overline{BD}$ is an altitude to $\overline{AC}$, $\overline{AE}$ is an altitude to $\overline{BC}$, $\overline{CE} \cong \overline{CD}$.
Prove: $\triangle AEC \cong \triangle BDC$.

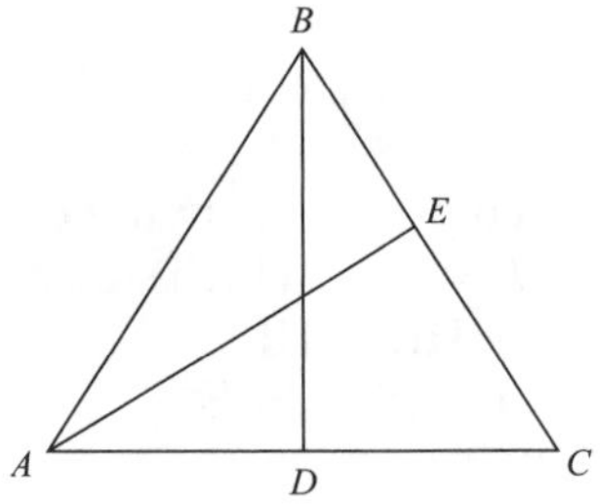

Exercise 9–10

11. Given: $\overline{AB} \cong \overline{BC}$, $\angle A \cong \angle C$.
Prove: $\triangle AEB \cong \triangle CDB$

12. Given: $\overline{AE} \perp \overline{BC}$, $\overline{CD} \perp \overline{AB}$, $\overline{BD} \cong \overline{BE}$.
Prove: $\triangle ABE \cong \triangle CBD$.

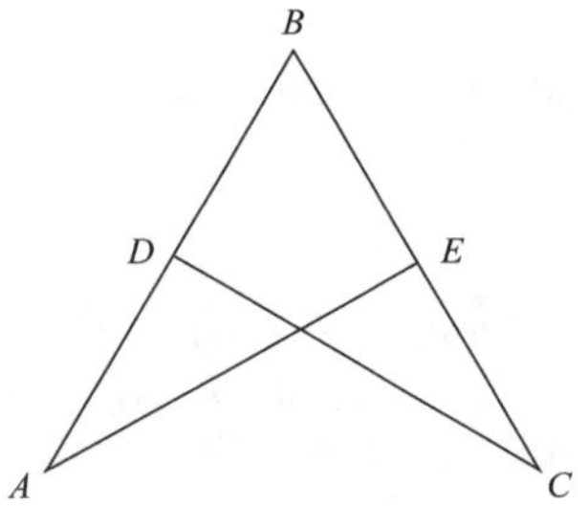

Exercise 11–12

13. Given: $\overline{DE} \cong \overline{AE}, \overline{BE} \cong \overline{CE},$
$\angle 1 \cong \angle 2$
Prove: $\triangle DBC \cong \triangle ACB.$

14. Given: $\overline{AB} \perp \overline{BE}, \overline{DC} \perp \overline{CE},$
$\angle 1 \cong \angle 2, \overline{AB} \cong \overline{DC}.$
Prove: $\triangle ABC \cong \triangle DCB.$

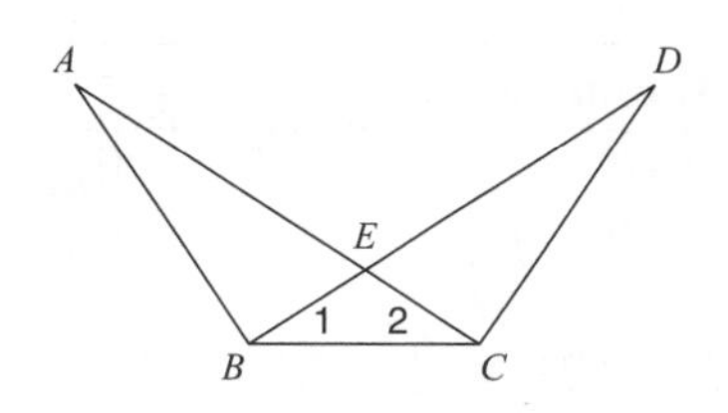

Exercise 13–14

3.2 APPLYING CONGRUENT TRIANGLES

KEY IDEAS

After two triangles are proved congruent, you may conclude that any remaining pair of corresponding parts not used in the proof must also be congruent.

Proving Segments or Angles Congruent

To prove that a pair of angles or pair of segments are congruent:

- Select a pair of triangles that contain these parts.
- Prove that the two triangles are congruent.
- Conclude that the required pair of parts are congruent because "Corresponding Parts of Congruent Triangles are Congruent." When used as a reason in a proof, this statement can be abbreviated as **CPCTC**.

Example 1

Given: $\overline{AC} \cong \overline{DB}, \overline{AB} \cong \overline{DC}.$
Prove: $\angle A \cong \angle D.$

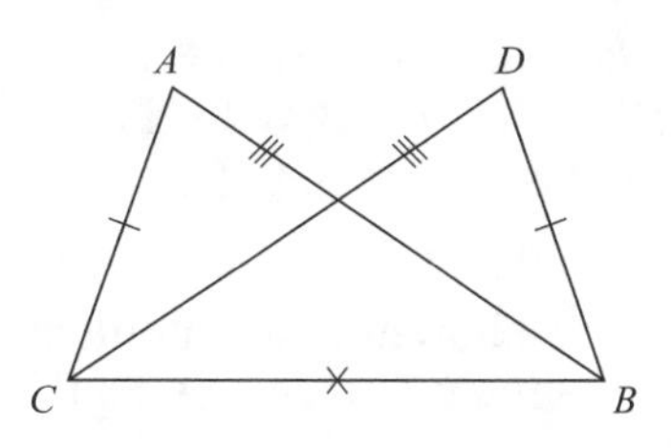

Solution: Reason backward from what you are required to prove by answering these questions:

- Which triangles contain $\angle A$ and $\angle D$ as parts? Since $\angle A$ is an angle of $\triangle ABC$ and D is an angle of $\triangle DBC$, you need to prove $\triangle ABC \cong \triangle DCB$.
- How can these triangles be proved congruent? After marking off the diagram, it is clear that $\triangle ABC \cong \triangle DCB$ by SSS $\cong$ SSS. Because the triangles are congruent, corresponding angles such as $\angle A$ and $\angle D$ are congruent.

Proof

Statement		Reason
1. $\overline{AC} \cong \overline{DB}$.	Side	1. Given.
2. $\overline{AB} \cong \overline{DC}$.	Side	2. Given.
3. $\overline{BC} \cong \overline{BC}$.	Side	3. Reflexive property.
4. $\triangle ABC \cong \triangle DCB$.		4. SSS ≅ SSS.
5. $\angle A \cong \angle D$.		5. CPCTC.

Proving Statements Lacking Diagrams

To prove a theorem that has the form "If . . . then . . . ," draw your own diagram using the hypothesis in the *if* clause as the Given and the conclusion in the *then* clause as the Prove. Consider the Perpendicular Bisector Theorem.

> **Theorem: Perpendicular Bisector Theorem**
> If a point is on the perpendicular bisector of a line segment, then it is equidistant from the endpoints of the line segment.

To prove the Perpendicular Bisector Theorem, first draw a diagram in which line ℓ is the perpendicular bisector of $\overline{AB}$ and point P is any point on line ℓ. You are required to prove $AP = BP$. Here is the set-up and outline of the proof.

Given: Line $\ell \perp \overline{AB}$ at M, $\overline{AM} \cong \overline{BM}$, point P is any point on line ℓ.
Prove: $AP = BP$.

Solution: Using SAS ≅ SAS, $\triangle AMP \cong \triangle BMP$. By CPCTC, $\overline{AP} \cong \overline{BP}$, so $AP = BP$.

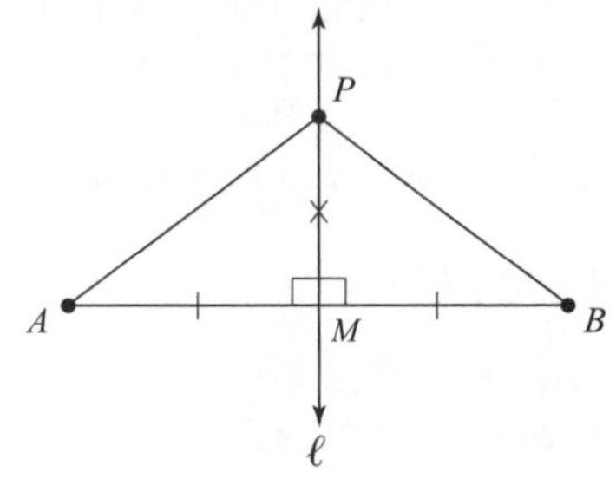

When you show that a statement is true by outlining the key details of the proof in sentence form, as in this example, you are providing an *informal* proof. A two-column proof is considered a "formal" proof. Example 2 asks you to prove the next theorem informally.

> **Theorem: Angle Bisector Theorem**
> If a point is on the bisector of an angle, then it is equidistant from the sides of the angle.

Example 2

Give an informal proof of the Angle Bisector Theorem.

Solution: Draw a diagram in which line $\overrightarrow{BD}$ is the bisector of $\angle B$ and point P is any point on $\overrightarrow{BD}$. From P, draw $\overline{PA}$ perpendicular to one side of the angle and $\overline{PC}$ perpendicular to the other side. The lengths of the perpendicular segments represent the distances that you are required to prove are equal. As part of an informal proof, identify the Given and the Prove.

Given: $\overrightarrow{BD}$ bisects $\angle ABC$, $\overline{PA} \perp \overrightarrow{BA}$, $\overline{PC} \perp \overrightarrow{BC}$

Prove: $PA = PC$.

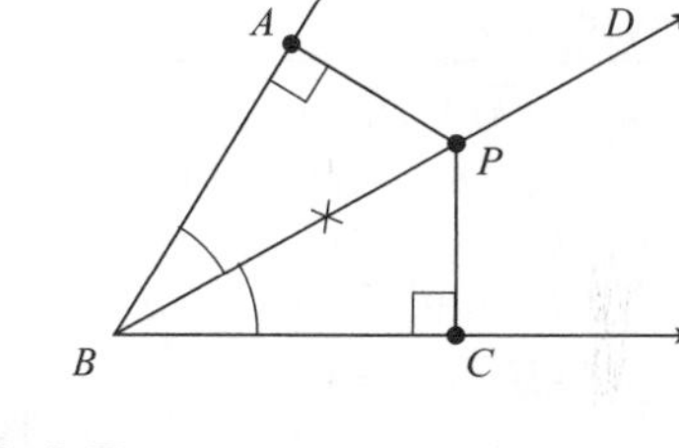

Informal Proof:

- $\triangle PAB \cong \triangle PCB$ by AAS $\cong$ AAS.
- By CPCTC, $\overline{PA} \cong \overline{PC}$, so $PA = PC$.

Proving a Segment Has a Special Property

To prove that a segment or ray is

- A *median* of a triangle, prove that it divides the side to which it is drawn into two congruent segments. In Figure 3.5, to prove $\overline{BX}$ is a median, use congruent triangles to prove $\overline{AX} \cong \overline{CX}$.
- An *angle bisector*, prove that it divides an angle into two congruent angles. To prove $\overline{BX}$ is an angle bisector, use congruent triangles to prove $\angle 3 \cong \angle 4$.

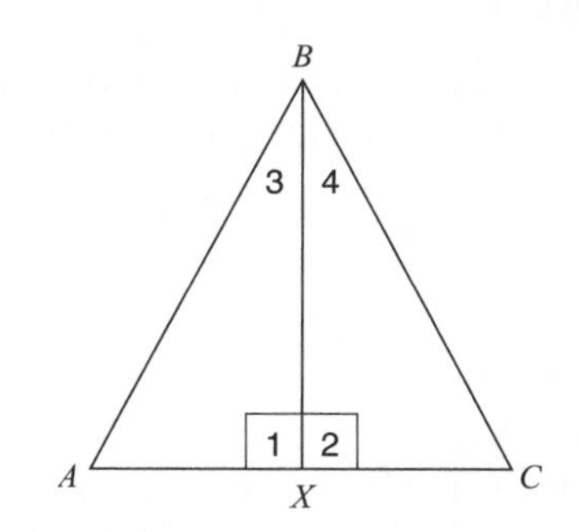

Figure 3.5 Proving a special property.

- An *altitude* of a triangle, prove that it forms congruent adjacent angles with the side that it intersects. To prove $\overline{BX}$ is an altitude, show that $\overline{BX} \perp \overline{AC}$ by using congruent triangles to demonstrate that $\angle 1 \cong \angle 2$.

Example 3

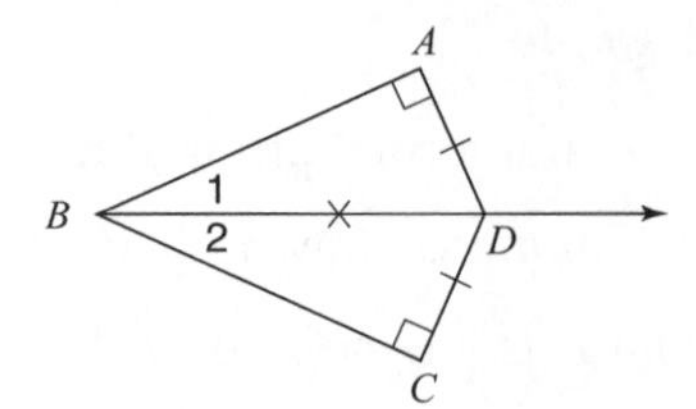

Given: $\overline{BA} \perp \overline{CD}, \overline{BC} \perp \overline{CD},$
$\overline{AD} \cong \overline{CD}.$

Prove: $\overrightarrow{BD}$ bisects $\angle ABC$.

Solution: Show $\angle 1 \cong \angle 2$ by proving that $\triangle BAD \cong \triangle BCD$ using the Hy-Leg Theorem.

Proof

Statement		**Reason**
1. $\overline{BA} \perp \overline{AD}, \overline{BC} \perp \overline{CD}$.		1. Given.
2. $\angle A$ and $\angle C$ are right angles.		2. Perpendicular lines intersect to form right angles.
3. $\triangle BAD$ and $\triangle BCD$ are right triangles.		3. A triangle that contains a right angle is a right triangle.
4. $\overline{BD} \cong \overline{BD}$.	Hyp	4. Reflexive property.
5. $\overline{AD} \cong \overline{CD}$.	Leg	5. Given.
6. $\triangle BAD \cong \triangle BCD$.		6. HL $\cong$ HL.
7. $\angle 1 \cong \angle 3$.		7. CPCTC.
8. $\overrightarrow{BD}$ bisects $\angle ABC$.		8. A ray that divides an angle into two congruent angles bisects the angle.

Proving Lines Parallel

You may be able to prove a pair of lines parallel by using congruent triangles to show that a pair of alternate interior angles or corresponding angles are congruent.

Example 4

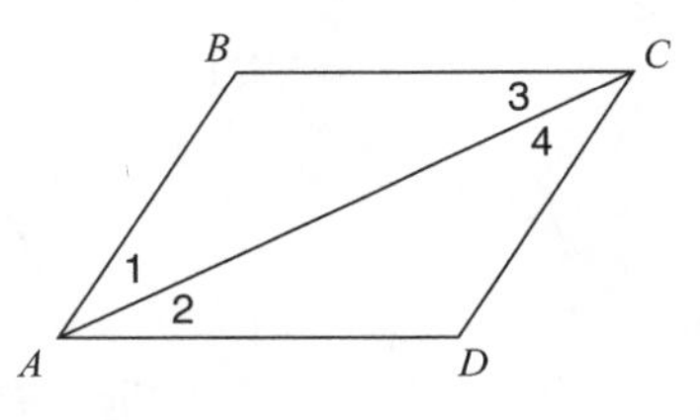

In the accompanying diagram of quadrilateral $ABCD$, $\overline{AB} \parallel \overline{CD}$, and $\overline{AB} \cong \overline{CD}$. Prove that $\overline{BC} \parallel \overline{AD}$.

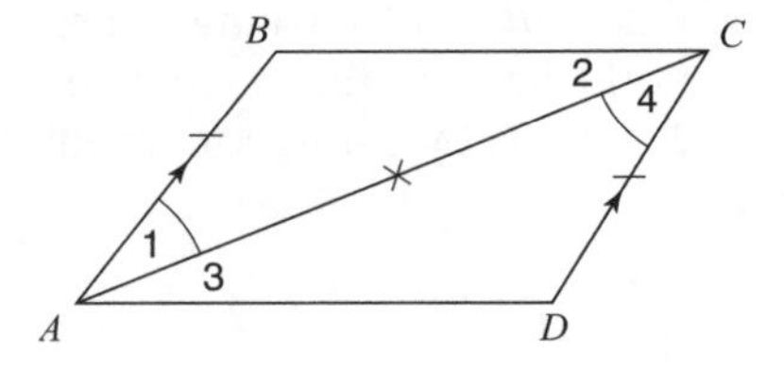

Solution: $\overline{BC} \parallel \overline{AD}$ if $\angle 2 \cong \angle 3$. To prove $\angle 2 \cong \angle 3$, show $\triangle ABC \cong \triangle CDA$ by SAS $\cong$ SAS.

Proof

Statement		Reason
1. $\overline{AB} \cong \overline{CD}$.	Side	1. Given.
2. $\overline{AB} \parallel \overline{CD}$.		2. Given.
3. $\angle 1 \cong \angle 4$.	Angle	3. If two parallel lines are cut by a transversal, alternate interior angles are congruent.
4. $\overline{AC} \cong \overline{AC}$.	Side	4. Reflexive property.
5. $\triangle ABC \cong \triangle CDA$.		5. SAS $\cong$ SAS.
6. $\angle 2 \cong \angle 3$.		6. CPCTC.
7. $\overline{BC} \parallel \overline{AD}$.		7. If two lines are cut by a transversal such that a pair of alternate interior angles are congruent, the lines are parallel.

Line Perpendicular To A Plane

A given line is perpendicular to a plane if and only if they intersect and the given line is perpendicular to all lines in the plane that contain the point of intersection. If in Figure 3.6, $\overleftrightarrow{AX} \perp$ plane P, then any line in the plane that contains point X is perpendicular to $\overleftrightarrow{AX}$ at X. Thus, $\overleftrightarrow{AX} \perp \overleftrightarrow{RX}$, $\overleftrightarrow{AX} \perp \overleftrightarrow{SX}$, $\overleftrightarrow{AX} \perp \overleftrightarrow{TX}$, and so forth.

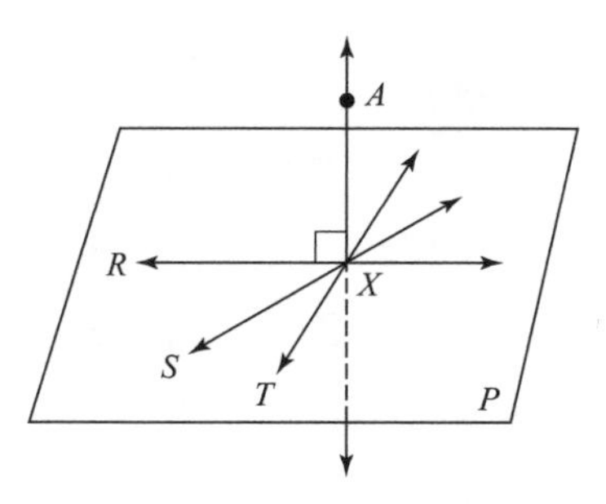

Figure 3.6 $\overleftrightarrow{AX}$ perpendicular to plane P.

Example 5

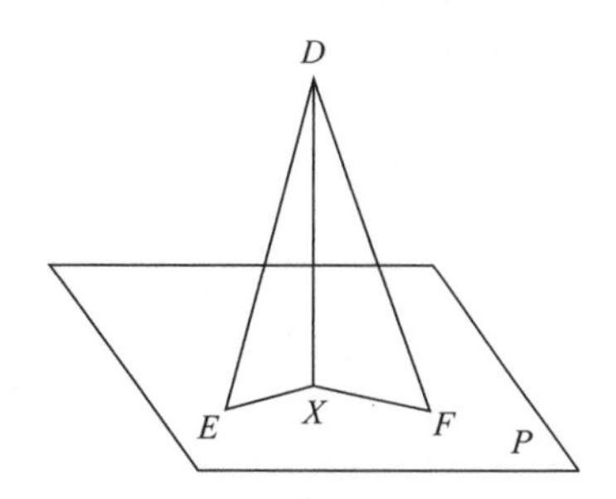

Given: $\overline{DX} \perp$ plane P, $\angle DEX \cong \angle DFX$.
Prove: $\overline{DE} \cong \overline{DF}$.

Solution: Angles DXE and DXF are right angles so $\triangle DXE \cong \triangle DXF$ by the AAS theorem.

Proof

Statement		Reason
1. $\angle DEX \cong \angle DFX$.	Angle	1. Given.
2. $\overline{DX} \perp$ plane P.		2. Given
3. $\overline{DX} \perp \overline{EX}$ and $\overline{DX} \perp \overline{FX}$.		3. Definition of a line perpendicular to a plane.
4. $\angle DXE$ and $\angle DXF$ are right angles.		4. Perpendicular lines intersect to form right angles.
5. $\angle DXE \cong \angle DXF$.	Angle	5. All right angles are congruent.
6. $\overline{DX} \cong \overline{DX}$.	Side	6. Given.
7. $\triangle DXE \cong \triangle DXF$.		7. AAS $\cong$ AAS.
8. $\overline{DE} \cong \overline{DF}$.		8. CPCTC.

Check Your Understanding of Section 3.2

1. Given: $\overline{AB} \cong \overline{AC}$, $\overline{BD} \cong \overline{CD}$.
Prove: $\angle B \cong \angle D$.

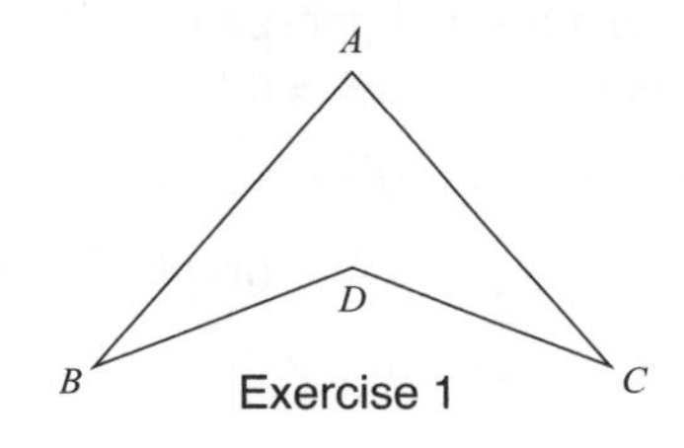

Exercise 1

2. Given: $\overline{MP} \cong \overline{ST}$, $\overline{PL} \cong \overline{RT}$,
$\overline{MP} \parallel \overline{ST}$.
Prove: $\overline{RS} \parallel \overline{LM}$.

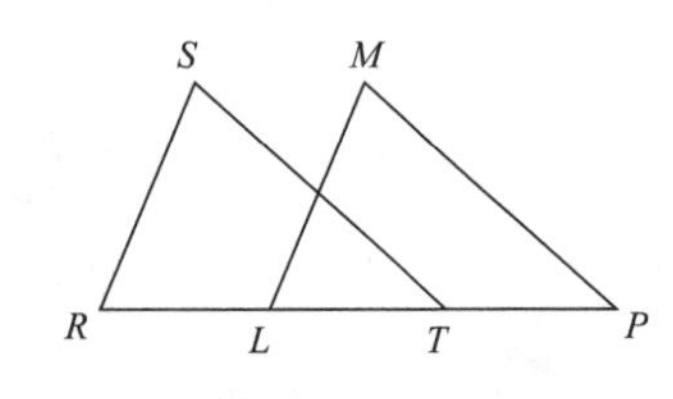

Exercise 2

3. In the accompanying diagram, $\overline{HK}$ bisects $\overline{IL}$ and $\angle H \cong \angle K$.
Prove: $\overline{IL}$ bisects $\overline{HK}$.

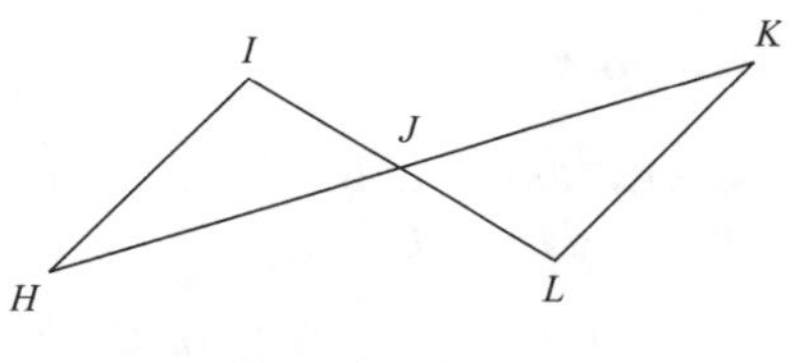

Exercise 3

4. In the accompanying diagram, ADC is equilateral and $\angle ADB \cong \angle CDB$. Write an informal proof that shows $\overline{BD}$ bisects ABC.

5. Given: $\overline{BE} \perp \overline{AC}$, $\overline{DF} \perp \overline{AC}$, $\overline{AB} \cong \overline{CD}$, $\overline{BE} \cong \overline{DF}$.
Prove: $\overline{AB} \parallel \overline{CD}$.

6. Given: $\overline{DF} \parallel \overline{BE}$, $\overline{DF} \cong \overline{BE}$, $\overline{AE} \cong \overline{CF}$.
Prove: $\overline{AD} \parallel \overline{BC}$.

7. Given: S is the midpoint of $\overline{RT}$, $\overline{RP} \cong \overline{SW}$, $\overline{RP} \parallel \overline{SW}$.
Prove: $\overline{SP} \parallel \overline{TW}$.

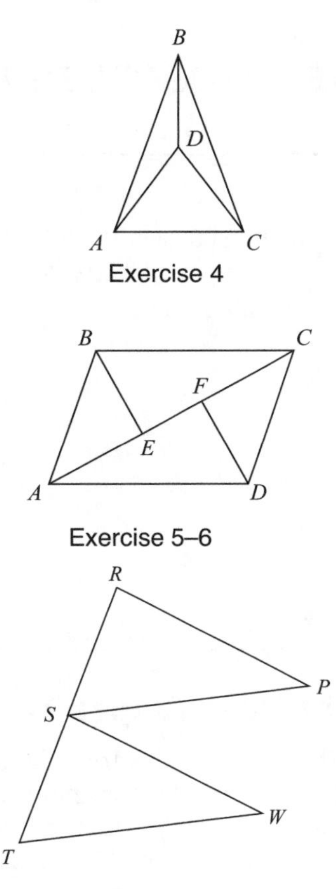

Exercise 4

Exercise 5–6

Exercise 7

8. Given: $\overline{TL} \perp \overline{RS}$, $\overline{SW} \perp \overline{RT}$, $\overline{TL} \cong \overline{SW}$
Prove: $\overline{SL} \cong \overline{TW}$.

9. Given: $\overline{TL}$ is the altitude to $\overline{RS}$, $\overline{SW}$ is the altitude to $\overline{RT}$, $\overline{RS} \cong \overline{RT}$
Prove: $\overline{RW} \cong \overline{RL}$.

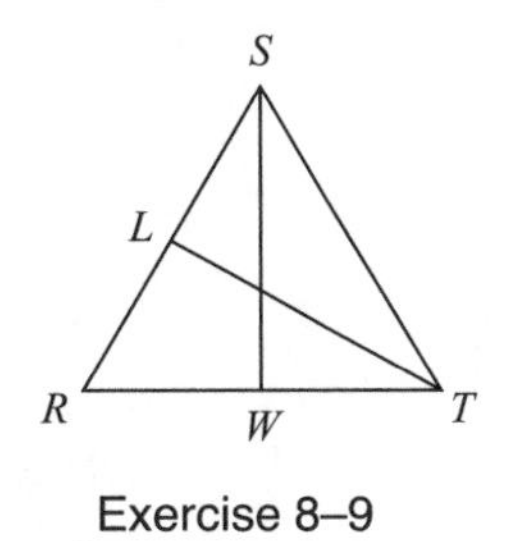

Exercise 8–9

10. In the accompanying diagram of $\triangle ABC$, $\overline{CM}$ is the median to $\overline{AB}$, $\overline{CM}$ is extended to point P so that $\overline{CM} \cong \overline{MP}$, and $\overline{AP}$ is drawn. Write an explanation or informal proof that shows $\overline{AP} \parallel \overline{CB}$.

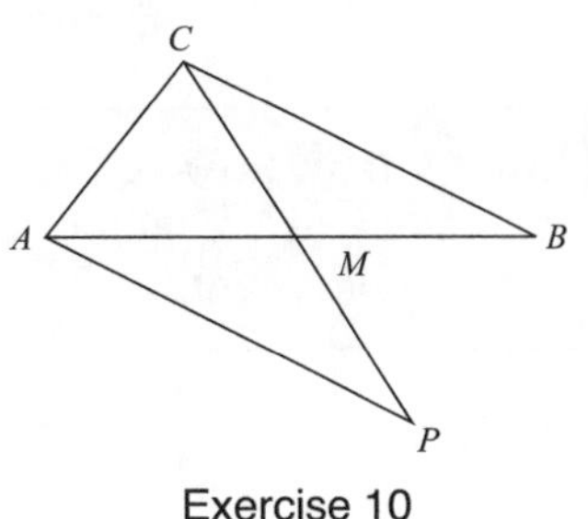

Exercise 10

3.3 ISOSCELES TRIANGLE THEOREMS

KEY IDEAS

In an isosceles triangle, the base angles are congruent. Conversely, if two angles of a triangle are congruent, then the sides opposite them are congruent.

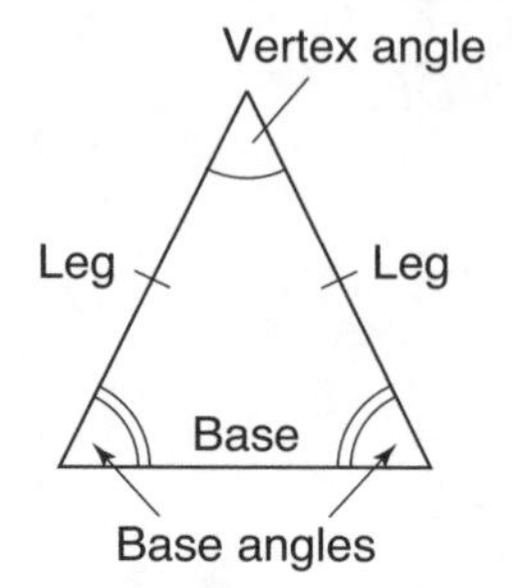

The Base Angles Theorem

When an isosceles triangle is folded along the bisector of the vertex angle, the two parts of the triangle coincide. The two triangles formed are congruent since $\overline{AB} \cong \overline{AC}$, $\angle 1 \cong \angle 2$, and $\overline{AD} \cong \overline{AD}$ so $\triangle ADB \cong \triangle ADC$ by SAS $\cong$ SAS. Thus, $\angle B \cong \angle C$ (CPCTC).

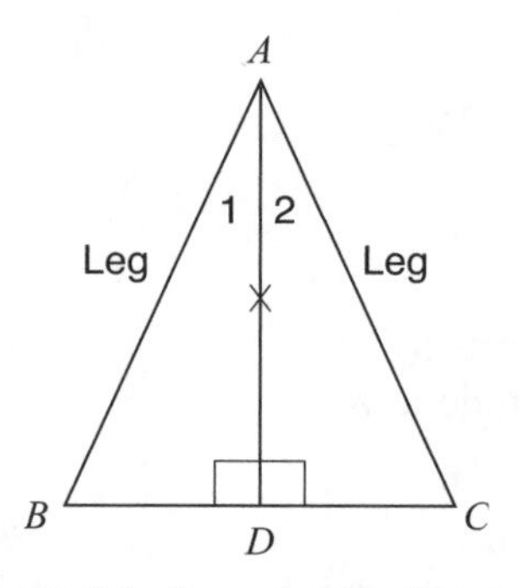

Figure 3.6 Base Angles Theorem.

Theorem: Base Angles Theorem
If two sides of a triangle are congruent, then the angles opposite these sides are congruent.

MATH FACTS

- An equilateral triangle is equiangular:

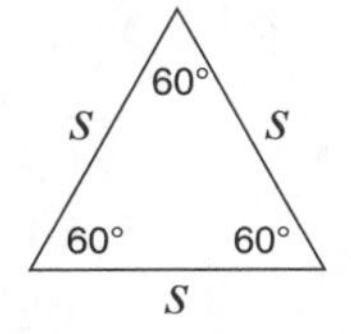

- In an isosceles triangle, the bisector of the vertex angle is also the median and the altitude to the base. In Figure 3.6, $\overline{AD} \perp \overline{BC}$ and $\overline{BD} \cong \overline{CD}$.

Example 1

The degree measure of a base angle of an isosceles triangle exceeds twice the degree measure of the vertex angle by 15. Find the degree measure of the vertex angle.

Solution: If x represent the degree measure of the vertex angle, then $2x + 15$ represents the measure of each base angle.

$$\begin{aligned}(2x + 15) + (2x + 15) + x &= 180\\ 5x + 30 &= 180\\ 5x &= 180 - 30\\ x &= \frac{150}{5}\\ &= \mathbf{30}\end{aligned}$$

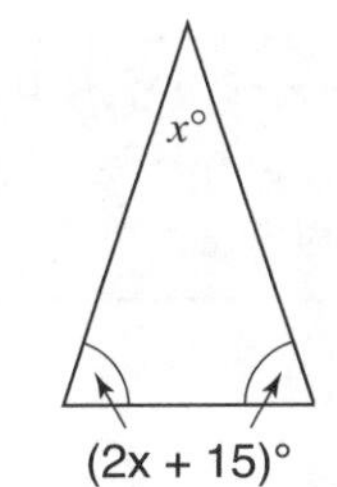

Example 2

Given: $\overline{AB} \cong \overline{BC}$, $\overline{EF} \cong \overline{GF}$,
$\overline{AE} \cong \overline{CG}$.
Prove: $\overline{GJ} \cong \overline{EK}$.

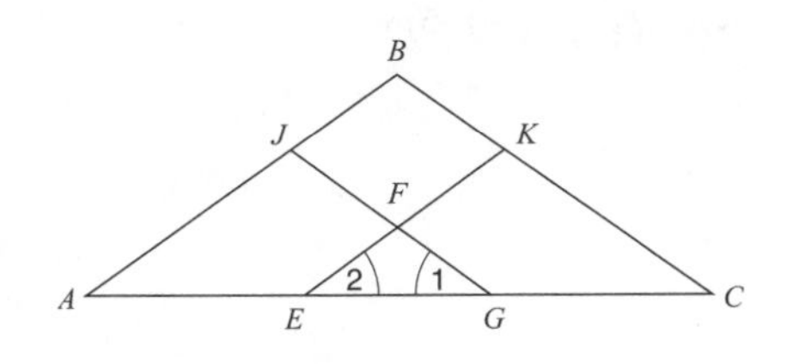

Solution: Mark off the diagram.

- Because $\overline{GJ}$ is a side of $\triangle AJG$ and $\overline{EK}$ is a side of $\triangle CKE$, show that these triangles are congruent.
- $\angle A \cong \angle C$ (base angles theorem), $\overline{AG} \cong \overline{CE}$ (addition), and $\angle 1 \cong \angle 2$ (base angles theorem).
- $\triangle AJG \cong \triangle CKE$ by ASA $\cong$ ASA, so $\overline{GJ} \cong \overline{EK}$ by CPCTC.

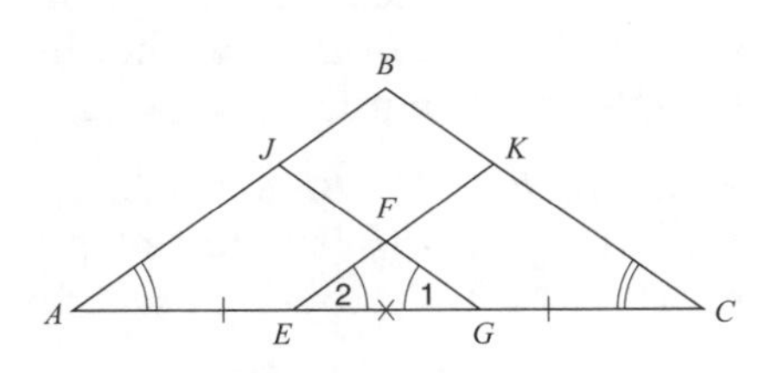

Proof

Statement		Reason
1. $\overline{AB} \cong \overline{BC}$.		1. Given.
2. $\angle A \cong \angle C$.	Angle	2. If two sides of a triangle are congruent, then angles opposite them are congruent.
3. $AE = CG$.		3. Given.
4. $EG = EG$.		4. Reflexive property.
5. $\underbrace{AE + EG}_{AG} = \underbrace{CG + EG}_{CE}$		5. Addition property of equality.
6. $\overline{AG} \cong \overline{CE}$.	Side	6. Substitution property.
7. $\overline{EF} \cong \overline{GF}$.		7. Given.
8. $\angle 1 \cong \angle 2$.	Angle	8. Same as Reason 3.
9. $\triangle AJG \cong \triangle CKE$.		9. ASA $\cong$ ASA.
10. $\overline{GJ} \cong \overline{EK}$.		10. CPCTC.

The Converse of the Base Angles Theorem

Although the converse of a true statement is not necessarily true, the converse of the Base Angles theorem is true.

> **Theorem: Converse of Base Angles Theorem**
> If two angles of a triangle are congruent, then the sides opposite them are congruent.

Example 3

Given: $\overline{CEA} \cong \overline{CDB}$, $\angle 1 \cong \angle 2$.
Prove: $\overline{PE} \cong \overline{PD}$.

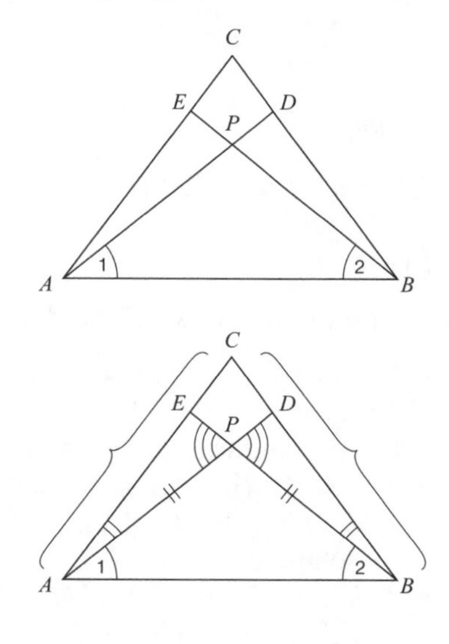

Solution: Since $\overline{PE}$ is a side of $\triangle EPA$ and $\overline{PD}$ is a side of $\triangle DPB$, show these triangles are congruent.

- In $\triangle APB$, $\overline{AP} \cong \overline{BP}$ and $\angle EPA \cong \angle DPB$.
- $\angle EAP \cong \angle DBP$ (subtraction property).
- $\triangle EPA \cong \triangle DPB$ by ASA $\cong$ ASA.

Proof

Statement		Reason
1. $\overline{CEA} \cong \overline{CDB}$.		1. Given.
2. $\angle CAB \cong \angle CBA$.		2. If two sides of a triangle are congruent, then angles opposite them are congruent.
3. $\angle 1 \cong \angle 2$.		3. Given.
4. $\underbrace{m\angle CAB - m\angle 1}_{m\angle EAP} = \underbrace{m\angle CBA - m\angle 2}_{m\angle DBP}$		4. Subtraction property of equality.
5. $\angle EAP \cong \angle DBP$.	Angle	5. Substitution property.
6. $\overline{AP} \cong \overline{BP}$.	Side	6. If two angles of a triangle are congruent, then sides opposite them are congruent.
7. $\angle EPA \cong \angle DPB$.	Angle	7. Vertical angles are congruent.
8. $\triangle EPA \cong \triangle DPB$.		8. ASA $\cong$ ASA.
9. $\overline{PE} \cong \overline{PD}$.		9. CPCTC.

Example 4

Write an informal proof that shows that if two altitudes of a triangle are congruent, then the triangle is isosceles.

Solution: Draw a diagram and identify the Given and the Prove.

Given: $\overline{CD}$ is the altitude to $\overline{AB}$,
$\overline{AE}$ is the altitude to $\overline{BC}$,
$\overline{CD} \cong \overline{AE}$.
Prove: $\triangle ABC$ is isosceles.

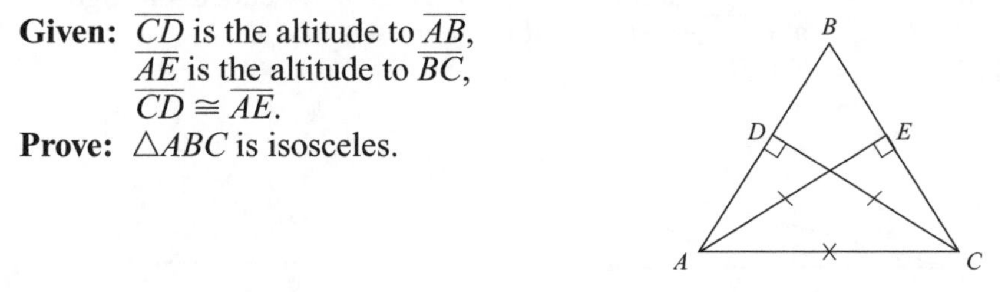

Informal Proof:

- Because $\overline{AC} \cong \overline{AC}$ (Hy) and $\overline{CD} \cong \overline{AE}$ (Leg), right $\triangle ADC \cong$ right $\triangle CEA$ by HL $\cong$ HL.
- $\angle BAC \cong \angle BCA$ (CPCTC) $\Rightarrow \overline{AB} \cong \overline{BC}$ so $\triangle ABC$ is isosceles.

Check Your Understanding of Section 3.3

A. *Multiple Choice*

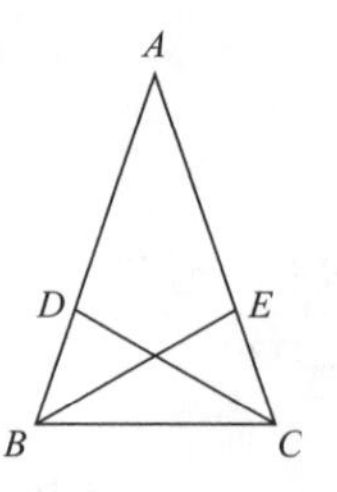

1. In the accompanying diagram of $\triangle ABC$, $\overline{AB} \cong \overline{AC}$, $BD = \frac{1}{3}BA$, and $CE = \frac{1}{3}CA$. Triangle EBC can be proved congruent to triangle DCB by

(1) SAS $\cong$ SAS
(2) ASA $\cong$ ASA
(3) SSS $\cong$ SSS
(4) HL $\cong$ HL

2. In isosceles triangle CAT, the measure of the vertex angle is one-half the measure of one of the base angles. Which statement is true about $\triangle CAT$?
(1) $\triangle CAT$ is an equilateral triangle.
(2) $\triangle CAT$ is an acute triangle.
(3) $\triangle CAT$ is a right triangle.
(4) $\triangle CAT$ is an obtuse triangle.

3. In the accompanying diagram of $\triangle CAB$, $\overline{AB}$ is extended through D, $m\angle CBD = 30$, and $\overline{AB} \cong \overline{BC}$. What is the measure of $\angle A$?
(1) 15 (2) 30 (3) 75 (4) 150

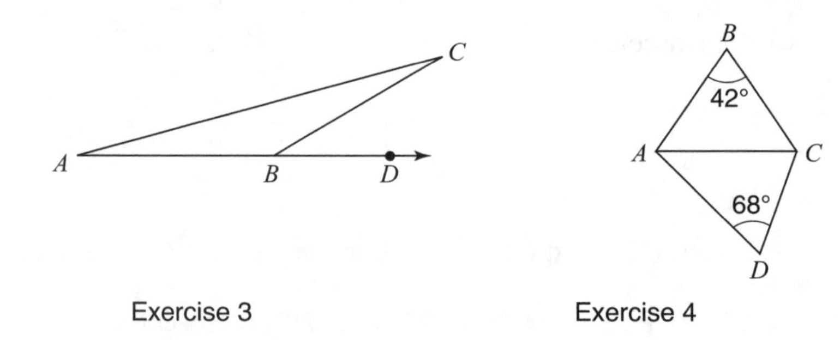

Exercise 3

Exercise 4

4. In the accompanying diagram, $\triangle ABC$ and $\triangle ACD$ are isosceles triangles with $\overline{AB} \cong \overline{BC}$ and $\overline{AD} \cong \overline{CD}$. If $m\angle ABC = 42$ and $m\angle ADC = 68$, what is $m\angle BAD$?
(1) 115 (2) 120 (3) 125 (4) 130

5. The measure of a base angle of an isosceles triangle is four times the measure of the vertex angle. The degree measure of the vertex angle is
(1) 20 (2) 30 (3) 36 (4) 135

6. Which statements describe properties of an isosceles triangle?
I. The median drawn to the base bisects the vertex angle.
II. The altitude drawn to the base bisects the base.
III. The vertex angle can never be obtuse.
(1) I and III only
(2) II and III only
(3) I and II only
(4) I, II, and III

7. In the accompanying diagram, $\triangle ABC$ and $\triangle DBE$ are isosceles triangles with $\overline{AB} \cong \overline{BC}$ and $\overline{DB} \cong \overline{BE}$. If $m\angle DBE = 26$ and $m\angle BCA = 65$, what is $m\angle ABD$?
(1) 9
(2) 12
(3) 18
(4) 24

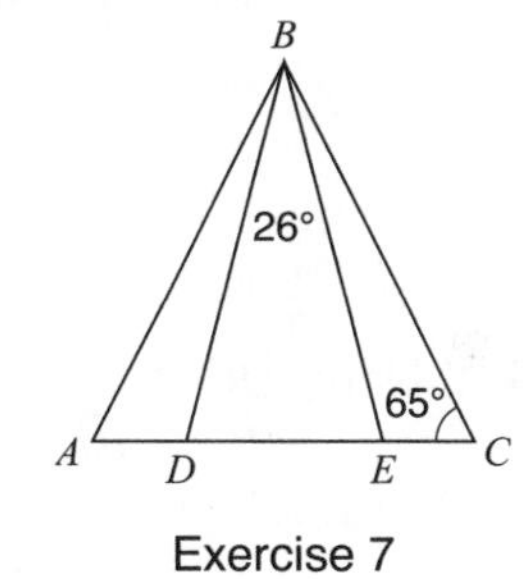

Exercise 7

8. If the median is drawn from the vertex angle to the base of an isosceles triangle, what is the probability that its length will be greater than the length of the altitude drawn to the base?
(1) 1 (2) 0 (3) $\frac{1}{2}$ (4) $\frac{1}{3}$

B. *Show how you arrived at your answer.*

9. Vertex angle A of isosceles triangle ABC measures 20 more than three times $m\angle B$. Find $m\angle C$.

10. In the accompanying figure, $\overline{AB} \cong \overline{AC} \cong \overline{AD} \cong \overline{BD}$. If $m\angle ACB = 7x - 12$, find the value of x.

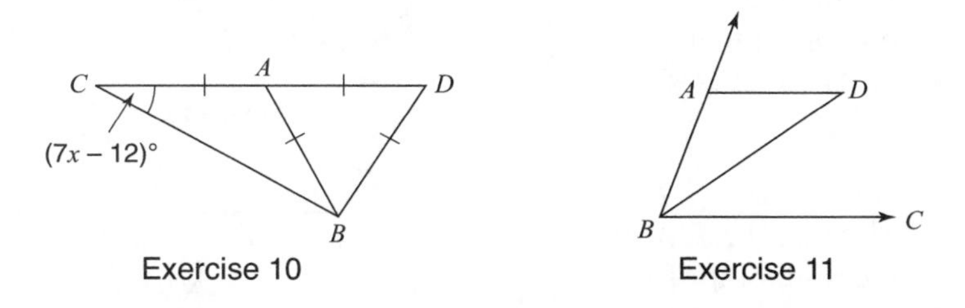

Exercise 10 Exercise 11

11. In the accompanying diagram, $\overline{BD}$ bisects acute angle ABC and $\overline{AD} \parallel \overrightarrow{BC}$. Classify $\triangle ABD$ as equilateral, scalene, isosceles, or right? Justify your answer.

C. Write a proof.

12. Given: $\angle 2 \cong \angle 4$, $\angle BDA \cong \angle BDC$.
Prove: $\triangle ABC$ is isosceles.

13. Given: $\angle 1 \cong \angle 3$, $\overline{AB} \cong \overline{BC}$.
Prove: $\triangle ADC$ is isosceles.

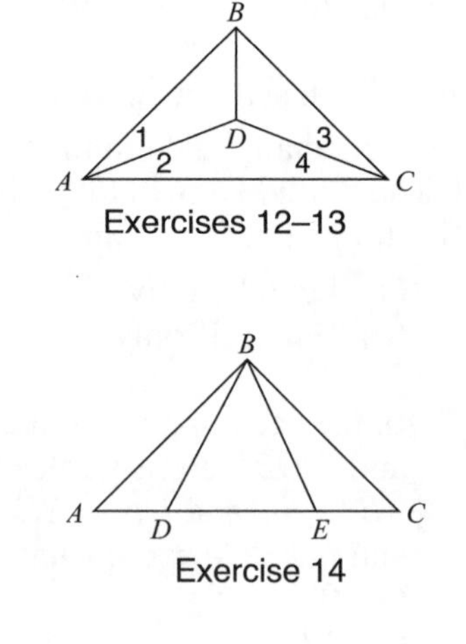

Exercises 12–13

14. Given: $\overline{BD} \cong \overline{BE}$, $\overline{AE} \cong \overline{CD}$.
Prove: $\triangle ABC$ is isosceles.

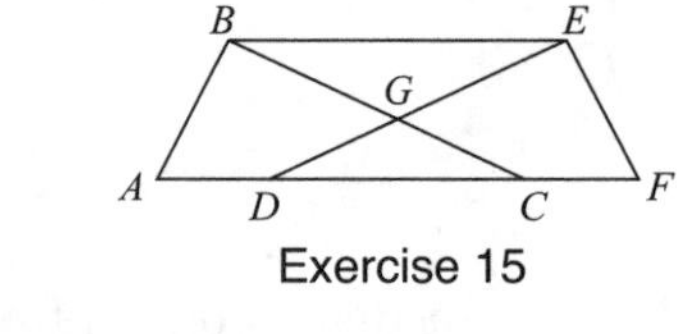

Exercise 14

15. Given: $\overline{DG} \cong \overline{CG}$, $\overline{AD} \cong \overline{FC}$,
$\overline{BC} \cong \overline{ED}$.
Prove: $\angle B \cong \angle E$.

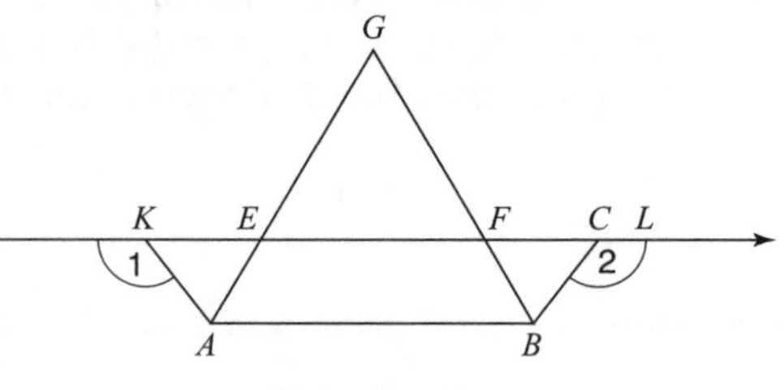

Exercise 15

16. Given: $\overline{GE} \cong \overline{GF}$,
$\angle 1 \cong \angle 2$,
$\overline{DE} \cong \overline{CF}$.
Prove: $\triangle DAE \cong \triangle CBF$.

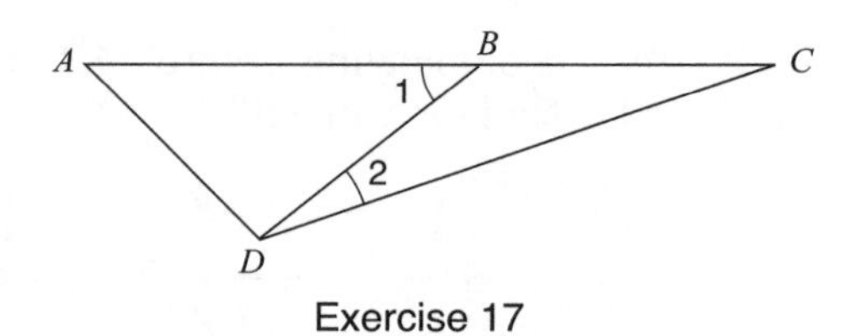

Exercise 16

17. In the accompanying diagram of $\triangle ACD$, $\overline{ABC}$, and $\overline{AD} \cong \overline{DB} \cong \overline{BC}$. Write an explanation or an informal proof that shows $m\angle A = 2m\angle C$.

Exercise 17

18. Given: $\overleftrightarrow{ARB} \parallel \overleftrightarrow{CST}$, $\overline{RT}$ bisects $\angle BRS$, M is the midpoint of $\overline{RT}$.
Prove: $\overline{SM} \perp \overline{RT}$.

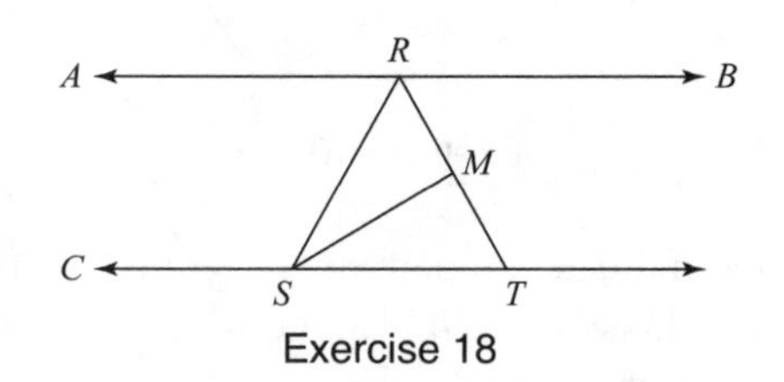

Exercise 18

19. In the accompanying diagram of $\triangle RST$, $\overline{RB} \perp \overline{ST}$, $\overline{SA} \perp \overline{RT}$, and $\overline{RB} \cong \overline{SA}$.
Prove: $\overline{TB} \cong \overline{TA}$.

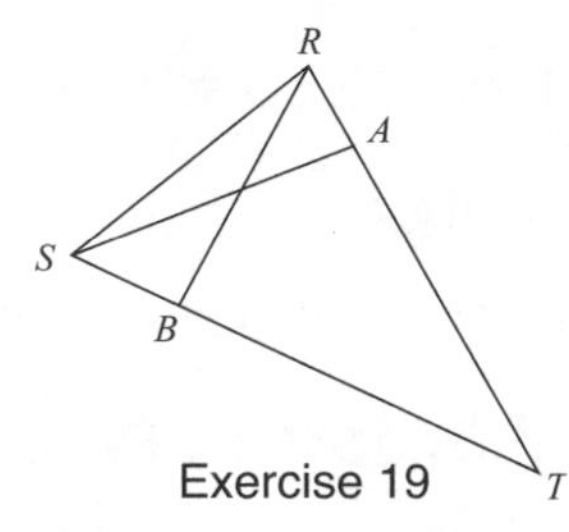

Exercise 19

20. Write an informal proof that shows that the altitudes drawn to the legs of an isosceles triangle are congruent.

3.4 USING PAIRS OF CONGRUENT TRIANGLES

KEY IDEAS

In order to reach a desired conclusion, it may be necessary to prove *two* pairs of triangles congruent. Proving one pair of triangles congruent provides a pair of congruent parts that are needed to prove a second pair of triangles congruent.

Double Congruence Proofs

When trying to prove a pair of triangles congruent, you may eventually realize that an additional pair of congruent parts are needed. Upon closer examination, you may find that the lacking parts are corresponding parts of two other triangles that can easily be proved congruent. After proving the second pair of triangles congruent, you can return to the original pair of triangles and use these congruent parts to help prove them congruent.

Example 1

Given: $\overline{AB} \cong \overline{CB}$, E is the midpoint of $\overline{AC}$.
Prove: $\triangle AED \cong \triangle CED$.

Solution: After observing that $\overline{AE} \cong \overline{CE}$ and $\overline{ED} \cong \overline{ED}$, you should realize that an additional pair of congruent parts is needed. If it happens to be the case that $\angle AED \cong \angle CED$, then $\triangle AED \cong \triangle CED$ by SAS $\cong$ SAS. As these angles are corresponding parts of $\triangle AEB \cong \triangle CEB$, these triangles must be proved congruent before the required pair of triangles can be proved congruent.

- Step 1: Prove $\triangle AEB \cong \triangle CEB$.
  $\overline{AB} \cong \overline{CB}$ (Given) and $\overline{BE} \cong \overline{BE}$ so $\triangle AEB \cong \triangle CEB$ by SSS $\cong$ SSS. Thus, $\angle AED \cong \angle CED$ by CPCTC.

- Step 2: Prove $\triangle AED \cong \triangle CED$.
  Use the fact that $\angle AED \cong \angle CED$ from Step 1, to prove $\triangle AED \cong \triangle CED$ by SAS $\cong$ SAS.

The actual two-column proof is left for you to complete.

Informal Proofs in Paragraph Form

Some of the earlier examples have used informal proof outlines rather than formal two-column proofs. Mathematicians often find it convenient to summarize the key steps of a mathematical proof in one or more easy-to-read paragraphs. Regardless of the style of proof, the statements in a proof should be logically organized, clearly written, and lead to the desired conclusion using valid mathematical reasoning. The next example illustrates from a student's point of view how to write an informal proof in paragraph form.

Example 2

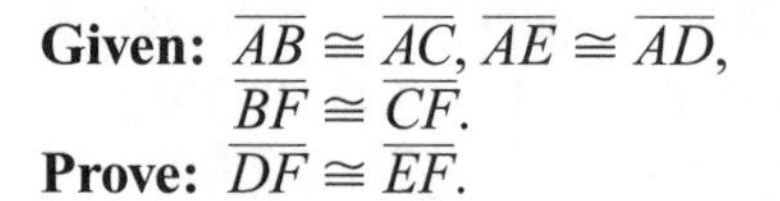

Given: $\overline{AB} \cong \overline{AC}$, $\overline{AE} \cong \overline{AD}$, $\overline{BF} \cong \overline{CF}$.
Prove: $\overline{DF} \cong \overline{EF}$.

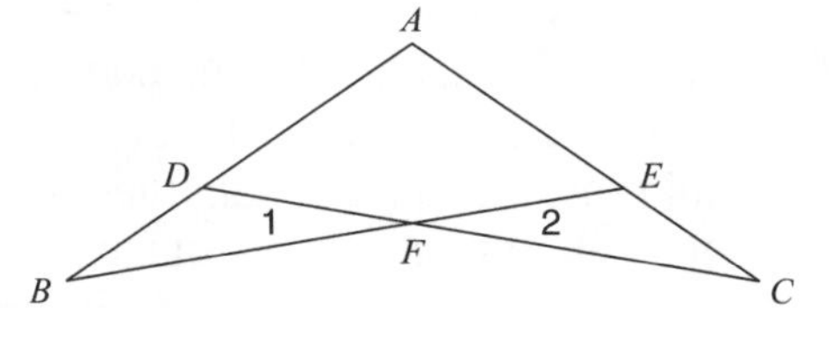

Solution:

Make a plan . . .

To prove $\overline{DF} \cong \overline{EF}$, I need to prove $\triangle BFD \cong \triangle CFE$ since these are the triangles that contain these segments as sides. I know that $\angle 1 \cong \angle 2$ and $\overline{BF} \cong \overline{CF}$. I need to find another pair of congruent parts. If I also knew that $\angle B \cong \angle C$, then I would be able to prove the required pair of triangles congruent using the ASA Postulate. Because overlapping triangles ABE and ADC contain angles B and C, I must first prove them congruent.

Carry out the plan . . .

Paragraph Proof

After marking off the diagram with the Given, I can see that $\triangle ABE \cong \triangle ACD$ by SAS $\cong$ SAS because $\overline{AB} \cong \overline{AC}$, $\angle A \cong \angle A$, and $\overline{AE} \cong \overline{AD}$.

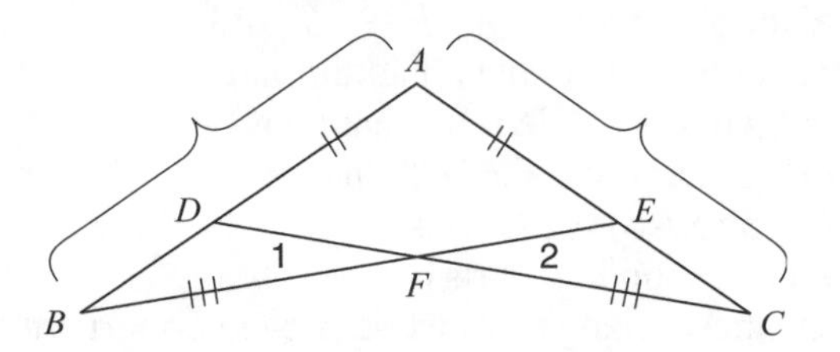

I now know that $\angle B \cong \angle C$ by CPCTC. From the Given, $\overline{BF} \cong \overline{CF}$. Also, $\angle 1 \cong \angle 2$ because they are vertical angles. This means that $\triangle BFD \cong \triangle CFE$. The reason is ASA $\cong$ ASA. By CPCTC, $\overline{DF} \cong \overline{EF}$.

Check Your Understanding of Section 3.4

Write a proof.

1. Given: $\overline{RS} \cong \overline{SL}$, $\overline{RT} \perp \overline{LT}$, $\overline{RS} \cong \overline{RT}$
Prove: a. $\triangle RLS \cong \triangle RLT$.
b. $\overline{WL}$ bisects $\angle SWT$.

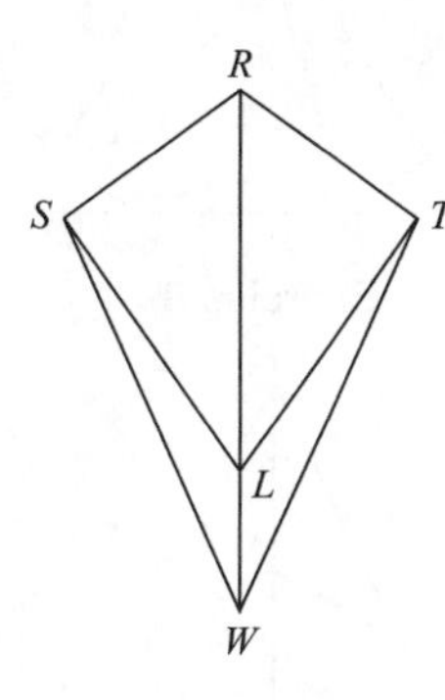

Exercise 1

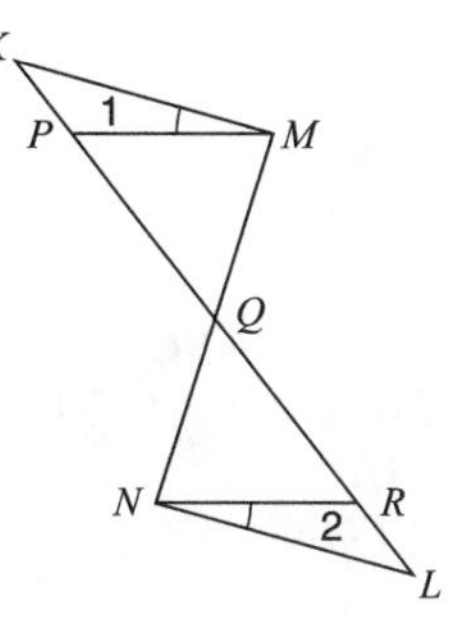

Exercise 2

2. Given: $\overline{KPQRL}$, $\overline{MQN}$, $\overline{KL}$, and $\overline{MN}$ bisect each other at Q.
$\angle 1 \cong \angle 2$
Prove: $\overline{PM} \cong \overline{NR}$

3. Given: Quadrilateral $ABCD$,
$\overline{BD}$ intersects $\overline{AC}$ at E,
$\overline{AB} \cong \overline{AD}$, $\angle 1 \cong \angle 2$.
Prove: a. $\triangle ABE \cong \triangle ADE$.
b. $\angle 3 \cong \angle 4$.

4. Given: Quadrilateral $ABCD$,
$\overline{BD}$ intersects $\overline{AC}$ at E, $\overline{AC}$ bisects $\angle BAD$ and $\angle BCD$.
Prove: $\angle 1 \cong \angle 2$.

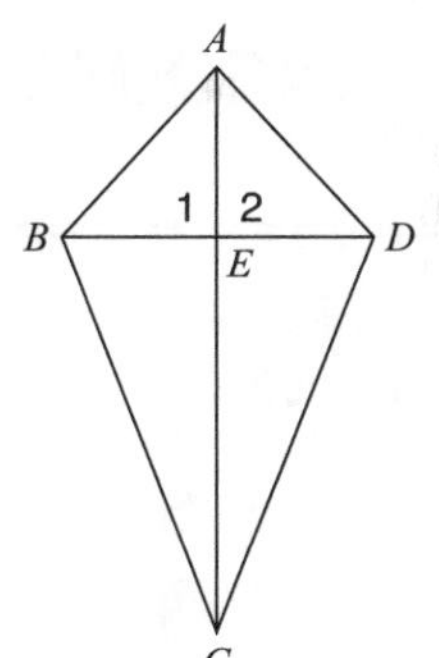

Exercises 3–4

5. Given: $\angle FAC \cong \angle FCA$,
$\overline{FD} \perp \overline{AB}$, $\overline{FE} \perp \overline{BC}$.
Prove: $\overline{BF}$ bisects $\angle DBE$.

6. Given: $\overline{BD} \cong \overline{BE}$, $\overline{FD} \cong \overline{FE}$.
Prove: $\triangle AFC$ is isosceles.

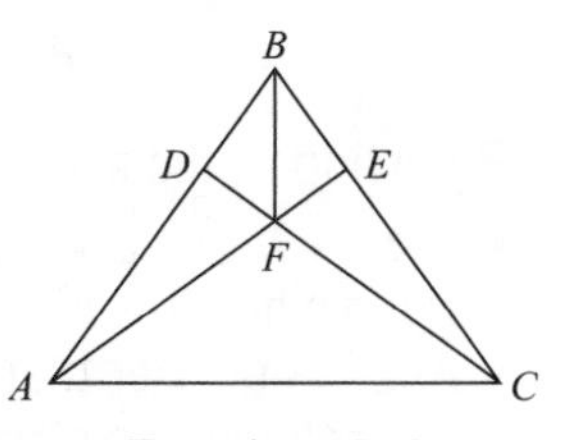

Exercises 5–6

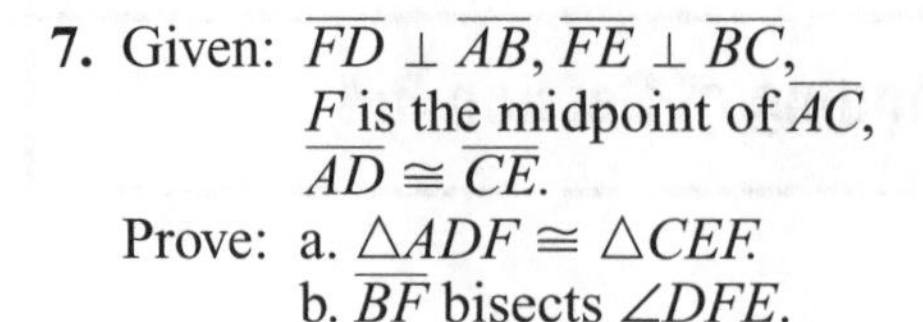

7. Given: $\overline{FD} \perp \overline{AB}$, $\overline{FE} \perp \overline{BC}$,
F is the midpoint of $\overline{AC}$,
$\overline{AD} \cong \overline{CE}$.
Prove: a. $\triangle ADF \cong \triangle CEF$.
b. $\overline{BF}$ bisects $\angle DFE$.

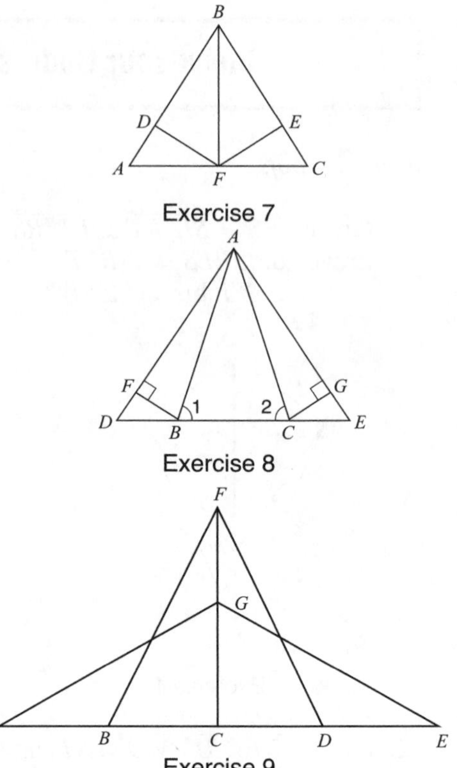

Exercise 7

Exercise 8

Exercise 9

8. Given: $\overline{AB} \cong \overline{AC}$, $\overline{BD} \cong \overline{CE}$,
$\overline{BF} \perp \overline{AD}$, $\overline{CG} \perp \overline{AE}$.
Prove: $\overline{DF} \cong \overline{EG}$.

9. Given: $\overline{ABCDE}$, $\angle FBC \cong \angle FDC$,
$\overline{CGF}$ bisects $\angle BFD$,
and $\overline{AB} \cong \overline{ED}$.
Prove: $\triangle ACG \cong \triangle ECG$.

3.5 INEQUALITIES IN A TRIANGLE

KEY IDEAS

The inequality symbol $<$ means "is less than" and $>$ means "is greater than."

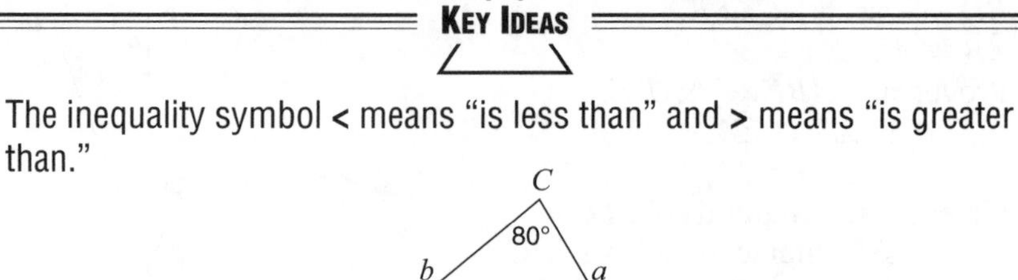

Referring to the accompanying figure:

- $c < a + b$, $b < a + c$, and $a < b + c$.
- $b > a$ since $m\angle B > m\angle A$.
- $m\angle 1$ is greater than $m\angle A$ and $m\angle C$.
- c is the longest side of the triangle because it is opposite the greatest angle.

Algebraic Inequality Facts

If a, b, and c represent positive numbers, then:

- Transitive Property: If $a > b$ and $b > c$, then $a > c$.
- Substitution Property: If $a > b$ and $b = c$, then $a > c$.
- Comparison Property: If $S = a + b$, then $S > a$ and $S > b$.
- Addition Property: If $a > b$, then $a + c > b + c$.

The transitive, substitution, and addition properties work for both the "is greater than" (>) and "is less than" (<) inequality relations.

Testing Numbers as Possible Side Lengths

Suppose points A, B, and C are the three vertices of a triangle, as shown in Figure 3.7. Since the shortest distance between two points is a straight line, AC must be less than $AB + BC$.

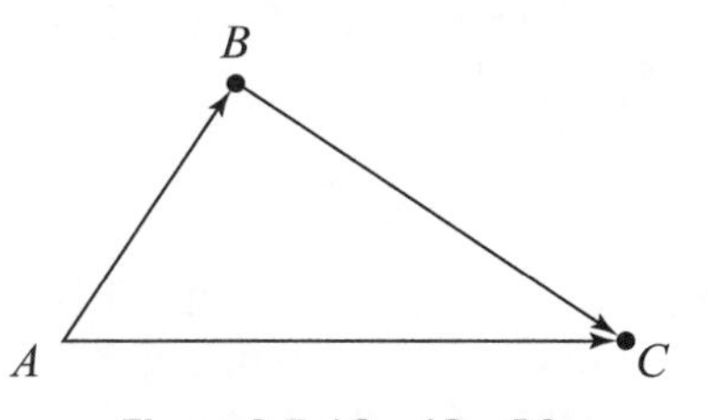

Figure 3.7 $AC < AB + BC$.

> **Theorem: Triangle Inequality Theorem**
> The length of each side of a triangle is less than the sum of the lengths of the other two sides.

To determine whether a set of three positive numbers can represent the lengths of the sides of a triangle, check that *each* of the three numbers is less than the sum of the other two numbers.

Example 1

Which of the following sets of numbers cannot represent the lengths of the sides of a triangle?

(1) {9, 40, 41} (2) {7, 7, 3} (3) (4, 5, 1} (4) {6, 6, 6}

Solution: Examine each choice in turn. For choice (3), $4 < 5 + 1$ and $1 < 4 + 5$, but 5 is *not* less than $4 + 1$.

The correct choice is **(3)**.

Side Length Limits

In a triangle, the range of possible lengths of each side depends on the lengths of the other two sides. In Figure 3.8, the lengths of two sides of the given triangle are 4 and 9. You already know that x, the length of the third side, is less than $4 + 9$. It is also true that the length of any side of a triangle is greater than the difference in the lengths of the other two sides. Thus, $x > 9 - 4$.

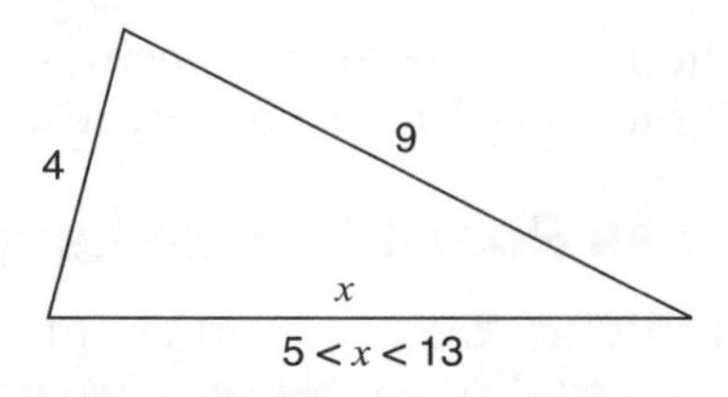

Figure 3.8 Side length limits.

Because $x > 5$ and $x < 13$, x is between 5 and 13 which is expressed by the "sandwich" inequality, $5 < x < 13$.

Example 2

The shortest distance between city A and city B is 150 miles. The shortest distance between city B and city C is 350 miles. Which could be the shortest distance, in miles, between city A and city C?

(1) 175 (2) 200 (3) 250 (4) 300

Solution: The shortest distances between the three cities can be represented by the lengths of the sides of a triangle whose vertices are the three cities.

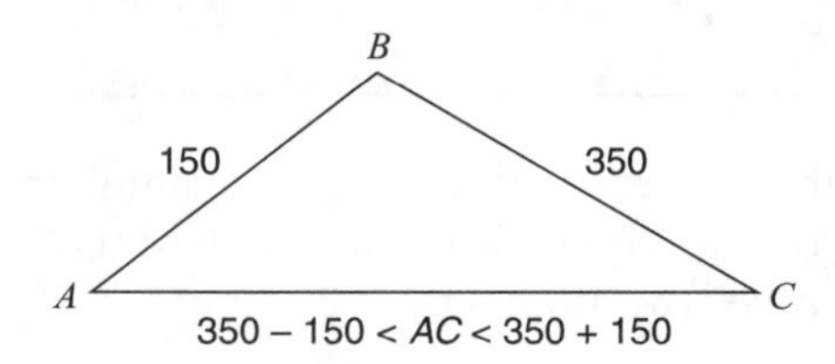

- Because the length of each side of a triangle is less than the sum of the lengths of the other two sides, $AC < 150 + 350$, so $AC < 500$.
- The length of each side of a triangle is also greater than the difference between the lengths of the other two sides. Hence, $AC > 350 - 150$, so $AC > 200$.
- Hence, $200 < AC < 500$. The smallest distance in this range among the answer choices is 250.

The correct choice is **(3)**.

Example 3

Devon wants to construct a triangular-shaped metal sculpture in which the lengths of two of the sides are 3 inches and 8 inches. How many different triangular sculptures can be constructed if the length of the third side must be an integer length?

Solution: The number of inches in the length of the third side must be greater than $8 - 3 = 5$ *and* less than $8 + 3 = 11$. Because the length of the third side must be an integer between 5 and 11, the length of the third side can be 6, 7, 8, 9, or 10 inches. Hence, Devon can construct **5** different triangular-shaped sculptures such that two of the sides measure 3 inches and 8 inches.

Exterior Angle Inequality

In Figure 3.9, $\angle 1$ is an exterior angle of $\triangle ABC$. Since $m\angle 1 = a + c$, $m\angle 1$ is greater than either of the two quantities on the right side of the equation:

$$m\angle 1 > a \text{ and } m\angle 1 > c.$$

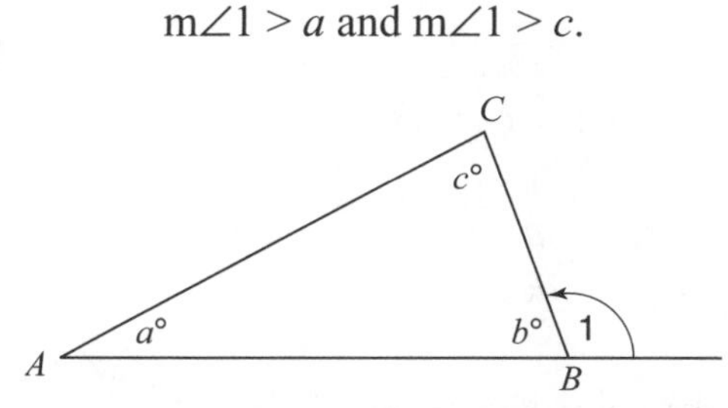

Figure 3.9 Exterior Angle Inequality Theorem.

> **Theorem: Exterior Angle Inequality Theorem**
> In a triangle, the measure of an exterior angle is greater than the measure of either nonadjacent (remote) interior angle.

Noncongruent Sides Imply Noncongruent Angles

The inverses of the Base Angles Theorem and its converse are true. Furthermore, the greater of the two angles of a triangle lies opposite the longer of the two facing sides, as illustrated in Figure 3.10.

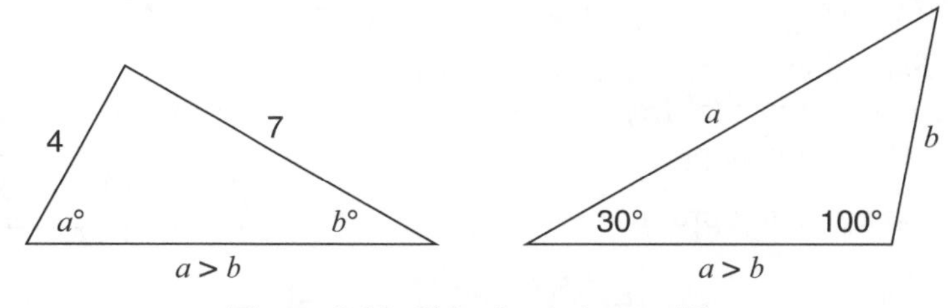

Figure 3.10 Side–Angle inequalities.

Theorem: Side–Angle Inequality Theorems

- **Theorem**: If two sides of a triangle are *not* congruent, the angles opposite them are *not* congruent, and the greater angle is opposite the longer side.

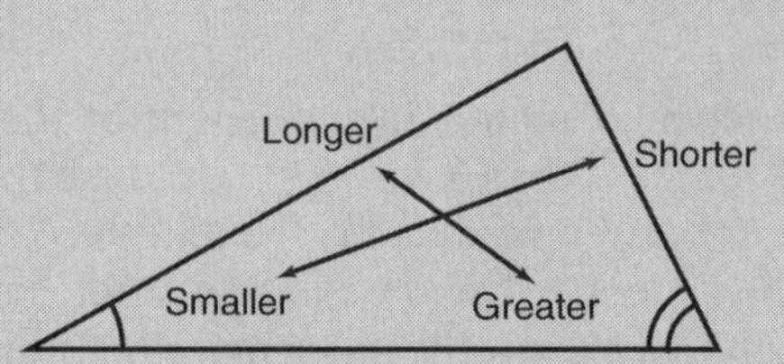

- **Theorem**: If two angles of a triangle are *not* congruent, the sides opposite them are *not* congruent, and the longer side is opposite the greater angle.

Example 4

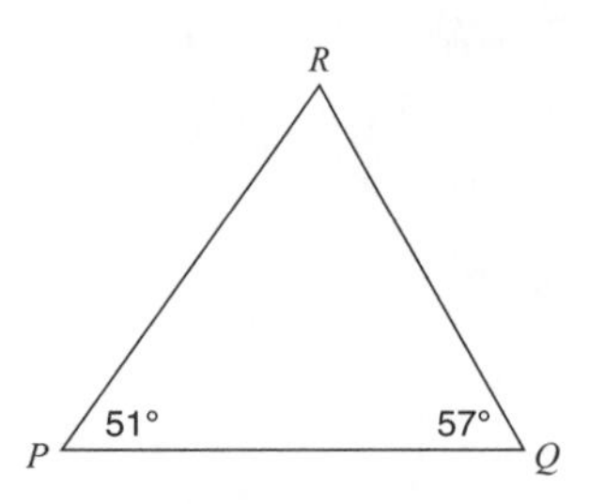

In $\triangle PQR$, $m\angle P = 51$ and $m\angle Q = 57$.

Which expression is true?
(1) $QR > PQ$ (3) $PQ = QR$
(2) $PR > PQ$ (4) $PQ > QR$

Solution: Find the degree measure of $\angle R$:

$$\begin{aligned} m\angle P + m\angle Q + m\angle R &= 180 \\ 51 + 57 + m\angle R &= 180 \\ m\angle R &= 180 - 108 \\ &= 72 \end{aligned}$$

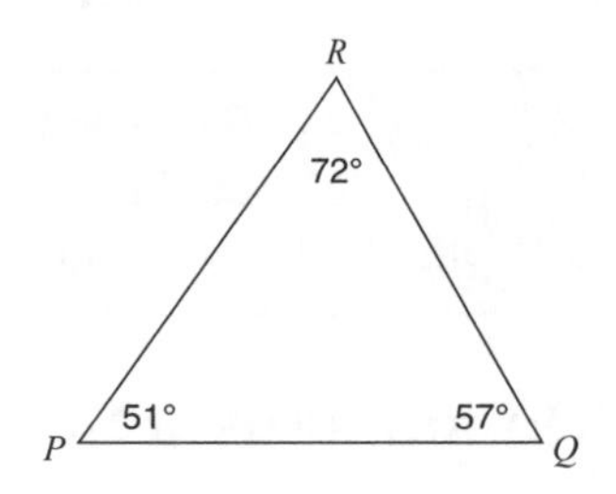

Since $\angle R$ is the largest angle of $\triangle PQR$, the side opposite, $\overline{PQ}$, is the longest side. Hence, $PQ > QR$. The correct choice is **(4)**.

Example 5

Given: $\overline{BD}$ bisects $\angle ABC$.
Prove: $AB > AD$.

Solution: To prove $AB > AD$, show that $m\angle 3$ (the angle opposite $\overline{AB}$) is greater than $m\angle 1$ (the angle opposite $\overline{AD}$).

See the accompanying proof.

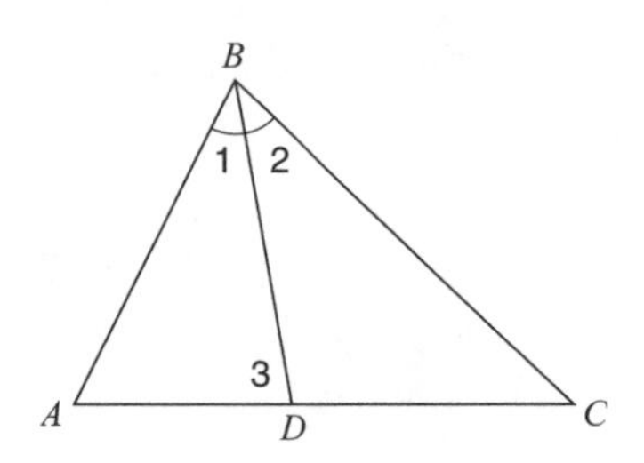

Proof

Statement	Reason
1. m∠3 > m∠2.	1. The degree measure of an exterior angle of a triangle is greater than the degee measure of either nonadjacent interior angle.
2. $\overline{BD}$ bisects ∠*ABC*.	2. Given.
3. m∠1 = m∠2.	3. A bisector divides an angle into two angles having the same degree measure.
4. m∠3 > m∠1.	4. Substitution property of inequalities.
5. *AB* > *AD*.	5. If two angles of a triangle are not equal in degree measure, then the sides opposite are not equal and the longere side is opposite the larger angle.

Check Your Understanding of Section 3.5

A. *Multiple Choice.*

1. In △*ABC*, *D* is a point on $\overline{AC}$ such that $\overline{BD}$ bisects ∠*ABC*. If m∠*ABC* = 60 and m∠*C* = 70, then
(1) *AC* > *AB* (2) *BD* > *BC* (3) *AD* > *BD* (4) *AB* > *AD*

2. Which set can *not* represent the lengths of the sides of a triangle?
(1) {4, 5, 6} (2) {5, 5, 11} (3) {7, 7, 12} (4) {8, 8, 8}

3. Sara is building a triangular pen for her pet rabbit. If two of the sides measure 8 feet and 15 feet, the length of the third side could be
(1) 13 ft (2) 7 ft (3) 3 ft (4) 23 ft

4. In △*ABC*, *D* is a point on $\overline{AC}$, m∠*A* > m∠*B*, and *E* is a point on $\overline{BC}$ such that $\overline{CE} \cong \overline{CD}$. Which statement is always true?
(1) m∠*CDE* > m∠*A*
(2) m∠*B* > m∠*CED*
(3) *EB* > *AD*
(4) *EB* = *AD*

5. In isosceles triangle *ABC*, $\overline{AC} \cong \overline{BC}$ and point *D* is between *A* and *B* on base $\overline{AB}$. If $\overline{CD}$ is drawn, which statement is *always* true?
(1) *AB* > *CD*
(2) *CD* < *AC*
(3) m∠*A* > m∠*ADC*
(4) m∠*B* > m∠*BDC*

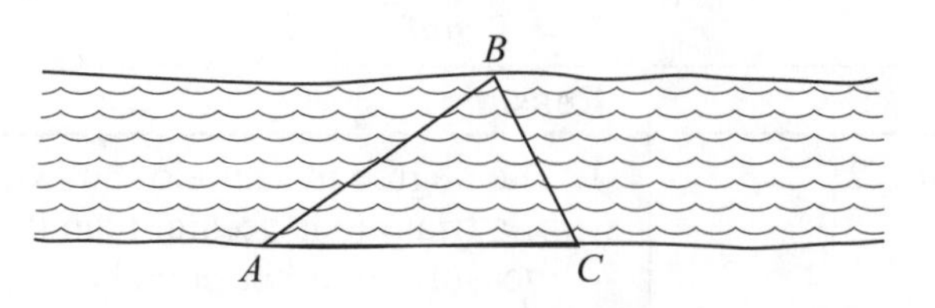

6. On the banks of a river, surveyors marked locations *A, B,* and *C*. The measure of $\angle ACB$ is 70° and the measure of $\angle ABC$ is 65°. Which expression shows the relationship between the lengths of the sides of this triangle?

(1) $AB < BC < AC$ (3) $BC < AC < AB$
(2) $BC < AB < AC$ (4) $AC < AB < BC$

7. In $\triangle NYC$, $m\angle N = 45$ and $m\angle Y = 70$. Which statement best describes $\triangle NYC$?

(1) $\overline{NY}$ is the longest side.
(2) $\overline{NC}$ is the longest side.
(3) $\overline{YC}$ is the longest side.
(4) $\overline{YC}$ and $\overline{NC}$ are the longest sides and are congruent.

8. If $\frac{1}{4}$ and $\frac{1}{2}$ represents the lengths of two sides of an isosceles triangle, then the perimeter of the triangle is

(1) 1 only (2) $\frac{5}{4}$ only (3) 1 or $\frac{5}{4}$ (4) $\frac{3}{4}$ only

9. If *D* is any point on side $\overline{AB}$ of equilateral triangle *ABC*, which statement is always true?

(1) $CD > DB$ (3) $m\angle A > m\angle ADB$
(2) $AB < CD$ (4) $m\angle B > m\angle BDC$

10. A box contains one 2-inch rod, one 3-inch rod, one 4-inch rod, and one 5-inch rod. What is the maximum number of different triangles that can be made using the full lengths of the rods as sides?

(1) 1 (2) 2 (3) 3 (4) 4

11. In the accompanying diagram, $\overline{QR} \cong \overline{QS}$. It is always true that
(1) $m\angle 1 > m\angle 2$ (3) $m\angle 3 > m\angle 4$
(2) $m\angle 1 > m\angle 4$ (4) $m\angle 4 > m\angle 2$

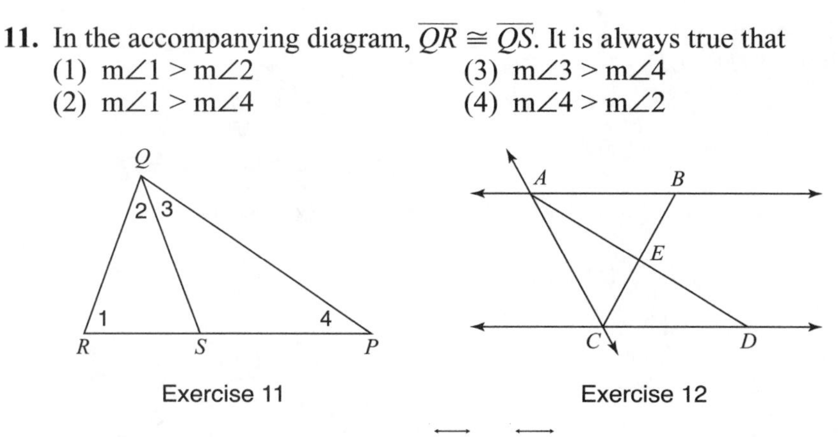

Exercise 11 Exercise 12

12. In the accompanying diagram, $\overleftrightarrow{AB} \parallel \overleftrightarrow{CD}$, $m\angle BAC < m\angle ACD$. If $\overline{CB}$ bisects $\angle ACD$ and $\overline{AD}$ bisects $\angle BAC$, then which statement is true?
(1) $AC = AB$ (3) $AE < CE$
(2) $AD < CD$ (4) $m\angle DAC > m\angle BCD$

B. *Show or explain how you arrived at your answer.*

13. José wants to build a triangular pen for his pet rabbit. He has three lengths of boards already cut that measure 7 feet, 8 feet, and 16 feet. Explain why José cannot construct a pen in the shape of a triangle with sides of 7 feet, 8 feet, and 16 feet.

14. In $\triangle SAT$, $m\angle A = 53$ and $m\angle T = 76$. Which side of $\triangle SAT$ is the shortest? Give a reason for your answer.

C. *Write a proof.*

15. Given: $\triangle ABC$, $\overline{ACF}$, $\overline{DEF}$, $\overline{AB} \cong \overline{BC}$.
Prove: $DF > AD$.

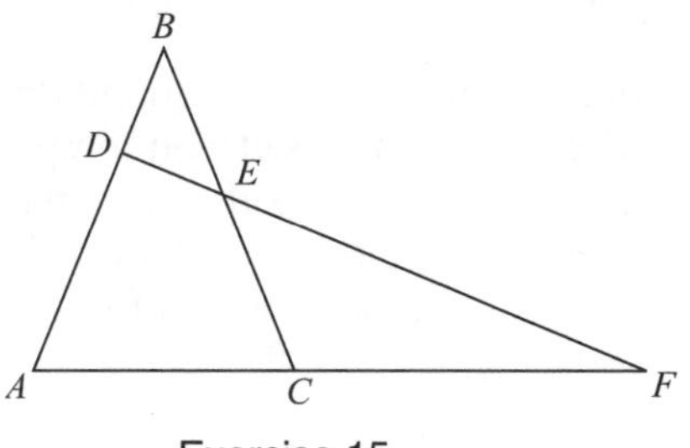

Exercise 15

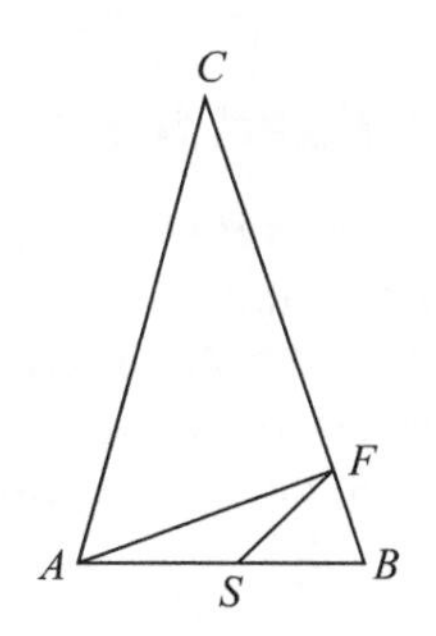

Exercise 16

16. Given: $\triangle ABC$, $\overline{BFC}$, $\overline{ASB}$, $\overline{AC} \cong \overline{BC}$.
Prove: $AF > FS$.

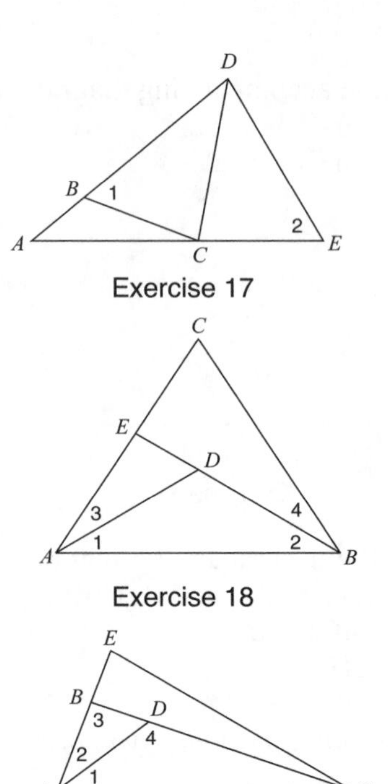

17. Given: $m\angle 1 = m\angle 2$.
Prove: $AD > ED$.

Exercise 17

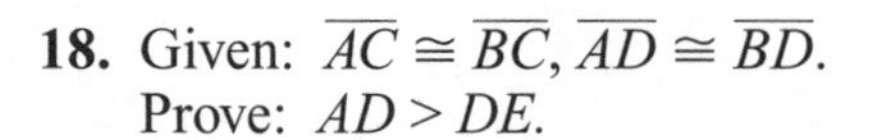

18. Given: $\overline{AC} \cong \overline{BC}, \overline{AD} \cong \overline{BD}$.
Prove: $AD > DE$.

Exercise 18

19. Given: $\overline{CB}$ is drawn to side $\overline{AE}$ of $\triangle AEC$, $\overline{AD}$ bisects $\angle CAB$, $AD > BD$.
Prove: $AC > DC$.

Exercise 19

3.6 PROVING STATEMENTS INDIRECTLY

KEY IDEAS

To prove a statement indirectly, prove its opposite cannot be true. An indirect proof is usually needed when the statement you are required to prove involves the word *not*.

The Need for an Indirect Proof

Sometimes it is too difficult or even impossible to prove a statement directly. In such situations, it may be helpful to see what happens if you temporarily assume the statement you must prove is *not* true. If you discover that this assumption leads to an impossible situation, then you may conclude that the required statement must be true as this is the only other possibility.

MATH FACTS

How to Write an Indirect Proof

- Step 1: Assume the *opposite* of what you need to prove is true.
- Step 2: Show that this assumption leads to a contradiction of a known fact.
- Step 3: Conclude that the only remaining possibility, the statement you were required to prove, is true.

An indirect proof is sometimes referred to as a "proof by contradiction." This type of proof can be extended to the situation in which there are more than two possibilities by accounting for all such possibilities and then showing that all but one produces some type of contradiction.

Example 1

Alex tells his friend that it is *not* possible to construct a triangle with two obtuse angles. Prove that Alex is correct.

Solution: Write an indirect proof:

- Step 1: Assume it is possible to construct a triangle with two obtuse angles.
- Step 2: If the triangle has two obtuse angles, then the sum of the measures of the two obtuse angles is greater than 180. But this contradicts the theorem that states the sum of the measures of the three angles of a triangle is 180.
- Step 3: Because of this contradiction, the original assumption is false. Thus, it is *not* possible to construct a triangle with two obtuse angles as this is the only other possibility.

Writing an Indirect Proof

An indirect proof lends itself to a paragraph format.

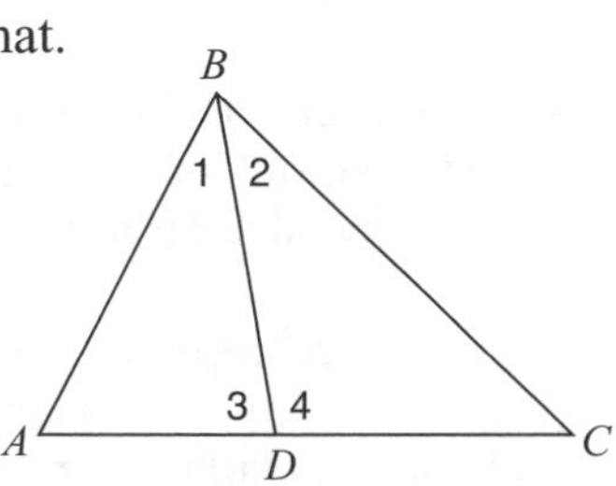

Example 2

Given: $\triangle ABC$ is scalene, $\angle 1 \cong \angle 3$.
Prove: $\overline{BD}$ is *not* perpendicular to $\overline{AC}$.

Solution: Write an indirect proof.

- Step 1: Assume $\overline{BD} \perp \overline{AC}$.
- Step 2: It follows that $\angle 3 \cong \angle 4$. Since $\triangle ADB \cong \triangle CDB$ by ASA $\cong$ ASA, $\overline{AB} \cong \overline{BC}$ by CPCTC. But this contradicts the Given fact that $\triangle ABC$ is scalene.
- Step 3: Because of this contradiction, the original assumption is false. The only remaining possibility, $\overline{BD}$ is *not* perpendicular to $\overline{AC}$, must be true.

Check Your Understanding of Section 3.6

1. In the accompanying diagram of $\triangle ABT$, $\overline{CBTD}$ is drawn and $\overline{AB} \perp \overline{CD}$. Write an explanation or an informal proof that shows $\overline{AT}$ is *not* perpendicular to $\overline{CD}$.

Exercise 1

2. Given: $\angle 1 \cong \angle 2$,
$\overline{BD}$ does not bisect $\triangle ABC$.
Prove: $\overline{AB} \not\cong \overline{BC}$.

Exercise 2

3. Given: $\triangle JAM$ is scalene, $\overline{JK}$ and $\overline{ML}$ are altitudes.
Prove: $\overline{JK} \not\cong \overline{ML}$.

Exercise 3

4. Given scalene triangle ABC. Point M the midpoint of side $\overline{AB}$. Write an explanation or an informal proof that shows the two segments drawn from M perpendicular to the other two sides of the triangle are *not* congruent.

5. In the accompanying diagram, $\overline{BC} \parallel \overline{AD}$ and $\triangle ABC$ is *not* isosceles. Write an explanation or an informal proof that shows $\overline{AC}$ does *not* bisect $\angle BAD$.

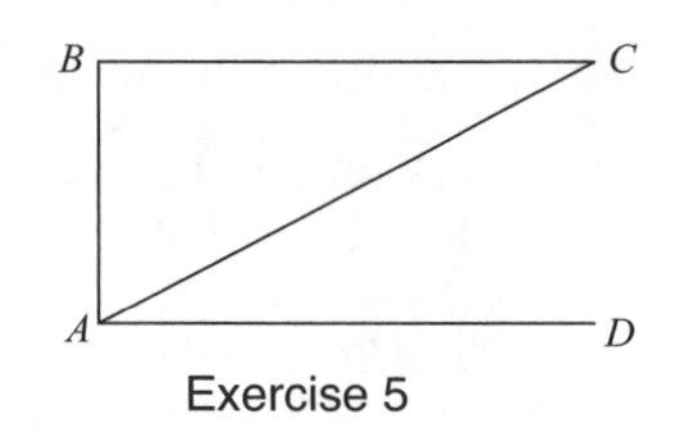

Exercise 5

6. Given: $\overline{AB} \cong \overline{AC}$, $\overline{BD} \not\cong \overline{CD}$.
Prove: $\overline{BE} \not\cong \overline{EC}$.

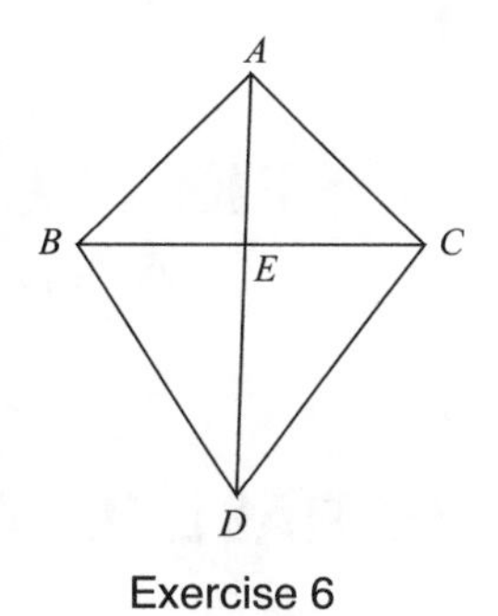

Exercise 6

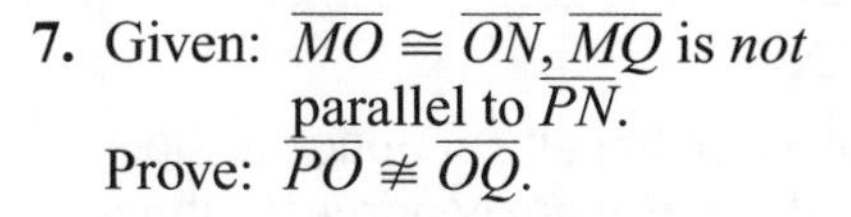

7. Given: $\overline{MO} \cong \overline{ON}$, $\overline{MQ}$ is *not* parallel to $\overline{PN}$.
Prove: $\overline{PO} \not\cong \overline{OQ}$.

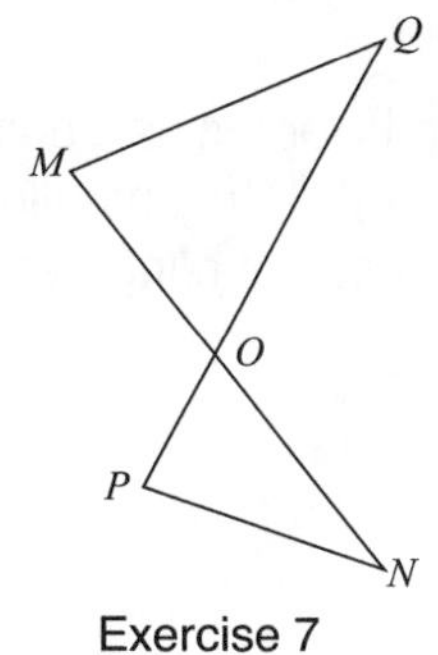

Exercise 7

8. Given: $AC > AB$, $\overline{DE} \cong \overline{CE}$.
Prove: $\overline{AB}$ is *not* parallel to $\overline{DE}$.

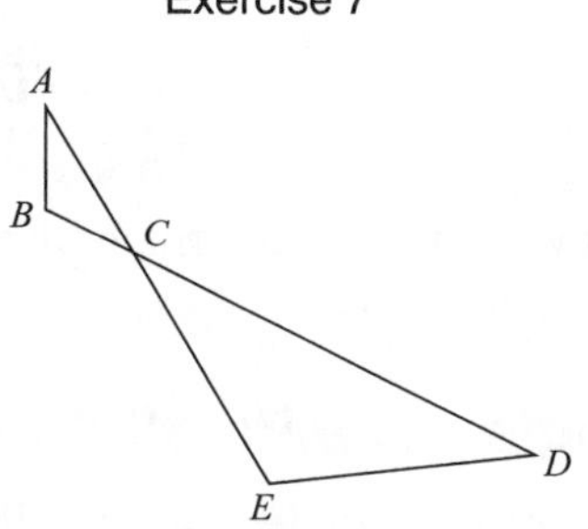

Exercise 8

CHAPTER 4
SPECIAL QUADRILATERALS AND COORDINATES

4.1 PARALLELOGRAMS

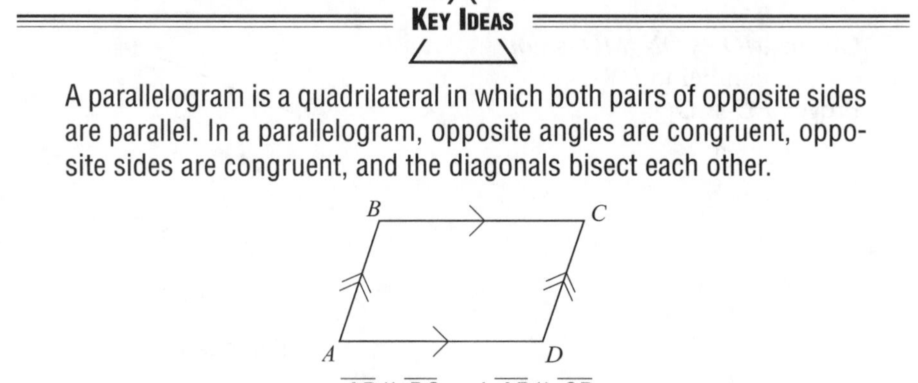

A parallelogram is a quadrilateral in which both pairs of opposite sides are parallel. In a parallelogram, opposite angles are congruent, opposite sides are congruent, and the diagonals bisect each other.

$\overline{AD} \parallel \overline{BC}$ and $\overline{AB} \parallel \overline{CD}$

Conversely, if a quadrilateral has any of these special properties, then it is a parallelogram.

Angles of a Parallelogram

Consecutive angles of a parallelogram are same side interior angles formed by parallel lines and, as a result, are supplementary. Because supplements of the same angle are congruent, $\angle A \cong \angle C$ and $\angle B \cong \angle D$. Thus, opposite angles of a parallelogram are congruent.

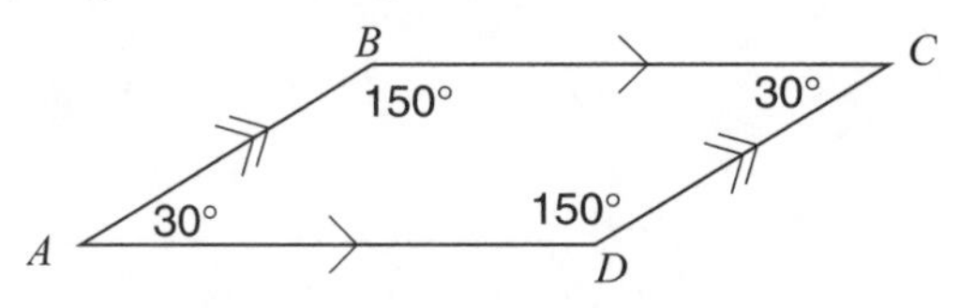

Figure 4.1 Angles of a parallelogram.

Theorems: Angles of a Parallelogram

- **Theorem:** If a quadrilateral is a parallelogram, then consecutive angles are supplementary.
- **Theorem:** If a quadrilateral is a parallelogram, then opposite angles are congruent.

Sides of a Parallelogram

You can prove the opposite sides of a parallelogram are congruent by proving a diagonal separates the parallelogram into two congruent triangles. In Figure 4.2, $\triangle BAD \cong \triangle DCB$ by AAS $\cong$ AAS. Then $\overline{AD} \cong \overline{CD}$ and $\overline{AD} \cong \overline{BC}$ by CPCTC.

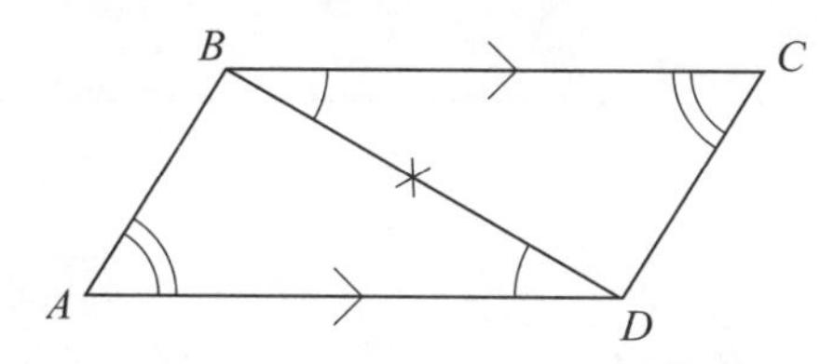

Figure 4.2 Opposites sides of parallelogram *ABCD* are congruent.

Example 1

Prove: The diagonals of a parallelogram bisect each other.

Solution: See the accompanying proof.

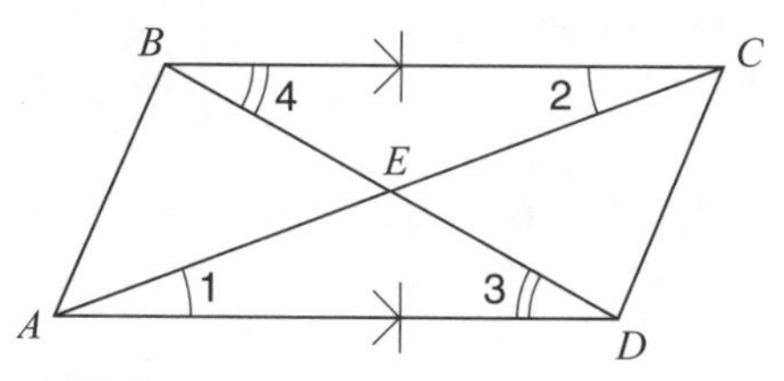

Given: *ABCD* is a parallelogram.
Prove: $\overline{AC}$ and $\overline{BD}$ bisect each other at *E*.

Solution: Prove $\triangle AED \cong \triangle CEB$ by ASA $\cong$ ASA.

Proof

Statement		**Reason**
1. $\overline{AD} \parallel \overline{BC}$.		1. Opposite sides of a parallelogram are parallel.
2. $\angle 1 \cong \angle 2$.	Angle	2. If two lines are parallel, then alternate interior angles are congruent.
3. $\overline{AD} \cong \overline{BC}$.	Side	3. Opposite sides of a parallelogram are congruent.
4. $\angle 3 \cong \angle 4$.	Angle	4. Same as #2.
5. $\triangle AED \cong \triangle CEB$.		5. ASA $\cong$ ASA.
6. $\overline{AE} \cong \overline{CE}$ and $\overline{DE} \cong \overline{EB}$.		6. CPCTC.
7. Point *E* is the midpoint of $\overline{AC}$ and $\overline{DB}$.		7. If a point divides a segment into two congruent segments, then the point is midpoint of that segment.
8. Diagonals $\overline{AC}$ and $\overline{BD}$ bisect each other at *E*.		8. If two line segments intersect at their midpoints, then they bisect each other.

Theorems: Sides and Diagonals of a Parallelogram

- **Theorem:** If a quadrilateral is a parallelogram, then opposite sides are congruent.
- **Theorem:** If a quadrilateral is a parallelogram, then the diagonals bisect each other.

Example 2

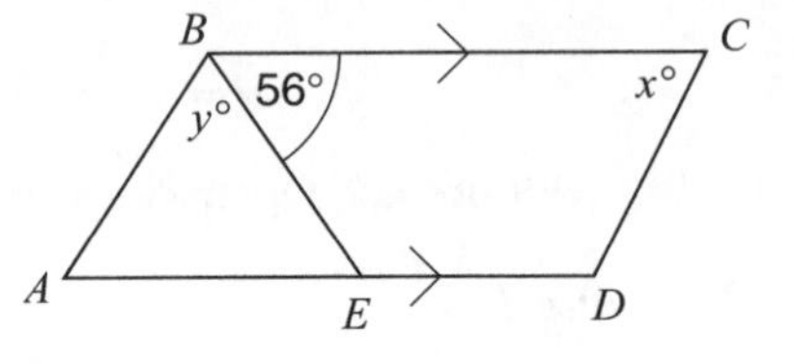

In the accompanying figure, $ABCD$ is a parallelogram. If $\overline{EB} \cong \overline{AB}$ and $m\angle CBE = 56°$, find the values of x and y ?

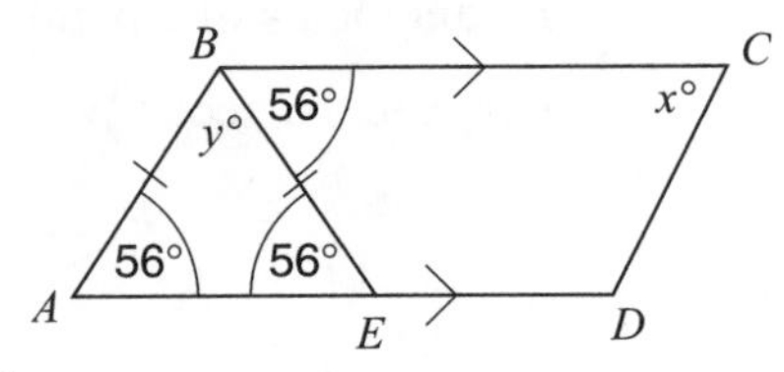

Solution: Since alternate interior angles formed by parallel lines have equal measures, $m\angle AEB = m\angle CBE = 56$.

- Because $\overline{EB} \cong \overline{AB}$, $m\angle A = m\angle AEB = 56$.
- Opposite angles of a parallelogram are congruent so $m\angle A = m\angle C = 56$. Hence, $\boldsymbol{x = 56}$.
- Since consecutive angles of a parallelogram are supplementary:

$$56 + (y + 56) = 180$$
$$y + 112 = 180$$
$$y = 180 - 112$$
$$\boldsymbol{y = 68}$$

Example 3

In parallelogram $ABCD$, $m\angle B = 5x - 43$ and $m\angle D = 3x - 7$. What is $m\angle A$?

Solution: First find the value of x.

- Since angles B and D are opposite angles of a parallelogram, they have equal measures:

$$m\angle B = m\angle D$$
$$5x - 43 = 3x - 7$$
$$5x - 3x = 43 - 7$$
$$2x = 36$$
$$\frac{2x}{2} = \frac{36}{2}$$
$$x = 18$$

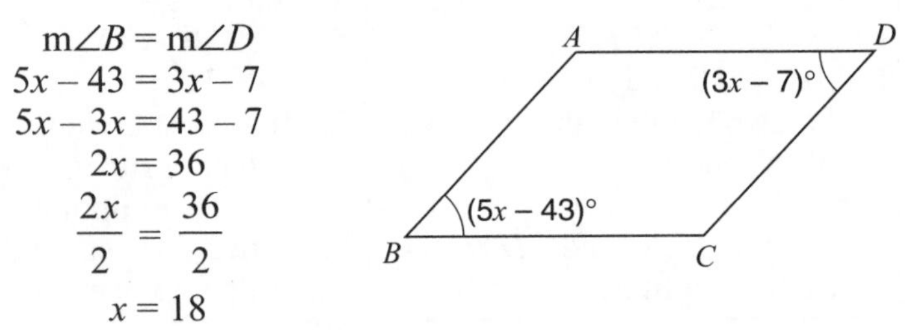

- Find m$\angle B$:

$$\begin{aligned} m\angle B &= 5x - 43 \\ &= 5(18) - 43 \\ &= 90 - 43 \\ &= 47 \end{aligned}$$

- Consecutive angles of a parallelogram are supplementary:

$$\begin{aligned} m\angle A &= 180 - m\angle B \\ &= 180 - 47 \\ &= \mathbf{133} \end{aligned}$$

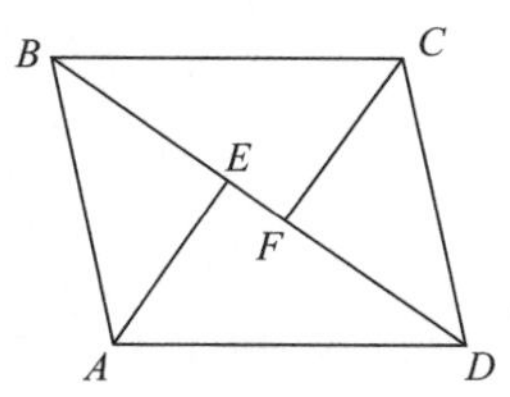

Example 4

Given: $ABCD$ is a parallelogram,
$\overline{AE} \perp \overline{BD}$, $\overline{CF} \perp \overline{BD}$.
Prove: $\overline{BE} \cong \overline{DF}$.

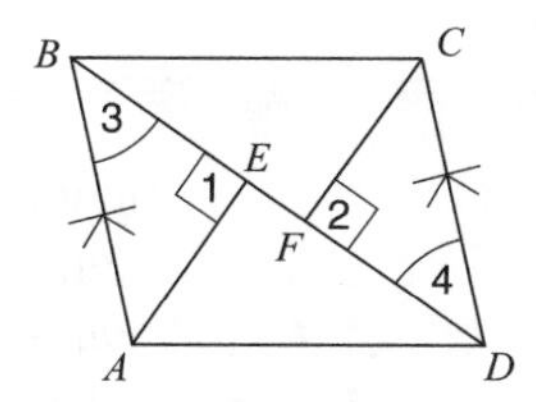

Solution: Prove the triangles that contain $\overline{BE}$ and $\overline{DF}$ as sides are congruent. By marking off the diagram, you know that $\triangle AEB \cong \triangle CFD$ (AAS $\cong$ AAS).

Proof

Statement		**Reason**
1. $ABCD$ is a parallelogram.		1. Given.
2. $\overline{AB} \parallel \overline{CD}$.		2. Opposite sides of a parallelogram are parallel.
3. $\angle 3 \cong \angle 4$.	Angle	3. If two parallel lines are cut by a transversal, alternate interior angles are congruent.
4. $\overline{AE} \perp \overline{BD}$, $\overline{CF} \perp \overline{BD}$.		4. Given.
5. $\angle 1 \cong \angle 2$.	Angle	5. Perpendicular lines form right angles and all right angles are congruent.
6. $\overline{AB} \cong \overline{CD}$.	Side	6. Opposite sides of a parallelogram are congruent.
7. $\triangle AEB \cong \triangle CFD$.		7. AAS $\cong$ AAS.
8. $\overline{BE} \cong \overline{DF}$.		8. CPCTC.

Proving That a Quadrilateral Is a Parallelogram

If a quadrilateral is a parallelogram, then

- Opposite sides are parallel.
- Opposite angles are congruent.
- Opposite sides are congruent.
- The diagonals bisect each other.

The converses of these statements are true and each represents a method for proving that a quadrilateral is a parallelogram. In addition, a quadrilateral is a parallelogram if the one pair of sides are both parallel and congruent, as in Figure 4.3.

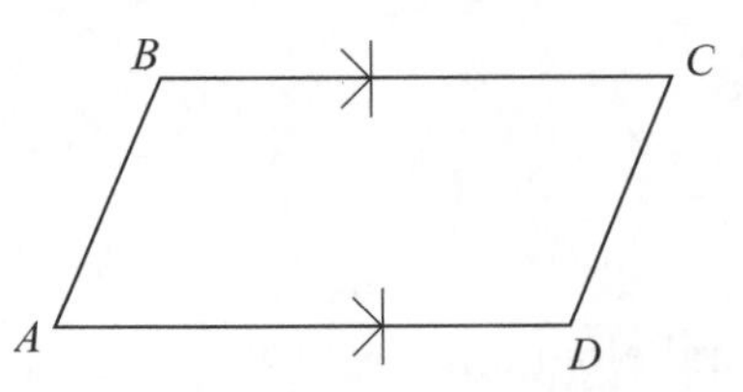

Figure 4.3 If $\overline{AD} \parallel \overline{BC}$ and $\overline{AD} \cong \overline{BC}$, then *ABCD* is a parallelogram.

MATH FACTS

Proving a Quadrilateral Is a Parallelogram

To prove that a quadrilateral is a parallelogram, show that it has any *one* of the following properties:

- Both pairs of opposite sides are parallel.
- Both pairs of opposite angles are congruent.
- Both pairs of opposite sides are congruent.
- Diagonals bisect each other.
- One pair of sides are both parallel *and* congruent.

Example 5

Given: Quadrilateral *ABCD*,
$\angle 1 \cong \angle 2$, $\overline{BD}$ bisects $\overline{AC}$ at *E*.
Prove: *ABCD* is a parallelogram.

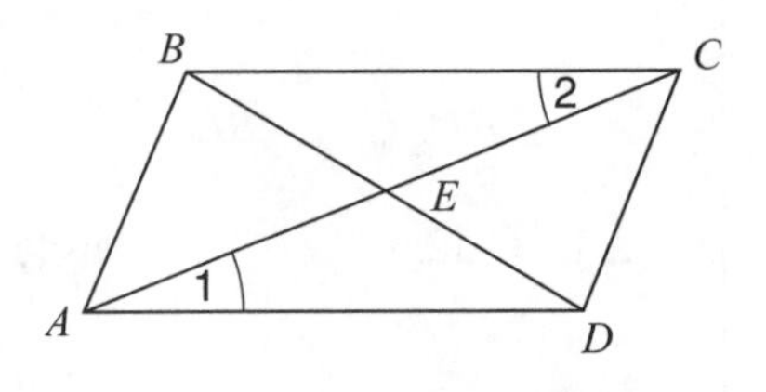

Solution: Since $\angle 1 \cong \angle 2$, $\overline{AD} \parallel \overline{BC}$. Thus, $\triangle BAD \cong \triangle DCB$ by ASA $\cong$ ASA. It follows that $\overline{AD} \cong \overline{BC}$. Because $\overline{AD}$ and $\overline{BC}$ are both parallel and congruent, *ABCD* is a parallelogram.

Proof

Statement		Reason
1. $\angle 1 \cong \angle 2$.	Angle	1. Given.
2. $\overline{AD} \parallel \overline{BC}$.		2. If two lines are cut by a transversal such that a pair of alternate interior angles are congruent, then the lines are parallel.
3. $\overline{BD}$ bisects $\overline{AC}$ at E.		3. Given.
4. $\overline{AE} \cong \overline{CE}$.	Side	4. A bisector of a line segment divides the segment into two congruent segments.
5. $\angle AED \cong \angle CEB$.	Angle	5. Vertical angles are congruent.
6. $\triangle BAD \cong \triangle DCB$		6. ASA $\cong$ ASA.
7. $\overline{AD} \cong \overline{BC}$.		7. CPCTC.
8. $ABCD$ is a parallelogram.		8. If one pair of sides of a quadrilateral are both parallel and congruent, then the quadrilateral is a parallelogram.

Example 6

Given: $ABCD$ is a parallelogram, $\overline{BE} \perp \overline{AC}$, $\overline{DF} \perp \overline{AC}$.
Prove: $BEDF$ is a parallelogram.

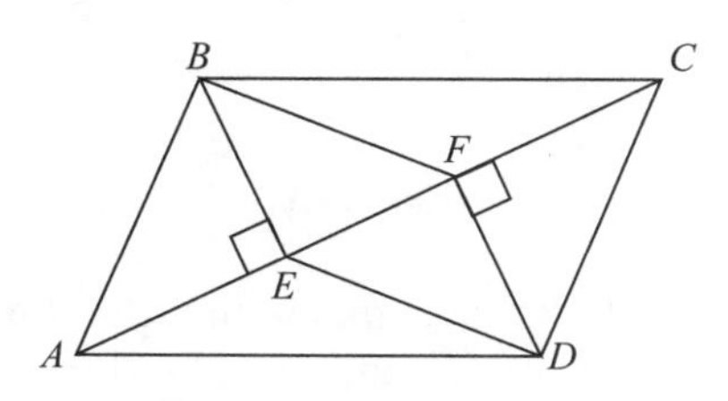

Solution: Mark off the diagram. $\overline{BE} \parallel \overline{DF}$ since the segments are perpendicular to the same line. $\triangle AEB \cong \triangle CFD$ by AAS $\cong$ AAS so $\overline{BE} \cong \overline{DF}$. $\therefore BEDF$ is a parallelogram. The two-column proof is left for you to complete.

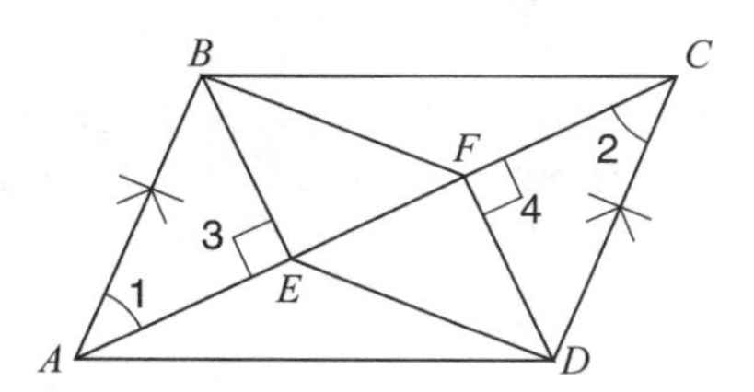

Check Your Understanding of Section 4.1

A. Multiple Choice

1. Which statement about a parallelogram is *not always* true?
(1) Diagonals are congruent.
(2) Opposite sides are parallel and congruent.
(3) Opposite angles are congruent.
(4) Consecutive angles are supplementary.

2. If quadrilateral $ABCD$ is a parallelogram, which statement *must* be true?
(1) $\overline{AC} \perp \overline{BD}$
(2) $\overline{AC} \cong \overline{BD}$
(3) $\overline{AC}$ bisects $\angle DAB$ and $\angle BCD$
(4) $\overline{AC}$ and $\overline{BD}$ have the same midpoint

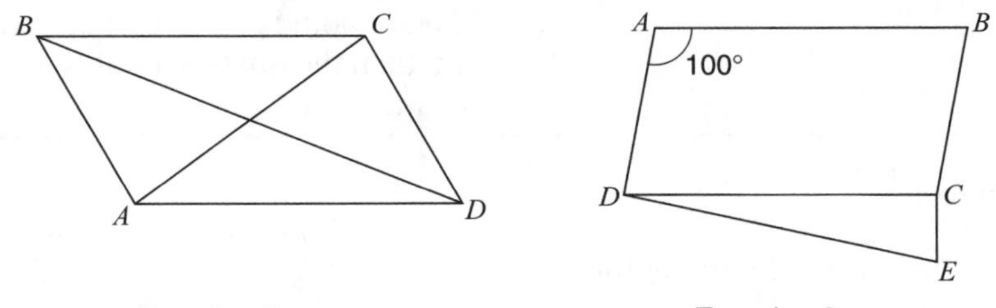

Exercise 2

Exercise 3

3. In the accompanying diagram of parallelogram $ABCD$, $\overline{EC} \perp \overline{DC}$, $\angle B \cong \angle E$, and $m\angle A = 100$. What is $m\angle CDE$?
(1) 10 (2) 20 (3) 30 (4) 80

4. In the accompanying diagram of parallelogram $MATH$, $m\angle T = 100$ and $\overline{SH}$ bisects $\angle MHT$. What is $m\angle HSA$?
(1) 80 (2) 100 (3) 120 (4) 140

Exercise 4

Exercise 5

5. In the accompanying diagram of parallelogram $ABCD$, $\overline{DE} \perp \overline{AC}$, $m\angle DCA = 40$, and $m\angle ADE = 70$. What is $m\angle ABC$?
(1) 100 (2) 110 (3) 120 (4) 140

6. In the accompanying diagram of parallelogram $ABCD$, m$\angle A = x$, m$\angle DEF = y$, and m$\angle DFE = z$. Which statement is correct?
(1) $x < y + z$
(2) $x = y + z$
(3) $x > y + z$
(4) $360 - x = y + z$

7. In parallelogram *MATH*, the degree measure of $\angle T$ exceeds two times the degree measure of $\angle H$ by 30. What is the degree measure of the largest angle of the parallelogram?
(1) 50 (2) 60 (3) 120 (4) 130

8. In parallelogram *TRIG*, m$\angle R = 2x + 19$ and m$\angle G = 4x - 17$. What is m$\angle T$?
(1) 48 (2) 55 (3) 125 (4) 132

9. If two sides of parallelogram *GEOM* are selected at random, what is the probability that the two sides are *not* congruent?
(1) $\frac{1}{2}$ (2) $\frac{1}{3}$ (3) $\frac{2}{3}$ (4) $\frac{1}{4}$

10. In the accompanying diagram of parallelogram $ABCD$, $AB > BC$. If altitude $\overline{AE}$ is drawn to side $\overline{DC}$, which statement *must* be true?
(1) $BC < AE$
(2) $EC = BC$
(3) $AE < AB$
(4) $BC > DC$

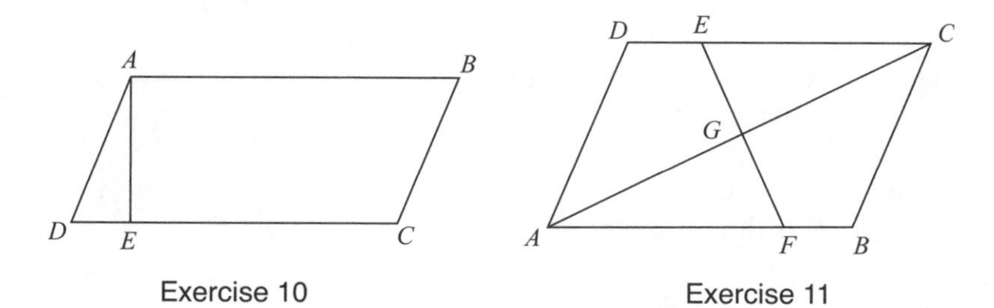

Exercise 10

Exercise 11

11. In the accompanying diagram of parallelogram $ABCD$, $\overline{DE} \cong \overline{BF}$. What is the most direct method that could be used to prove $\triangle EGC \cong \triangle FGA$?
(1) HL $\cong$ HL
(2) SAS $\cong$ SAS
(3) AAS $\cong$ AAS
(4) SSA $\cong$ SSA

B. *Show or explain how you arrived at your answer.*

12. In the accompanying diagram of parallelogram $ABCD$, $m\angle B = 5x$ and $m\angle C = 2x + 12$. Find the measure of $\angle D$.

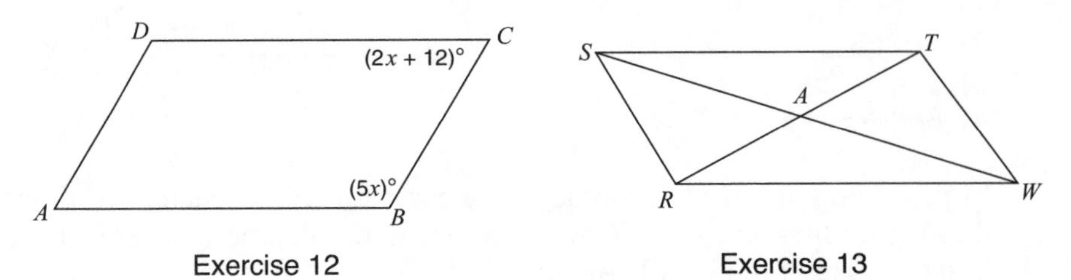

Exercise 12

Exercise 13

13. In parallelogram $RSTW$, diagonals $\overline{RT}$ and $\overline{SW}$ intersect at point A. If $SA = x - 5$ and $AW = 2x - 37$, what is the length of $\overline{SW}$?

C. *Write a proof.*

14. Given: $ABCD$ is a parallelogram, B is the midpoint of $\overline{AE}$.
Prove: $\overline{EF} \cong \overline{FD}$.

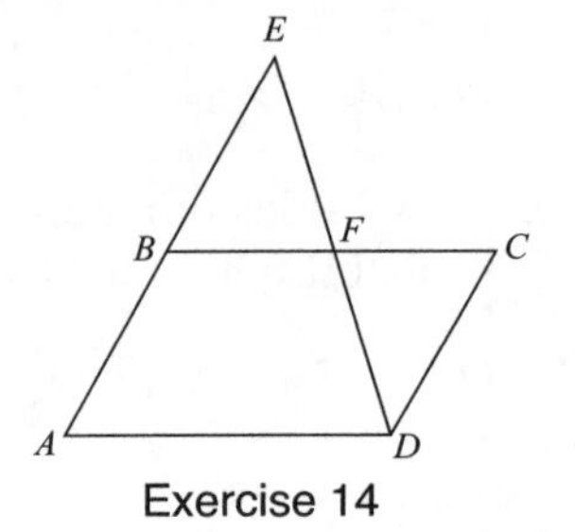

Exercise 14

15. Given: $ABCD$ is a parallelogram, $\overline{BF} \cong \overline{DG}$.
Prove: $\overline{EF} \cong \overline{HG}$.

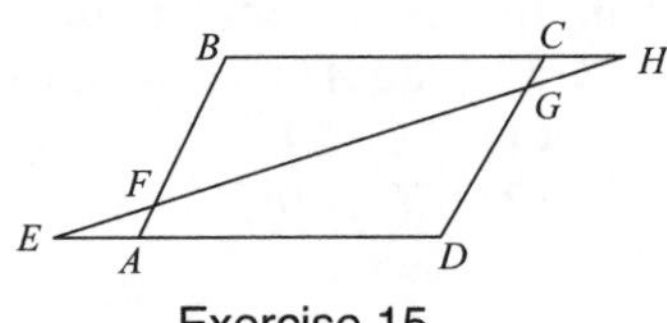

Exercise 15

16. Given: $ABCD$ is a parallelogram, $\overline{DK}$ bisects $\angle ADC$, $\overline{BK} \cong \overline{AB}$.
Prove: a. $\overline{CK} \cong \overline{CD}$.
b. K is the midpoint of $\overline{BC}$.

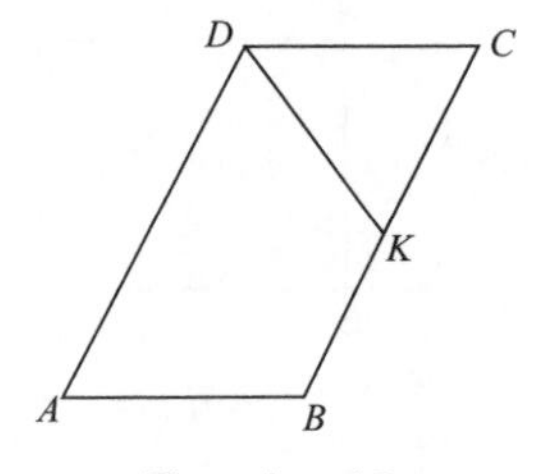

Exercise 16

17. Given: $ABCD$ is a parallelogram, $AD > DC$.
Prove: $m\angle 3 > m\angle 1$.

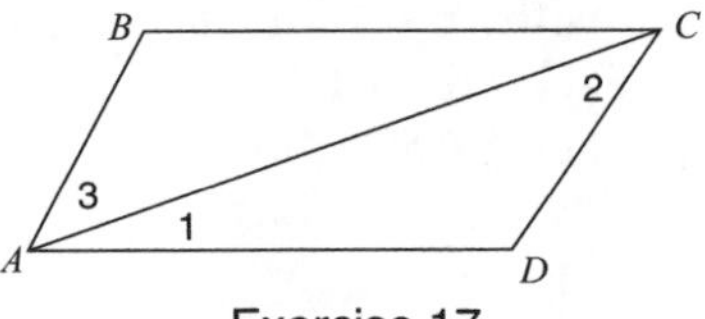

Exercise 17

18. Given: Parallelogram $ABCD$, diagonal $\overline{AC}$, and $\overline{ABE}$.
Prove: $m\angle 1 > m\angle 2$.

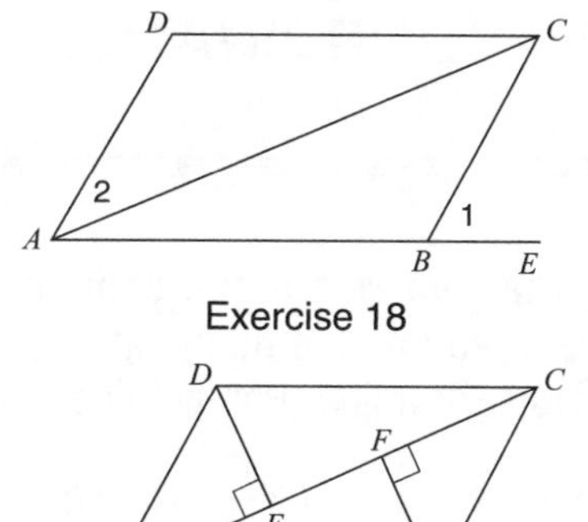

Exercise 18

19. Given: Quadrilateral $ABCD$, diagonal $\overline{AEFC}$, $\overline{DE} \perp \overline{AC}$, $\overline{BF} \perp \overline{AC}$, $\overline{AE} \cong \overline{CF}$, $\overline{DE} \cong \overline{BF}$.
Prove: $ABCD$ is a parallelogram.

Exercise 19

20. Given: $BMDL$ is a parallelogram, $\overline{AL} \cong \overline{CM}$.
Prove: $ABCD$ is a parallelogram.

21. Given: $ABCD$ is a parallelogram, $\angle ABL \cong \angle CDM$.
Prove: $BMDL$ is a parallelogram.

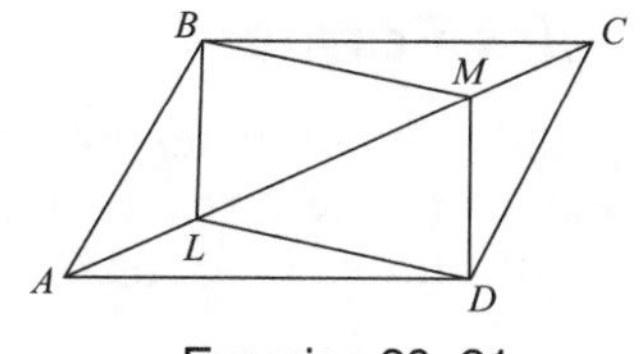

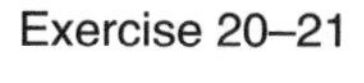

Exercise 20–21

22. Given: $\overline{NJ}$ and $\overline{DU}$ bisect each other at K, N is the midpoint of $\overline{DA}$, J is the midpoint of $\overline{QU}$.
Prove: $QUAD$ is a parallelogram.

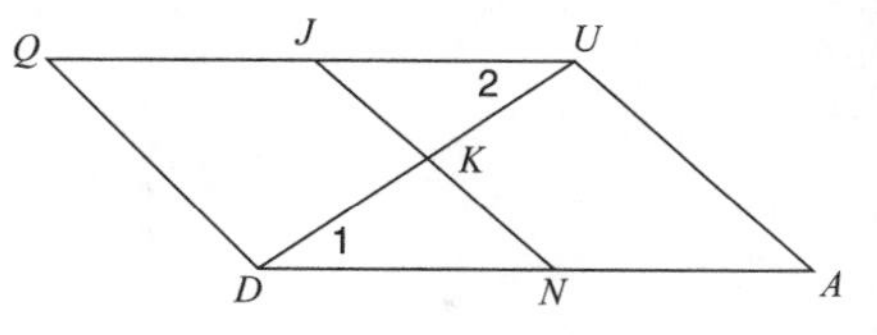

Exercise 22

23. Given: $ABCD$ is a quadrilateral, $\overline{FG} \cong \overline{EG}$, $\overline{AG} \cong \overline{CG}$, $\angle B \cong \angle D$.
Prove: a. $\overline{BC} \cong \overline{DA}$.
b. $ABCD$ is a parallelogram.

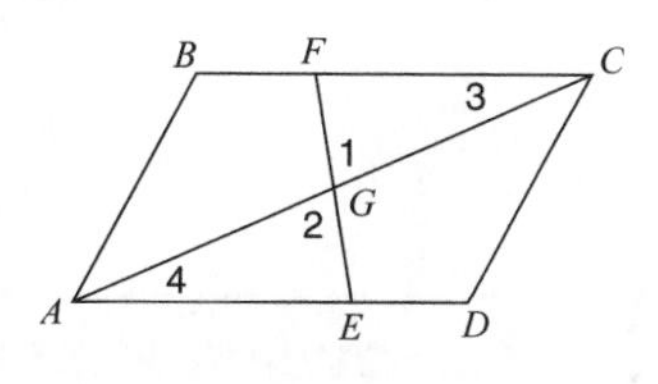

Exercise 23

4.2 COORDINATE FORMULAS

KEY IDEAS

When working in the coordinate plane, convenient formulas can be used to find the midpoint and length of a line segment as well as the slope of a line. Parallel lines have the same slope. Pairs of numbers such as $\frac{3}{4}$ and $-\frac{4}{3}$ are negative reciprocals because their product is −1. Perpendicular lines have slopes that are negative reciprocals. Coordinate formulas can be used to prove geometric relationships.

Midpoint Formula

The *x*– and *y*–coordinates of the midpoint *M* of a line segment are equal to the averages of the corresponding coordinates of the endpoints of the segment, as illustrated in Figure 4.4.

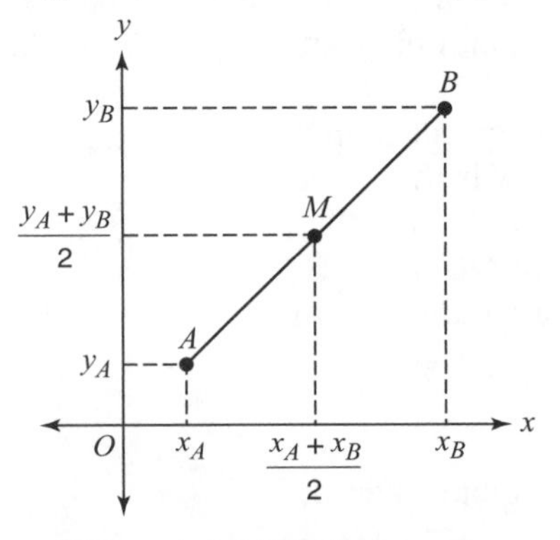

Figure 4.4 Midpoint formula.

Theorem: Midpoint Formula

If $A = (x_A, y_A)$, $B = (x_B, y_B)$, and M is the midpoint of $\overline{AB}$, then

$$M(x,y) = \left(\frac{x_A + x_B}{2}, \frac{y_A + y_B}{2}\right)$$

If the endpoints of $\overline{AB}$ are $A(3, 2)$ and $B(11, 8)$, then the midpoint M is at $(\bar{x}, \bar{y})$, where

$$\bar{x} = \frac{3+11}{2} = \frac{14}{2} = 7 \quad \text{and} \quad \bar{y} = \frac{2+8}{2} = \frac{10}{2} = 5$$

The notation $\bar{x}$ is sometimes used to indicate the average of two x–values. Similarly, $\bar{y}$ represents the average of two y–values.

Example 1

In $\triangle ABC$, median $\overline{CD}$ intersects side $\overline{AB}$ at point D. If the coordinates of point A are (4, 8) and the coordinates of point B are (10, –2), find the coordinates of point D.

Solution: Since median $\overline{CD}$ intersects side $\overline{AB}$ at point D, point D is the midpoint of $\overline{AB}$:

$$D(\bar{x}, \bar{y}) = \text{Midpoint of } \overline{AB}$$
$$= \left(\frac{4+10}{2}, \frac{8+(-2)}{2}\right)$$
$$= \left(\frac{14}{2}, \frac{6}{2}\right)$$
$$= \mathbf{(7, 3)}$$

Example 2

Given: The coordinates of the vertices of quadrilateral $ABCD$ are $A(-1,-3)$, $B(0, 3)$, $C(5, 5)$, and $D(4,-1)$. Prove that $ABCD$ is a parallelogram.

Solution: Find and then compare the midpoints of the two diagonals.

- The midpoint of diagonal $\overline{AC}$ is $\left(\frac{-1+5}{2}, \frac{-3+5}{2}\right) = (2, 1)$.
- The midpoint of diagonal $\overline{BD}$ is $\left(\frac{0+4}{2}, \frac{3+(-1)}{2}\right) = (2, 1)$.
- Since $\overline{AC}$ and $\overline{BD}$ have the same midpoint, the diagonals of quadrilateral $ABCD$ bisect each other, and, as a result, $ABCD$ is a parallelogram.

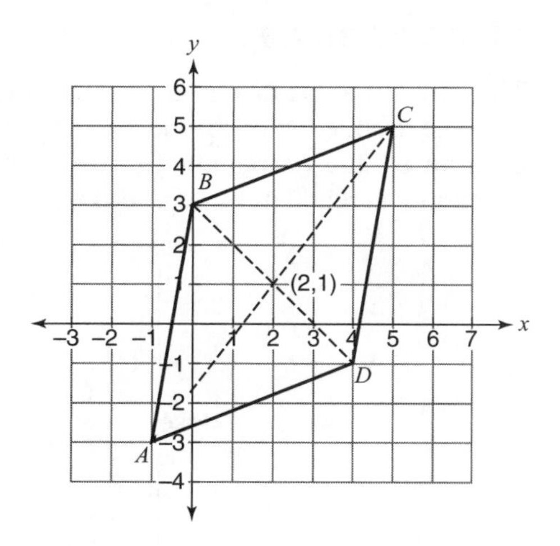

Example 3

Given that $F(2,1)$, $A(10, 7)$, $M(7, 11)$, and $E(x, y)$ are the vertices of a parallelogram. What are the coordinates of point E?

Solution: Find the midpoint of each diagonal.

- The midpoint of diagonal $\overline{FM}$ is $\left(\dfrac{2+7}{2}, \dfrac{1+11}{2}\right) = \left(\dfrac{9}{2}, 6\right)$.
- The midpoint of diagonal $\overline{AE}$ is $\left(\dfrac{10+x}{2}, \dfrac{7+y}{2}\right)$.
- Since *FAME* is a parallelogram, its diagonals have the same midpoint. Thus,

$$\left(\frac{10+x}{2}, \frac{7+y}{2}\right) = \left(\frac{9}{2}, 6\right)$$

Set the two x–coordinates equal and set the two y–coordinates equal:

$$\frac{10+x}{2} = \frac{9}{2} \quad \text{and} \quad \frac{7+y}{2} = 6$$

Multiply each side of each equation by 2:

$$\overset{1}{\cancel{2}}\left(\frac{10+x}{\cancel{2}}\right) = \overset{1}{\cancel{2}}\left(\frac{9}{\cancel{2}}\right) \qquad \overset{1}{\cancel{2}}\left(\frac{7+y}{\cancel{2}}\right) = 2(6)$$

$$10 + x = 9 \qquad 7 + y = 12$$

$$x = -1 \qquad y = 5$$

The coordinates of point E are **(–1, 5)**.

Slope Formula

The slope of a line is a number that represents its steepness. To find the slope of a line, divide the difference in the y–coordinates (Δy) of any two points on the line by the corresponding difference in their x–coordinates (Δx). The notation y is read "delta y" and Δx is read "delta x." See Figure 4.5.

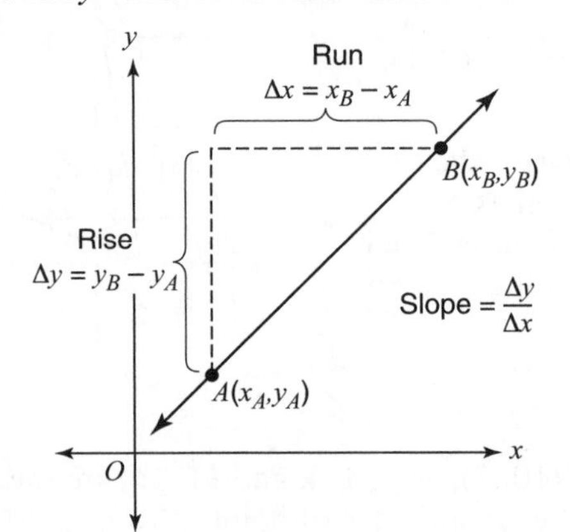

Figure 4.5. Slope as "rise over run."

Theorem: Slope Formula
If $A = (x_A, y_A)$ and $B = (x_B, y_B)$ are two points on a nonvertical line and m is the slope of that line, then

$$m = \frac{\Delta y}{\Delta x} = \frac{y_B - y_A}{x_B - x_A}$$

Because the order in which two numbers are subtracted matters, the order in which the points are taken when subtracting coordinates in the denominator of the slope fraction must be the same as in the numerator of the slope fraction. Thus, it would *not* be correct to perform the subtractions in the order indicated by

$$\frac{y_B - y_A}{x_A - x_B} \text{ or } \frac{y_A - y_B}{x_B - x_A}$$

Special Slope Relationships

Parallel lines have equal slopes and perpendicular lines have slopes that are negative reciprocals so that their product is –1.

Theorem: Slopes of Parallel and Perpendicular Lines
Let m represent the slope of $\overleftrightarrow{AB}$ and n represent the slope of $\overleftrightarrow{CD}$.

If $\overleftrightarrow{AB} \parallel \overleftrightarrow{CD}$, then $m = n$. Conversely, if $m = n$, then $\overleftrightarrow{AB} \parallel \overleftrightarrow{CD}$.

If $\overleftrightarrow{AB} \perp \overleftrightarrow{CD}$, then $m \times n = -1$. Conversely, if $m \times n = -1$, then $\overleftrightarrow{AB} \perp \overleftrightarrow{CD}$.

Example 4

Prove by means of slope that quadrilateral $ABCD$ from Example 1 is a parallelogram.

Solution: Find and then compare the slopes of the opposite sides of the quadrilateral with vertices $A(-1, -3)$, $B(0, 3)$, $C(5, 5)$, and $D(4, -1)$.

- Find the slopes of opposite sides $\overline{AB}$ and $\overline{CD}$:

$$\text{Slope of } \overline{AB} = \frac{\Delta y}{\Delta x} = \frac{3-(-3)}{0-(-1)} = \frac{3+3}{0+1} = 6$$

$$\text{Slope of } \overline{CD} = \frac{\Delta y}{\Delta x} = \frac{-1-5}{4-5} = \frac{-6}{-1} = 6$$

Because slope of $\overline{AB}$ = slope of $\overline{CD}$, $\overline{AB} \parallel \overline{CD}$.

- Find the slopes of opposite sides $\overline{BC}$ and $\overline{AD}$:

$$\text{Slope of } \overline{BC} = \frac{\Delta y}{\Delta x} = \frac{5-3}{5-0} = \frac{2}{5}$$

$$\text{Slope of } \overline{AD} = \frac{\Delta y}{\Delta x} = \frac{-1-(-3)}{4-(-1)} = \frac{-1+3}{4+1} = \frac{2}{5}$$

Since slope of $\overline{BC}$ = slope of $\overline{AD}$, $\overline{BC} \parallel \overline{AD}$.

- Both pairs of opposite sides of quadrilateral *ABCD* are parallel, and, as a result, *ABCD* is a parallelogram.

Example 5

The coordinates of the vertices of $\triangle PQR$ are $P(-1, -1)$, $Q(1, -2)$, and $R(3, 2)$. Prove by means of slope that $\triangle PQR$ is a right triangle.

Solution: If two sides of $\triangle PQR$ have slopes that are negative reciprocals, then those sides are perpendicular, and, consequently, $\triangle PQR$ is a right triangle.

$$\text{Slope of } \overline{PQ} = \frac{\Delta y}{\Delta x} = \frac{-2-(-1)}{1-(-1)} = \frac{-2+1}{1+1} = -\frac{1}{2}$$

$$\text{Slope of } \overline{PR} = \frac{\Delta y}{\Delta x} = \frac{2-(-1)}{3-(-1)} = \frac{2+1}{3+1} = \frac{3}{4}$$

$$\text{Slope of } \overline{QR} = \frac{\Delta y}{\Delta x} = \frac{2-(-2)}{3-1} = \frac{2+2}{2} = \frac{4}{2} = 2$$

Compare the slopes of the three sides. The slopes of $\overline{PQ}$ and $\overline{QR}$ are negative reciprocals since $\left(-\frac{1}{2}\right) \times 2 = -1$.

Hence, $\overline{PQ} \perp \overline{QR}$ so $\triangle PQR$ is a right triangle.

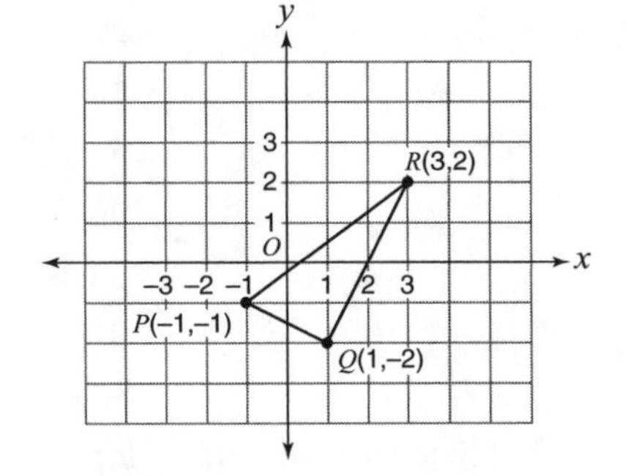

Example 6

Given: The coordinates of the vertices of $\triangle ABC$ are $A(-1, 2)$, $B(7, 0)$, and $C(1, -6)$.
Prove: a. Point $D(4, -3)$ lies on $\overline{BC}$.
b. $\overline{AD}$ is the perpendicular bisector of $\overline{BC}$.

Solution:

a. Points B, C, and D are collinear if the slopes of any two segments determined by these points are equal.

$$\text{Slope of } \overline{BC} = \frac{\Delta y}{\Delta x} = \frac{-6-0}{1-7} = \frac{-6}{-6} = 1$$

$$\text{Slope of } \overline{BD} = \frac{\Delta y}{\Delta x} = \frac{-3-0}{4-7} = \frac{-3}{-3} = 1$$

Since slope of $\overline{BC}$ = slope of $\overline{BD}$, point D lies on $\overline{BC}$.

b. To prove $\overline{AD}$ is the perpendicular bisector of $\overline{BC}$, you need to show two things: (1) $\overline{AD} \perp \overline{BC}$ and (2) D is the midpoint of $\overline{BC}$.

- From part a, slope of $\overline{BC} = 1$.
- Slope of $\overline{AD} = \frac{\Delta y}{\Delta x}$

$$= \frac{-3-2}{4-(-1)} = \frac{-5}{5} = -1$$

- Since the slopes of $\overline{BC}$ and $\overline{AD}$ are negative reciprocals, $\overline{AD} \perp \overline{BC}$.

- Find the midpoint of $\overline{BC}$:

$$\left(\frac{7+1}{2}, \frac{0+(-6)}{2}\right) = \left(\frac{8}{2}, \frac{-6}{2}\right) = (4, -3)$$

- Since the coordinates of D are given as $(4, -3)$, D is the midpoint of $\overline{BC}$.

Since $\overline{AD} \perp \overline{BC}$ and D is the midpoint of $\overline{BC}$, $\overline{AD}$ is the perpendicular bisector of $\overline{BC}$.

MATH FACTS

Proving Relationships Using Slope

- To prove two lines are *parallel*, use the slope formula to show that their slopes are equal.
- To prove two lines are *perpendicular*, use the slope formula to find their slopes. Then show that the product of the two slopes is –1.
- To prove three points are *collinear*, show that the slopes of any two of the segments determined by these points are equal.

Distance Formula

The distance between two points can be calculated using the **distance formula**.

Theorem: Distance Formula
If $A = (x_A, y_A)$ and $B = (x_B, y_B)$, then

$$AB = \sqrt{(\Delta x)^2 + (\Delta y)^2}$$

where $\Delta x = x_B - x_A$ and $\Delta y = y_B - y_A$. The distance AB also represents the length of $\overline{AB}$.

To find the length of the segment whose endpoints are $A(3, 2)$ and $B(11, 8)$, use the distance formula where $\Delta x = 11 - 3 = 8$ and $\Delta y = 8 - 2 = 6$:

$$\begin{aligned} AB &= \sqrt{(\Delta x)^2 + (\Delta y)^2} \\ &= \sqrt{8^2 + 6^2} \\ &= \sqrt{100} \\ &= 10 \end{aligned}$$

Example 7

Prove by means of the distance formula that quadrilateral $ABCD$ from Example 1 is a parallelogram.

Solution: Compare the lengths of the opposite sides of the quadrilateral with vertices $A(-1, -3)$, $B(0, 3)$, $C(5, 5)$, and $D(4, -1)$:

- Find the lengths of opposites sides $\overline{AB}$ and $\overline{CD}$:

$$\begin{aligned} AB &= \sqrt{(\Delta x)^2 + (\Delta y)^2} \\ &= \sqrt{(0-(-1))^2 + (3-(-3))^2} \\ &= \sqrt{1^2 + (3+3)^2} \\ &= \sqrt{1+36} \\ &= \sqrt{37} \end{aligned}$$

$$\begin{aligned} CD &= \sqrt{(\Delta x)^2 + (\Delta y)^2} \\ &= \sqrt{(4-5)^2 + (-1-5)^2} \\ &= \sqrt{(-1)^2 + (-6)^2} \\ &= \sqrt{1+36} \\ &= \sqrt{37} \end{aligned}$$

Thus, $\overline{AB} \cong \overline{CD}$.

- Find the lengths of opposite sides $\overline{BC}$ and $\overline{AD}$:

$$\begin{aligned} BC &= \sqrt{(\Delta x)^2 + (\Delta y)^2} \\ &= \sqrt{(5-0)^2 + (5-3)^2} \\ &= \sqrt{5^2 + 2^2} \\ &= \sqrt{25+4} \\ &= \sqrt{29} \end{aligned}$$

$$\begin{aligned} AD &= \sqrt{(\Delta x)^2 + (\Delta y)^2} \\ &= \sqrt{(4-(-1))^2 + (-1-(-3))^2} \\ &= \sqrt{(4+1)^2 + (-1+3)^2} \\ &= \sqrt{5^2 + 2^2} \\ &= \sqrt{25+4} \\ &= \sqrt{29} \end{aligned}$$

Hence, $\overline{BC} \cong \overline{AD}$.

- Both pairs of opposite sides of quadrilateral *ABCD* are congruent and, as result, *ABCD* is a parallelogram.

MATH FACTS

To prove a quadrilateral is a parallelogram, use

- The *midpoint formula* to show that the diagonals have the same midpoint and, as a result, bisect each other; or
- The *slope formula* to show that both pairs of opposite sides have the same slope and, as a result, are parallel; or
- The *distance formula* to show that both pairs of opposite sides have the same length and, as a result, are congruent; or
- The *slope and distance formulas* to show that the same pair of sides are both parallel and congruent.

Check Your Understanding of Section 4.2

A. Multiple Choice

1. The vertices of parallelogram *STAR* are *S*(0, 0), *T*(6, –1), *A*(4, 2), and *R*(–2, 3). What are the coordinates of the point of intersection of the diagonals?

(1) $\left(3, -\frac{1}{2}\right)$ (2) (2,1) (3) $\left(-1, \frac{3}{2}\right)$ (4) $\left(5, \frac{1}{2}\right)$

2. The median to side $\overline{AB}$ of $\triangle ABC$ is $\overline{CM}$. If the coordinates of *M* are (–2, 5) and the coordinates of *A* are (4,7), what are the coordinates of *B*?

(1) (1,6) (2) (2,12) (3) (–8,6) (4) (–8,3)

3. The slope of $\overline{AB}$ is $\frac{3}{4}$. Points *A*, *B*, and *C* are collinear. If the coordinates of point *A* are (2, 5), the coordinates of point *C* could be

(1) (6, 8) (2) (5,9) (3) (–1,1) (4) (6,2)

4. If the points (3, 5), (4, 2), and (5, *k*) are collinear, the value of *k* is

(1) 1 (2) 0 (3) –1 (4) –2

5. Given: points *A*(0, 0), *B*(3, 2), and *C*(–2, 3). Which statement is true?

(1) $\overline{AB} \parallel \overline{AC}$
(2) $\overline{AB} \perp \overline{AC}$
(3) $AB > BC$
(4) $\overline{BC} \perp \overline{CA}$

6. Given: Points *A*(1, 2), *B*(4, 5), and *C*(6, 7). Which statement is true?

(1) $\overline{AB} \cong \overline{BC}$.
(2) $\overline{AB} \perp \overline{BC}$.
(3) Points *A*, *B*, and *C* are collinear.
(4) The slope of $\overline{AC} = -1$.

7. The distance between points (4*a*, 3*b*) and (3*a*, 2*b*) is

(1) $a^2 + b^2$
(2) $\sqrt{a^2 + b^2}$
(3) $a + b$
(4) $\sqrt{a + b}$

8. The midpoint of $\overline{AB}$ is *M*, the coordinates of *A* are (*a*, *b*), and the coordinates of *B* are (*a* + 4, 5*b*). What are the coordinates of point *M*?

(1) (2, 2*b*)
(2) (*a* + 2, 3*b*)
(3) (2*a* + 4, 6*b*)
(4) $\left(\frac{a+4}{2}, \frac{5b}{2}\right)$

B. *Show or explain how you arrived at your answer.*

9. Parallelogram $GAME$ has vertices $G(2,–2)$, $A(p, q)$, $M(9, 3)$, and $E(3, 2)$. Find the coordinates of point A.

10. The diagonals of parallelogram $S(1,1)$, $T(–2, 3)$, $A(0, k)$, and $R(3,–5)$ intersect at point E. If the length of $\overline{EA}$ is represented as $3x – 19$, find the values of k and x.

11. Prove that $M(–2,–1)$, $A(6, 2)$, $T(7, 7)$, and $H(–1, 4)$ are the vertices of a parallelogram.

12. Triangle CAT has vertices $C(–2, 6)$, $A(6, 4)$, and $T(0,–2)$. Point $S(3, 1)$ is on side $\overline{AT}$.
Prove by means of coordinate geometry that
a. $\triangle CAT$ is isosceles.
b. $\overline{CS}$ is the perpendicular bisector of $\overline{AT}$.

13. Prove that $A(–2, 8)$, $B(6, 6)$, $C(0, 12)$, and $D(–4, 14)$ are *not* the vertices of a parallelogram.

14. The vertices of $\triangle NYS$ are $N(–2, –1)$, $Y(0, 10)$, and $S(10, 5)$.
a. If the coordinates of point T are $(4, 2)$, prove that points N, T, and S are collinear.
b. $\overline{YT}$ is both a median and an altitude to side $\overline{NS}$.

15. Given: $A(5, 7)$, $B(11, –1)$, and $C(3, 3)$.
a. Prove that $\triangle ABC$ is a right triangle.
b. Prove that the length of the median drawn to the hypotenuse is one-half the length of the hypotenuse.

16. The vertices of $\triangle ABC$ are $A(–1, 2)$, $B(3, 8)$, and $C(5,–2)$.
a. Prove that $\triangle ABC$ is a right triangle.
b. Prove that point M, the midpoint of the hypotenuse, is equidistant from all three vertices of the triangle.

4.3 SPECIAL PARALLELOGRAMS

KEY IDEAS

A parallelogram may have four congruent angles, four congruent sides, or both of these special properties.

Rectangle

A **rectangle** is a parallelogram with four right angles. In addition to all of the properties of a parallelogram, a rectangle has congruent diagonals, as shown in Figure 4.6.

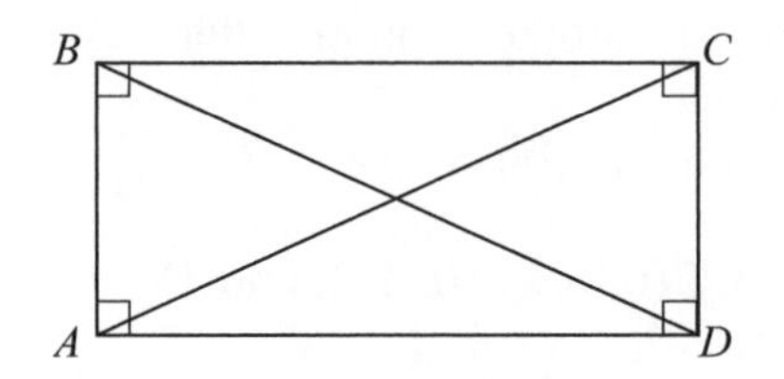

Figure 4.6 In rectangle *ABCD*, $\overline{AC} \cong \overline{BD}$.

Example 1

Given: Rectangle *ABCD*, $\overline{EG} \cong \overline{EF}$.
Prove: $\overline{AG} \cong \overline{DF}$.

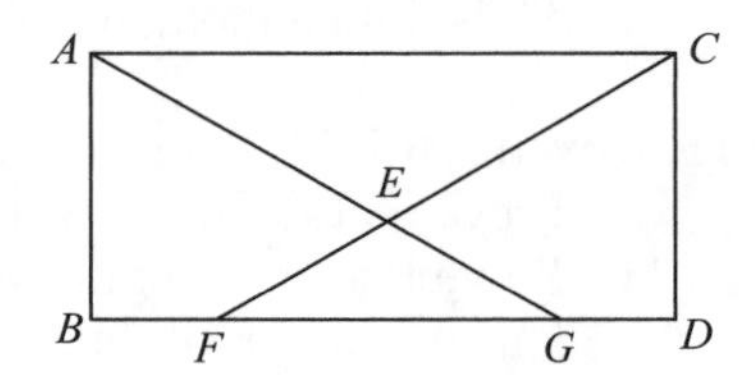

Solution: Mark off the diagram with the Given. Prove $\triangle ABG \cong \triangle DCF$ by AAS $\cong$ AAS.

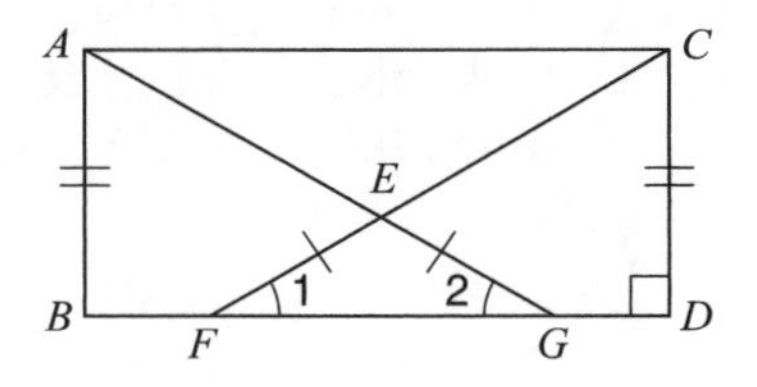

Proof

Statement		Reason
1. Rectangle *ABCD*.		1. Given.
2. Angles *B* and *C* are right angles.		2. A rectangle contains four right angles.
3. $\angle B \cong \angle C$.	Angle	3. All right angles are congruent.
4. $\overline{EF} \cong \overline{EG}$.		4. Given.
5. $\angle 1 \cong \angle 2$.	Angle	5. If two sides of a triangle are congruent, the angles opposite these sides are congruent.
6. $\overline{AB} \cong \overline{CD}$.	Side	6. Opposite sides of a rectangle are congruent.
7. $\triangle ABG \cong \triangle DCF$		7. AAS ≅ AAS.
8. $\overline{AG} \cong \overline{DF}$.		8. Corresponding sides of congruent triangles are congruent.

Proving That a Quadrilateral Is a Rectangle

To prove that a quadrilateral is a rectangle, show that the quadrilateral is a parallelogram that contains a right angle or is a parallelogram with congruent diagonals. To prove a parallelogram is *not* a rectangle, show that it does *not* contain a right angle or does *not* have congruent diagonals.

Example 2

Given: $R(-5, 2)$, $E(7, 6)$, $C(8, 3)$, and $T(-4, -1)$. Prove that *RECT* is a rectangle.

Solution: Compare the slopes of the sides of *RECT*.

$$\text{Slope of } \overline{RE} = \frac{\Delta y}{\Delta x} = \frac{6-2}{7-(-5)} = \frac{4}{12} = \frac{1}{3}$$

$$\text{Slope of } \overline{EC} = \frac{\Delta y}{\Delta x} = \frac{3-6}{8-7} = \frac{-3}{1} = -3$$

$$\text{Slope of } \overline{CT} = \frac{\Delta y}{\Delta x} = \frac{-1-3}{-4-8} = \frac{-4}{-12} = \frac{1}{3}$$

$$\text{Slope of } \overline{RT} = \frac{\Delta y}{\Delta x} = \frac{-1-2}{-4-(-5)} = \frac{-3}{-4+5} = -3$$

- Because lines with the same slope are parallel, $\overline{RE} \parallel \overline{CT}$ and $\overline{EC} \parallel \overline{RT}$. Hence, *RECT* is a parallelogram.

- The slopes of $\overline{RE}$ and $\overline{EC}$ are negative reciprocals since $\left(\frac{1}{3}\right) \times (-3) = -1$. Because lines with slopes that are negative reciprocals are perpendicular, $\overline{RE} \perp \overline{EC}$.
- *RECT* is a rectangle since it is a parallelogram that contains a right angle.

Rhombus

A **rhombus** is a parallelogram with four congruent sides. A rhombus has

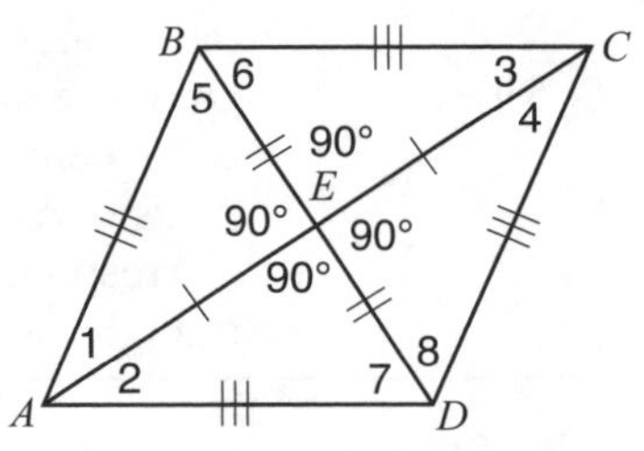

Figure 4.7 Properties of a rhombus.

- All the properties of a parallelogram.
- Perpendicular diagonals. In Figure 4.7, $\overline{AC} \perp \overline{BD}$. Because a rhombus is a parallelogram, $\overline{AE} \cong \overline{EC}$ and $\overline{BE} \cong \overline{ED}$. Therefore, each diagonal of a rhombus is the perpendicular bisector of the other diagonal.
- Diagonals that bisect opposite pairs of angles:

$$\angle 1 \cong \angle 2 \text{ and } \angle 3 \cong \angle 4;\ \angle 5 \cong \angle 6 \text{ and } \angle 7 \cong \angle 8.$$

Square

A **square** is a rectangle with four congruent sides, as shown in Figure 4.8. In addition, a square has

- All the properties of a parallelogram.
- All the properties of a rectangle.
- All the properties of a rhombus.

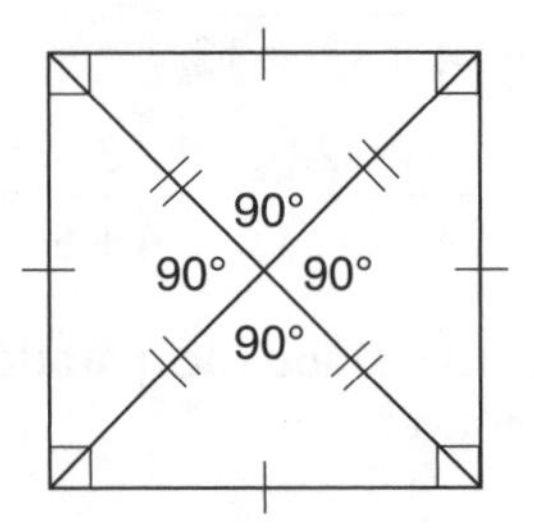

Figure 4.8 Properties of a square.

Proving That a Parallelogram Is a Rhombus

To prove that a quadrilateral is a rhombus, simply show that all four sides have the same length. To prove that a parallelogram is a rhombus, show that any one of the following statements is true:

- A pair of adjacent sides are congruent.
- The diagonals intersect at right angles.
- The diagonals bisect opposite pairs of angles.

Use the inverse of any of these three properties to prove that a parallelogram is *not* a rhombus.

Proving That a Parallelogram Is a Square

To prove that a parallelogram is a square, show that one of the following statements is true:

- The parallelogram is a rectangle with a pair of congruent adjacent sides.
- The parallelogram is a rhombus that contains a right angle.

To prove that a rectangle is *not* a square, show that a pair of adjacent sides are not congruent. To prove that a rhombus is *not* a rectangle or square, show that it does not contain a right angle.

Example 3

The coordinates of the vertices of quadrilateral *DRAW* are $D(0, -4)$, $R(8, -3)$, $A(4, 4)$, and $W(-4, 3)$. Show by means of coordinate geometry that quadrilateral *DRAW* is a rhombus but *not* a square.

Solution: Prove that quadrilateral *DRAW* is a rhombus by using the distance formula to show that all four sides have the same length.

$$\begin{aligned} DR &= \sqrt{(\Delta x)^2 + (\Delta y)^2} \\ &= \sqrt{(8-0)^2 + (-3-(-4))^2} \\ &= \sqrt{(8)^2 + (-3+4)^2} \\ &= \sqrt{64+1} \\ &= \sqrt{65} \end{aligned}$$

$$\begin{aligned} WA &= \sqrt{(\Delta x)^2 + (\Delta y)^2} \\ &= \sqrt{(-4-4)^2 + (3-4)^2} \\ &= \sqrt{(-8)^2 + (-1)^2} \\ &= \sqrt{64+1} \\ &= \sqrt{65} \end{aligned}$$

$$\begin{aligned} RA &= \sqrt{(\Delta x)^2 + (\Delta y)^2} \\ &= \sqrt{(4-8)^2 + (4-(-3))^2} \\ &= \sqrt{(-4)^2 + (7)^2} \\ &= \sqrt{16+49} \\ &= \sqrt{65} \end{aligned}$$

$$\begin{aligned} DW &= \sqrt{(\Delta x)^2 + (\Delta y)^2} \\ &= \sqrt{(-4-0)^2 + (3-(-4))^2} \\ &= \sqrt{(-4)^2 + (7)^2} \\ &= \sqrt{16+49} \\ &= \sqrt{65} \end{aligned}$$

Since the length of each side is $\sqrt{65}$, $DR = RA = WA = DW$ so *DRAW* is a rhombus. To prove *DRAW* is *not* a square, use the slope formula to show that one of its angles is not a right angle.

$$\begin{aligned} \text{Slope of } \overline{DR} &= \frac{\Delta y}{\Delta x} \\ &= \frac{-3-(-4)}{8-0} \\ &= \frac{-3+4}{8} \\ &= \frac{1}{8} \end{aligned}$$

$$\begin{aligned} \text{Slope of } \overline{RA} &= \frac{\Delta y}{\Delta x} \\ &= \frac{4-(-3)}{4-8} \\ &= \frac{7}{-4} \\ &= -\frac{7}{4} \end{aligned}$$

Since the slopes of $\overline{DR}$ and $\overline{RA}$ are *not* negative reciprocals, $\overline{DR} \not\perp \overline{RA}$ so $\angle DRA$ is not a right angle. Because rhombus *DRAW* does *not* contain four right angles, it is *not* a square.

Check Your Understanding of Section 4.3

A. Multiple Choice

1. A parallelogram must be a rectangle if its diagonals
(1) bisect each other
(2) bisect the angles to which they are drawn
(3) are perpendicular to each other
(4) are congruent

2. The diagonals of a rhombus do not always
(1) bisect each other
(2) bisect the angles to which they are drawn
(3) intersect at right angles
(4) have the same length

3. In rhombus *PQRS*, diagonals $\overline{PR}$ and $\overline{QS}$ intersect at *T*. Which statement is always true?
(1) Quadrilateral *PQRS* is a square.
(2) Triangle *RTQ* is a right triangle.
(3) Triangle *PQS* is equilateral.
(4) Diagonals $\overline{PR}$ and $\overline{QS}$ are congruent.

4. Which statements describe the properties of the diagonals of a rectangle?
I. The diagonals are congruent.
II. The diagonals are perpendicular.
III. The diagonals bisect each other.
(1) II and III only
(2) I and II only
(3) I and III only
(4) I, II, and III

5. In the accompanying diagram of rectangle *ABCD*, $m\angle BAC = 3x + 4$ and $m\angle ACD = x + 28$. What is $m\angle CAD$?
(1) 12
(2) 37
(3) 40
(4) 50

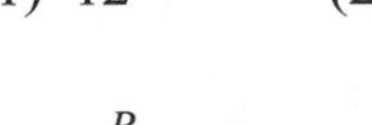

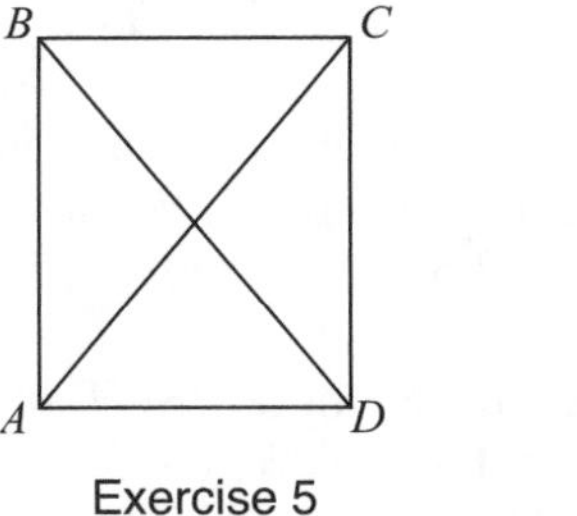

Exercise 5

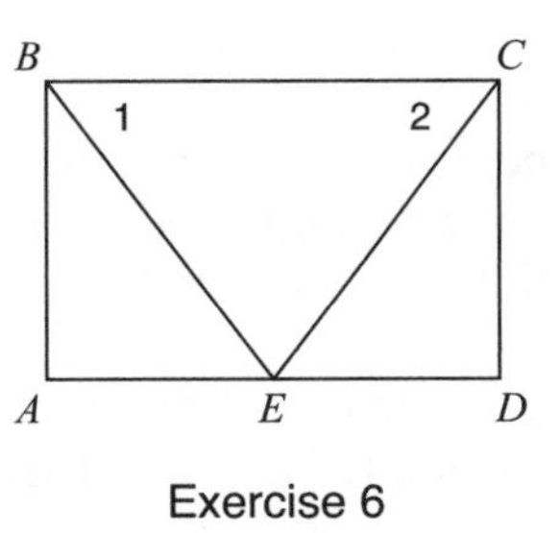

Exercise 6

6. In the accompanying diagram of rectangle *ABCD*, $\angle 1 \cong \angle 2$. Which of the following does *not* represent a method that could be used to prove $\triangle EAB \cong \triangle EDC$?
(1) SSS ≅ SSS
(2) HL ≅ HL
(3) SAS ≅ SAS
(4) ASA ≅ ASA

7. If *p* represents "Diagonals are congruent" and *q* represents "All sides are congruent," then for which figure is the statement $p \vee q$ not always true?
(1) rectangle
(2) rhombus
(3) square
(4) parallelogram

8. In quadrilateral $WXYZ$, $\overline{WZ} \cong \overline{XY}$ and $\overline{WZ} \parallel \overline{XY}$. Which statement *must* be true?
(1) The diagonals bisect the angles of the quadrilateral.
(2) The diagonals bisect each other.
(3) The diagonals are congruent.
(4) The diagonals are perpendicular.

9. In the accompanying diagram of rhombus $QRST$, diagonals $\overline{QS}$ and $\overline{RT}$ intersect at M. Which statement *must* be true?
(1) Triangle QRM is an isosceles right triangle.
(2) $\triangle QRM \cong \triangle SRM$.
(3) Triangle QRM is an obtuse triangle.
(4) $\overline{QS} \cong \overline{RT}$.

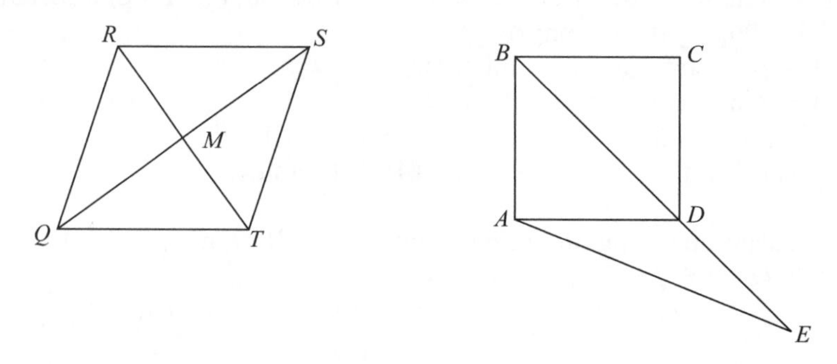

Exercise 9

Exercise 10

10. In the accompanying diagram, $ABCD$ is a square, diagonal $\overline{BD}$ is extended through D to E, $\overline{AD} \cong \overline{DE}$, and $\overline{AE}$ is drawn. What is $m\angle BAE$?
(1) 22.5 (2) 45.0 (3) 112.5 (4) 135.0

11. The coordinates of three of the vertices of rectangle $RECT$ are $R(-1, 1)$, $E(3,1)$, and $C(3,5)$. What are the coordinates of vertex T?
(1) $(-5, 3)$ (2) $(-1, 5)$ (3) $(1, 3)$ (4) $(3, -5)$

B. *Show or explain how you arrived at your answer.*

12. In rhombus $ABCD$, the length, in inches, of $\overline{AB}$ is $3x + 2$ and $\overline{BC}$ is $x + 12$. Find the number of inches in the length of $\overline{DC}$.

13. Prove that $M(-2, -1)$, $A(1, 6)$, $T(8, 3)$, and $H(5, -4)$ are the vertices of a square.

14. Prove that $A(-3, 6)$, $B(6, 0)$, $C(9, -9)$, and $D(0, -3)$ are the vertices of a parallelogram but *not* a rhombus.

15. Given: $A(-2, 2)$, $B(6, 5)$, $C(4, 0)$, and $D(-4, -3)$. Prove by means of coordinate geometry that $ABCD$ is a parallelogram but *not* a rectangle.

16. Prove that $R(2, 1)$, $E(10, 7)$, $C(7, 11)$, and $T(-1, 5)$ are the vertices of a rectangle but *not* a square.

17. The vertices of quadrilateral $DEFG$ are $D(3, 2)$, $E(7, 4)$, $F(9, 8)$, and $G(5,6)$. Prove by means of coordinate geometry that $DEFG$ is a rhombus.

18. Jim is experimenting with a new drawing program on his computer. He created quadrilateral $TEAM$ with coordinates $T(-2, 3)$, $E(-5, -4)$, $A(2, -1)$, and $M(5, 6)$. Jim believes that he has created a rhombus but *not* a square. Prove by means of coordinate geometry that Jim is correct.

19. a. Prove that $A(-2, 8)$, $B(6, 6)$, $C(0, 12)$, and $D(-4, 14)$ are *not* the vertices of a paralellogram.

b. Prove that the quadrilateral formed by joining the midpoints of the sides of quadrilateral $ABCD$ is a paralellogram.

C. *Write a proof.*

20. Given: Quadrilateral $DRUM$ is a rectangle, $\angle A \cong \angle Q$.
Prove: Quadrilateral $QUAD$ is a parallelogram

21. Given: Rhombus $ABCD$, $\overline{BL} \cong \overline{CM}$, $\overline{AL} \cong \overline{BM}$.
Prove: $ABCD$ is a square.

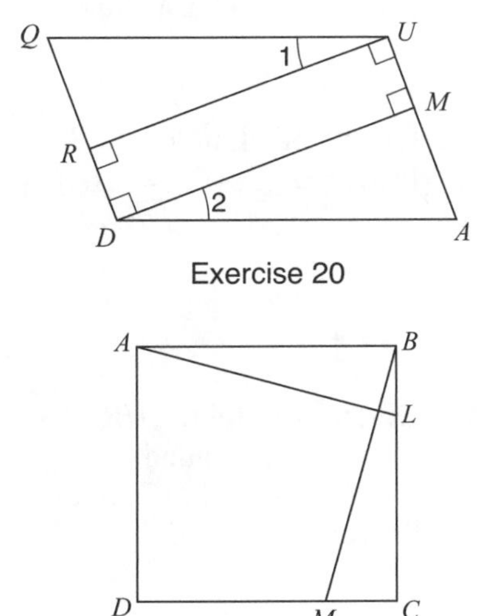

Exercise 20

Exercise 21

22. Given: Rhombus $RSTV$, $\overline{VTX}$, $\overline{STW}$, $\overline{SX}$, $\overline{VW}$, $\angle RSX \cong \angle RVW$.
Prove: $\overline{TX} \cong \overline{TW}$

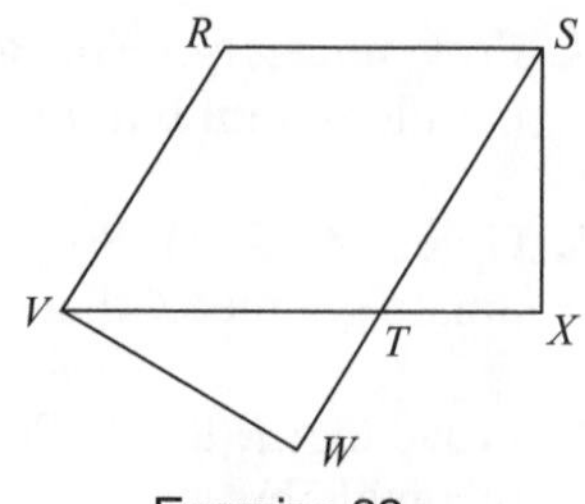

Exercise 22

23. In the accompanying diagram of parallelogram $ABCD$, $m\angle 1 > m\angle 2$. Write an explanation or an informal proof that shows $ABCD$ is *not* a rectangle.

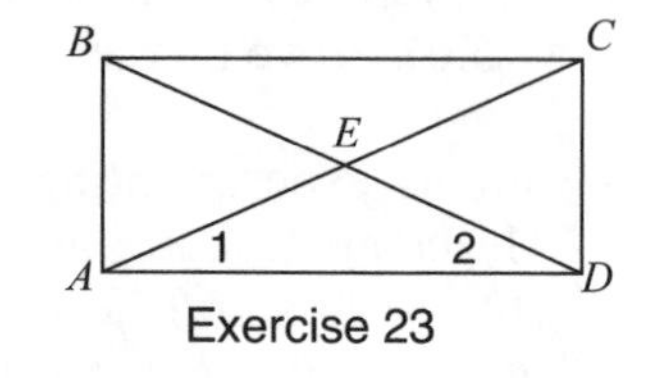

Exercise 23

24. Given: Square $ABCD$, $\angle 1 \cong \angle 2$.
Prove: $\overline{BE} \cong \overline{DF}$

25. Given: Rectangle $ABCD$, $\angle 1 \cong \angle 2$, $\angle BEF \cong \angle DFE$.
Prove: $ABCD$ is a square.

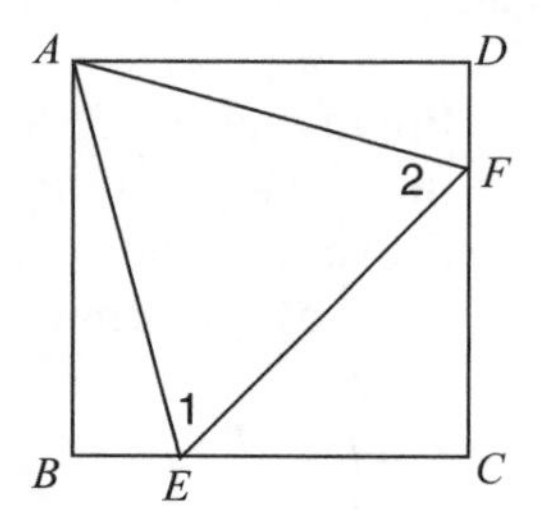

Exercises 24–25

26. Given: Rectangle $ABCD$, $\overline{AP} \cong \overline{DN}$.
Prove: a. $\triangle ABP \cong \triangle DCN$.
b. $\overline{AE} \cong \overline{DE}$.

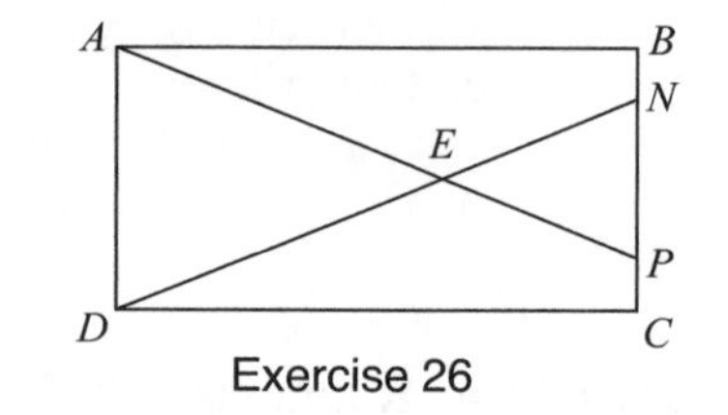

Exercise 26

27. Given: Rhombus $ABCD$, diagonal $\overline{AC}$ is extended through C to E, $\overline{BE}$ and $\overline{DE}$ are drawn.
Prove: $\overline{BE} \cong \overline{DE}$.

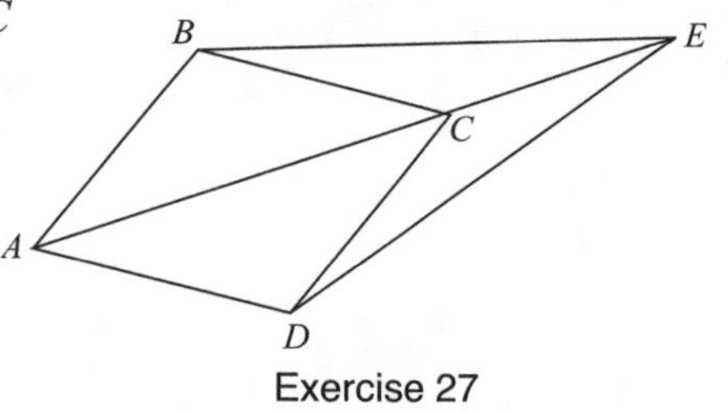

Exercise 27

28. In right triangle ABC, $\overline{CD}$ is the median to hypotenuse $\overline{AB}$. If $\overline{CD}$ is extended its own length through D to E and $\overline{EA}$ and $\overline{EB}$ are drawn, prove that $AEBC$ is a rectangle.

4.4 TRAPEZOIDS

KEY IDEAS

A **trapezoid**, shown in the accompanying figure, is a quadrilateral that has exactly one pair of parallel sides, called **bases**. The nonparallel sides of a trapezoid are the **legs**.

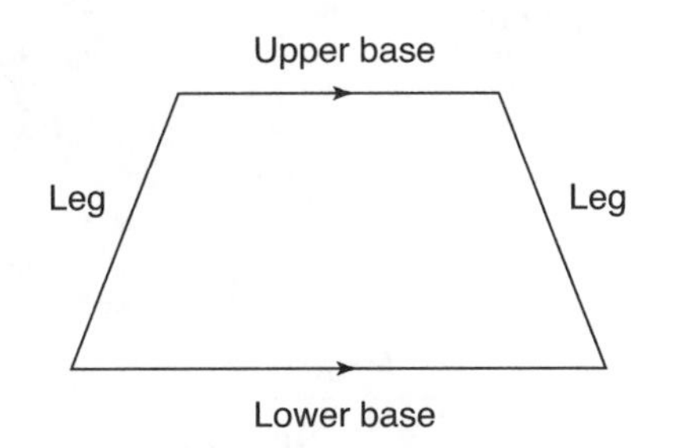

Properties of Trapezoids

An **isosceles trapezoid** has the following additional properties (see Figure 4.9):

- The legs are congruent:

$$\overline{AB} \cong \overline{CD}$$

- The upper and lower base angles are congruent:

$$\angle A \cong \angle D \quad \text{and} \quad \angle B \cong \angle C.$$

- The diagonals are congruent:

$$\overline{AC} \cong \overline{DB}$$

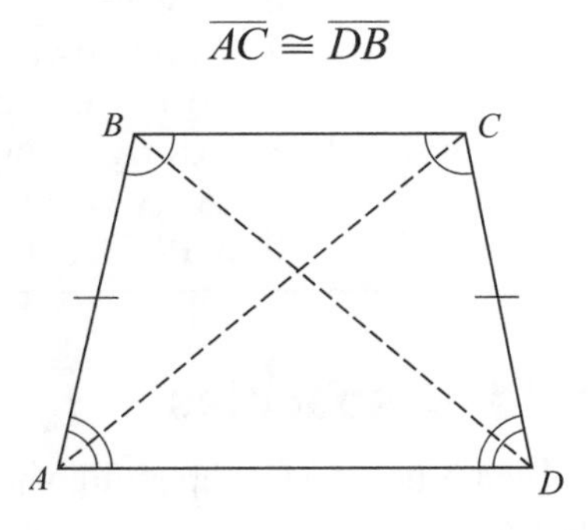

Figure 4.9 Isosceles trapezoid.

Example 1

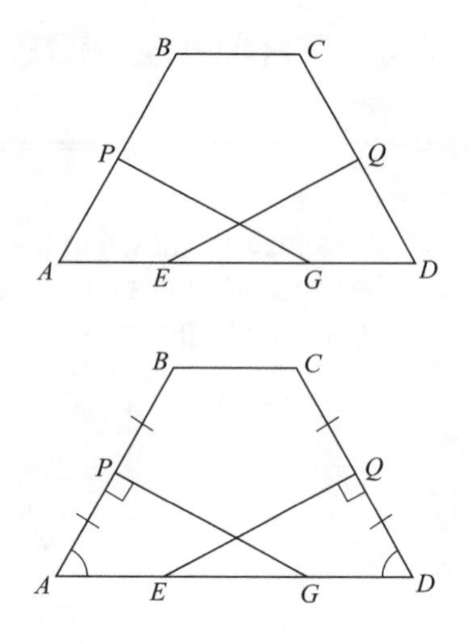

Given: Isosceles trapezoid with $\overline{BC} \parallel \overline{AD}$, $\overline{GP} \perp \overline{AB}$, $\overline{EQ} \perp \overline{CD}$, points P and Q are midpoints of $\overline{AB}$ and $\overline{CD}$, respectively.
Prove: $\overline{GP} \cong \overline{EQ}$.

Solution: Mark off the diagram with the Given. $\triangle APG \cong \triangle EQD$ by ASA $\cong$ ASA.

Proof

Statement	Reason
1. Isosceles trapezoid with $\overline{BC} \parallel \overline{AD}$.	1. Given.
2. $\angle AD \cong \angle D$. Angle	2. Base angles of an isosceles trapezoid are congruent.
3. P and Q are midpoints of $\overline{AB}$ and $\overline{CD}$, respectively.	3. Given.
4. $\overline{AB} \cong \overline{DC}$.	4. The legs of an isosceles trapezoid are congruent.
5. $\overline{AP} \cong \overline{DQ}$. Side	5. Halves of congruent segments are congruent.
6. $\overline{GP} \perp \overline{AB}$ and $\overline{EQ} \perp \overline{CD}$.	6. Given.
7. $\angle APG$ and $\angle DQE$ are right angles.	7. Perpendicular lines intersect to form right angles.
8. $\angle APG \cong \angle DQE$. Angle	8. All right angles are congruent.
9. $\triangle APG \cong \triangle EQD$.	9. ASA $\cong$ ASA.
10. $\overline{GP} \cong \overline{EQ}$.	10. CPCTC.

Proving a Trapezoid is Isosceles

To prove that a trapezoid is an isosceles trapezoid, show that one of the following statements is true:

- The legs are congruent.
- A pair of base angles are congruent.
- The diagonals are congruent.

To prove that a trapezoid is *not* an isosceles trapezoid, show that the inverse of one of its special properties is true. For example, a trapezoid is *not* isosceles if its diagonals are not congruent.

Example 2

The coordinates of the vertices of quadrilateral $ABCD$ are $A(-2, 0)$, $B(10, 3)$, $C(5, 7)$, and $D(1, 6)$. Prove that quadrilateral $ABCD$ is a trapezoid but *not* an isosceles trapezoid.

Solution: A trapezoid has exactly one pair of parallel sides. To prove that the $ABCD$ is a trapezoid, show by means of slope that one pair of sides is parallel ***and*** the other pair of sides is *not* parallel.

$$\text{Slope of } \overline{AB} = \frac{\Delta y}{\Delta x} = \frac{3-0}{10-(-2)} = \frac{3}{12} = \frac{1}{4}$$

$$\text{Slope of } \overline{BC} = \frac{\Delta y}{\Delta x} = \frac{7-3}{5-10} = \frac{4}{-5} = -\frac{4}{5}$$

$$\text{Slope of } \overline{CD} = \frac{\Delta y}{\Delta x} = \frac{6-7}{1-5} = \frac{-1}{-4} = \frac{1}{4}$$

$$\text{Slope of } \overline{AD} = \frac{\Delta y}{\Delta x} = \frac{6-0}{1-(-2)} = \frac{6}{3} = 2$$

$\overline{AB} \parallel \overline{CD}$

$\overline{BC} \nparallel \overline{AD}$

Since $\overline{AB} \parallel \overline{CD}$ *and* $\overline{BC} \nparallel \overline{AD}$, quadrilateral $ABCD$ is a trapezoid.

To prove that $ABCD$ is *not* an isosceles trapezoid, use the distance formula to show its diagonals have unequal lengths and, as a result, are not congruent.

$$\begin{aligned} AC &= \sqrt{(\Delta x)^2 + (\Delta y)^2} \\ &= \sqrt{(5-(-2))^2 + (7-0)^2} \\ &= \sqrt{7^2 + 7^2} \\ &= \sqrt{49+49} \\ &= \sqrt{98} \end{aligned}$$

$$\begin{aligned} BD &= \sqrt{(\Delta x)^2 + (\Delta y)^2} \\ &= \sqrt{(1-10)^2 + (6-3)^2} \\ &= \sqrt{(-9)^2 + 3^2} \\ &= \sqrt{81+9} \\ &= \sqrt{90} \end{aligned}$$

Since $AC \neq BD$, trapezoid $ABCD$ does *not* have congruent diagonals so it is *not* an isosceles trapezoid.

Check Your Understanding of Section 4.4

A. Multiple Choice

1. If the diagonals of a parallelogram are perpendicular but *not* congruent, then the parallelogram is
(1) a rectangle
(2) a rhombus
(3) a square
(4) an isosceles trapezoid

2. If a quadrilateral is selected at random from the set, {rectangle, rhombus, square, trapezoid}, what is the probability that the quadrilateral has diagonals that bisect each other?
(1) 1 (2) $\frac{3}{4}$ (3) $\frac{1}{2}$ (4) $\frac{1}{4}$

3. In a certain quadrilateral, two opposite sides are parallel, and the other two opposite sides are *not* congruent. This quadrilateral could be a
(1) rhombus
(2) parallelogram
(3) square
(4) trapezoid

4. Which quadrilateral must have diagonals that are congruent and perpendicular?
(1) rhombus
(2) square
(3) isosceles trapezoid
(4) parallelogram

5. In the accompanying diagram of isosceles trapezoid $ABCD$, $\overline{AB} \parallel \overline{DC}$ and diagonals $\overline{DB}$ and $\overline{AC}$ intersect at E. Which statement is *not* true?
(1) $\overline{AB} \cong \overline{BD}$
(2) $\angle CDB \cong \angle DBA$
(3) $\triangle ADC \cong \triangle ABC$
(4) $\triangle CBA \cong \triangle DAB$

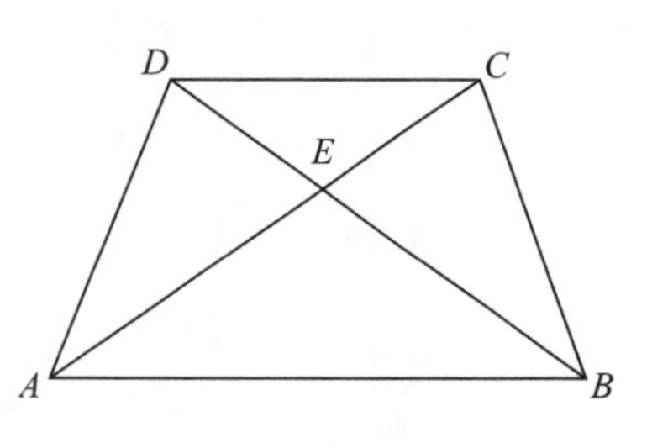

Exercise 5

B. Show or explain how you arrived at your answer.

6. The coordinates of quadrilateral $JKLM$ are $J(1, -2)$, $K(13, 4)$, $L(6, 8)$, and $M(-2, 4)$. Prove that quadrilateral $JKLM$ is a trapezoid but *not* an isosceles trapezoid.

7. The coordinates of the vertices of quadrilateral $JAKE$ are $J(0, 3a)$, $A(3a, 3a)$, $K(4a, 0)$, and $E(-a, 0)$. Prove that $JAKE$ is an isosceles trapezoid.

8. Given points $A(-2, 3)$, $B(1, 0)$, $C(7, 6)$, and $D(0, 5)$.
  a. Prove that quadrilateral $ABCD$ is a trapezoid.
  b. If points B, $E(h, k)$, and C are collinear, find the values of h and k such that points A, B, E, and D are the vertices of a parallelogram.
  c. Prove that $ABED$ is a rectangle.

C. *Write a proof.*

9. Given: Trapezoid $ABCD$, $\overline{AD} \parallel \overline{BC}$, $\overline{BE}$ and $\overline{CF}$ are altitudes drawn to $\overline{AD}$, $\overline{AE} \cong \overline{DF}$.
Prove: Trapezoid $ABCD$ is isosceles.

Exercise 9

10. Given: Isosceles trapezoid $RSTW$.
Prove: $\triangle RPW$ isosceles.

Exercise 10

11. Given: Trapezoid $ABCD$, $\overline{EF} \cong \overline{EG}$, $\overline{AF} \cong \overline{DG}$, $\overline{BG} \cong \overline{CF}$.
Prove: Trapezoid $ABCD$ is isosceles.

Exercise 11

12. Given: Isosceles trapezoid $ABCD$, $\angle 1 \cong \angle 2$.
Prove: $BKDC$ is a parallelogram.

Exercise 12

13. Given: Trapezoid $ROSE$ with $\overline{OS} \parallel \overline{RE}$, diagonals $\overline{RS}$ and $\overline{EO}$ intersecting at point M. Prove the diagonals do *not* bisect each other.

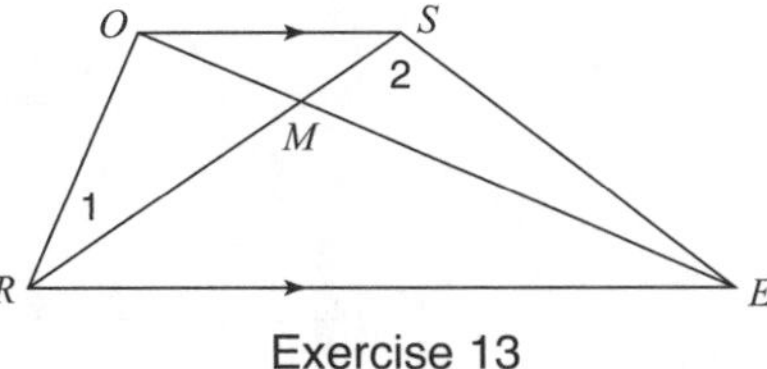

Exercise 13

4.5 MIDPOINT THEOREMS

KEY IDEAS

The special properties of parallelograms can be used to help prove relationships involving the midpoints of the sides of triangles and quadrilaterals.

Median–Hypotenuse Theorem

In Figure 4.10, since M is the midpoint of $\overline{AC}$,

$$AM = \frac{1}{2}AC$$

Because the diagonals of a rectangle are congruent, $\overline{AC} \cong \overline{BD}$. By substitution,

$$AM = \frac{1}{2}BD$$

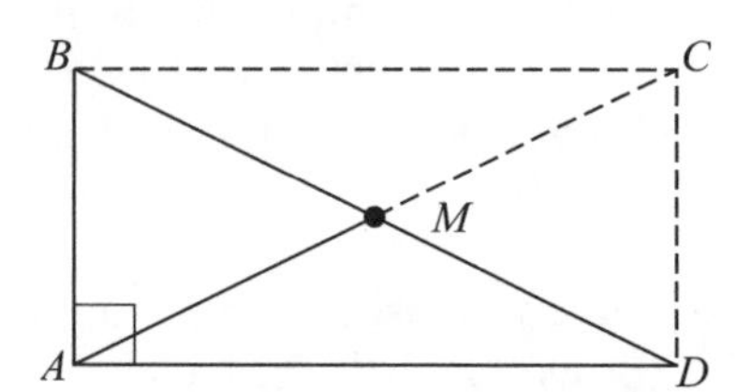

Figure 4.10 Median-Hypotenuse Theorem.

In right triangle BAD, $\overline{AM}$ represents the median to hypotenuse $\overline{BD}$.

Theorem: Median–Hypotenuse
The length of the median drawn to the hypotenuse of a right triangle is one-half of the length of the hypotenuse.

Midpoints of a Triangle

A **midsegment** of a triangle is a segment whose endpoints are the midpoints of two sides of a triangle. A midsegment is half the length and parallel to the side opposite it. In Figure 4.11, $\overline{DE} \parallel \overline{AC}$ and $DE = \frac{1}{2}AC$.

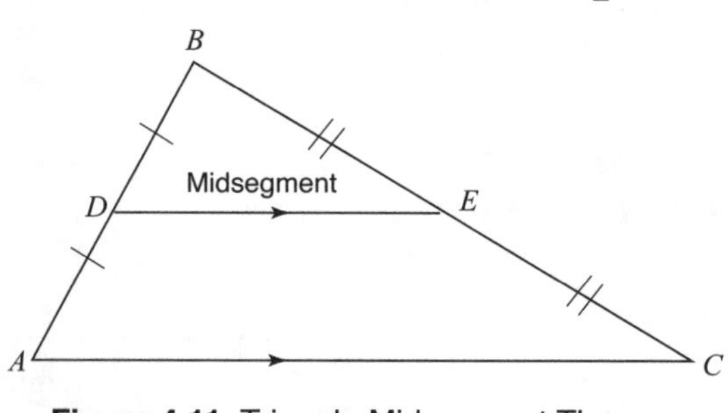

Figure 4.11 Triangle Midsegment Theorem.

Theorem: Triangle Midsegment Theorem
The line segment connecting the midpoints of two sides of a triangle is parallel to the third side and one-half of its length.

You are asked to prove this theorem in Exercise 7 at the end of this lesson.

Perimeter of the Midsegment Triangle

The triangle formed by joining the midpoints of its three sides is called the **midsegment triangle**. In Figure 4.12, $\triangle RST$ is the midsegment triangle.

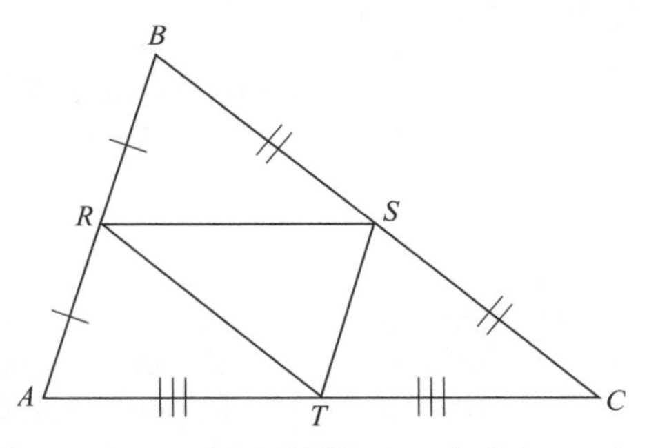

Figure 4.12 The perimeter of $\triangle RST$ is one-half the perimeter of $\triangle ABC$.

Since each side of $\triangle RST$ is one-half the length of the opposite side, the perimeter of the midsegment triangle is one-half the perimeter of the original triangle. For example, if the lengths of the three sides of $\triangle ABC$ in Figure 4.12 are 5 in, 8 in, and 11 in, then the perimeter of $\triangle RST$ is

$$\frac{1}{2}(5 + 8 + 11) = \frac{1}{2}(24 \text{ in}) = 12 \text{ in}$$

Theorem: Perimeter of the Midsegment Triangle
The perimeter of the midsegment triangle is one-half the perimeter of the original triangle.

Midpoints of a Quadrilateral

If you draw any quadrilateral and connect the midpoints of consecutive sides, another quadrilateral is formed. The new quadrilateral is always a parallelogram.

Theorem: Midsegment Quadrilateral
If the midpoints of a quadrilateral are joined consecutively, the quadrilateral that is formed is a parallelogram.

Here is the proof:

Given: Quadrilateral $ABCD$, points J, K, L, and M are midpoints of sides $\overline{AB}$, $\overline{BC}$, $\overline{CD}$, and $\overline{AD}$, respectively.
Prove: $JKLM$ is a parallelogram.

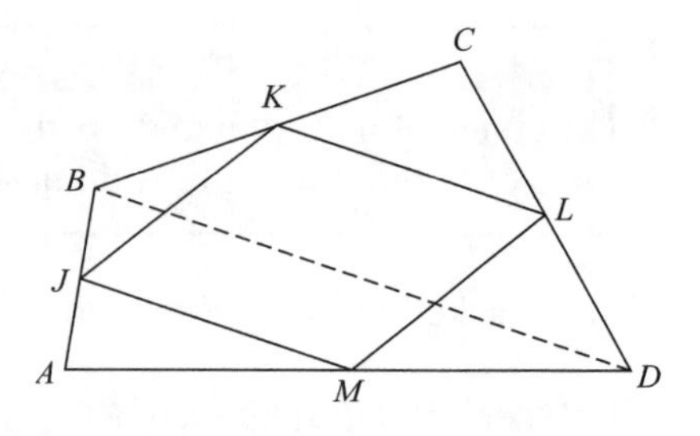

Separate the quadrilateral into two triangles by drawing diagonal $\overline{BD}$. Then use the Triangle Midsegments Theorem:

- In $\triangle BAD$, $\overline{JM} \parallel \overline{BD}$ and $JM = \frac{1}{2}BD$. Similarly, in $\triangle BCD$, $\overline{KL} \parallel \overline{BD}$ and $KL = \frac{1}{2}BD$.
- Hence, $\overline{JM} \parallel \overline{KL}$ (both lines are parallel to the same line) and $\overline{JM} \cong \overline{KL}$ (both segments are one-half the length of the same segment).
- $JKLM$ is a parallelogram since one pair of opposite sides are both parallel and congruent.

Median of a Trapezoid

The **median** of a trapezoid is the segment whose endpoints are the midpoints of its legs. In Figure 4.13, if $\overline{LM}$ is the median of trapezoid $ABCD$, then

- $\overline{LM}$ is parallel to the two bases, and
- LM is one-half the sum of the lengths of the two bases.

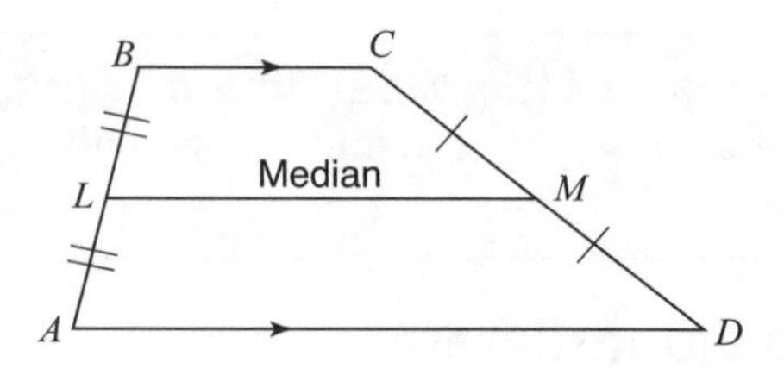

Figure 4.13 Median of a trapezoid.

The proof is based on drawing $\overline{BM}$, extending it so that it meets the extension of $\overline{AD}$ at E, and then applying the Triangle Midsegment Theorem.

Given: $\overline{LM}$ is the median of trapezoid $ABCD$.
Prove: a. $\overline{LM} \parallel \overline{AD}$ and $\overline{LM} \parallel \overline{BC}$.

b. $LM = \frac{1}{2}(AD + BC)$.

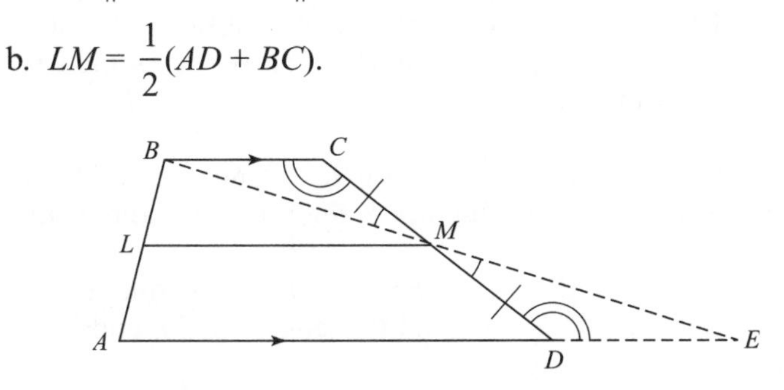

Paragraph Proof

- Since $\triangle EMD \cong \triangle BMC$ by ASA $\cong$ ASA, $\overline{EM} \cong \overline{BM}$ and $\overline{DE} \cong \overline{BC}$.
- In $\triangle ABE$, L is the midpoint of $\overline{AB}$ and M is the midpoint of $\overline{BE}$. Since $\overline{LM}$ is a midsegment of $\triangle ABE$, $\overline{LM} \parallel \overline{AD}$. As $\overline{AD} \parallel \overline{BC}$, $\overline{LM} \parallel \overline{BC}$.
- Because of the Triangle Midsegment Theorem,

$$LM = \frac{1}{2}(AE) = \frac{1}{2}(AD + \mathbf{DE})$$

and, by substitution, $LM = \frac{1}{2}(AD + BC)$

Theorem: Median of a Trapezoid
The median of a trapezoid is parallel to the bases, and its length is one-half the sum of the lengths of the bases.

Check Your Understanding of Section 4.5

A. Multiple Choice.

1. In the accompanying figure, equilateral triangle ABC has a perimeter of 18. Points R, S, and T are midpoints of the sides of triangle ABC. What is the length of $\overline{RS}$?
(1) 6 (2) 2 (3) 3 (4) 4

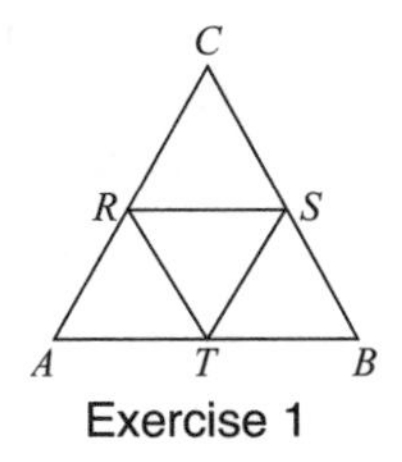

Exercise 1

2. How many congruent triangles are formed by connecting the midpoints of the three sides of a scalene triangle?
(1) 1 (2) 2 (3) 3 (4) 4

B. *Show or explain how you arrived at your answer.*

3. The length of the median drawn to the hypotenuse of a right triangle is represented by $3x - 7$, while the hypotenuse is represented by $5x - 4$. Find the length of the median.

4. The length of the sides of a triangle are 9, 40, and 41. Find the perimeter of the triangle formed by joining the midpoints of the three sides.

5. In $\triangle RST$, E is the midpoint of $\overline{RS}$, and F is the midpoint of $\overline{ST}$. If $EF = 5y - 1$ and $RT = 7y + 10$, find the lengths of $\overline{EF}$ and $\overline{RT}$.

C. *Write a proof.*

6. Prove the Triangle Midsegment Theorem on page 147 using the accompanying figure. Hint: Extend $\overline{DE}$ its own length so that $\overline{DE} \cong \overline{EF}$. Draw $\overline{CF}$. Prove $ADFC$ is a parallelogram.

Given: D and E are the midpoints of $\overline{AB}$ and $\overline{BC}$, respectively.

Prove: a. $\overline{DE} \parallel \overline{AC}$.

b. $DE = \frac{1}{2}AC$.

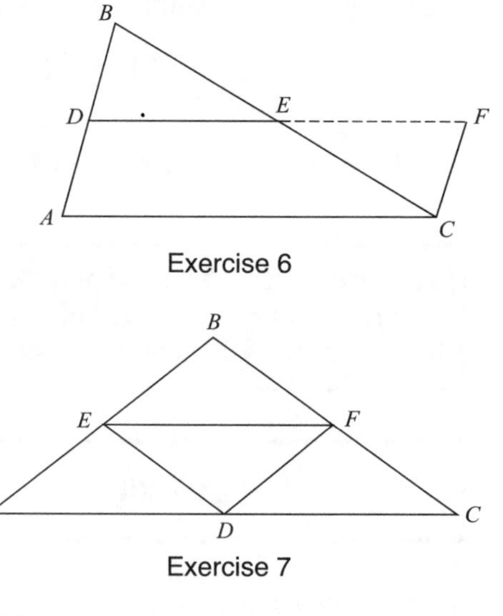

Exercise 6

Exercise 7

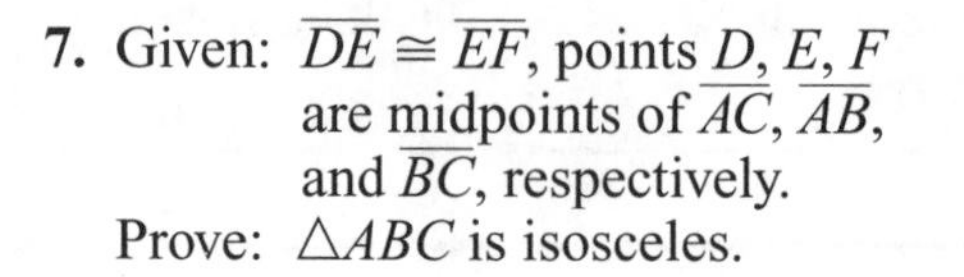

7. Given: $\overline{DE} \cong \overline{EF}$, points D, E, F are midpoints of $\overline{AC}$, $\overline{AB}$, and $\overline{BC}$, respectively.

Prove: $\triangle ABC$ is isosceles.

8. Given: Trapezoid $ABCD$, median $\overline{LM}$, P is midpoint of $\overline{AD}$, $\overline{LP} \cong \overline{MP}$.

Prove: Trapezoid $ABCD$ is isosceles.

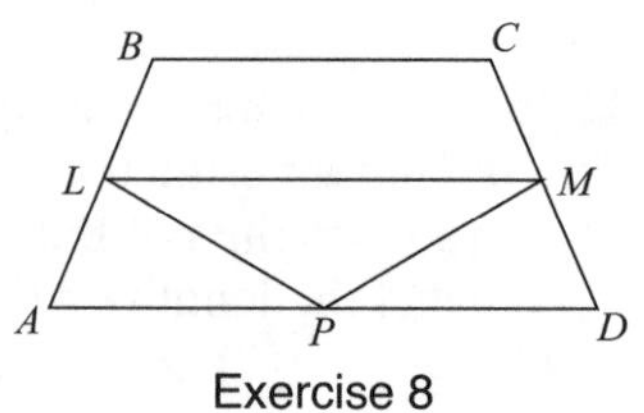

Exercise 8

9. Given: $RSTW$ is a parallelogram, B is the midpoint of $\overline{SW}$, C is the midpoint of $\overline{ST}$.
Prove: $WACT$ is a parallelogram.

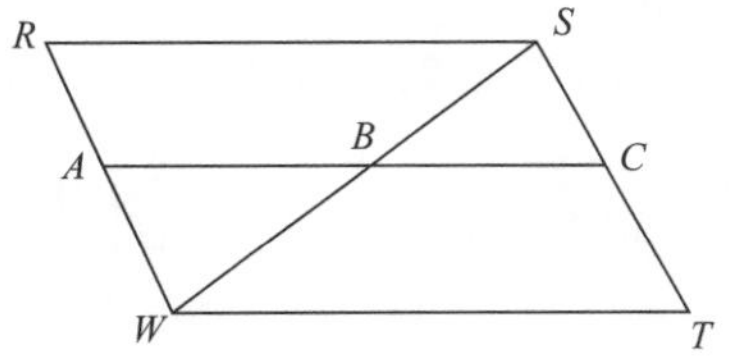

Exercise 9

10. Given: $ABCD$ is a rhombus, E, F, G, and H are midpoints of $\overline{AX}$, $\overline{BX}$, $\overline{CX}$, and $\overline{DX}$, respectively.
Prove: $EFGH$ is a rhombus.

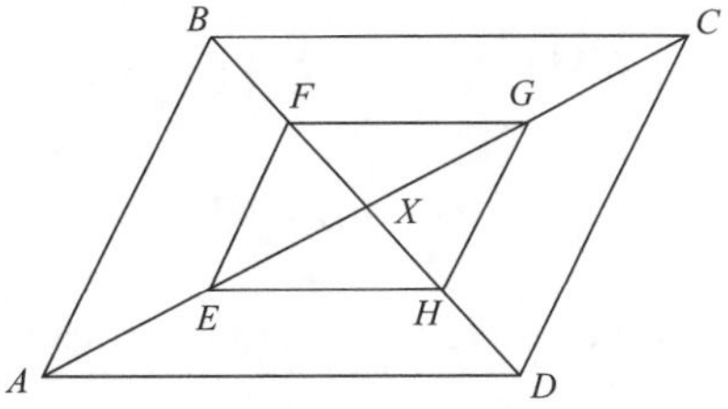

Exercise 10

11. Prove that if the midpoints of the sides of a rectangle are joined consecutively, the resulting quadrilateral is a rhombus.

4.6 GENERAL COORDINATE PROOFS

KEY IDEAS

In a **coordinate proof** a figure is positioned in the coordinate plane using vertices with generalized variable coordinates. One or more coordinate geometry formulas are then used to show that the geometric relationship you are required to prove is true. Theorems that are difficult to prove deductively may be easier to prove using general coordinates.

Proving a Theorem Using Coordinates

When variables are used as coordinates instead of specific numbers, generalized relationships can be proved using coordinate formulas.

Example 1

Write a coordinate proof.

Given: The vertices of right triangle ABC are $A(2a, 0)$, $B(0, 2b)$, and $C(0, 0)$.
Prove: The median drawn to the hypotenuse of right triangle ABC is one-half its length.

Solution: The midpoint, M, of hypotenuse $\overline{AB}$ is

$$M\left(\frac{0+2a}{2}, \frac{2b+0}{2}\right) = M(a, b).$$

- Find the length of median $\overline{CM}$:

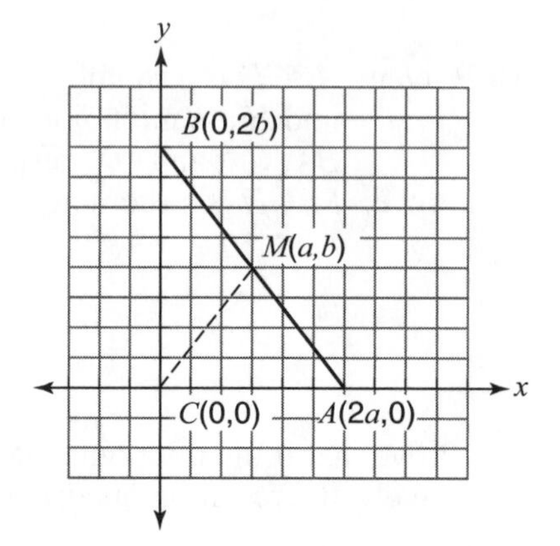

$$CM = \sqrt{(a-0)^2 + (b-0)^2}$$
$$= \sqrt{a^2 + b^2}$$

- Find the length of hypotenuse $\overline{AB}$:

$$AB = \sqrt{(2a-0)^2 + (0-2b)^2}$$
$$= \sqrt{4a^2 + 4b^2}$$
$$= \sqrt{4(a^2 + b^2)}$$
$$= 2\sqrt{a^2 + b^2}$$

- Compare the lengths of $\overline{CM}$ and $\overline{AB}$: $AB = 2\sqrt{a^2 + b^2} = 2(CM)$ or, equivalently, $CM = \frac{1}{2}AB$.

Because general coordinates were used for the vertices of the right triangle, the median property is true for all right triangles.

Positioning Figures Using General Coordinates

If the general coordinates of a figure are not given, place the figure in the coordinate plane following these guidelines:

- Use the origin as one of the vertices while confining the figure to Quadrant I.
- Use one or both of the coordinate axes as sides of the figure.
- Make use of the defining properties of the figure. For example, position square $ABCD$ so that A is at the origin, adjacent sides $\overline{AB}$ and $\overline{AD}$ each coincide with a different coordinate axis, and all four sides have the same length.
- If you anticipate using the midpoint formula, assign general coordinates that are divisible by 2.

Example 2

Prove using the methods of coordinate geometry that the line segment whose endpoints are the midpoints of two sides of a triangle is parallel to the third side of the triangle and one-half of its length.

Solution: Position the triangle so that one of its vertices is at the origin and another is on the x–axis, as shown in the accompanying figure.

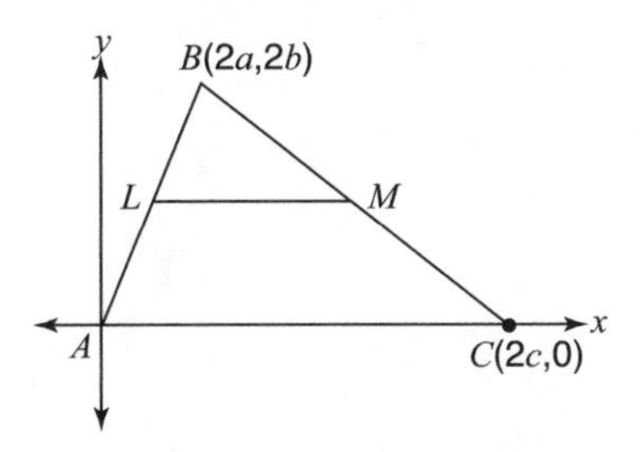

- The vertices of $\triangle ABC$ are $A(0, 0)$, $B(2a, 2b)$, and $C(2c, 0)$. Choosing $2a$, $2b$, and $2c$ rather than a, b, and c will eliminate the need to work with fractions when applying the midpoint formula.
- Find the coordinates of points L and M, the midpoints of $\overline{AB}$ and $\overline{BC}$, respectively:

$$L(x,y) = L\left(\frac{0+2a}{2}, \frac{0+2b}{2}\right) \qquad M(x,y) = M\left(\frac{2a+2c}{2}, \frac{2b+0}{2}\right)$$

$$= L(a, b) \qquad\qquad = M(a + c, b)$$

  As points L and M have the same y–coordinate, $\overline{LM}$ is a horizontal segment. Since $\overline{AC}$ lies on the x–axis, it is also a horizontal segment. Because horizontal segments are parallel, $\overline{LM} \parallel \overline{AC}$.
- The length of a horizontal segment is the difference in the x–coordinates of its endpoints. Hence, $LM = (a + c) - a = c$ and $AC = 2c - 0 = 2c$. Thus,

$$LM = \frac{1}{2}AC.$$

Check Your Understanding of Section 4.6

Write a coordinate proof for each exercise.

1. In the accompanying diagram of $ABCD$, where $a \neq b$, prove $ABCD$ is an isosceles trapezoid.

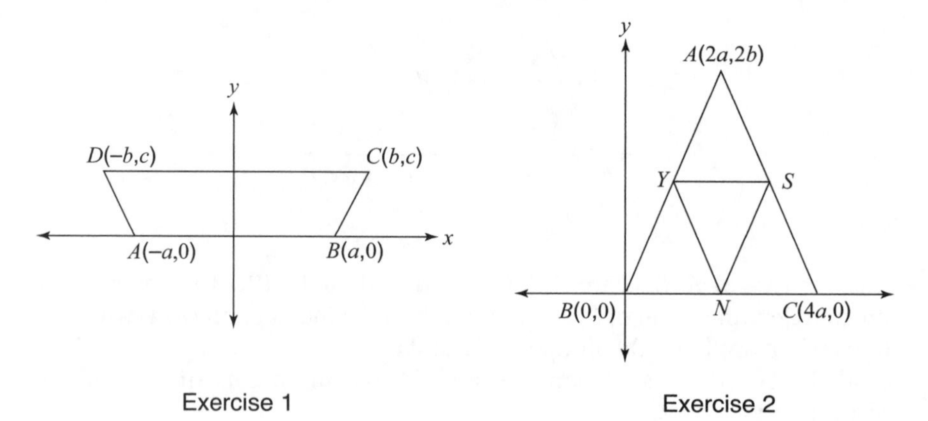

Exercise 1 Exercise 2

2. In the accompanying diagram, $\triangle ABC$ is isosceles. Prove the triangle formed by joining the midpoints of isosceles triangle ABC is also isosceles.

3. Quadrilateral $QRST$ has vertices $Q(a, b)$, $R(0, 0)$, $S(c, 0)$, and $T(a + c, b)$. Prove $QRST$ is a parallelogram.

4. Prove the diagonals of a rectangle are congruent.

5. The coordinates of the vertices of $\triangle ABC$ are $A(-2r, 0)$, $B(0,2s)$, and $C(2r,0)$.
Prove: a. $\triangle ABC$ is isosceles.
b. The medians drawn to the legs are congruent.

6. The coordinates of the vertices of square $ABCD$ are $A(0, 0)$, $B(0, t)$, $C(x, y)$, and $D(t, 0)$.
a. Express the coordinates of C in terms of t.
b. Prove the diagonals of square $ABCD$ are congruent and perpendicular.

7. The coordinates of the vertices of quadrilateral $MATH$ are $M(0, 0)$, $A(r, t)$, $T(s, t)$, and $H(s - r, 0)$. Prove $MATH$ is a parallelogram but *not* a rectangle.

8. The vertices of isosceles trapezoid $ABCD$ with bases $\overline{BC}$ and $\overline{AD}$ are $A(0, 0)$, $B(b, c)$, $C(h, k)$, and $D(a, 0)$.
a. Express h and k in terms of a, b, or c.
b. Prove the diagonals of $ABCD$ are congruent.

9. The coordinates of the vertices of parallelogram $ABCD$ are $A(0, 0)$, $B(b, y)$, $C(a + b, y)$, and $D(a, 0)$.
a. If $ABCD$ is a rhombus, express y in terms of a and b.
b. Prove that the diagonals of rhombus $ABCD$ are perpendicular to each other.

10. The vertices of quadrilateral $ABCD$ are $A(0, 0)$, $B(s, 0)$, $C(t + s, s)$, and $D(t, s)$. If $s > 0$, and $t > 0$, show by means of coordinate geometry, stating reasons for your conclusion, that $ABCD$ is a parallelogram but *not* a rhombus.

CHAPTER 5

RATIO, PROPORTION, AND SIMILARITY

5.1 RATIO AND PROPORTION

KEY IDEAS

A **ratio** is a comparison by division of two quantities measured in the same units. The ratio of x and y can be written in any one of the following three ways:

$$\frac{x}{y}, \quad x:y, \quad x \text{ to } y$$

provided $y \neq 0$. An equation that states that two ratios are equal is called a **proportion**. To find an unknown member of a proportion, cross-multiply and solve for the unknown term.

Ratios and Variables

The ratio of two quantities does not indicate the actual amounts of each quantity. If you know that M is the midpoint of $\overline{AB}$, you can conclude that the ratio of AM to AB is 1:2 without knowing the length of $\overline{AB}$. If you are told that the ratio of Bill's age to Glen's age is 3:2, then their actual ages could be 3 and 2, 6 and 4, 9 and 6, and so forth. In each case, the possible ages are multiples of the base values of 3 and 2.

MATH FACTS

If two quantities are in the ratio $a{:}b$, then the two quantities can be represented by ax and bx, respectively. If the ratio of Bill's age to Glen's age is 3:2, then their ages can be expressed as $3x$ and $2x$, respectively.

Example 1

If the degree measures of two complementary angles are in the ratio of 2 to 13, what is the degree measure of the smaller angle?

Solution: If $2x$ = the measure of the smaller angle, then $13x$ = the measure of the larger angle. Since the two angles are complementary, the sum of their measures is 90.

$$\begin{aligned} 2x + 13x &= 90 \\ 15x &= 90 \\ \frac{15x}{15} &= \frac{90}{15} \\ x &= 6 \\ 2x &= 2(6) = \mathbf{12} \\ \text{and } 13x &= 13(6) = \mathbf{78} \end{aligned}$$

Example 2

The measures of the three angles of a triangle are in the ratio of 2:3:7. What is the measure of the largest angle of the triangle?

Solution: Represent the measures of the three angles of the triangle by $2x$, $3x$, and $7x$. Since the sum of the measures of the angles of a triangle is 180:

$$\begin{aligned} 2x + 3x + 7x &= 180 \\ 12x &= 180 \\ \frac{12x}{12} &= \frac{180}{12} \\ x &= 15 \\ \text{and } 7x &= 7(15) = \mathbf{105}. \end{aligned}$$

$(2x)°$ $(7x)°$ $(3x)°$

Terms of a Proportion

The proportion $\frac{a}{b} = \frac{c}{d}$ is read, "a is to b as c is to d." The same proportion may also be written in the form $a : b = c : d$. The two inside terms, b and c, are the **means**. The two outside terms, a and d, are the **extremes**. In the proportion $\frac{12}{9} = \frac{4}{3}$ or, equivalently, $12 : 9 = 4 : 3$, 9 and 4 are the *means*, and 12 and 3 are the *extremes*. Because the proportion is true, $9 \times 4 = 12 \times 3$. In any true proportion, **the product of the means is equal to the product of the extremes** or, more simply, the cross-products are equal.

Solving a Proportion

To solve a proportion that contains a variable, set the cross-products equal. Then solve for the variable. If $\frac{3}{4} = \frac{x}{7}$, then $4x = 21$ and $x = \frac{21}{4}$.

Example 3

Solve the proportion $\frac{n-3}{4} = \frac{5n+19}{3}$. Check your answer.

Solution: Set the cross-products equal:

$$4(5n + 19) = 3(n - 3)$$
$$20n + 76 = 3n - 9$$
$$20n - 3n = -76 - 9$$
$$17n = -85$$
$$\frac{17n}{17} = \frac{-85}{17}$$
$$n = \mathbf{-5}$$

CHECK:

$$\frac{n-3}{4} = \frac{5n+19}{3}$$

Set $n = -5$: $\frac{-5-3}{4} \overset{?}{=} \frac{5(-5)+19}{3}$

$$\frac{-8}{4} \overset{?}{=} \frac{-6}{3}$$
$$-2 = -2 \checkmark$$

Properties of Proportions

Sometimes a proportion needs to be replaced by another proportion that has a different but equivalent form.

- Property 1: The terms of a proportion can be inverted:

  $\frac{a}{b} = \frac{c}{d}$ and $\frac{b}{a} = \frac{d}{c}$ are equivalent

- Property 2: Opposite terms of a proportion can be interchanged:

  $\frac{a}{b} = \frac{c}{d}$ and $\frac{a}{c} = \frac{b}{d}$ are equivalent

  $\frac{a}{b} = \frac{c}{d}$ and $\frac{d}{b} = \frac{c}{a}$ are equivalent

- Property 3: On each side of a proportion, the denominator can be added to the numerator:

  $\frac{a}{b} = \frac{c}{d}$ and $\frac{a+b}{b} = \frac{c+d}{d}$ are equivalent

Check Your Understanding of Section 5.1

A. *Multiple Choice*

1. The diagonals of rectangle *DATE* intersect at *K*. What is the ratio of *DK* to *AE*?
(1) 1:1 (2) 2:1 (3) 1:2 (4) 1:3

2. The measures of the angles of a triangle are in the ratio 2:3:5. The triangle is
(1) an obtuse triangle
(2) an acute triangle
(3) a right triangle
(4) an isosceles triangle

3. Points *L, I, N*, and *E* are collinear such that *M* is the midpoint of $\overline{LE}$, *I* is the midpoint of $\overline{LM}$, and *N* is the midpoint of $\overline{ME}$. What is the ratio of *IN* to *LE*?
(1) 1:2 (2) 2:3 (3) 1:3 (4) 1:4

4. The ratio of two supplementary angles is 3:6. What is the measure of the smaller angle?
(1) 10 (2) 20 (3) 30 (4) 60

5. One angle of a triangle measures 30°. If the measures of the other two angles are in the ratio 3:7, the measure of the largest angle of the triangle is
(1) 15 (2) 45 (3) 105 (4) 126

6. If the ratio of the measures of the interior angles of a quadrilateral is 2:3:4:6, what is the measure of the smallest angle of the quadrilateral?
(1) 12 (2) 24 (3) 36 (4) 48

7. If the measures of the angles of a triangle are in the ratio 3:4:5, the measure of an exterior angle of the triangle can *not* be
(1) 165 (2) 135 (3) 120 (4) 105

B. *Show or explain how you arrived at your answer.*

8–11. Solve each proportion and check your answer.

8. $\frac{2}{3} = \frac{2-k}{12}$

9. $\frac{7y-5}{3} = \frac{9y}{4}$

10. $\frac{4}{11} = \frac{x+6}{2x}$

11. $\frac{10-h}{5} = \frac{7-h}{2}$

12. If the acute angles of a right triangle are in the ratio of 5 to 13, find the measure of the smaller angle.

13. In isosceles triangle ABC, the ratio of the measure of vertex angle A to the measure of $\angle B$ is 2:5. Find $m\angle C$.

14. The measures of two complementary angles are in the ratio of 2 to 3. What is the measure of the smaller angle?

15. Two parallel lines are cut by a transversal so that the measures of a pair of interior angles on the same side of the transversal are in the ratio of 4:11. Find the measure of the smaller of these angles.

16. Points A, B, P, and M are collinear such that M is the midpoint of $\overline{AB}$ and P is between points A and B. The ratio of AP to PB is 3 to 5. If $AM = 28$, what is the length of $\overline{AP}$?

17. In an isosceles triangle, the measures of the vertex angle and a base angle are in the ratio of 7 to 4. What is the measure of the vertex angle?

18. The measures of the angles of a pentagon are in the ratio of 2:3:5:7:7. What is the measure of the smallest angle of the pentagon.

5.2 SIMILAR TRIANGLES

KEY IDEAS

When a photograph is enlarged, the original photograph and the enlarged image are *similar*. Two figures are **similar** if they have the same shape, but not necessarily the same size. To prove two *triangles* are similar, it is sufficient to prove that only two pairs of corresponding angles are congruent. The symbol for "is similar to" is ~.

Drawing Conclusions from Similar Figures

Two *polygons* are **similar** if their corresponding angles are congruent and the lengths of their corresponding sides have the same ratio and, as a result, "are in proportion."

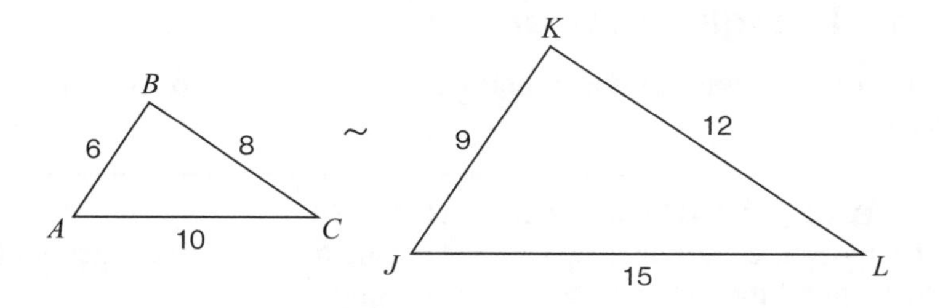

Figure 5.1 Given: $\triangle ABC \sim \triangle JKL$.

If two triangles are similar, as in Figure 5.1, you may conclude that

- Corresponding angles are congruent:

$$\angle A \cong \angle J, \cong \angle B \cong \angle K, \text{ and } \angle C \cong \angle L$$

- The lengths of corresponding sides are in proportion:

$$\frac{AB}{JK} = \frac{BC}{KL} = \frac{AC}{JL}$$

The **similarity ratio** or **ratio of similitude** between the two triangles in Figure 5.1 is $\frac{2}{3}$ since the ratio of any pair of corresponding sides is 2 to 3:

$$\frac{AB}{JK} = \frac{6}{9} = \frac{2}{3}, \frac{BC}{KL} = \frac{8}{12} = \frac{2}{3}, \text{ and } \frac{AC}{JL} = \frac{10}{15} = \frac{2}{3}$$

Example 1

The lengths of the sides of a triangle are 9, 15, and 21. If the length of the shortest side of a similar triangle is 12, find the length of its longest side.

Solution: Assume x represents the length of the longest side of the larger triangle. Since the lengths of corresponding sides of similar triangles are in proportion:

$$\frac{\text{Side in smaller } \triangle}{\text{Corresponding side in larger } \triangle} = \frac{9}{12} = \frac{21}{x}$$

$$9x = 252$$

$$\frac{9x}{9} = \frac{252}{9}$$

$$x = \mathbf{28}.$$

Proving Triangles Similar

The shape of a triangle is determined when the measures of two of its angles are given.

> **Postulate: AA Postulate of Similarity**
> Two triangles are similar if two angles of one triangle are congruent to the corresponding angles of the other triangle.

Example 2

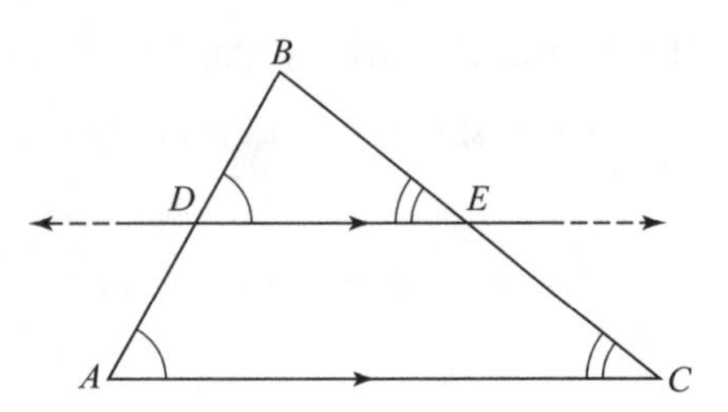

Given: $\overleftrightarrow{DE} \parallel \overline{AC}$.

Prove: $\triangle DBE \sim \triangle ABC$.

Solution: Mark off the diagram with the two pairs of congruent corresponding angles formed by the parallel lines.

Proof

Statement		Reason
1. $\overline{DE} \parallel \overline{AC}$.		1. Given.
2. $\angle BDE \cong \angle A$.	Angle	2. If two lines are parallel, corresponding angles are congruent.
3. $\angle BED \cong \angle C$.	Angle	3. Same as Reason 2.
4. $\triangle DBE \sim \triangle ABC$.		4. AA Postulate of Similarity.

This example proves the next theorem.

> **Theorem**
> A line parallel to one side of a triangle and intersecting the other two sides forms a triangle similar to the original triangle.

Example 3

In the accompanying diagram of $\triangle SRT$, $\overline{LM} \parallel \overline{RT}$. If $SL = 4$, $LR = 3$, and $RT = 21$, what is the length of $\overline{LM}$?

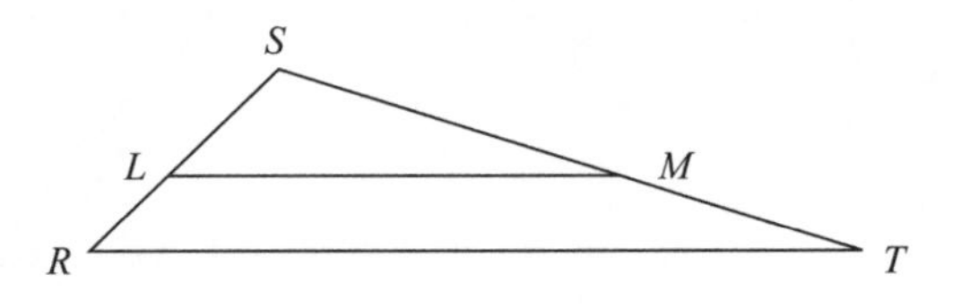

Solution: Because $\overline{LM} \parallel \overline{RT}$, $\triangle SLM \sim \triangle SRT$ so the lengths of the corresponding sides of the two triangles are in proportion:

$$\frac{SL}{SR} = \frac{LM}{RT}$$

$$\frac{4}{4+3} = \frac{LM}{21}$$

$$\frac{4}{7} = \frac{LM}{21}$$

$$7(LM) = 4 \times 21$$

$$\frac{7(LM)}{7} = \frac{84}{7}$$

$$LM = 12$$

The length of $\overline{LM}$ is **12**.

Proving a Proportion

Once two triangles are proved congruent, you may conclude that a pair of corresponding sides are congruent. In much the same way, after proving two triangles are similar, you can write a proportion involving the lengths of corresponding sides citing as a reason, "Lengths of corresponding sides of similar triangles are in proportion."

Example 4

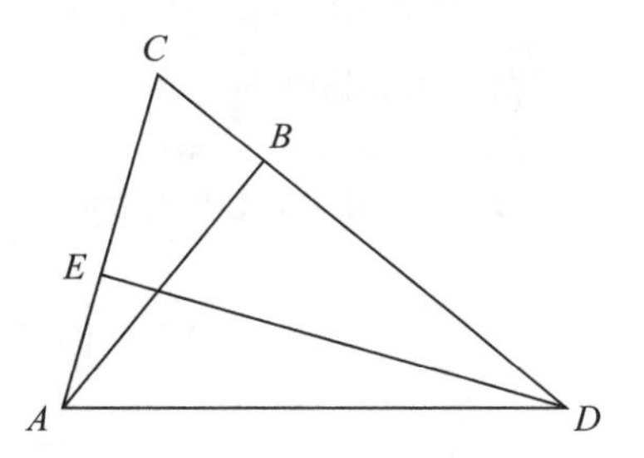

Given: $\overline{DE} \perp \overline{AC}$, $\overline{AB} \perp \overline{CD}$.

Prove: $\frac{EC}{BC} = \frac{ED}{AB}$.

Solution: Find the triangles whose sides are the segments in the Prove. Then prove these triangles are similar.

- Read across the proportion to determine the triangles that you need to prove are similar:

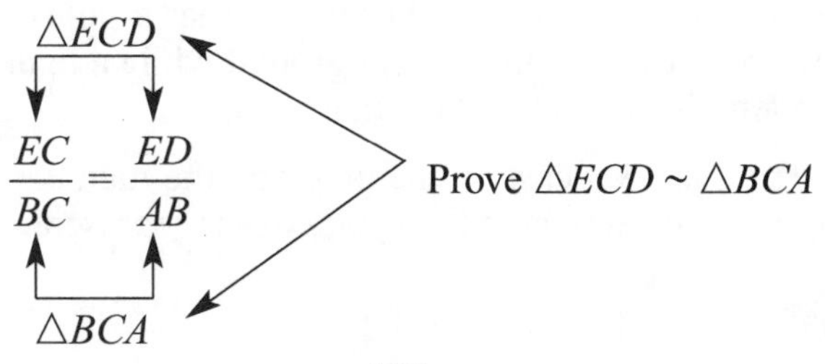

- Mark off on the diagram the two pairs of congruent angles that can be used to prove the two triangles similar.
- Write the proof.

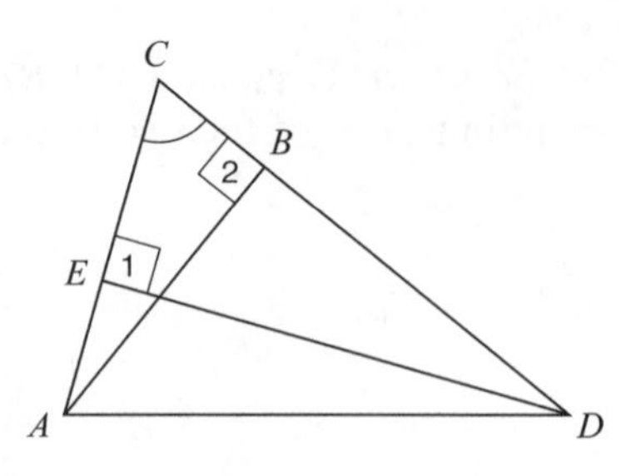

Proof

Statement		Reason
1. $\overline{DE} \perp \overline{AC}, \overline{AB} \perp \overline{CD}$.		1. Given.
2. $\angle 1$ and $\angle 2$ are right angles.		2. Perpendicular lines intersect to form right angles.
3. $\angle 1 \cong \angle 2$.	Angle	3. All right angles are congruent.
4. $\angle C \cong \angle C$.	Angle	4. Reflexive property of congruence.
5. $\triangle ECD \sim \triangle BCA$.		5. AA postulate of similarity.
6. $\dfrac{EC}{BC} = \dfrac{ED}{AB}$.		6. Lengths of corresponding sides of similar triangles are in proportion.

MATH FACTS

If the original proportion in the Prove had been written as $\dfrac{EC}{ED} = \dfrac{BC}{AB}$, then reading *across* the bottom of the proportion (*E–D–A–B*) would not tell you the vertices of the triangles that you need to prove similar. When this happens, try reading *down* each ratio rather than across:

$$\frac{EC}{ED} = \frac{BC}{AB}$$

$$\triangle ECD \sim \triangle BCA$$

Example 5

At a certain time during the day, light falls so that a pole 10 feet in height casts a shadow 15 feet in length on level ground while a man casts a shadow that is 9 feet in length. How tall is the man?

Solution: The shadows cast by the pole and the man can be represented as legs of right triangles in which the hypotenuses represent the rays of light.

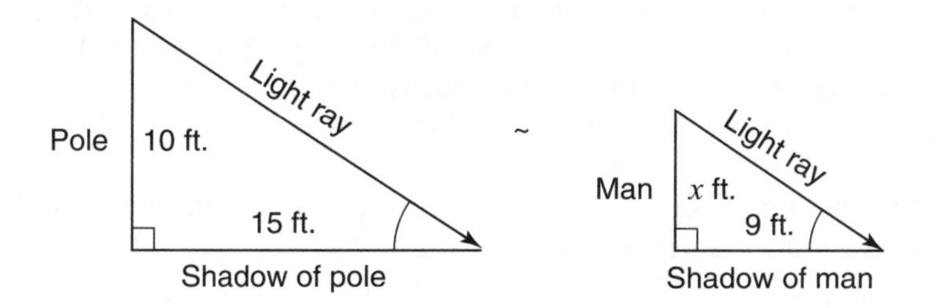

Assume that in each right triangle the ray of light makes the same angle with the ground. Because the two right triangles are similar, the lengths of their corresponding sides are in proportion, as shown below.

$$\frac{\text{Height of pole}}{\text{Height of man}} = \frac{\text{Shadow of pole}}{\text{Shadow of man}}$$

$$\frac{10}{x} = \frac{15}{9}$$

$$15x = 90$$

$$\frac{15x}{15} = \frac{90}{15}$$

$$x = 6$$

The man is **6 feet** tall.

Check Your Understanding of Section 5.2

A. Multiple Choice.

1. The length of the shortest side of a triangle is 12, and the length of the shortest side of a similar triangle is 5. If the longest side of the first triangle is 15, what is the longest side of the larger triangle?
(1) 2.4 (2) 6.25 (3) 24 (4) 36

2. A person 5 feet tall is standing near a tree 30 feet high. If the length of the person's shadow is 3 feet, what is the length of the shadow of the tree?
(1) 50 (2) 24 (3) 18 (4) 12

3. On a scale drawing of a new school playground, a triangular area has sides with lengths of 8 centimeters, 15 centimeters, and 17 centimeters. If the triangular area located on the playground has a perimeter of 120 meters, what is the length of its longest side?
(1) 24 m (2) 40 m (3) 45 m (4) 51 m

4. Jordan and Missy are standing together in the schoolyard. Jordan, who is 6 feet tall, casts a shadow that is 54 inches long. At the same time, Missy casts a shadow that is 45 inches long. How tall is Missy?
(1) 38 in (2) 86.4 in (3) 5 ft (4) 5 ft 6 in

5. The sides of $\triangle ABC$ are 2, 3, and 4. Which set of numbers could represent the sides of a triangle similar to $\triangle ABC$?
(1) {5, 6, 7} (3) {12, 13, 14}
(2) {6, 9, 16} (4) {20, 30, 40}

6. The accompanying diagram shows two similar triangles.

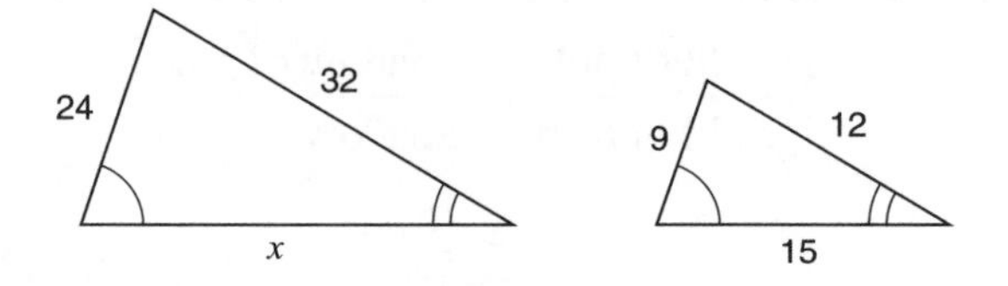

Which proportion could be used to solve for x?

(1) $\frac{x}{24} = \frac{9}{15}$ (2) $\frac{24}{9} = \frac{15}{x}$ (3) $\frac{32}{x} = \frac{12}{15}$ (4) $\frac{32}{12} = \frac{15}{x}$

7. In the accompanying diagram, $\angle B$ and $\angle E$ are right angles, $AB = 3$, $BC = 4$, and $CD = 20$. What is the perimeter of $\triangle AED$?
(1) 22 (2) 36 (3) 45 (4) 48

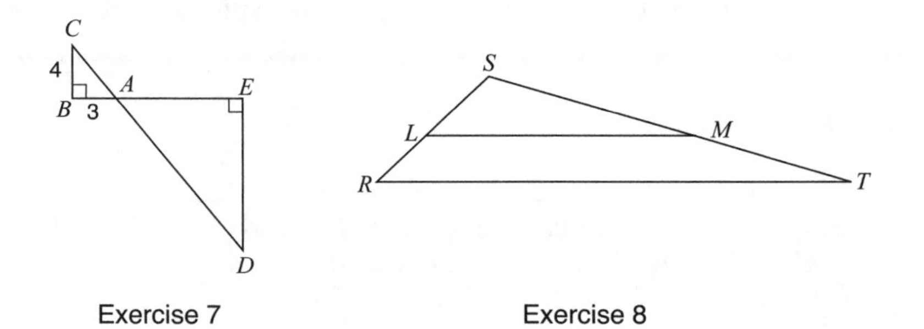

Exercise 7 Exercise 8

8. In the accompanying diagram of $\triangle SRT$, $\overline{LM} \parallel \overline{RT}$. If $SM = 12$, $MT = 8$, and $RS = 15$. What is the length of LR?
(1) 9 (2) 6 (3) 3 (4) 4

9. At a certain time during the day, light falls so that a flagpole 12 feet in height casts a shadow 9 feet in length on level ground, while a man casts a shadow that is 4 feet in length. How tall is the man?
(1) 5 ft 4 in (2) 5 ft 3 in (3) 5 ft 2 in (4) 5.0 ft

10. The corresponding altitudes of two similar triangles are 6 and 14. If the perimeter of the smaller triangle is 21, what is the perimeter of the larger triangle?
(1) 9 (2) 36 (3) 49 (4) 64

11. In the accompanying diagram of $\triangle ABC$, D is a point on $\overline{AC}$, $\overline{AB}$ is extended to E, and $\overline{DE}$ is drawn so that $\triangle ADE \sim \triangle ABC$. If $m\angle C = 30$ and $m\angle A = 70$, what is $m\angle ADE$?
(1) 30 (2) 70 (3) 80 (4) 100

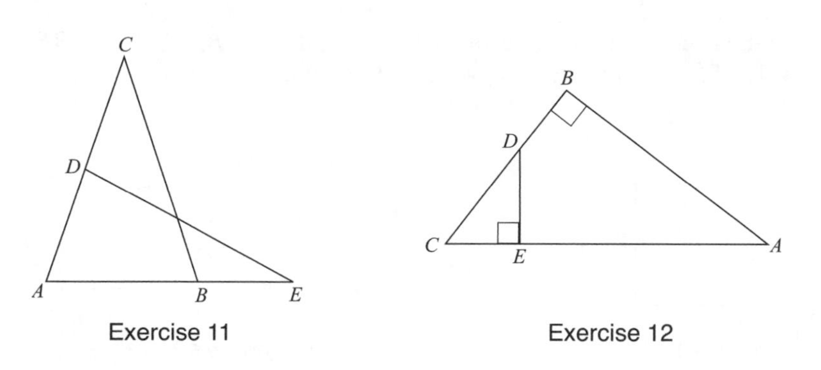

Exercise 11

Exercise 12

B. *Show or explain how you arrived at your answer.*

12. In $\triangle ABC$, $\overline{AB} \perp \overline{BC}$ and $\overline{DE} \perp \overline{CE}$. If $DE = 8$, $CD = 10$, and $CA = 30$, find AB.

13. The lengths of the sides of a triangle are 14, 18, and 20. If the length of the longest side of a similar triangle is 25, what is the perimeter of this triangle?

14. An image of a building in a photograph is 6 centimeters wide and 11 centimeters tall. If the image is similar to the actual building and the actual building is 174 meters wide, how tall is the actual building in meters?

15. If a boy 5 feet 6 inches tall casts a shadow 6 feet long, what is the length of the shadow cast by a tree that is 11 feet high?

16. In the accompanying diagram of $\triangle ABC$, $\overline{DE} \perp \overline{AB}$, $DE = 8$, $CD = 12$, and $DA = 3$. Find the length of $\overline{AB}$.

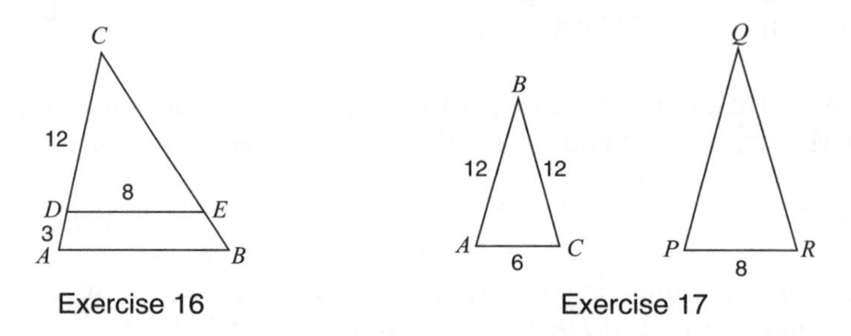

Exercise 16 Exercise 17

17. In the accompanying diagram, $\triangle ABC$ is similar to $\triangle PQR$, $AC = 6$, $AB = BC = 12$, and $PR = 8$. Find the perimeter of $\triangle PQR$.

18. In the accompanying diagram of $\triangle ABC$, $\overline{AC} \perp \overline{CB}$, $\overline{AE} \perp \overline{EF}$, $BF = 8$, $FA = 12$, $FE = 9$, and $BC = x$. What is the value of x ?

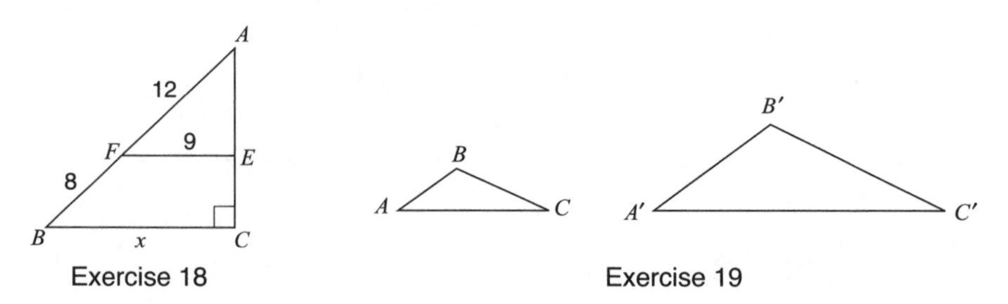

Exercise 18 Exercise 19

19. In the accompanying diagram, $\triangle ABC \sim \triangle A'B'C'$ and $A'B'=4$.

a. If AC is 2 more than AB, and $A'C'$ is 6 more than AB, find the length of $\overline{AB}$.

b. If the length of an altitude in $\triangle A'B'C'$ is h units, express in terms of h the length of the corresponding altitude in $\triangle ABC$.

c. Using the results from part a, determine the smallest possible integral value of BC. Give a reason for your answer.

20. In the accompanying figure, angles D and B are right angles. If $BC = 80$ meters, $DE = 15$ meters, and $BD = 171$ meters.

a. Write a proportion that could be used to find the length of AB and explain why the proportion is valid.

b. Solve the proportion written in part a.

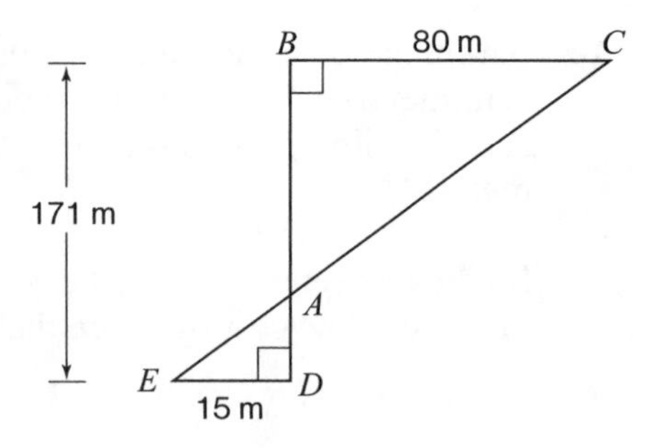

C. *Write a proof.*

21. In the accompanying diagram of parallelogram $STAR$, $\overline{TK} \perp \overline{SR}$, and $\overline{TL} \perp \overline{AR}$.
Prove: $\triangle SKT \sim \triangle ALT$.

22. Given: $ABCD$ is a rectangle, $\overline{CF} \perp \overline{BD}$.
Prove: $\dfrac{EC}{AD} = \dfrac{AB}{BD}$.

23. Given: Paralellogram $ABCD$, $\overline{AC}$ bisects $\angle FAB$, $\overline{AG} \cong \overline{AE}$.
Prove: $\triangle AHG \sim \triangle CHD$.

24. Given: $\overline{MN} \parallel \overline{AT}$, $\angle 1 \cong \angle 2$.
Prove: $\dfrac{NT}{AT} = \dfrac{RN}{RT}$.

25. Given: $\overline{SR} \cong \overline{SQ}$, $\overline{RQ}$ bisects $\angle SRW$.
Prove: $\dfrac{SQ}{RW} = \dfrac{SP}{PW}$.

26. Given: $\overline{MC} \perp \overline{JK}$, $\overline{PM} \perp \overline{MQ}$, $\overline{TP} \cong \overline{TM}$.
Prove: $\dfrac{PM}{MC} = \dfrac{PQ}{MK}$.

27. Given: T is the midpoint of $\overline{PQ}$, $\overline{MP} \cong \overline{MQ}$, $\overline{JK} \parallel \overline{MQ}$.
Prove: $\dfrac{PM}{JK} = \dfrac{TQ}{JT}$.

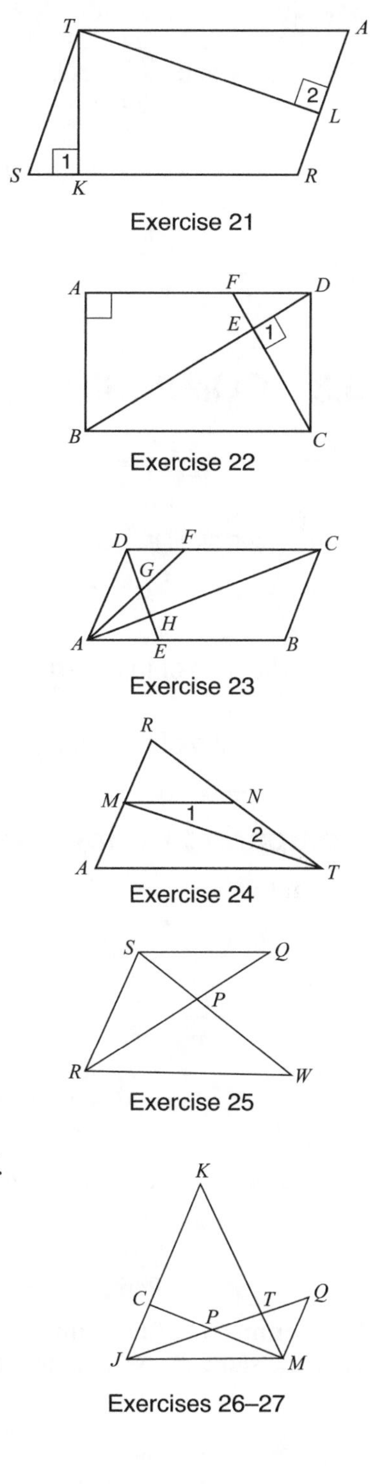

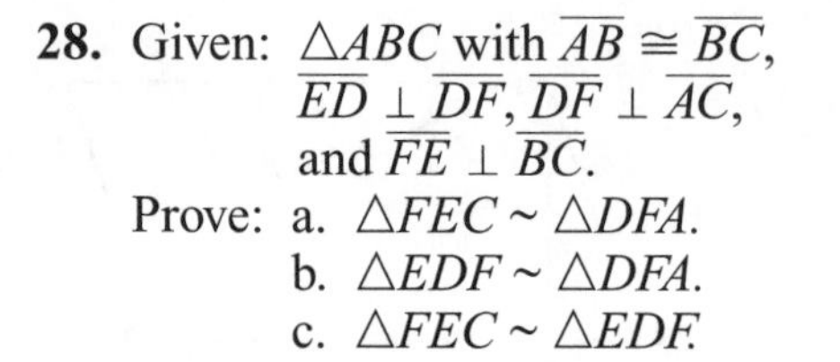

28. Given: $\triangle ABC$ with $\overline{AB} \cong \overline{BC}$,
$\overline{ED} \perp \overline{DF}$, $\overline{DF} \perp \overline{AC}$,
and $\overline{FE} \perp \overline{BC}$.
Prove: a. $\triangle FEC \sim \triangle DFA$.
b. $\triangle EDF \sim \triangle DFA$.
c. $\triangle FEC \sim \triangle EDF$.

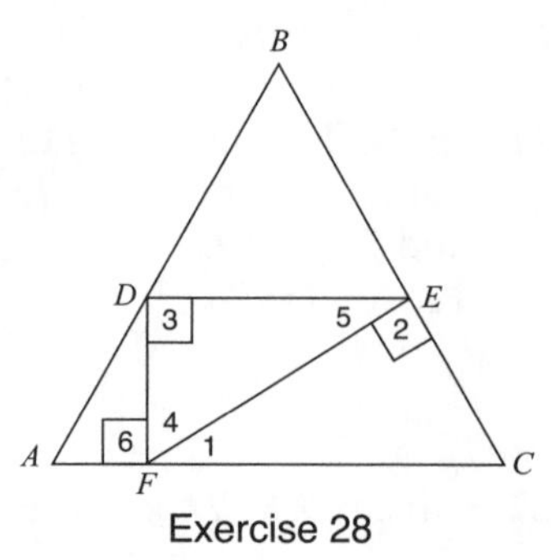

Exercise 28

5.3 COMPARING SPECIAL RATIOS

KEY IDEAS

If in two similar triangles the ratio of the lengths of a pair of corresponding sides is $\frac{a}{b}$, then

- The ratio of their perimeters and corresponding altitudes is $\frac{a}{b}$.
- The ratio of their areas is $\left(\frac{a}{b}\right)^2$.

Comparing Perimeters of Similar Triangles

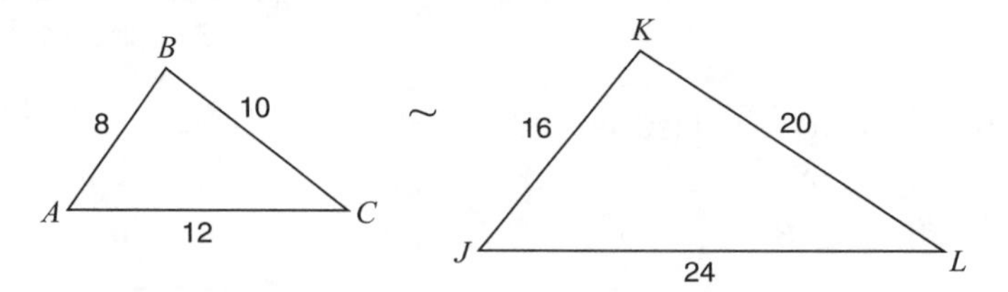

Figure 5.2 Comparing perimeters of similar triangles.

In Figure 5.2, the similarity ratio is 1 to 2 or $\frac{1}{2}$.

$$\frac{\text{Perimeter } \triangle ABC}{\text{Perimeter } \triangle JKL} = \frac{8+10+12}{16+20+24} = \frac{30}{60} = \frac{1}{2}$$

The perimeters of two similar triangles, as well as any two similar polygons, have the same ratio as the lengths of any pair of corresponding sides.

Example 1

The lengths of the sides of a triangle are 8 cm, 11 cm, and 17 cm. If the length of the shortest side of a similar triangle is 12 cm, find the perimeter of the larger triangle.

Solution: Assume x represents the perimeter of the larger triangle.

$$\frac{\text{Side in smaller } \triangle}{\text{Corresponding side in larger } \triangle} = \frac{\text{Perimeter of smaller } \triangle}{\text{Perimeter of larger } \triangle}$$

$$\frac{8}{12} = \frac{36}{x}$$

$$8x = 432$$

$$\frac{8x}{8} = \frac{432}{8}$$

$$x = 54$$

The perimeter of the larger triangle is **54 cm**.

Comparing Altitudes in Similar Triangles

In the accompanying diagram, $\triangle ABC \sim \triangle DEF$, h is the length of the altitude to side $\overline{AC}$, and h' is the length of the altitude to side $\overline{DF}$.

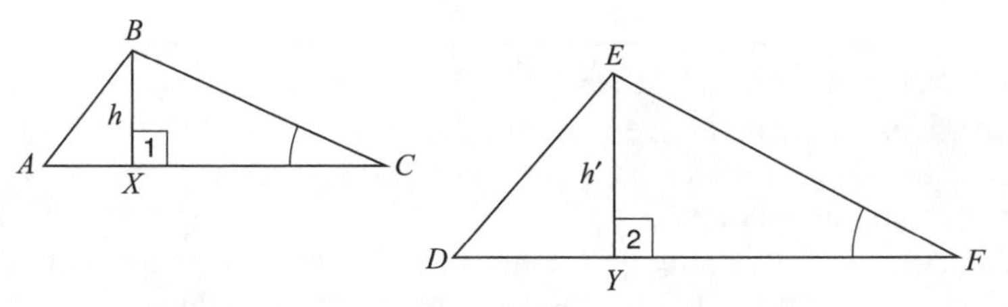

Figure 5.3 Comparing altitudes of similar triangles.

- Since the two triangles are similar, $\angle C \cong \angle F$.
- $\angle 1$ and $\angle 2$ are right angles so they are congruent.
- $\triangle BXC \sim \triangle EYF$ and, as a result, $\frac{h}{h'} = \frac{BC}{EF}\left(= \frac{AB}{DE} = \frac{AC}{DF}\right)$.

This proves that the lengths of corresponding *altitudes* of similar triangles have the same ratio as the lengths of any pair of corresponding sides.

Comparing Areas of Similar Triangles

It is not necessary to calculate the areas of two similar triangles in order to be able to compare them. Referring to Figure 5.3:

$$\frac{\text{Area of } \triangle ABC}{\text{Area of } \triangle DEF} = \frac{\frac{1}{2}(AC)h}{\frac{1}{2}(DF)h'} = \frac{AC}{DF} \cdot \frac{h}{h'}$$

- Since the lengths of corresponding altitudes of similar triangles have the same ratio as the lengths of any pair of corresponding sides, you can substitute $\frac{AC}{DF}$ for $\frac{h}{h'}$:

$$\begin{aligned}\frac{\text{Area of } \triangle ABC}{\text{Area of } \triangle DEF} &= \frac{AC}{DF} \cdot \frac{h}{h'} \\ &= \left(\frac{AC}{DF}\right)^2 \\ &= \left(\frac{\text{Side of } \triangle ABC}{\text{Corresponding side of } \triangle DEF}\right)^2\end{aligned}$$

Theorems: Comparing Parts of Similar Triangles

- **Theorem**: If two triangles are similar, perimeters and the lengths of corresponding altitudes have the same ratio as the lengths of any pair of corresponding sides.
- **Theorem**: If two triangles are similar, the ratio of their areas is the same as the *square* of the ratio of the lengths of any pair of corresponding sides.

Example 2

In the accompanying figure, $\angle B \cong \angle D$, $AC = 10$ cm, and $EC = 15$ cm. If the area of $\triangle ABC$ is 32 cm^2, what is the area of $\triangle EDC$?

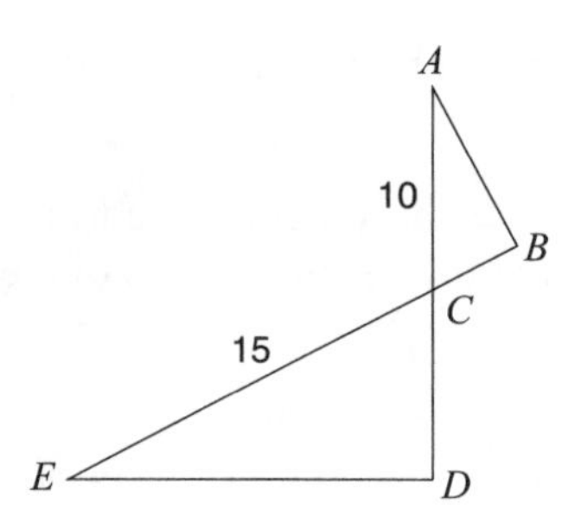

Solution: Because $\triangle ABC \sim \triangle CDE$, the ratio of their areas is equal to the square of the ratio of the lengths of corresponding sides $\overline{AC}$ and $\overline{EC}$. If x represents the area of $\triangle EDC$, then

$$\frac{\text{Area of } \triangle ABC}{\text{Area of } \triangle EDC} = \left(\frac{AC}{EC}\right)^2$$

$$\frac{32}{x} = \left(\frac{10}{15}\right)^2$$

$$\frac{32}{x} = \left(\frac{2}{3}\right)^2$$

$$\frac{32}{x} = \frac{4}{9}$$

$$4x = 288$$

$$\frac{4x}{4} = \frac{288}{4}$$

$$x = 72$$

The area of $\triangle EDC$ is **72 cm²**.

Check Your Understanding of Section 5.3

A. Multiple Choice.

1. On a scale drawing of a new school playground, a triangular area has sides with lengths of 8 centimeters, 15 centimeters, and 17 centimeters. If the triangular area located on the playground has a perimeter of 120 meters, what is the length of its longest side?
(1) 24 m (2) 40 m (3) 45 m (4) 51 m

2. The lengths of the sides of a triangular flag are 30 in, 40 in, and 50 in. What is the length of the shortest side of a flag with a similar triangular shape that has a perimeter of 360 inches?
(1) 120 in (2) 90 in (3) 30 in (4) 150 in

3. Craig's sailboat has two sails that are similar triangles. The larger sail has sides of 10 feet, 24 feet, and 26 feet. If the shortest side of the smaller sail measures 6 feet, what is the perimeter of the *smaller* sail?
(1) 15 ft (2) 36 ft (3) 60 ft (4) 100 ft

4. The corresponding altitudes of two similar triangles are 6 and 14. If the perimeter of the smaller triangle is 21, what is the perimeter of the larger triangle?
(1) 9 (2) 36 (3) 49 (4) 64

5. If the midpoints of the sides of a triangle are connected, the area of the triangle formed is what part of the area of the original triangle?
(1) $\frac{1}{4}$ (2) $\frac{1}{16}$ (3) $\frac{3}{8}$ (4) $\frac{1}{2}$

6. In $\triangle RST$, points X, Y, and Z are the midpoints of sides $\overline{RS}$, $\overline{ST}$, and $\overline{TR}$, respectively. If X, Y, and Z are connected to form triangle XYZ, which statement must be true?
(1) $\triangle XYZ \cong \triangle RST$.
(2) The perimeter of $\triangle XYZ$ is equal to the perimeter of $\triangle RST$.
(3) $\triangle XYZ \sim \triangle RST$.
(4) The area of $\triangle XYZ$ is equal to one-half the area of $\triangle RST$.

B. *Show or explain how you arrived at your answer.*

7. The lengths of the sides of two similar rectangular billboards are in the ratio 5:4. If 250 square feet of material is needed to cover the larger billboard, how much material, in square feet, is needed to cover the smaller billboard?

8. In the accompanying diagram, $\overline{PQ} \parallel \overline{RT}$, $QT = 8$ in, and $ST = 20$ in. If the area of $\triangle RST$ is 125 in^2, what is the number of square inches in the area of $\triangle PSQ$?

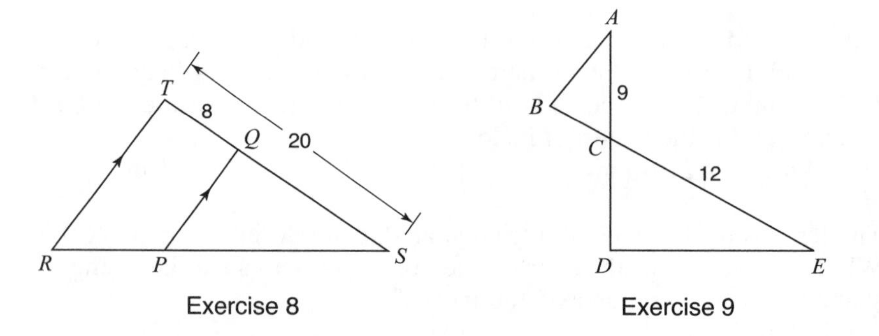

Exercise 8

Exercise 9

9. In the accompanying diagram, $\angle B \cong \angle D$, $AC = 9$ in, and $EC = 12$ in. If the area of $\triangle ABC$ is 208 in^2, what is the area of $\triangle EDC$?

5.4 SIMILARITY THEOREMS

KEY IDEAS

There are two additional methods for proving triangles similar. Using similar triangles, some special proportions can be proved.

SAS Similarity Theorem

Two triangles have the same shape if they have one pair of angles congruent and the sides forming those angles are in proportion. If in Figure 5.4, $\frac{AB}{RS} = \frac{AC}{RT}$ and $\angle A \cong \angle R$, then $\triangle ABC \sim \triangle RST$.

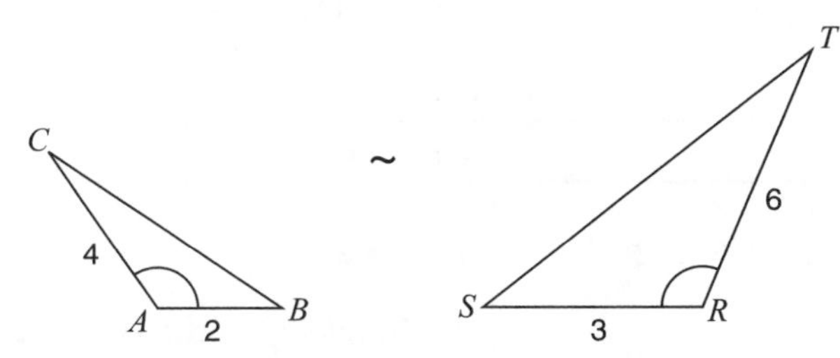

Figure 5.4 SAS Similarity Theorem.

> **Theorem: SAS Similarity Theorem**
> If the lengths of two pairs of corresponding sides of two triangles are in proportion and their included angles are congruent, then the triangles are similar [SAS ~ SAS].

SSS Similarity Theorem

Two triangles have the same shape when the lengths of their corresponding sides have the same ratio. In Figure 5.5, if $\frac{AB}{RS} = \frac{AC}{RT} = \frac{BC}{ST}$, then $\triangle ABC \sim \triangle RST$.

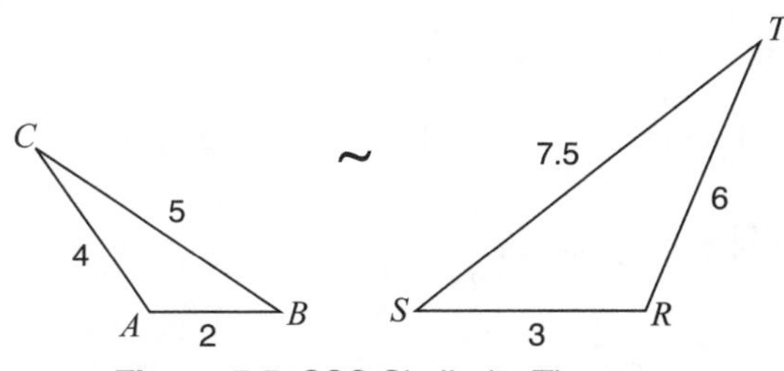

Figure 5.5 SSS Similarity Theorem.

> **Theorem: SSS Similarity Theorem**
> If the lengths of corresponding sides of two triangles are in proportion, then the triangles are similar [SSS ~ SSS].

Example 1

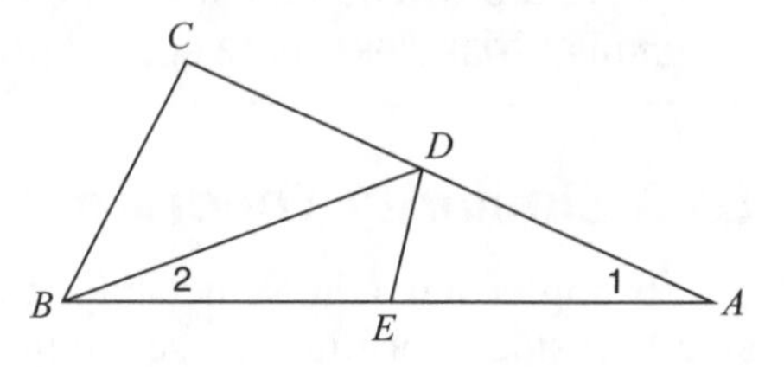

Given: $\frac{BD}{AC} = \frac{AE}{AB}$, $\angle 1 \cong \angle 2$.

Prove: $\triangle DAE \sim \triangle CAB$

Solution: In $\triangle BDA$, $BD = AD$ since the base angles are congruent. Substituting AD for BD in the given proportion allows the required triangles to be proved similar using the SAS Similarity Theorem.

Proof

Statement	**Reason**
1. $\frac{BD}{AC} = \frac{AE}{AB}$, $\angle 1 \cong \angle 2$.	1. Given.
2. $BD = AD$.	2. Converse of the Base Angles Theorem.
3. $\frac{AD}{AC} = \frac{AE}{AB}$.	3. Substitution property.
4. $\angle 1 \cong \angle 1$. (Included Angle)	4. Reflexive property of congruence.
5. $\triangle DAE \sim \triangle CAB$.	5. SAS ~ SAS.

Dividing Sides Proportionally

In Figure 5.6, $\overline{JK}$ and $\overline{LM}$ are **divided proportionally** because $\frac{6}{2} = \frac{9}{3}$.

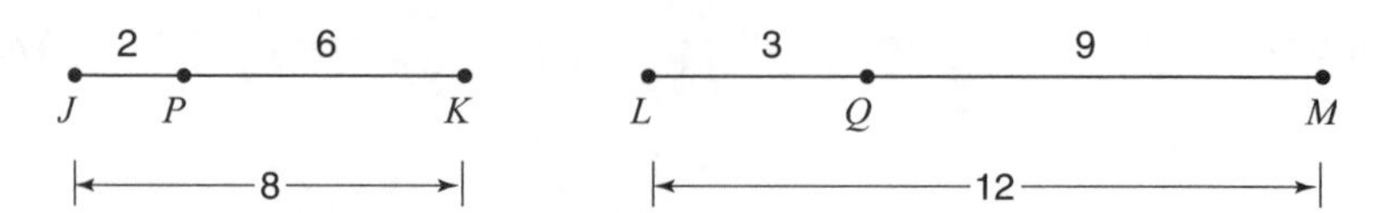

Figure 5.6 Segments divided proportionally.

Some equivalent proportions are $\frac{2}{6} = \frac{3}{9}$, $\frac{2}{8} = \frac{3}{12}$, and $\frac{6}{8} = \frac{9}{12}$.

If $\overline{XY} \parallel \overline{RT}$ in Figure 5.7, then $\overline{XY}$ divides the two sides it intersects proportionally such that $\frac{b}{a} = \frac{d}{c}$. Equivalent proportions include

$$\frac{a}{b} = \frac{c}{d} \qquad \frac{a}{p} = \frac{c}{k} \qquad \frac{b}{p} = \frac{d}{k}$$

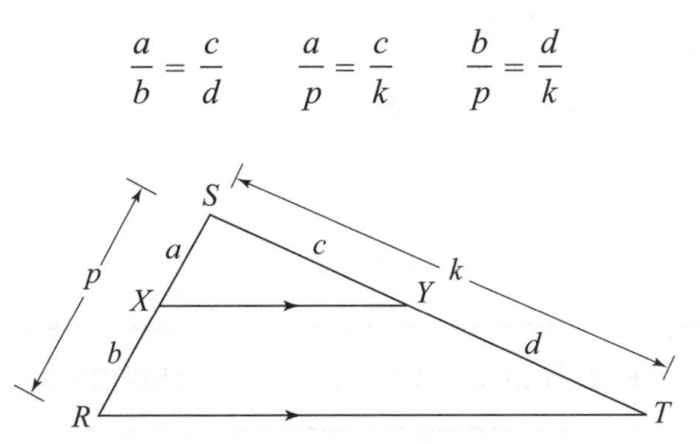

Figure 5.7 $\overline{XY}$ divides $\overline{RS}$ and $\overline{TS}$ proportionally.

> **Theorem: Splitting Sides of a Triangle Into Proportions**
> If a line parallel to one side of a triangle intersects the other two sides, then it divides those sides proportionally.

Corollary: If three parallel lines are cut by two transversals, then the segments intercepted on the transversals are in proportion. In the accompanying figure,

$$p \parallel q \parallel r \quad \Rightarrow \quad \frac{a}{b} = \frac{c}{d}$$

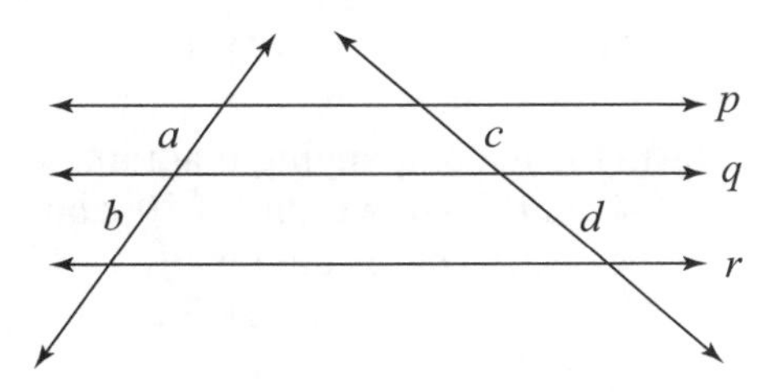

Example 2

In the accompanying figure, $\overline{DE} \parallel \overline{AB}$. If $CD = 12$, $CA = 20$, and $BE = 6$, what is the length of $\overline{CE}$?

Solution: Since $\overline{DE} \parallel \overline{AB}$, $\overline{DE}$ divides the sides it intersects proportionally:

$$\frac{AD}{CD} = \frac{BE}{CE}$$

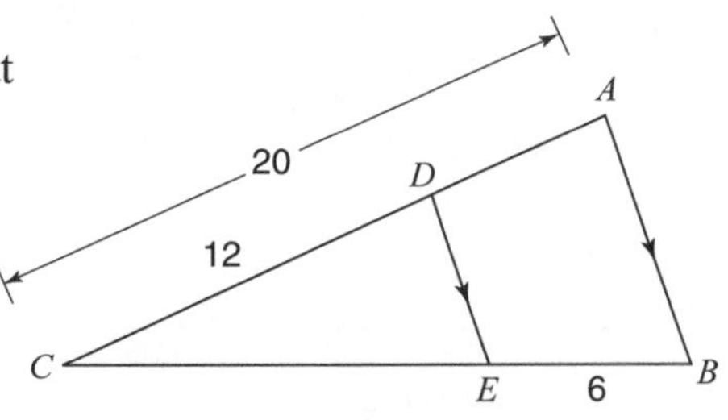

Since $CA = 20$, $AD = 20 - 12 = 8$.

$$\frac{8}{12} = \frac{6}{CE}$$

Set the cross-products equal:

$$8(CE) = 12 \times 6$$

$$\frac{8(CE)}{8} = \frac{72}{8}$$

$$\boldsymbol{CE = 9}$$

Check Your Understanding of Section 5.4

A. Multiple Choice.

1. In the accompanying diagram of $\triangle ABC$, $\overline{AC} \parallel \overline{DE}$, $AB = 10$, $BC = 15$, and $BD = 8$. What is the length of $\overline{EC}$?

(1) $5\frac{1}{3}$ (3) 3

(2) 2 (4) 12

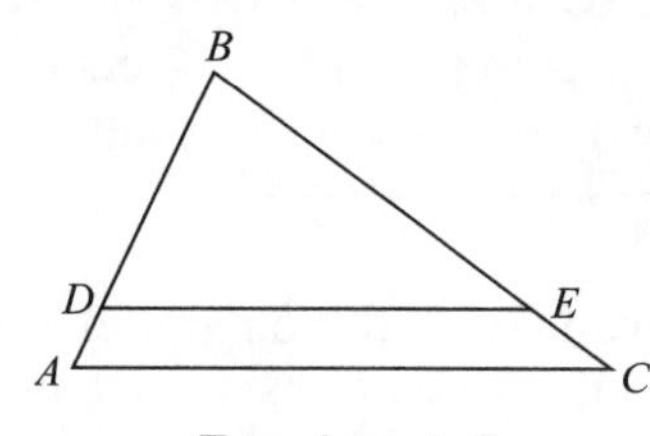

Exercises 1–2

2. In the accompanying diagram, $\overline{AC} \parallel \overline{DE}$, $AB = 6$, $BC = 10$, and $EC = 2$. What is the length of $\overline{BD}$?

(1) 5 (3) 3.6

(2) 4.8 (4) 1.2

3. In the accompanying diagram of $\triangle DEF$, $\overline{AB} \parallel \overline{DE}$, $AF = 4$, $DF = 16$, and $FE = 20$. What is the length of $\overline{FB}$?

(1) 5 (2) 5.4 (3) 6 (4) 8

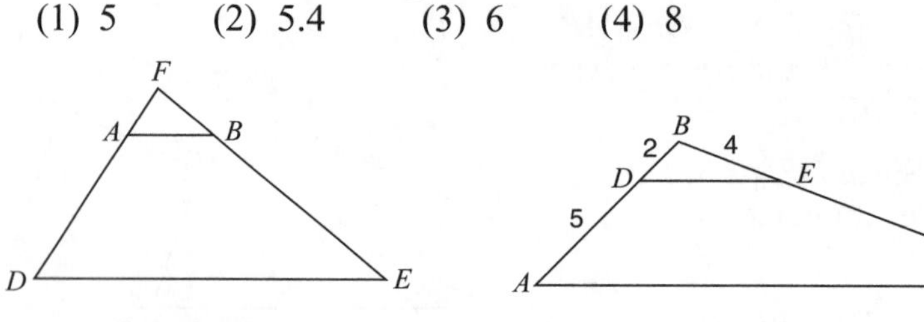

Exercise 3 Exercise 4

4. In the accompanying diagram of $\triangle ABC$, $\overline{DE} \parallel \overline{AC}$, $BD = 2$, $BE = 4$, and $DA = 5$. Find the length of $\overline{BC}$.
(1) 7 (2) 10 (3) 12 (4) 14

5. In the accompanying diagram of $\triangle ABC$, $\overline{DE} \parallel \overline{AB}$, $\overline{CFG} \perp \overline{AB}$, $\overline{CD} = 6$, $\overline{DA} = 4$, and $\overline{CF} = 5$. What is the length of $\overline{FG}$?

(1) $1\frac{1}{3}$ (3) $5\frac{1}{3}$

(2) $3\frac{1}{3}$ (4) $10\frac{1}{3}$

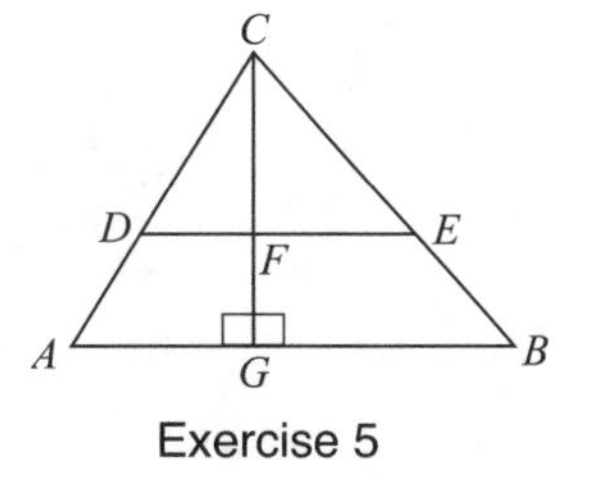

Exercise 5

B. *Show or explain how your arrived at your answer.*

6. In $\triangle ABC$, D is a point on $\overline{AB}$ and E is a point on $\overline{AC}$ such that $\overline{DE} \parallel \overline{BC}$. If $AB = 12$, $AC = 18$, and $AD = 4$, find the length of $\overline{AE}$.

7. In the accompanying diagram of $\triangle CAT$, $\overline{WG} \parallel \overline{AT}$, $TG = x$, $GC = x - 1$, $CW = x + 5$, and $WA = 2x + 6$. Find the length of $\overline{TG}$. [Only an algebraic solution will be accepted.]

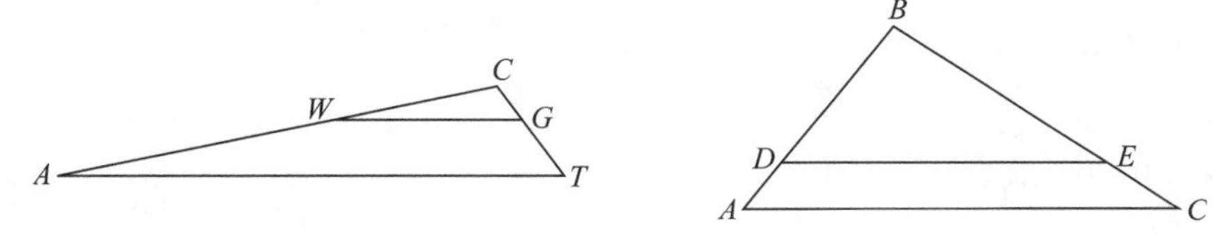

Exercise 7

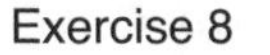
Exercise 8

8. In the accompanying diagram of $\triangle ABC$, $\overline{DE} \parallel \overline{AC}$, $AB = 18$, $BC = 24$, $AC = 30$, and EC exceeds AD by 1. Find the length of $\overline{DE}$.

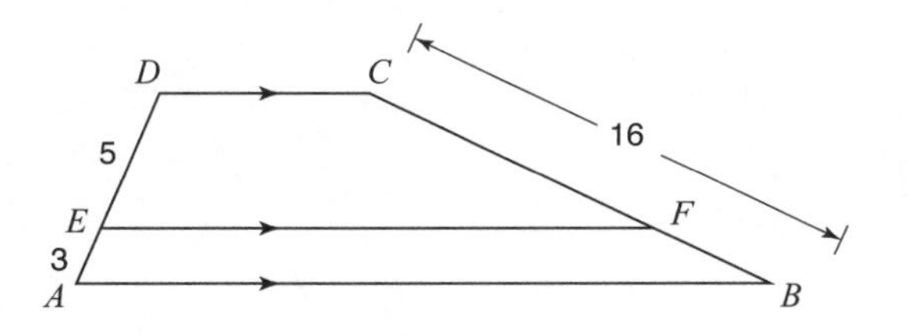

9. In the accompanying diagram, $\overline{AB} \parallel \overline{EF} \parallel \overline{DC}$. If $AE = 3$, $ED = 5$, and $CB = 16$, find FB.

10. In the accompanying diagram of $\triangle PRT$, $\overline{KG} \parallel \overline{PR}$. If $TP = 20$, $KP = 4$, and $GR = 7$, what is the length of $\overline{TG}$?

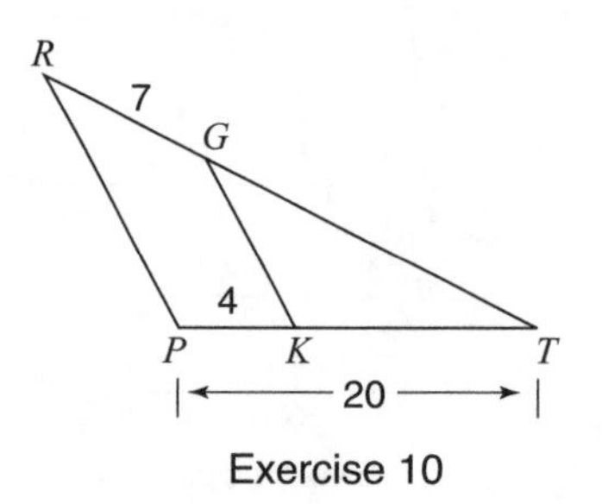

Exercise 10

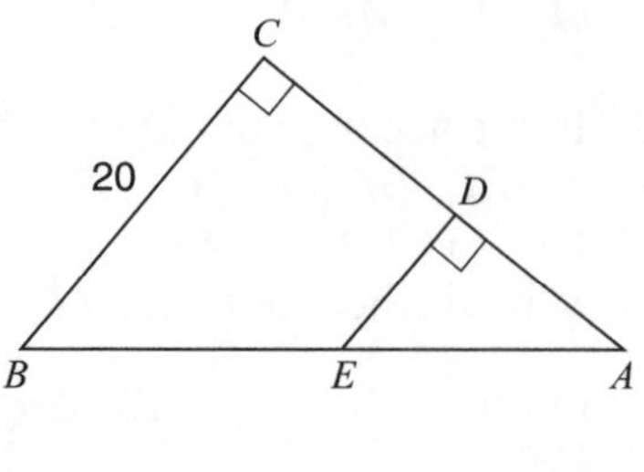

Exercise 11

11. In the accompanying diagram of right triangle ACB, C is a right angle and $\overline{ED} \perp \overline{AC}$. The length of $\overline{BC}$ is 20, CD is 3 less than AD, and DE is 3 more than AD. Find the length of $\overline{AD}$.

12. In the accompanying diagram, $\overline{PQ} \parallel \overline{RT}$, $QT = 3$, and $ST = 8$, $RT = 4$ and $PS = 7$, what is the perimeter of trapezoid $TRPQ$?

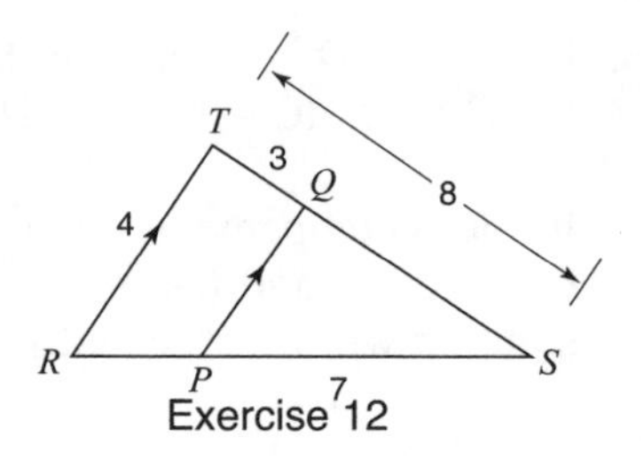

Exercise 12

C. Write a proof.

13. Given: $\dfrac{AC}{EC} = \dfrac{BC}{DC}$.

Prove: $\angle B \cong \angle D$.

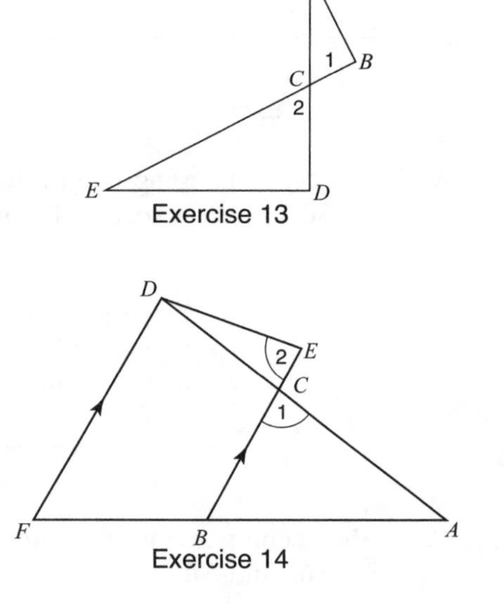

Exercise 13

Exercise 14

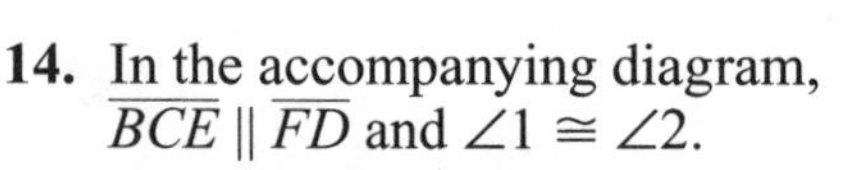

14. In the accompanying diagram, $\overline{BCE} \parallel \overline{FD}$ and $\angle 1 \cong \angle 2$.

Prove: $\dfrac{BF}{AF} = \dfrac{ED}{AD}$.

15. In the accompanying diagram, points R, S, and T are the midpoints of $\overline{JK}$, $\overline{KL}$, and $\overline{LJ}$, respectively. Write an informal proof that shows that $\angle 1 \cong \angle 2$.

16. In the accompanying diagram, $\overline{TK} \perp \overline{RS}$, $\overline{KH} \perp \overline{RT}$, $RH = 3$, $HT = 8$, $KH = 6$, and $SK = 5$. Write an explanation or an informal proof that shows that $\triangle RHK \sim \triangle SKT$.

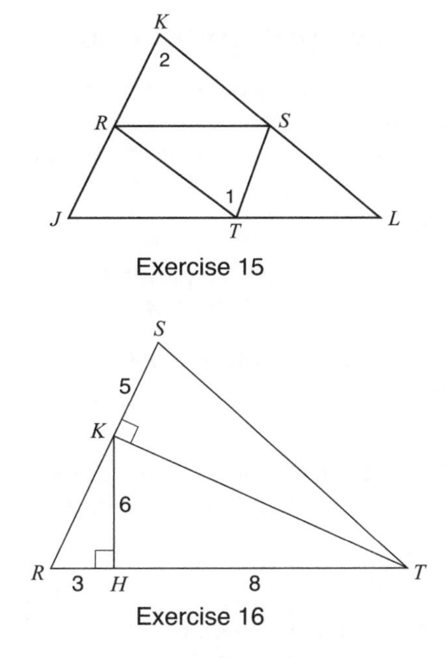

Exercise 15

Exercise 16

5.5 PROVING PRODUCTS OF SEGMENTS EQUAL

KEY IDEAS

To prove that the product of the lengths of two segments is equal to the product of the lengths of another pair of segments, work backwards from the two products to determine the pair of triangles that contain these sides. Then prove these triangles are similar.

Forming a Proportion from a Product

Since $\frac{3}{6} = \frac{1}{2}$, $6 \times 1 = 3 \times 2$. Starting with the product $6 \times 1 = 3 \times 2$, you can figure out an equivalent proportion by making the factors of either of the two products the means and the factors of the other product the extremes:

$$\frac{3}{6} = \frac{1}{2} \quad \text{or} \quad \frac{6}{3} = \frac{2}{1}$$

Proving Products Equal

To prove the products of two pairs of triangle side lengths are equal, write an equivalent proportion. From the proportion, determine the triangles that must be proved similar. Then prove those triangles similar in the usual way.

Example 1

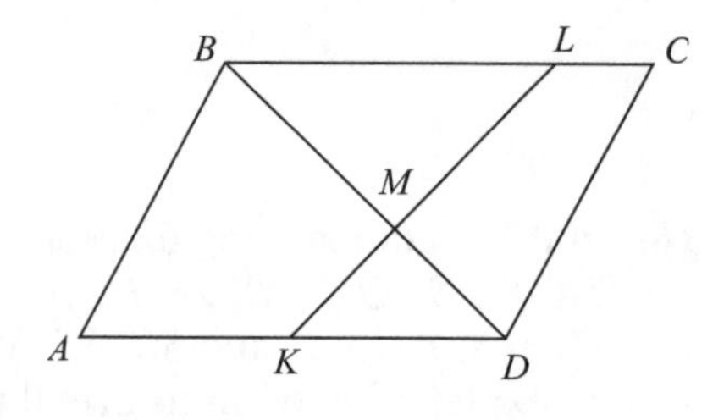

Given: $ABCD$ is a parallelogram.
Prove: $KM \times LB = LM \times KD$.

Solution: Make a plan.

- From the product $KM \times LB = LM \times KD$, form an equivalent proportion by making KM and LB the extremes, and LM and KD the means:

$$\frac{KM}{LM} = \frac{KD}{LB}$$

- Read across the proportion to determine the pair of triangles that you need to prove are similar:

$$\triangle KMD$$

$$\frac{KM}{LM} = \frac{KD}{LB}$$

$$\triangle LMB$$

- Mark off on the diagram the two pairs of congruent angles that can be used to prove the two triangles similar.

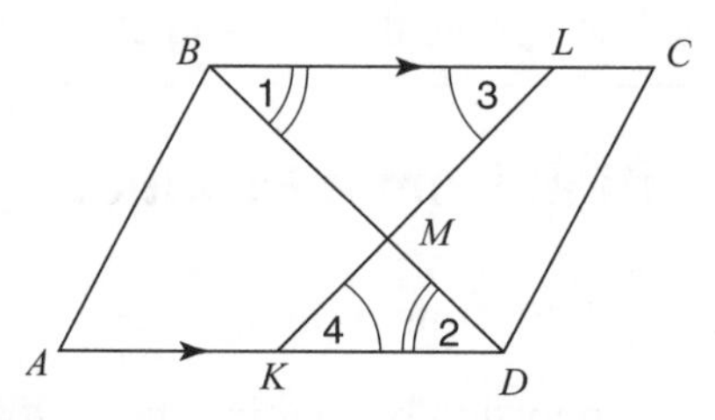

- Write the proof by reversing the steps taken in developing your plan: Prove the pair of triangles similar, form the required proportion, and then write the product you were asked to prove.

Proof

Statement	Reason
1. $ABCD$ is a parallelogram.	1. Given.
2. $\overline{AD} \parallel \overline{BC}$.	2. Opposite sides of a parallelogram are parallel.
3. $\angle 1 \cong \angle 2$ and $\angle 3 \cong \angle 4$.	3. If two lines are parallel, then alternate interior angles are congruent.
4. $\triangle KMD \sim \triangle LMB$.	4. AA postulate of similarity.
5. $\frac{KM}{LM} = \frac{KD}{LB}$.	5. The lengths of corresponding sides of similar triangles are in proportion.
6. $KM \times LB = LM \times KD$.	6. In a proportion, the product of the means is equal to the product of the extremes.

MATH FACTS

In developing the plan for the proof in Example 1, if you had formed the proportion $\frac{KM}{KD} = \frac{LM}{LB}$, reading across the top (K–M–L) would not give the vertices of a triangle. When this happens, try reading down each ratio rather than across.

Check Your Understanding of Section 5.5

Write a proof.

1. Given: $\overline{AF}$ bisects $\angle BAC$,
$\overline{BH}$ bisects $\angle ABC$,
$\overline{BC} \cong \overline{AC}$.
Prove: $AH \times EF = BF \times EH$.

2. Given: $\overline{XY} \parallel \overline{LK}$, $\overline{XZ} \parallel \overline{JK}$.
Prove: $JY \times ZL = XZ \times KZ$.

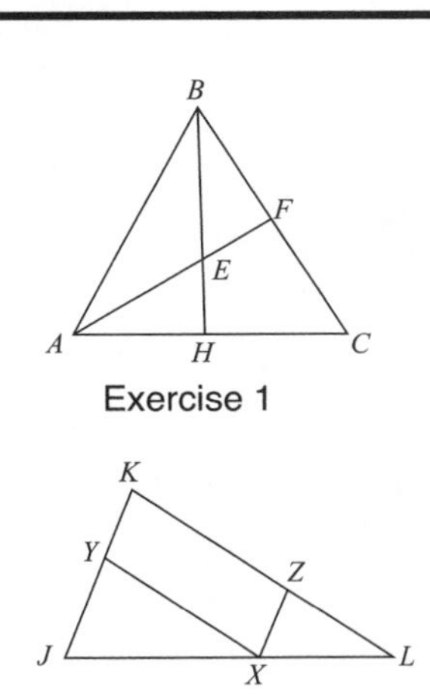

3. Given: $\overline{EF}$ is the median of trapezoid $ABCD$.
Prove: $EI \times GH = IH \times EF$.

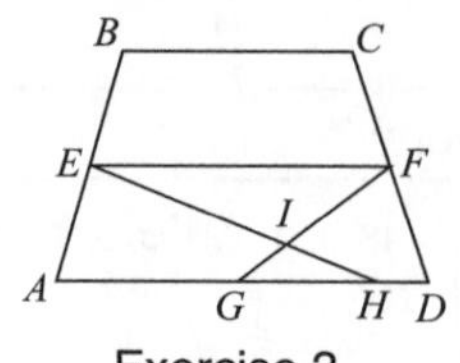

Exercise 3

4. Given: V is a point on $\overline{ST}$ such that $\overline{RVW}$ bisects $\angle SRT$, $\overline{TW} \cong \overline{TV}$.
Prove: $RW \times SV = RV \times TW$.

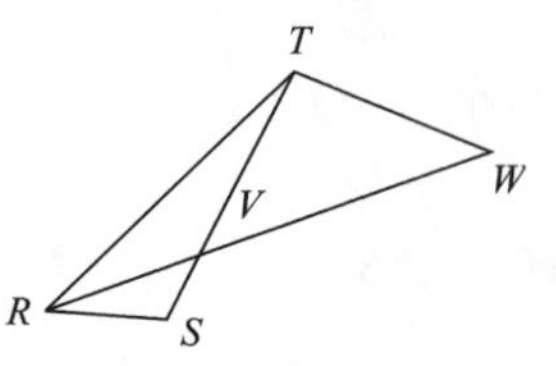

Exercise 4

5. Given: $\triangle ABC$ with $\overline{CDA}$, $\overline{CEB}$, $\overline{AFB}$, $\overline{DE} \parallel \overline{AB}$, $\overline{EF} \parallel \overline{AC}$, $\overline{CF}$ intersects $\overline{DE}$ at G.
Prove: a. $\triangle CAF \sim \triangle FEG$.
b. $DG \times GF = EG \times GC$

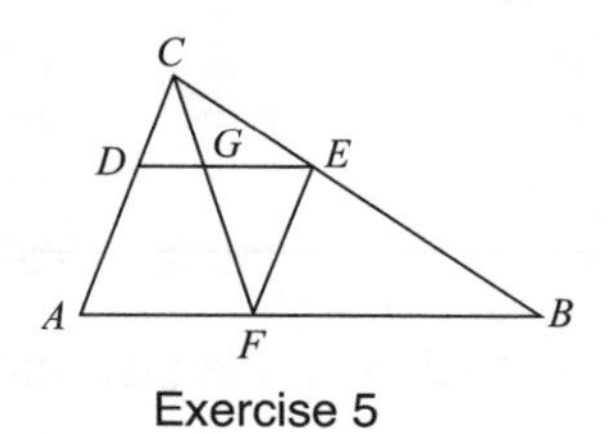

Exercise 5

6. Rosalie thinks she discovered a new theorem. She claims that "The product of the lengths of the legs of a right triangle is always equal to the product of the lengths of the hypotenuse and the altitude drawn to the hypotenuse." Prove or disprove Rosalie's theorem.

CHAPTER 6

RIGHT TRIANGLES AND TRIGONOMETRY

6.1 PROPORTIONS IN A RIGHT TRIANGLE

KEY IDEAS

Drawing an altitude to the hypotenuse of a right triangle creates pairs of similar right triangles from which useful proportions can be derived.

Mean Proportional

If the means of a proportion are the same, as in $\frac{8}{4} = \frac{4}{2}$, then either mean is called the **mean proportional** or **geometric mean** between the other two numbers. Thus, 4 is the *mean proportional* between 8 and 2. Equivalently, 4 is the *geometric mean* of 8 and 2.

Example 1

Find the mean proportional between:
a. 2 and 50 b. $6a$ and $8a^3$

Solution: Let x represent the mean proportional between the given pair of numbers.

a.
$$\frac{4}{x} = \frac{x}{25}$$
$$x \cdot x = 4 \cdot 25$$
$$x^2 = 100$$
$$x = \sqrt{100} = \mathbf{10}.$$

b.
$$\frac{6a}{x} = \frac{x}{8a^3}$$
$$x^2 = 48a^4$$
$$x = \sqrt{48} \cdot \sqrt{a^4}$$
$$= \sqrt{16} \cdot \sqrt{3} \cdot a^2$$
$$= \mathbf{4\sqrt{3}\,a^2}$$

Right Triangle Proportions

The altitude drawn to the hypotenuse of a right triangle separates it into two other right triangles that are similar to each other and to the original right triangle. In the accompanying figure of right triangle ABC,

- $\Delta\text{I} \sim \Delta ABC$ since the two right triangles have $\angle B$ in common.
- $\Delta\text{II} \sim \Delta ABC$ as the two right triangles have $\angle A$ in common.
- $\Delta\text{I} \sim \Delta\text{II}$ because two triangles similar to the same triangle are similar to each other (Transitive Property of Similarity).

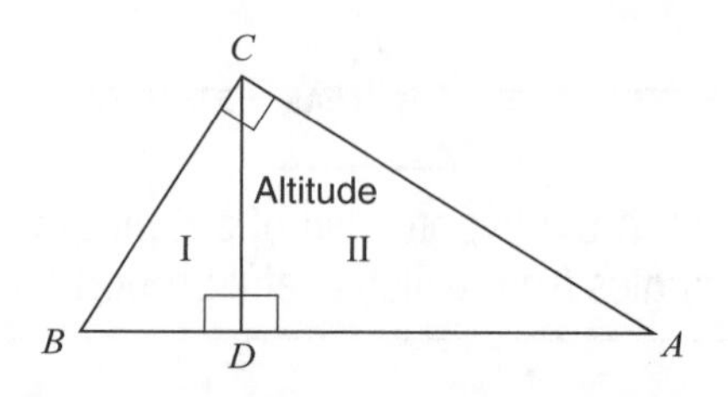

> **Theorem: Altitude–Hypotenuse Theorem**
> If the altitude is drawn to the hypotenuse of a right triangle, then the two right triangles that are formed are similar to each other and to the original right triangle.

Two important corollaries of this theorem tell how the segments formed on the hypotenuse are related to the legs and to the altitude on the hypotenuse.

Corollary	Proportions	Figure
Corollary 1: The altitude to the hypotenuse of a right triangle divides the hypotenuse so that either leg is the mean proportional between the hypotenuse and the segment of the hypotenuse adjacent to that leg.	$\frac{x}{a} = \frac{a}{c}$ and $\frac{y}{b} = \frac{b}{c}$	C, a, b, h, x, y, B, D, A, c
Corollary 2: The altitude to the hypotenuse of a right triangle is the mean proportional between the two segments along the hypotenuse.	$\frac{x}{h} = \frac{h}{y}$	

Example 2

In the accompanying diagram of right triangle ABC, $\overline{CD}$ is the altitude to hypotenuse $\overline{AB}$. Find the values of x and y.

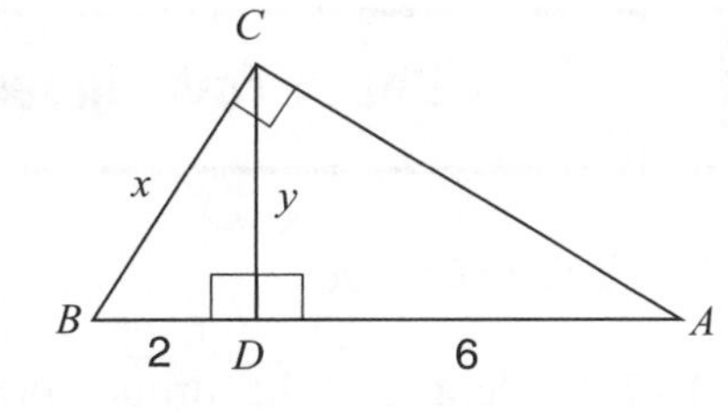

Solution: The length of hypotenuse $\overline{AB}$ is $2 + 6 = 8$.

Use Corollary 1 to find x:

$$\frac{BD}{x} = \frac{x}{AB}$$

$$\frac{2}{x} = \frac{x}{8}$$

$$x^2 = 16$$

$$x = \sqrt{16} = \mathbf{4}$$

Use Corollary 2 to find y:

$$\frac{BD}{y} = \frac{y}{AD}$$

$$\frac{2}{x} = \frac{x}{6}$$

$$x^2 = 12$$

$$x = \sqrt{4 \cdot 3} = \mathbf{2\sqrt{3}}$$

Example 3

In the accompanying diagram of right triangle JKL, $\overline{KN}$ is the altitude to hypotenuse $\overline{LJ}$. Find the values of x and y.

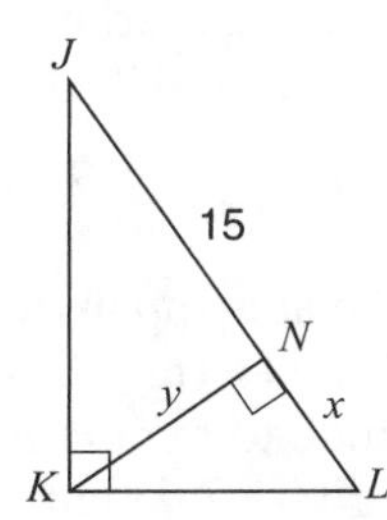

Solution: The length of hypotenuse $\overline{LJ}$ is represented by $x + 15$.

Use Corollary 1 to find x:

$$\frac{LN}{10} = \frac{10}{LJ}$$

$$\frac{x}{10} = \frac{10}{x+15}$$

$$x(x + 15) = 100$$

$$x^2 + 15x - 100 = 0$$

$$(x - 5)(x + 20) = 0$$

$$x - 5 = 0 \quad \text{or} \quad x + 20 = 0$$

$$\mathbf{x = 5} \quad \text{or} \quad x = -20 \leftarrow \text{Reject since } x \text{ must be positive}$$

Use Corollary 2 to find y:

$$\frac{LN}{y} = \frac{y}{NJ}$$

$$\frac{5}{x} = \frac{x}{15}$$

$$x^2 = 75$$

$$x = \sqrt{25 \cdot 3}$$

$$= \mathbf{5\sqrt{3}}$$

Check Your Understanding of Section 6.1

A. Multiple Choice

1. If the length of the altitude drawn to the hypotenuse of a right triangle is 10 inches, the number of inches in the lengths of the segments of the hypotenuse may be
(1) 5 and 20 (2) 2 and 5 (3) 3 and 7 (4) 50 and 50

2. In the accompanying diagram, $\triangle FUN$ is a right triangle, $\overline{UR}$ is the altitude to hypotenuse $\overline{FN}$, $UR = 12$, and the lengths of $\overline{FR}$ and $\overline{FN}$ are in the ratio 1:10. What is the length of $\overline{FR}$?

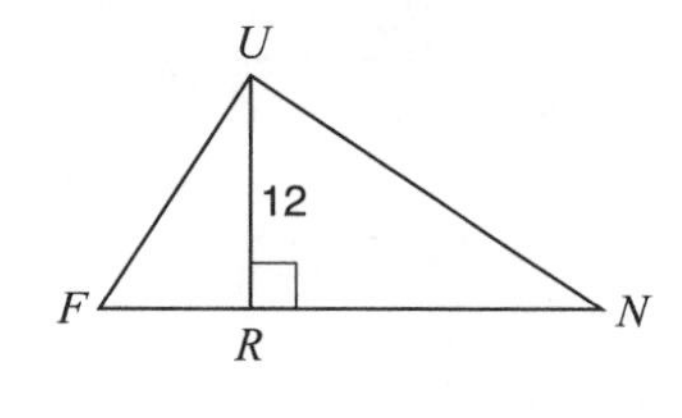

(1) 1 (3) 36
(2) 1.2 (4) 4

3. In the right triangle ABC, $m\angle C = 90$ and altitude $\overline{CD}$ is drawn to hypotenuse $\overline{AB}$. If $AD = 4$ and $DB = 5$, what is AC?

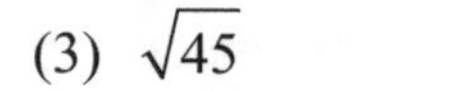

(1) $\sqrt{20}$ (2) 6 (3) $\sqrt{45}$ (4) 9

4. In the accompanying diagram, $\triangle RST$ is a right triangle, $\overline{SU}$ is the altitude to hypotenuse $\overline{RT}$. $RT = 16$, and $RU = 7$. What is the length of $\overline{ST}$?

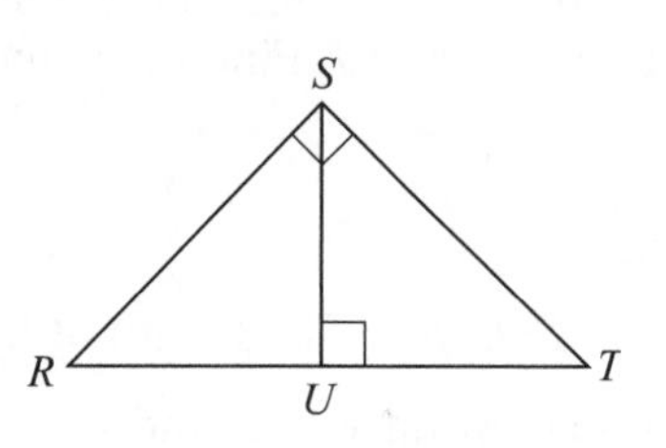

(1) $3\sqrt{7}$ (3) 9
(2) $4\sqrt{7}$ (4) 12

5. The altitude drawn to the hypotenuse of a right triangle divides the hypotenuse into segments of lengths 4 and 12. The length of the shorter leg of the right triangle is
(1) 8 (2) $\sqrt{20}$ (3) $\sqrt{48}$ (4) $\sqrt{192}$

6. What is the geometric mean of $\frac{1}{3}x$ and $27x^3$ when $x \neq 0$?
(1) $3x^2$ (2) $9x$ (3) $3x$ (4) $9x^2$

7. In the accompanying diagram, $\triangle RST$ is a right triangle, $\overline{SU}$ is the altitude to hypotenuse $\overline{RT}$, $RS = 8$, and the ratio of RU to UT is 1 to 3. What is the length of $\overline{RT}$?
(1) 16 (3) 24
(2) 20 (4) 32

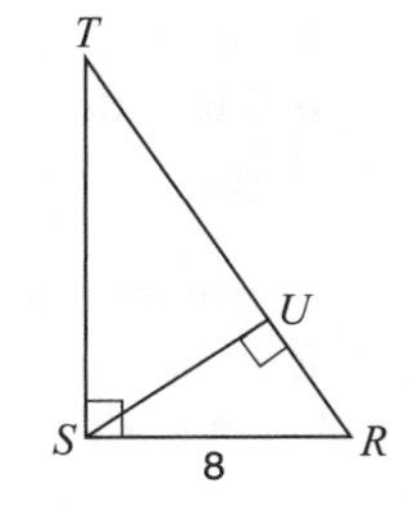

B. *Show or explain how you arrived at your answer.*

8. The altitude drawn to the hypotenuse of a right triangle divides the hypotenuse into two segments whose lengths are in the ratio of 1 to 4. If the length of the altitude is 8, find the length of the longer leg of the triangle.

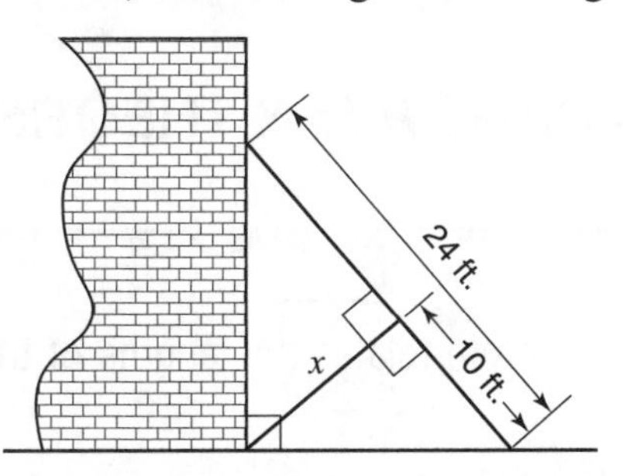

9. The accompanying diagram shows a 24-foot ladder leaning against a building. A steel brace extends from the ladder to the point where the building meets the ground. The brace forms a right angle with the ladder. If the steel brace is connected to the ladder at a point that is 10 feet from the foot of the ladder, find to the nearest tenth of a foot the length, x, of the steel brace.

10. In right triangle JKL, $\angle K$ is a right angle. Altitude $\overline{KH}$ intersects the hypotenuse at H in such a way that JH exceeds HL by 5. If $KH = 6$, find the length of the hypotenuse.

11. In right triangle ABC, the length of altitude $\overline{CD}$ to hypotenuse $\overline{AB}$ is 12. If the length of the longer segment of the hypotenuse exceeds the length of the shorter segment of the hypotenuse by 7, find the length of the hypotenuse.

12. In right triangle ABC, altitude $\overline{CD}$ is drawn to hypotenuse $\overline{AB}$. If AB is four times as great as AD and AC is 3 more than AD, find the length of altitude $\overline{CD}$.

13. In right triangle ABC, altitude $\overline{CD}$ is drawn to hypotenuse $\overline{AB}$, $AD = 12$, and DB is 3 less than CD. Find, in simplest radical form, the perimeter of triangle ABC.

14. In the accompanying diagram of right triangle RST, altitude $\overline{YS}$ is drawn to hypotenuse $\overline{RT}$, $RT = 20$, $TY < YR$, and $YS = 8$.

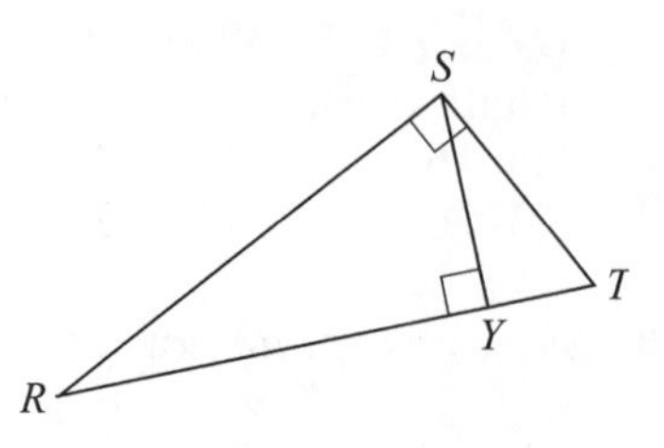

a. If x is used to represent the length of $\overline{TY}$, write a proportion that could be used to find x. Solve the proportion algebraically for x.

b. Find the length of $\overline{ST}$ in simplest radical form.

6.2 THE PYTHAGOREAN THEOREM

The Pythagorean Theorem relates the lengths of the three sides of a right triangle:

$$(\text{Leg } 1)^2 + (\text{Leg } 2)^2 = (\text{Hypotenuse})^2$$

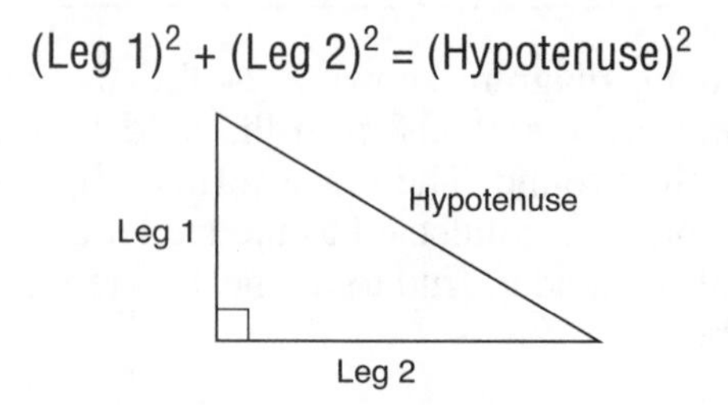

The Pythagorean Theorem can be proved using the proportions derived from the Altitude-Hypotenuse Theorem.

A Proof of the Pythagorean Theorem

Although there is more than one way of proving the Pythagorean Theorem, here is a proof that uses proportions in a right triangle.

Theorem: Pythagorean Theorem
In a right triangle, the sum of the squares of the lengths of the legs is equal to the square of the length of the hypotenuse ($a^2 + b^2 = c^2$).

Proof

Given: $\triangle ABC$, $\angle C$ is a right angle,
c is the length of the hypotenuse,
a and b are the lengths of the legs.
Prove: $a^2 + b^2 = c^2$.

Draw altitude $\overline{CD}$, and let $BD = m$, so $AD = c - m$, as shown in the accompanying diagram. Then apply Corollary 1:

$$\frac{m}{a} = \frac{a}{c} \quad \text{so} \quad a^2 = +\,mc$$

$$\frac{c-m}{b} = \frac{b}{c} \quad \text{so} \quad b^2 = c^2 - mc$$

Add the two equations.

$$a^2 + b^2 = c^2 \quad \leftarrow \text{Pythagorean Theorem}$$

Applying the Pythagorean Theorem

The Pythagorean Theorem can be used to find the length of *any* side of a right triangle when the lengths of the other two sides are known:

If $a = 3$ and $b = 5$, then

$$a^2 + b^2 = c^2$$
$$3^2 + 5^2 = c^2$$
$$9 + 25 = c^2$$
$$c^2 = 34$$
$$c = \sqrt{34}$$

If $a = \sqrt{13}$ and $c = 7$, then

$$a^2 + b^2 = c^2$$
$$(\sqrt{13})^2 + b^2 = 7^2$$
$$13 + b^2 = 49$$
$$b^2 = 49 - 13$$
$$b = \sqrt{36} = 6$$

Example 1

Ray wants to build a square garden in which the distance between opposite corners is *at least* 18.0 feet. What is the shortest possible side length of the square garden correct to the *nearest tenth of a foo*t ?

Solution: Use the Pythagorean Theorem where x represents the length of a side of the square garden:

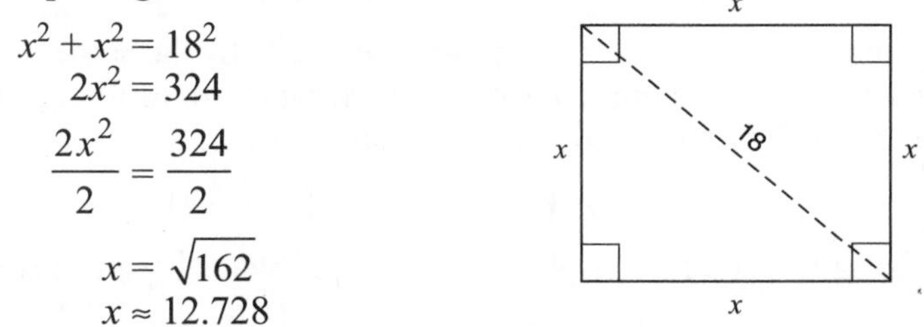

$$x^2 + x^2 = 18^2$$
$$2x^2 = 324$$
$$\frac{2x^2}{2} = \frac{324}{2}$$
$$x = \sqrt{162}$$
$$x \approx 12.728$$

Round up from 12.728 to **12.8 feet** in order for the diagonal length to be *at least* 18.0 feet.

Example 2

Katie hikes 5 miles north, 7 miles east, and then 3 miles north again. *To the nearest tenth of a mile*, how far, in a straight line, is Katie from her starting point?

Solution: The four key points on Katie's trip are labeled A through D in the accompanying diagram.

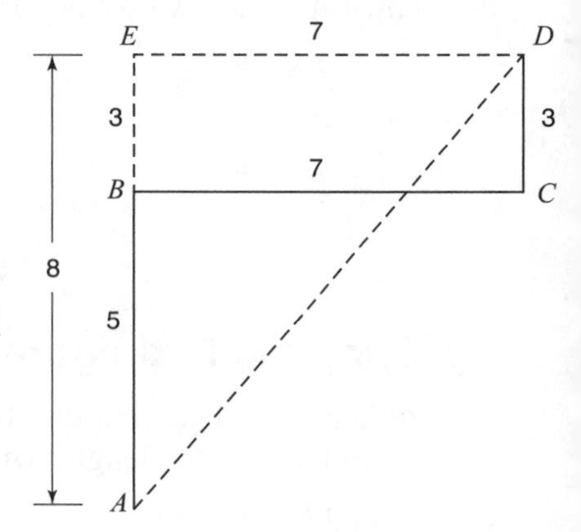

- To determine how far, in a straight line, Katie is from her starting point at A, find the length of $\overline{AD}$.
- Form a right triangle in which $\overline{AD}$ is the hypotenuse by completing rectangle $BCDE$ as shown in the accompanying diagram.
- Because opposite sides of a rectangle have the same length, $ED = BC = 7$, and $BE = CD = 3$. Thus, $AE = 5 + 3 = 8$.
- Since AD is the hypotenuse in right triangle AED:

$$
\begin{aligned}
(AD)^2 &= (AF)^2 + (FD)^2 \\
&= 8^2 + 7^2 \\
&= 64 + 4 \\
&= 113 \\
AD &= \sqrt{113} \approx 10.63
\end{aligned}
$$

Correct to the *nearest tenth of a mile*, Katie is **10.6** miles from her starting point.

Pythagorean Triples

A **Pythagorean triple** is a set of three positive integers $\{x, y, z\}$ that satisfy the relationship $x^2 + y^2 = z^2$. Here are some commonly encountered Pythagorean triples that you should memorize:

{3, 4, 5}, **{5, 12, 13}**, and **{8, 15, 17}**

Multiplying each member of a Pythagorean triple by the same whole number produces another Pythagorean triple. For example, multiplying each member of $\{3, 4, 5\}$ by 2 forms a 6–8–10 Pythagorean triple:

$$\{\underline{3} \times 4, \underline{4} \times 2, \underline{5} \times 2\} = \{6, 8, 10\}$$

Recognizing Pythagorean triples can make problems easier to solve.

Example 3

A rectangular yard measures 12 yards by 16 yards. What is the distance from one corner of the yard to the opposite corner?

Solution 1: Draw rectangle *ABCD*. The lengths of the sides of right triangle *ADC* are a multiple of the basic 3–4–5 Pythagorean triple since $12 = \underline{3} \times 4$ and $16 = \underline{4} \times 4$ so $AC = \underline{5} \times 4 = 20$.

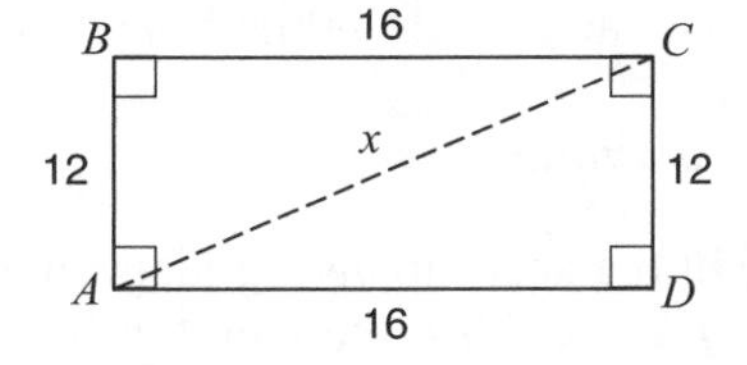

Solution 2: If you did not recognize that a Pythagorean triple was involved, use the Pythagorean theorem to find the length of leg $\overline{AB}$:

$$12^2 + 16^2 = (AC)^2$$
$$144 + 256 = (AC)^2$$
$$(AC)^2 = 400$$
$$AC = \sqrt{400} = 20$$

Area Applications

You may need to use the Pythagorean Theorem or a Pythagorean triple to find the length of a segment required in an area formula. A summary of area formulas is included in the section at the back of the book titled, "Some Geometric Relationships Worth Remembering."

Example 4

In the accompanying figure of rhombus *MATH*, diagonals $\overline{TM}$ and $\overline{AH}$ intersect at *K*, $AK = 6$, and each side measures 10. Find the area of rhombus *MATH*.

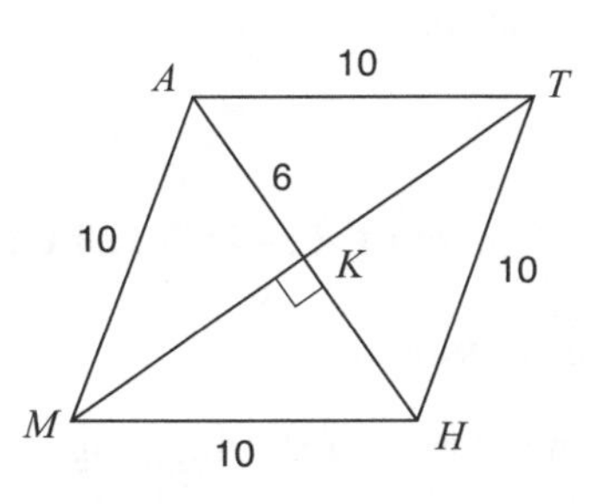

Solution: The area, *A*, of a rhombus is given by the formula,

$$A = \frac{1}{2}d_1d_2$$

where d_1 and d_2 are the lengths of its diagonals.

- The length of the sides of right triangle *MKA* form a 6–***8***–10 Pythagorean triple where $MK = 8$.
- Since the diagonals of a rhombus bisect each other,

$$d_1 = AH = 6 + 6 = 12 \text{ and } d_2 = MT = 8 + 8 = 16$$

- Use the area formula $A = \frac{1}{2}d_1d_2$ where $d_1 = 12$ and $d_2 = 16$:

$$A = \frac{1}{2}(12)(16) = 96$$

The area of rhombus *MATH* is **96** square units.

Example 5

In the accompanying diagram of isosceles triangle *JAM*, the length of base $\overline{JM}$ is 16 cm. If the perimeter of $\triangle JAM$ is 50 cm, find the area of the triangle.

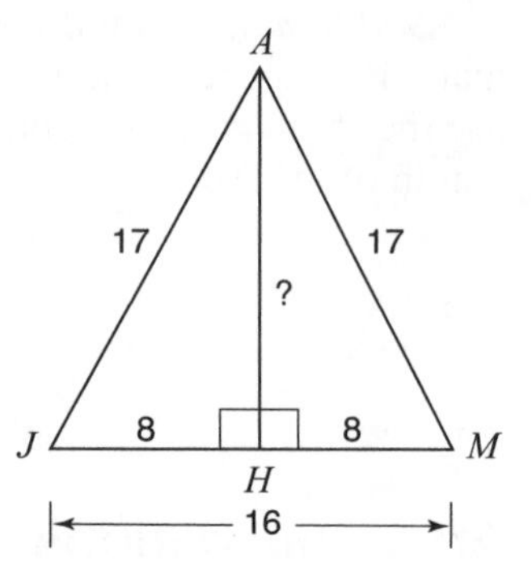

Solution: The sum of the lengths of the three sides is 50 cm, so the lengths of the two congruent legs add up to 50 – 16 = 34 cm. Therefore, the length of each leg is 17 cm. The altitude drawn to the base bisects the base so that an 8–***15***–17 right triangle is formed where *AH* =15. If you did not recognize that a Pythagorean triple was involved, you could have used the Pythagorean Theorem to find the unknown length.

$$\begin{aligned}\text{Area of } \triangle JAM &= \frac{1}{2} \times \text{base} \times \text{height}\\ &= \frac{1}{2}(16 \text{ cm} \times 15 \text{ cm})\\ &= 120 \text{ cm}^2\end{aligned}$$

The area of $\triangle JAM$ is **120 cm²**.

Example 6

In the accompanying diagram of isosceles trapezoid *TRAP*, $\overline{RA} \parallel \overline{TP}$, *RA* = 4, *TP* = 28, and *RT* = 13. What is the area of trapezoid *TRAP*?

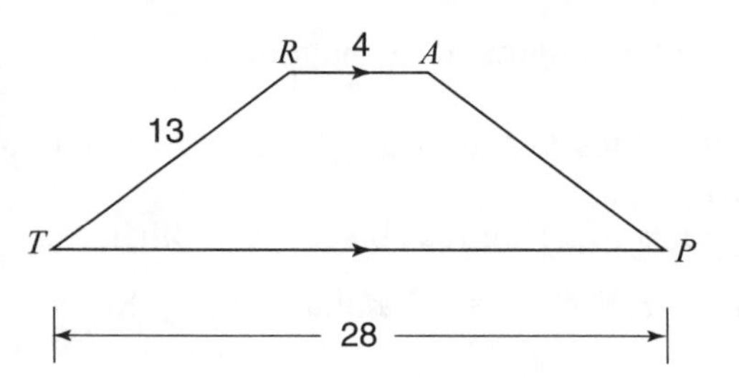

Solution: The area, A, of a trapezoid is given by the formula

$$A = \frac{1}{2}h(b_1 + b_2)$$

where b_1 and b_2 represent the lengths of the two bases and h is the altitude.

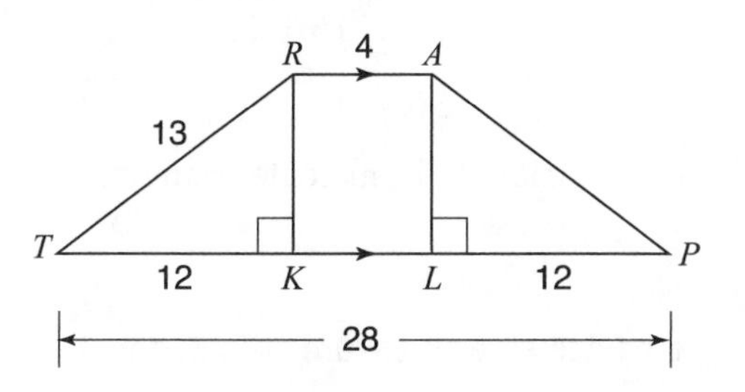

- Draw altitudes $\overline{RK}$ and $\overline{AL}$ thereby forming rectangle *RALK.* Since $KL = RA = 4$, $TK + LP = 28 - 4 = 24$.
- Because $\triangle TKR \cong \triangle PLA$, $TK = LP = \frac{1}{2}(24) = 12$.
- The lengths of the sides of right triangle *TKR* form a **5**–12–13 Pythagorean triple where altitude $RK = 5$.
- Use the area formula $A = \frac{1}{2}h(b_1 + b_2)$, where $h = 5$, $b_1 = 4$, and $b_2 = 28$:

$$\begin{aligned} A &= \frac{1}{2}(5)(4 + 28) \\ &= \frac{1}{2}(5)(32) \\ &= 80 \end{aligned}$$

The area of trapezoid *TRAP* is **80** square units.

Pythagorean Inequalities

You can tell whether a triangle is acute, right, or obtuse by comparing the squares of the lengths of its sides.

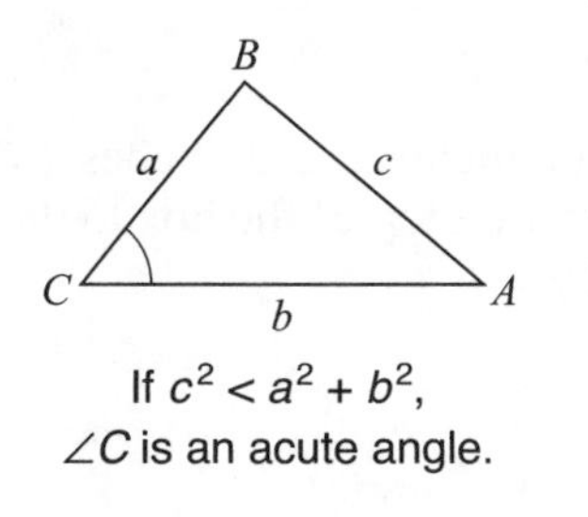

If $c^2 < a^2 + b^2$, $\angle C$ is an acute angle.

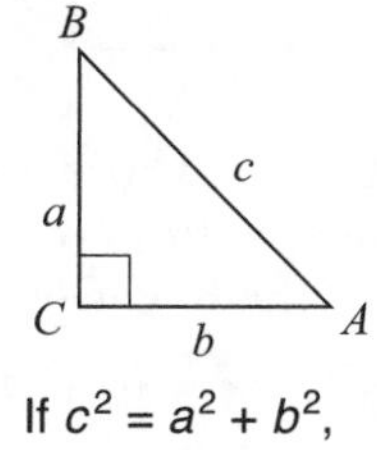

If $c^2 = a^2 + b^2$, $\angle C$ is a right angle.

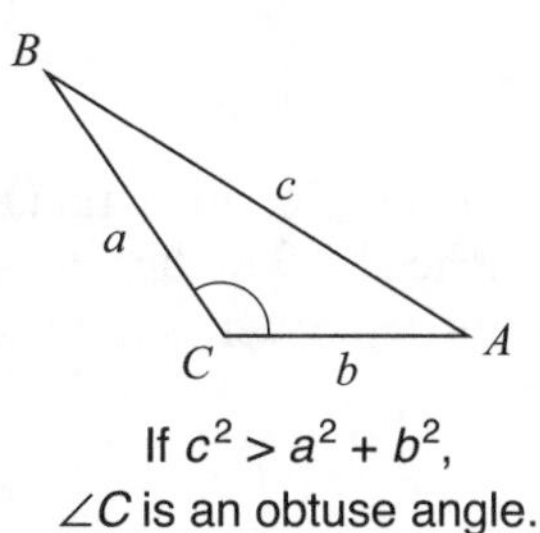

If $c^2 > a^2 + b^2$, $\angle C$ is an obtuse angle.

Given the lengths of a side of a triangle are 4, 5, and 6. Compare the square of the largest of the three numbers to the sum of the squares of the other two numbers:

$$6^2 \ \boxed{?} \ 4^2 + 5^2$$

$$6^2 \ \boxed{?} \ 16 + 25$$

$$36 < 41$$

Because $c^2 < a^2 + b^2$, the triangle is an acute triangle.

***Example* 7**

Classify the triangle whose sides measure $\sqrt{7}$, 3, and 4.

Solution: Compare the square of the largest of the three numbers to the sum of the squares of the other two numbers:

$$4^2 \ \boxed{?} \ (\sqrt{7})^2 + 3^2$$

$$16 \ \boxed{?} \ 7 \quad + 9$$

$$16 = 16$$

Because $c^2 = a^2 + b^2$, the triangle is a **right triangle**.

The Pythagorean Theorem In Rectangular Solids

The Pythagorean relationship may be extended to rectangular solids. In the accompanying figure,

$$x^2 + y^2 + z^2 = d^2$$

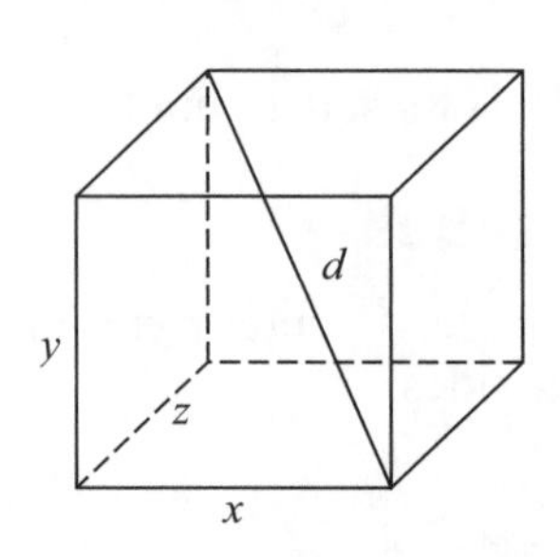

For example, in a rectangular box whose measurements are 5 inches by 6 inches by 9 inches, the distance, d, from one corner on top of the box to the opposite corner at the bottom of the box is given by

$$5^2 + 6^2 + 9^2 = d^2$$

so $25 + 36 + 81 = d^2$ and $d = \sqrt{142}$.

Check Your Understanding of Section 6.2

A. Multiple Choice

1. The lengths of the two legs of a right triangle are 2 and $\sqrt{3}$. What is the length of the hypotenuse?

(1) $4\sqrt{3}$ (2) 10 (3) $\sqrt{14}$ (4) 4

2. The length of the hypotenuse of a right triangle is $\sqrt{15}$ and the length of one leg is 3. What is the length of the other leg?

(1) 6 (2) 9 (3) $\sqrt{6}$ (4) $3\sqrt{2}$

3. What is the length of a diagonal of a square whose perimeter is 32?

(1) $\sqrt{2}$ (2) $2\sqrt{2}$ (3) $4\sqrt{2}$ (4) $8\sqrt{2}$

4. Which set of numbers do *not* represent the lengths of the sides of a right triangle?

(1) $\{9, 40, 41\}$ (3) $\{\sqrt{12}, 6, 7\}$

(2) $\{2, 4, 2\sqrt{5}\}$ (4) $\{1, 1, \sqrt{2}\}$

5. What is the length of a diagonal of a rectangle in which the lengths of two adjacent sides are $\sqrt{13}$ and 6?

(1) 7 (2) 19 (3) $5\sqrt{3}$ (4) $\sqrt{23}$

6. If the lengths of the diagonals of a rhombus are 6 and 8, the perimeter of the rhombus is

(1) 14 (2) 20 (3) 28 (4) 40

7. In rectangle *MATH*, $AT = 8$ and $TH = 12$. What is the length of diagonal $\overline{HA}$ to the *nearest tenth*?

(1) 14.4 (2) 11.7 (3) 8.9 (4) 7.2

8. What is the length of the altitude drawn to a side of an equilateral triangle whose perimeter is 60 cm?

(1) 10 cm (2) $10\sqrt{2}$ cm (3) $10\sqrt{3}$ cm (4) $10\sqrt{5}$ cm

9. A carpenter is building a rectangular deck with dimensions of 16 feet by 30 feet. To ensure that the adjacent sides form 90° angles, what should each diagonal measure?
(1) 26 ft (2) 30 ft (3) 34 ft (4) 46 ft

10. At 9:00 A.M. a car starts at point A and travels north for 1 hour at an average rate of 60 miles per hour. Without stopping, the car then travels east for 2 hours at an average rate of 45 mile per hour. At 12:00 P.M., what is the best approximation of the distance, in miles, of the car from point A?
(1) 100 (2) 105 (3) 108 (4) 115

11. If the length of each leg of an isosceles triangle is 17 and the base is 16, the length of the altitude to the base is
(1) 8 (2) $8\frac{1}{2}$ (3) 15 (4) $\sqrt{32}$

12. The lengths of the bases of an isosceles trapezoid are 6 centimeters and 12 centimeters. If the length of each leg is 5 centimeters, what is the area of the trapezoid?
(1) 18 cm^2 (2) 36 cm^2 (3) 45 cm^2 (4) 90 cm^2

B. *Show or explain how you arrived at your answer.*

13. A baseball diamond is in the shape of a square with a side length of 90 feet. What is the distance from home plate to second base, correct to the *nearest tenth of a foot*?

14. The cross section of an attic is in the shape of an isosceles trapezoid, as shown in the accompanying figure. If $AB = CD = 25$ feet, $BC = 20$ feet, and $AD = 68$ feet, what is the area of the cross section?

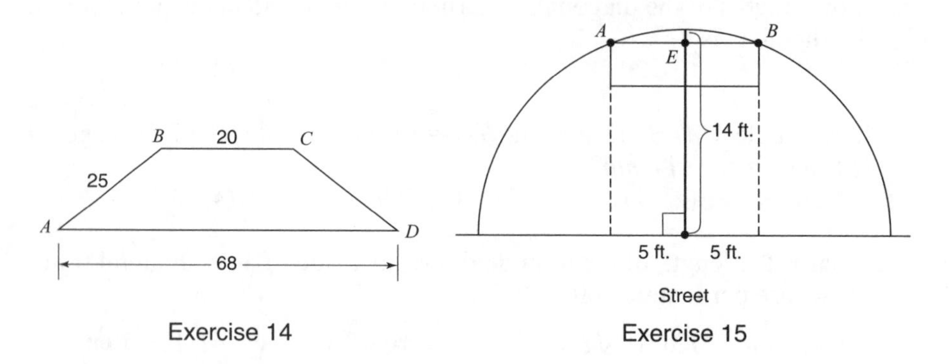

Exercise 14

Exercise 15

15. The accompanying diagram shows a semicircular arch over a street that has a radius of 14 feet. A banner is attached to the arch at points A and B in such a way that $AE = EB = 5$ feet. How many feet above the ground are these points of attachment for the banner? Estimate to the *nearest tenth of a foot*.

16. The perimeter of a rhombus is 100 centimeters and the length of the longer diagonal is 48 centimeters. Find the area of the rhombus.

17. The length and width of a rectangle are in the ratio of 3:4. If the length of the diagonal of the rectangle is 60, what are the length and width of the rectangle?

18. Two hikers started at the same location. One traveled 2 miles east and then 1 mile north. The other traveled 1 mile west and then 3 miles south. At the end of their hikes, how many miles apart were the two hikers?

19. To get from his high school to his home, Jamal travels 5.0 miles east and then 4.0 miles north. When Sheila goes to her home from the same high school, she travels 8.0 miles east and 2.0 miles south. What is the shortest distance, to the *nearest tenth* of a mile, between Jamal's home and Sheila's home?

20. In the accompanying diagram of right triangles ABD and DBC, $AB = 5$, $AD = 4$, and $CD = 1$. Find the length of $\overline{BC}$, to the *nearest tenth*.

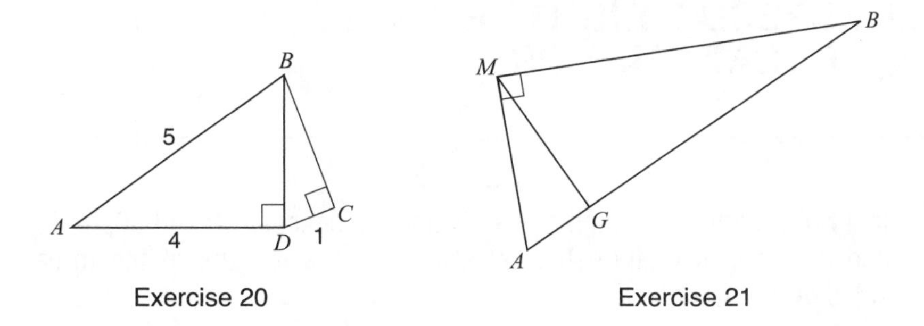

Exercise 20

Exercise 21

21. Town A is 8 miles from town C, town B is 15 miles from town C, and angle ACB is a right angle. On the straight road that connects towns A and B, a restaurant will be built at the point that is closest to town C. To the *nearest tenth of a mile*, find

a. The distance from town A to the restaurant
b. The distance from town C to the restaurant

22. Exits A and B are on a straight highway. A motel M is located off the highway such that $\angle AMB$ is a right angle. A gas station is to be built along the highway at a point G between A and B. Point G is chosen so that its distance from the motel is as short as possible. If $MG = 20$ miles and $AM = 25$ miles, how far is exit A from exit B?

23. A straw with negligible thickness is placed into a rectangular box that is 3 inches by 4 inches by 8 inches, as shown in the accompanying diagram. If the straw fits exactly into the box diagonally from the bottom left front corner to the top right back corner, how long is the straw, to the *nearest tenth of an inch*?

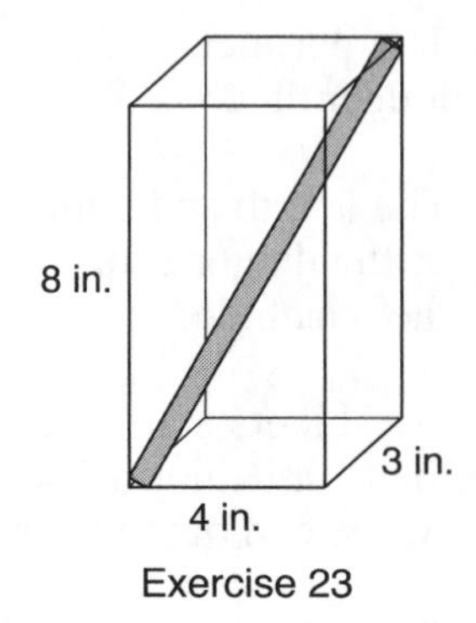

Exercise 23

24. A rectangular box has a square bottom that measures 5.0 inches on each side. A fly lands in one corner of the base of the box. It walks in a straight path along the diagonal of the base and then straight up an edge to the top of the box at which point it flies away. If the fly walks a total distance of 11.0 inches, find the height of the box to the *nearest tenth* of an inch.

6.3 SPECIAL RIGHT TRIANGLE RELATIONSHIPS

KEY IDEAS

In a right triangle in which the acute angles measure 30 and 60, or 45 and 45, special relationships exist among the lengths of the three sides of the triangle.

The 45–45 Right Triangle

If s represents the length of each of the congruent legs of an isosceles right triangle, as in Figure 6.1, then according to the Pythagorean Theorem,

$$(AB)^2 = x^2 + x^2 = 2x^2$$

so $AB = \sqrt{2x^2} = x\sqrt{2}$. The length of hypotenuse $\overline{AB}$ is $\sqrt{2}$ times the length of a leg.

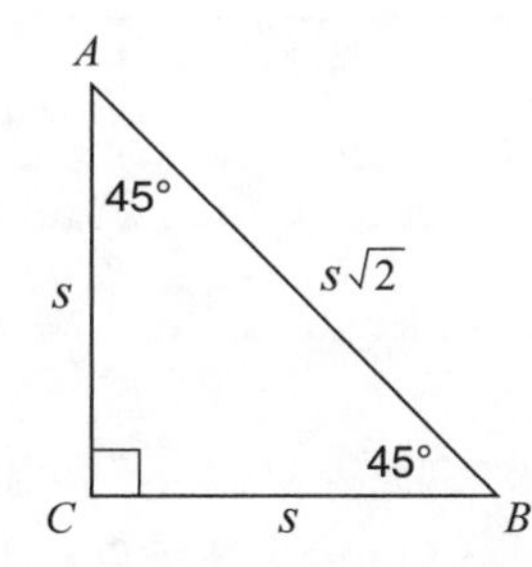

Figure 6.1 45–45–90 triangle relationships.

Theorem
In a 45–45 right triangle, the hypotenuse is $\sqrt{2}$ times the length of either leg.

The 30–60 Right Triangle

If s represents the length of each side of an equilateral triangle, then drawing an altitude to one of the sides divides both the vertex angle and the base into congruent parts, as shown in Figure 6.2. To find BD, apply the Pythagorean Theorem in right triangle ADB:

$$(BD)^2 + \left(\frac{s}{2}\right)^2 = s^2$$

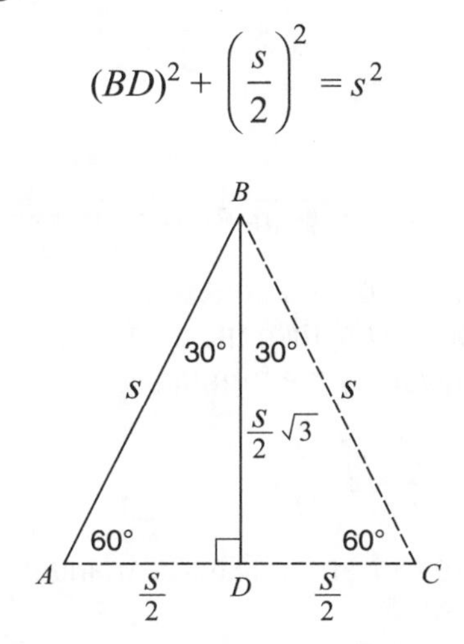

Figure 6.2 30–60 right triangle relationships.

so $(BD)^2 = \frac{3s^2}{4}$ and $BD = \frac{s}{2}\sqrt{3}$. In right triangle ADB, $\overline{AD}$ is the shorter leg, $\overline{BD}$ is the longer leg, and $\overline{AB}$ is the hypotenuse. Based on Figure 6.2, you can make generalizations about how these sides compare in length.

Theorem

In a 30–60 right triangle:

- The shorter leg is the side opposite the 30° angle and is one-half the length of the hypotenuse. In Figure 6.2, $AD = \frac{1}{2} \times AB$.
- The longer leg is the side opposite the 60° angle and is $\sqrt{3}$ times the length of the shorter leg. In Figure 6.2, $BD = AD \times \sqrt{3}$.

Area of an Equilateral Triangle

The area, A, of equilateral triangle ABC is

$$\begin{aligned} A &= \frac{1}{2} \times \text{base} \times \text{height} \\ &= \frac{1}{2} \times AC \times BD \\ &= \frac{1}{2} \times s \times \left(\frac{s}{2}\sqrt{3}\right) \\ &= \frac{s^2}{4}\sqrt{3} \end{aligned}$$

MATH FACTS

If the length of each side of an **equilateral triangle** is represented by s, then the area, A, of the triangle is given by the formula

$$A = \frac{s^2}{4}\sqrt{3}$$

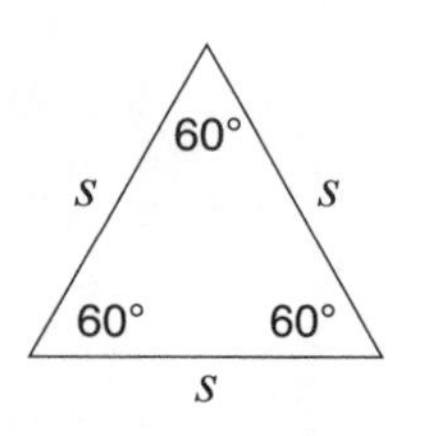

For example, the area of an equilateral triangle in which each side measures 6 inches is

$$A = \frac{6^2}{4}\sqrt{3} = \frac{36}{4}\sqrt{3} = 9\sqrt{3}\ \text{in}^2$$

Example 1

In $\triangle JKL$, $KJ = KL = 15$ inches, $m\angle JKL = 120$. Find the area of $\triangle JKL$ to the *nearest tenth* of a square inch.

Solution: Drop a perpendicular from K to $\overline{JL}$. Since $\triangle JKL$ is isosceles, the altitude bisects the vertex angle and the base.

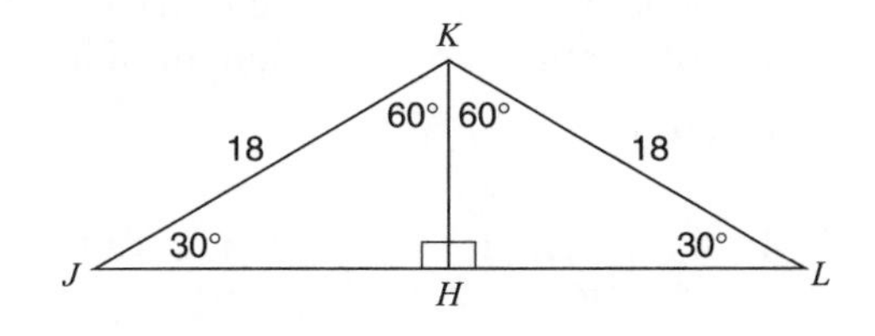

- $KH = \frac{1}{2} \times 18 = 9$.
- $JH = KH \times \sqrt{3} = 9\sqrt{3}$.
- $JL = 2 \times JH = 18\sqrt{3}$.

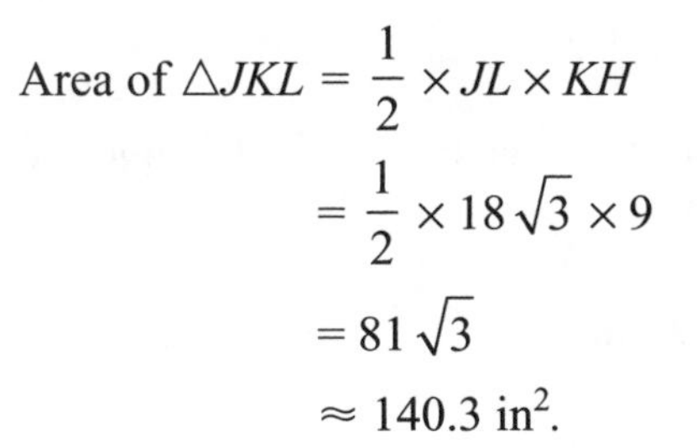

$$\begin{aligned} \text{Area of } \triangle JKL &= \frac{1}{2} \times JL \times KH \\ &= \frac{1}{2} \times 18\sqrt{3} \times 9 \\ &= 81\sqrt{3} \end{aligned}$$

Use a calculator: $\approx 140.3 \text{ in}^2$.

The area of $\triangle JKL$ to the *nearest tenth* of a square inch is **140.3 in²**.

Example 2

In isosceles trapezoid *ABCD*, the measure of a lower base angle is 45 and the length of the shorter base is 5 cm. If the length of an altitude is 7 cm, find the perimeter of the trapezoid.

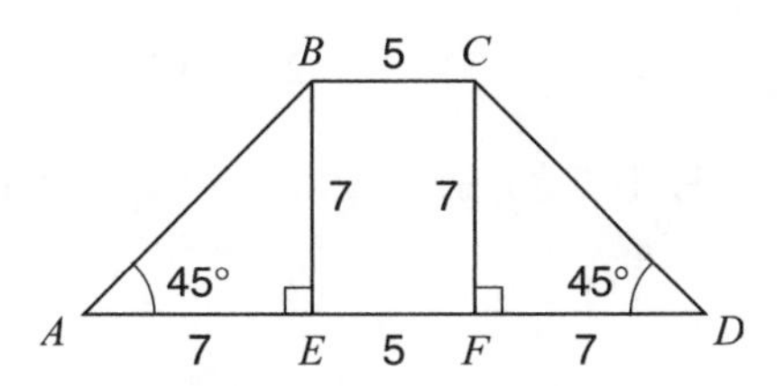

Solution: First find the length of the longer base. Drop altitudes from B and C, thereby forming two congruent right triangles, $\triangle AEB$ and $\triangle DFC$. In a 45–45 right triangle, the legs have the same length:

$$AE = BE = 7 \text{ cm} \quad \text{and} \quad DF = CF = 7 \text{ cm}$$

- Since quadrilateral $BEFC$ is a rectangle, $EF = BC = 5$ cm. Hence, $AD = 7 \text{ cm} + 5 \text{ cm} + 7 \text{ cm} = 19$ cm.
- In a 45–45 right triangle, the length of the hypotenuse is $\sqrt{2}$ times the length of a leg. Hence, $AB = CD = 7\sqrt{2}$.
- The perimeter of the trapezoid is the sum of the lengths of its four sides:

$$5 + 7\sqrt{2} + 19 + 7\sqrt{2} = 24 + 14\sqrt{2} \text{ cm}$$

The perimeter of the trapezoid is **$24 + 14\sqrt{2}$ cm**.

Check Your Understanding of Section 6.3

A. Multiple Choice

1. If the length of each side of a triangle is 4 centimeters, what is the number of centimeters in the length of an altitude to a side?

(1) $2\sqrt{3}$ (2) 2 (3) $4\sqrt{3}$ (4) 4

2. In right triangle ABC, $\angle C$ is a right angle and $m\angle B = 30$. What is the ratio of AC to BC?

(1) $2:\sqrt{3}$ (2) 1:2 (3) $1:\sqrt{3}$ (4) 1:1

3. What is the perimeter of an equilateral triangle if the length of an altitude is $5\sqrt{3}$?

(1) 15 (2) 30 (3) $\frac{15}{\sqrt{3}}$ (4) 45

4. The lengths of a pair of adjacent sides of a parallelogram are 6 and 15. If the measure of their included angle is 60, what is the length of the shorter diagonal of the parallelogram?

(1) $\sqrt{171}$ (2) $\sqrt{148}$ (3) $\sqrt{153}$ (4) $\sqrt{261}$

5. If the perimeter of an equilateral triangle is 15, the area of the triangle is

(1) $5\sqrt{3}$ (2) $\frac{15}{4}\sqrt{3}$ (3) $25\sqrt{3}$ (4) $\frac{25}{4}\sqrt{3}$

6. In a triangle, the measures of the angles are 30, 60, and 90. What is the ratio of the length of the side opposite the 60 degree angle to the length of the hypotenuse?

(1) $\sqrt{3}$ (2) $\frac{1}{2}$ (3) $\frac{1}{3}$ (4) $\frac{1}{\sqrt{3}}$

7. The vertex angle of an isosceles triangle measures 120°, and the length of one of the congruent legs is 12 cm. What is the number of centimeters in the length of the altitude drawn to the base of the triangle?

(1) $6\sqrt{3}$ (2) $\frac{12}{\sqrt{3}}$ (3) 6 (4) $12\sqrt{3}$

8. The accompanying diagram shows two cables of equal length supporting a pole. Both cables are 14 meters long, and they are anchored to points in the ground that are 14 meters apart. What is the exact height of the pole, in meters?

(1) 7 (3) $7\sqrt{3}$

(2) $7\sqrt{2}$ (4) 14

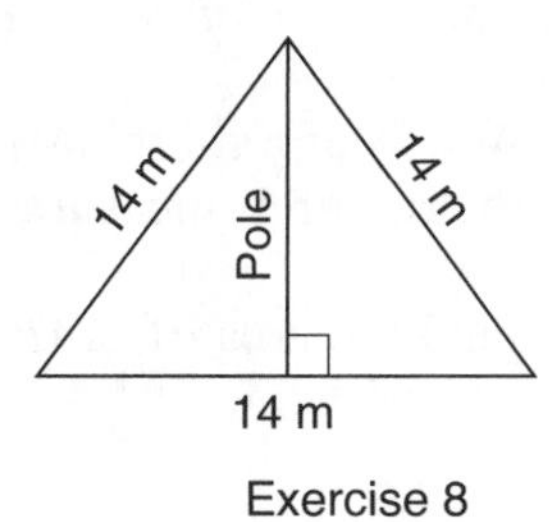

Exercise 8

9. In the accompanying diagram of rhombus *ABCD*, $\overline{AC}$ and $\overline{BD}$ are diagonals intersecting at point *M*. What is the perimeter of the rhombus if m$\angle DAB$ = 120 and *AC* = 12?

(1) 48 (2) $48\sqrt{3}$ (3) $24\sqrt{3}$ (4) 24

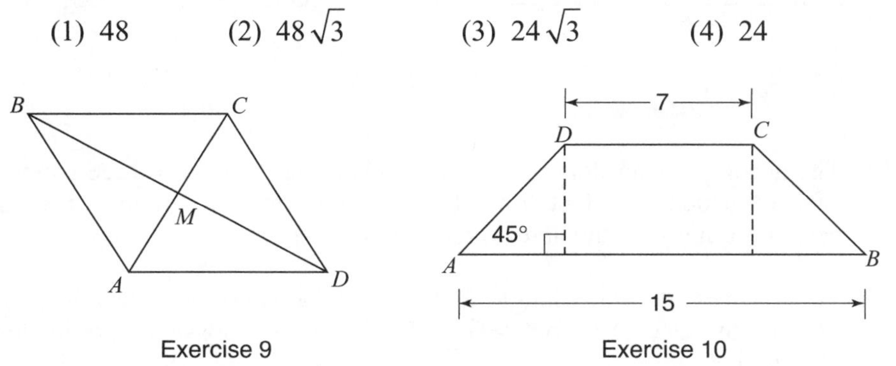

Exercise 9 Exercise 10

10. In the accompanying diagram, the lengths of the bases of isosceles trapezoid *ABCD* are 7 and 15. Each leg makes an angle of 45° with the longer base. What is the length of the altitude of the trapezoid?

(1) $4\sqrt{2}$ (2) $\frac{4}{\sqrt{2}}$ (3) $8\sqrt{2}$ (4) 4

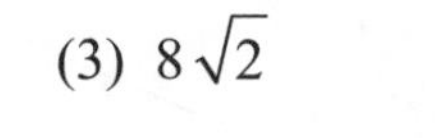

11. Points *A, B, C,* and *D* are midpoints of the sides of square *JETS*. If the area of *JETS* is 36, the area of *ABCD* is

(1) $9\sqrt{2}$ (3) 9

(2) $18\sqrt{2}$ (4) 18

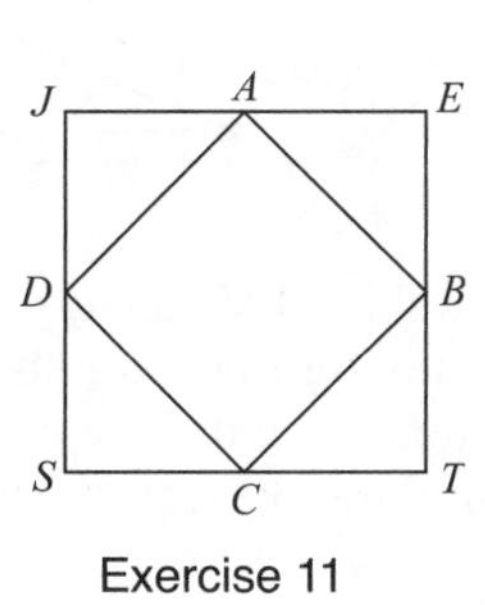

Exercise 11

B. *Show or explain how you arrived at your answer.*

12. A side of a rhombus measures 10 inches. If the measure of an angle of the rhombus measures 60, find the area of the rhombus.

13. Find the area of $\triangle ABC$ where $\overline{BD} \perp \overline{ADC}$, $\angle A = 45°$, $CD = 8$ cm, and $BC = 10$ cm.

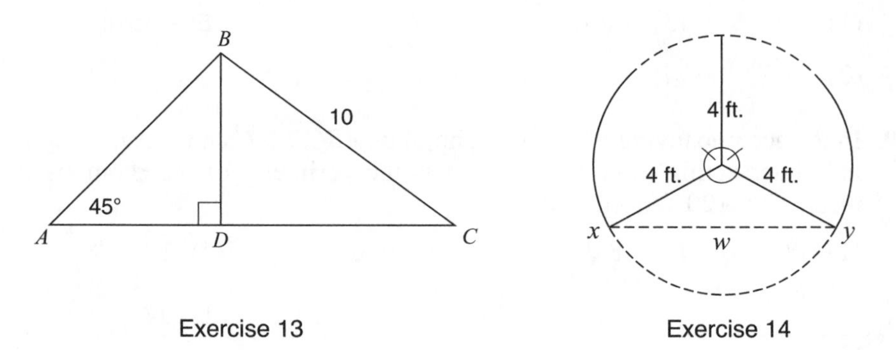

Exercise 13

Exercise 14

14. The accompanying diagram shows a revolving door with three panels, each of which is 4 feet long. What is the width, w, of the opening between x and y, to the *nearest tenth of a foot*?

15. The vertices of right triangle RAG with hypotenuse $\overline{AG}$ are $R(-2, 4)$, $A(7, 4)$, and $G(x, y)$. If $m\angle RAG = 45$, what are the possible coordinates of point G?

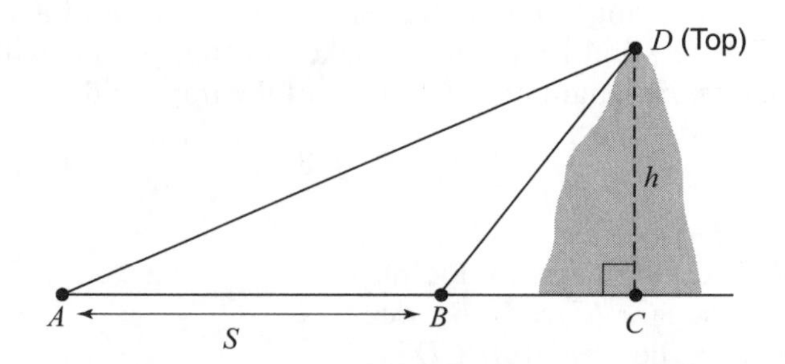

16. A ship at sea heads directly toward a cliff on the shoreline. The accompanying diagram shows the top of the cliff, D, sighted from two locations, A and B, separated by distance S. If $m\angle DAC = 30$, $m\angle DBC = 45$, and $h = 30$ feet, what is the distance, S, to the *nearest foot*?

6.4 REVIEWING TRIGONOMETRY

KEY IDEAS

A **trigonometric ratio** relates the measures of *two* sides and one of the acute angles of a *right* triangle. Trigonometric ratios can be used to find the measures of unknown parts of geometric figures.

THE TRIGONOMETRIC RATIOS

The three basic trigonometric ratios are sine, cosine, and tangent, which are defined as follows:

Sine ratio

$$\sin A = \frac{\text{Opposite side}}{\text{Hypotenuse}}$$

Cosine ratio

$$\cos A = \frac{\text{Adjacent side}}{\text{Hypotenuse}}$$

Tangent ratio

$$\tan A = \frac{\text{Opposite side}}{\text{Adjacent side}}$$

You should also remember these facts:

- The values of $\sin x$, $\cos x$, and $\tan x$ do not change in different right triangles in which the measure of acute angle x is the same. These right triangles are similar so that the ratios of the lengths of the sides used to form a given trigonometric ratio are the same.
- Since the hypotenuse is the longest side of a right triangle, the numerators of the sine and cosine ratio are always less than their denominators. Thus, the value of the sine or cosine of an acute angle is always a number between 0 and 1. The tangent ratio may have a value greater than 1.
- Values of trigonometric functions or the measures of angles obtained from a calculator should not be rounded off. Instead, you should use the full power/display of your calculator when performing calculations. Unless otherwise specified, rounding, if required, should be done only when the *final* answer is reached.

Solving a Right Triangle: Geometry Applications

When the measures of two parts of a right triangle are known, an appropriate trigonometric ratio or the Pythagorean Theorem can be used to solve for any of the remaining parts of the triangle.

Example 1

The lengths of the bases of an isosceles trapezoid are 7 and 29, and the length of each leg is 15. Find the measure of a base angle formed by the longer base and a leg, correct to the *nearest tenth* of a degree.

Solution: Draw an isosceles trapezoid that fits the conditions of the problem.

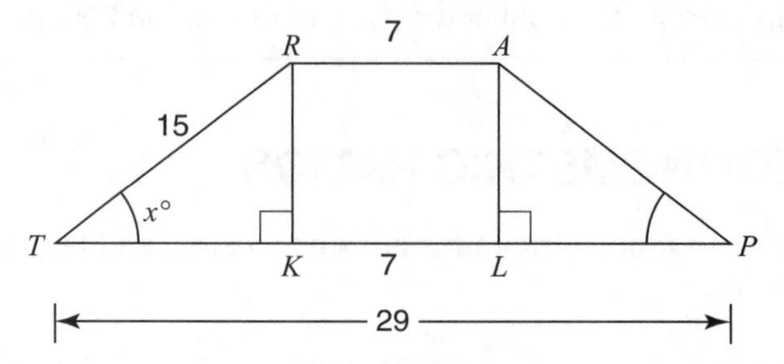

- Mark the diagram with the given dimensions where x represents the measure of $\angle RTK$.
- Drop altitudes, thereby forming rectangle $KRAL$ so that $KL = 7$ and $TK = PL$.

  As $TK + LP = 29 - 7 = 22$, $TK = \frac{1}{2}(22) = 11$.
- Decide which trigonometric ratio to use. Because the known lengths of the sides of right $\triangle RKT$ are *adjacent* to $\angle RTK$ and the *hypotenuse*, use the **cosine** ratio to find $\angle RTK$:

$$\cos x^\circ = \frac{\text{Adjacent side } (TK)}{\text{Hypotenuse } (RT)}$$

$$= \frac{11}{15}$$

- Solve for x:

$$x = \cos^{-1}\left(\frac{11}{15}\right)$$

- Use the full power/display of your calculator to find x:

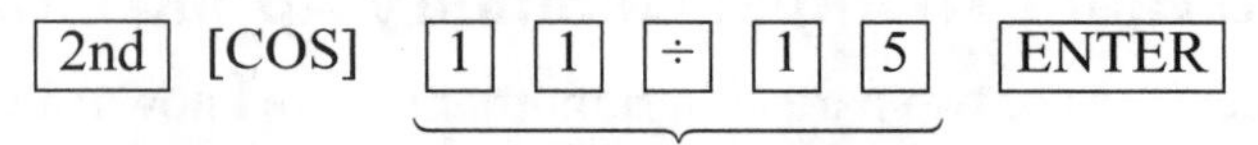

The calculator display window shows 42.83342807°. As the final step, round off the answer in the display window.

To the *nearest tenth of a degree*, the leg makes an angle of **42.8°** with the longer side.

Example 2

The length of the base of an isosceles triangle is 40 inches and the measure of the vertex angle 70°. Find the area of the triangle correct to the *nearest tenth* of a square inch.

Solution: Draw isosceles $\triangle ABC$ and the bisector of vertex angle B. Since the bisector of the vertex angle of an isosceles triangle is the perpendicular bisector of the base, $\triangle ADB$ is a right triangle and $AD = \frac{1}{2} \times 40 = 20$.

- If x represents the length of $\overline{BD}$, use the tangent ratio to find x:

$$\tan 35° = \frac{\text{Opposite side}}{\text{Adjacent side}}$$

$$\tan 35° = \frac{20}{x}$$

$$x = 20 \div \tan 35°$$

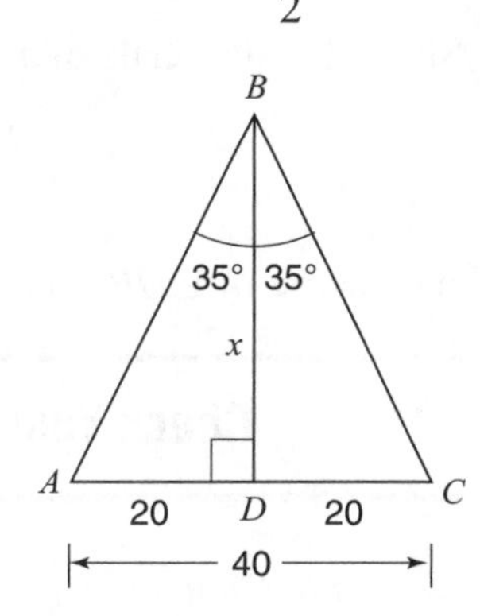

- Use the formula for the area of a triangle:

$$\text{Area of } \triangle ABC = \frac{1}{2} \times (BD) \times (AC)$$

$$= \frac{1}{2} \times (20 \div \tan 35°) \times (40)$$

$$= 400 \div$$

tan 35°

$$= 571.2592027$$

The area of the triangle, correct to the *nearest tenth* of a square inch, is **571.3 in^2**.

Example 3

The perimeter of a rhombus *RHOM* is 40 inches and the longer diagonal measures 16 inches. Find the measure of $\angle HRM$ correct to the *nearest tenth* of a degree.

Solution: The diagonals of a rhombus bisect each other and intersect at right angles. Since the perimeter of the rhombus is 40 inches, each side measures 10 inches.

- In right triangle REM, use the cosine ratio to find $m\angle ERM$:

$$\cos x° = \frac{\text{Adjacent side}}{\text{Hypotenuse}}$$

$$\cos x° = \frac{8}{10}$$

$$x° = \cos^{-1} 0.8$$

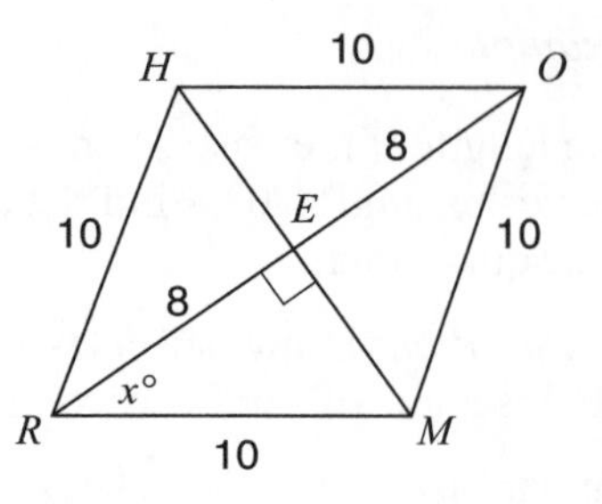

- Since the diagonals of a rhombus bisect its angles,

$$m\angle HRM = 2 \times (\cos^{-1} 0.8)$$
$$= 73.73979529$$

The measure of $\angle HRM$ is **73.7** to the *nearest tenth* of a degree.

Check Your Understanding of Section 6.4

A. Multiple Choice

1. In right $\triangle REM$, $\angle E$ is a right angle, $RE = 6$ and $EM = 8$. The value of $\cos R$ is
   (1) 0.25 (2) 0.6 (3) 0.8 (4) 0.75

2. In the accompanying diagram of right triangle RUN, $m\angle U = 90$, $m\angle N = 37$, and $RN = 21$. What is the length of $\overline{RU}$, expressed to the *nearest tenth*?
   (1) 12.6 (2) 15.8 (3) 16.8 (4) 34.9

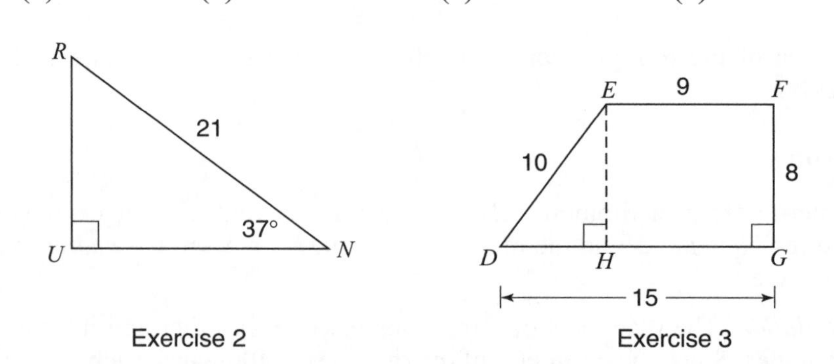

Exercise 2 Exercise 3

3. In the accompanying diagram, altitude $\overline{EH}$ is drawn in trapezoid $DEFG$, $DE = 10$, $EF = 9$, $FG = 8$, and $GD = 15$. What is $m\angle D$ to the *nearest degree*?
   (1) 37 (2) 53 (3) 60 (4) 80

4. The length of the base of an isosceles triangle is 30 inches and the measure of the vertex angle is 100. What is the length of the altitude drawn to the base correct to the *nearest inch*?
(1) 17.9 (2) 12.6 (3) 11.5 (4) 9.6

5. In the accompanying diagram of rhombus $ABCD$, $DB = 28$ and $AC = 20$ inches. What is m$\angle BAC$ to the *nearest degree*?
(1) 35.5 (3) 44.4
(2) 54.5 (4) 45.6

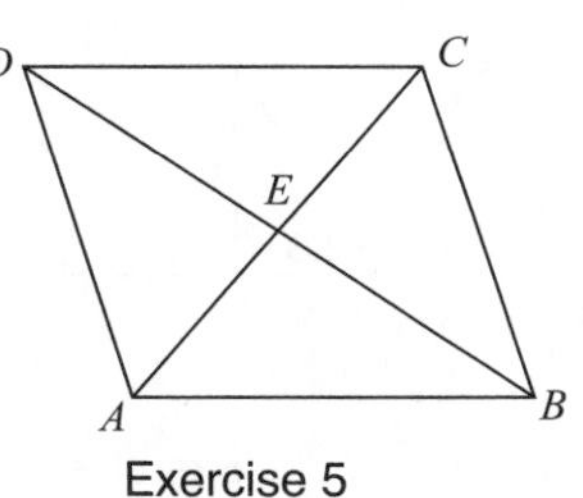

Exercise 5

B. *Show how you arrived at your answer.*

6. The lengths of a pair of adjacent sides of rectangle $JKLM$ are 7 and 18. Find to the *nearest tenth of a degree* the angle a diagonal makes with the shorter side of the rectangle.

7. In right triangle ABC, $\angle C$ is a right angle, $AB = 17$, and $AC = 15$. Find to the *nearest degree* the measure of the angle opposite the shorter leg of the right triangle.

8. In rectangle $ABCD$, diagonal $\overline{AC}$ makes an angle of 39° with the longer side. If the length of the diagonal is 20.0 inches, find the length of the longer side to the *nearest tenth* of an inch.

9. In the accompanying diagram, the bases of isosceles trapezoid $ABCD$ measure 10 and 18 and the length of a leg is 8. Find m$\angle A$.

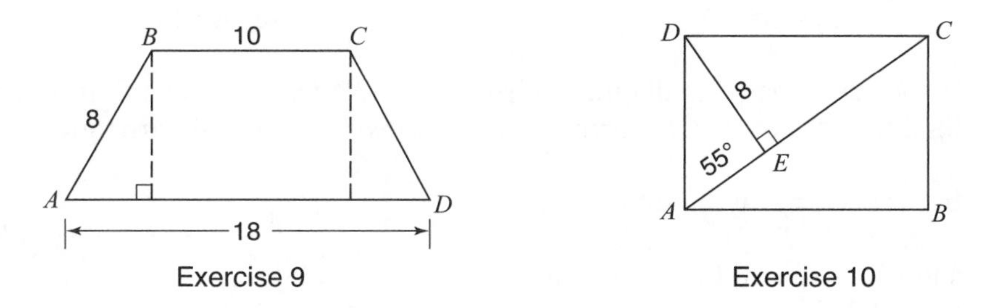

Exercise 9 Exercise 10

10. In the accompanying diagram of rectangle $ABCD$, diagonal $\overline{AC}$ is drawn, $DE = 8$. $\overline{DE} \perp \overline{AC}$, and m$\angle DAC = 55$. Find the area of rectangle $ABCD$ to the *nearest integer*.

11. In the accompanying diagram of isosceles trapezoid $ABCD$, $m\angle A = 53$, $DE = 6$, and $DC = 10$. Find the length of base $\overline{AEB}$ to the *nearest tenth* of an integer.

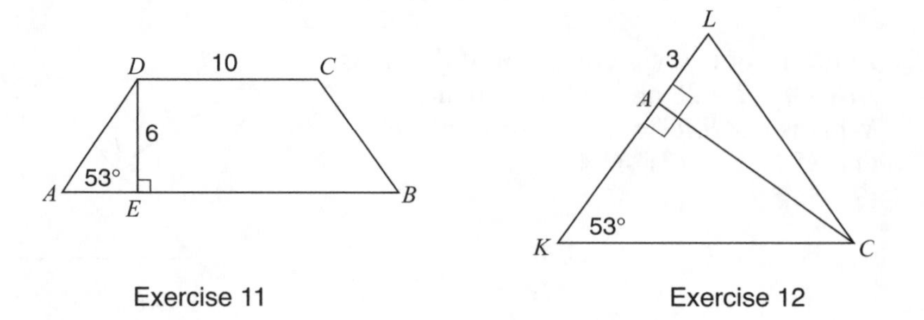

Exercise 11 Exercise 12

12. In the accompanying diagram of isosceles triangle KLC, $\overline{LK} \cong \overline{LC}$, $m\angle K = 53$, altitude $\overline{CA}$ is drawn to leg $\overline{LK}$, and $LA = 3.0$ cm. Find the number of square centimeters in the area of $\triangle KLC$ correct to the nearest integer.

13. In the accompanying diagram of right triangle ABD, $AB = 6$ and altitude BC divides hypotenuse $\overline{AD}$ into segments of lengths x and 9.

a. Find AC .

b. Using the answer from part a, find the measure of $\angle A$ to the *nearest tenth* of a degree.

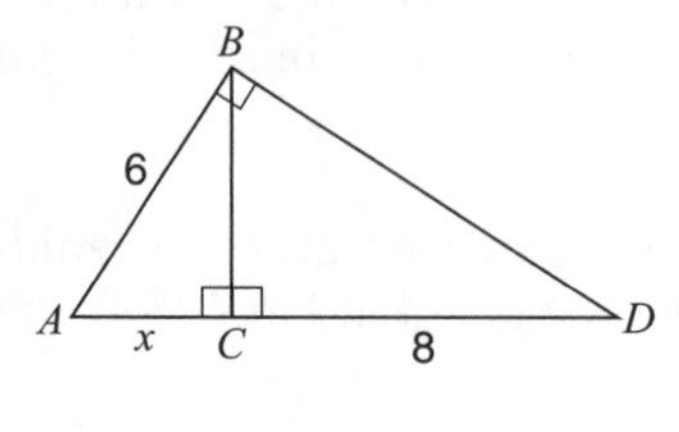

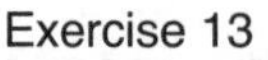

Exercise 13

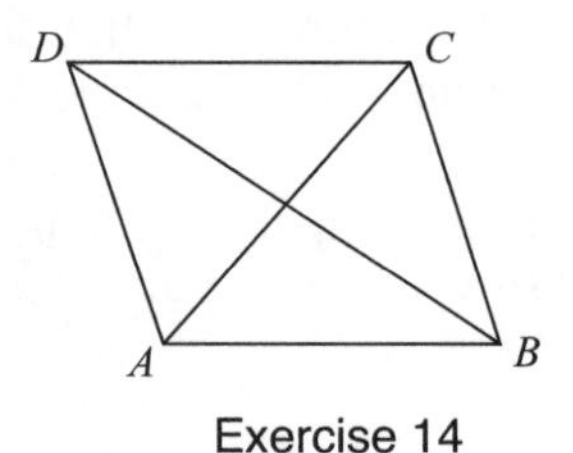

Exercise 14

14. In the accompanying diagram of rhombus $ABCD$, $AB = 10$ and $AC = 10$. Find the area of $ABCD$ correct to the *nearest tenth* of a square unit.

15. In the accompanying diagram of rhombus $RUDE$, $m\angle RUD = 42$ and $UE = 20$ cm. Find the area of $RUDE$ to the *nearest tenth* of a square centimeter.

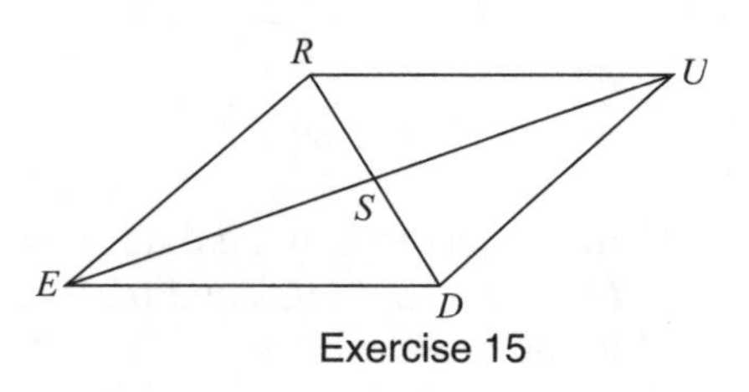

Exercise 15

16. In the accompanying diagram of parallelogram *MATH*, $\overline{AV} \perp \overline{MH}$, $AT = 18$, $VH = 10$, and $m\angle M = 42$.

a. Find the area of parallelogram *MATH correct* to the *nearest tenth* of a square unit.

b. Find the perimeter of parallelogram *MATH correct* to the *nearest tenth.*

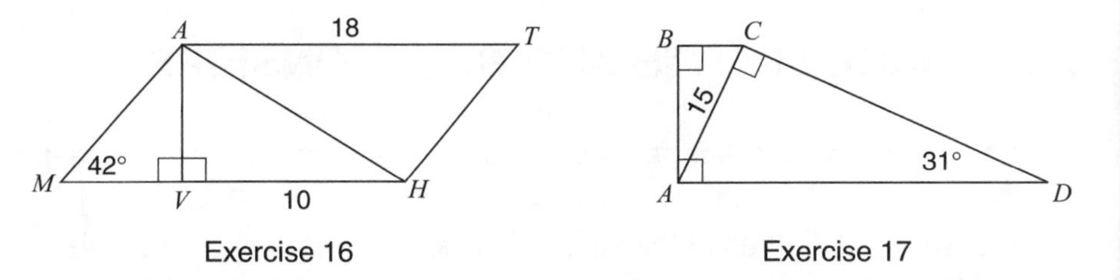

Exercise 16

Exercise 17

17. In the accompanying diagram of trapezoid *ABCD*, $\overline{AB} \perp \overline{BC}$, $\overline{BA} \perp \overline{AD}$, and $\overline{AC} \perp \overline{CD}$. If $AC = 15$ cm and $m\angle D = 31$, find the area of trapezoid *ABCD* to the *nearest square* centimeter.

CHAPTER 7
CIRCLES AND ANGLE MEASUREMENT

7.1 CIRCLE PARTS AND RELATIONSHIPS

KEY IDEAS

The two basic figures in the study of plane geometry are the triangle and the circle. A **circle** is a set of points in a plane at a fixed distance from a given point. The fixed distance is the **radius** of the circle and the given point is its **center**. A circle is named by its lettered center. If the center of a circle is point *O*, then that circle is referred to as circle *O*. A **semicircle** is one-half of a circle. **Congruent circles** are circles with congruent radii.

Points, Lines, and Segments of a Circle

If the distance from the center of a circle to a point *Y is less than* the radius, it is an **interior point** of the circle; if the distance from the center to point *X is greater than* the radius, then *X* is an **exterior point** of the circle.

A continuous set of points of the circle that trace out a curved portion of it is called an **arc**. An arc that is less than a semicircle is named by its two endpoints. The notation $\overset{\frown}{LM}$ refers to the *arc* whose endpoints are points *L* and *M*, as shown in Figure 7.1.

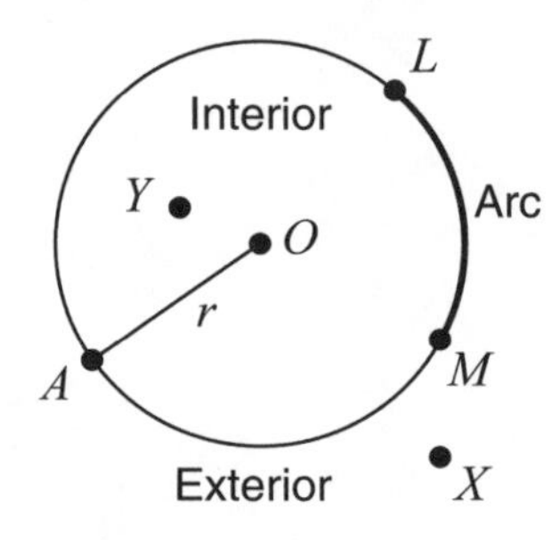

Figure 7.1 Interior and exterior points.

The accompanying table identifies some different types of lines and segments that contain points of a circle.

Special Line or Segment	Circle Diagram
A **chord** is a segment whose endpoints are points of the circle.	
A **diameter** is a chord that passes through the center of the circle.	
A **secant** is a line that intersects a circle at two points.	
A **tangent** is a line in the same plane as the circle that intersects it at exactly one point called the **point of tangency** or **point of contact.**	

Central Angles and Arcs

A **central angle** is an angle whose vertex is the center of a circle and whose sides contain radii of the circle. If radii $\overline{OA}$ and $\overline{OB}$ are the sides of a central angle that measures less than 180 degrees, as in Figure 7.2, then

- Points A and B and the points of the circle that are in the *interior* of the central angle form a **minor arc**.
- Points A and B and the points of the circle that are in the *exterior* of the central angle form a **major arc**.

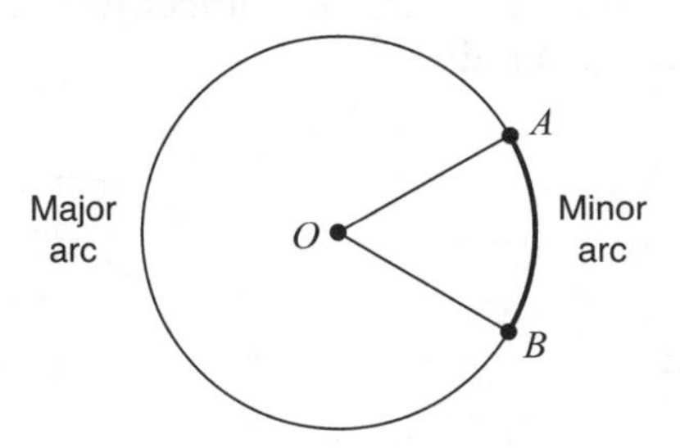

Figure 7.2 Classifying arcs.

If points A and B are the endpoints of a diameter, then each arc that is formed is a semicircle.

Naming and Measuring Arcs

To help distinguish minor arc AB from the major arc having the same two endpoints, a third letter is placed on the major arc. In Figure 7.3, $\overset{\frown}{APB}$ is the *major* arc with endpoints A and B.

- The **measure of a circle** is 360 degrees, and the **measure of a semicircle** is 180 degrees.
- The **measure of minor arc** *AB* is the measure of central angle *AOB* that intercepts it. Thus,

$$m\overset{\frown}{AB} = m\angle AOB = 50.$$

- The **measure of major arc** *APB* is 360 minus the degree measure of its minor arc. Hence, $m\overset{\frown}{APB} = 360 - 50 = 310$.

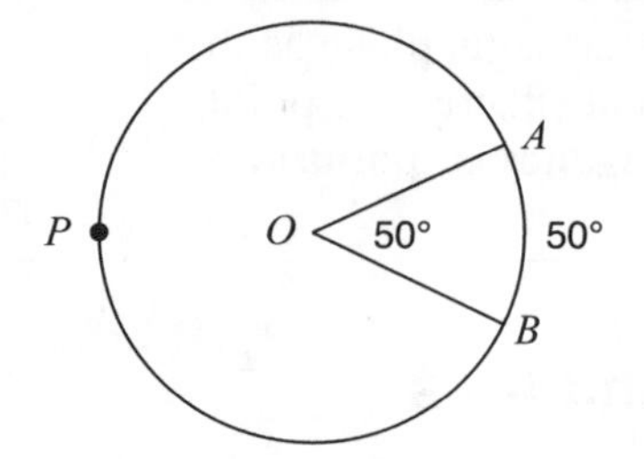

Figure 7.3 Degree Measure of Arcs.

Congruent Circles and Arcs

In the same or congruent circles, congruent central angles have **congruent arcs**, as shown in Figure 7.4. Although arcs *AB* and *CD* in Figure 7.5 have congruent central angles, their arcs are *not* congruent since they are in *concentric* rather than in congruent circles. **Concentric circles** are circles with the same center but unequal radii.

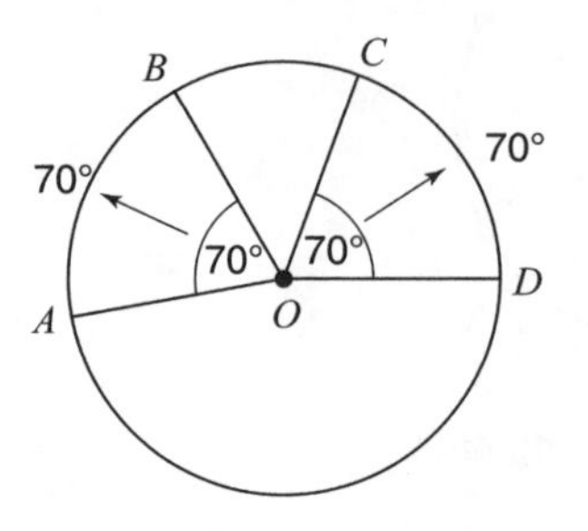

Figure 7.4 $\overset{\frown}{AB} \cong \overset{\frown}{CD}$.

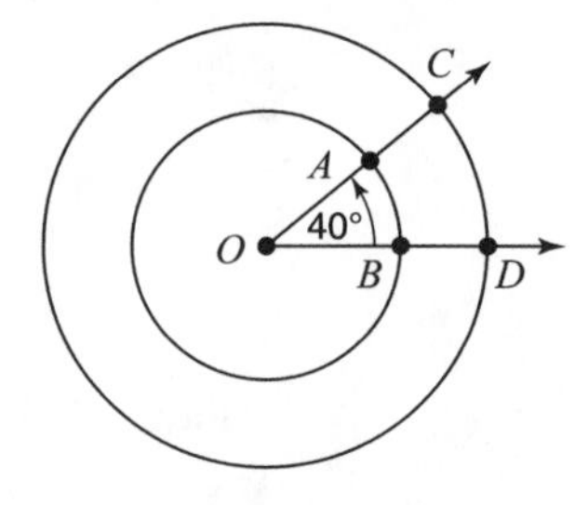

Figure 7.5 $\overset{\frown}{AB} \ncong \overset{\frown}{CD}$.

Congruent Arcs and Chords

In Figure 7.6, if chords $\overline{AB}$ and $\overline{CD}$ are congruent, then $\overset{\frown}{AB} \cong \overset{\frown}{CD}$. Conversely, if $\overset{\frown}{AB}$ and $\overset{\frown}{CD}$ are congruent, then $\overline{AB} \cong \overline{CD}$.

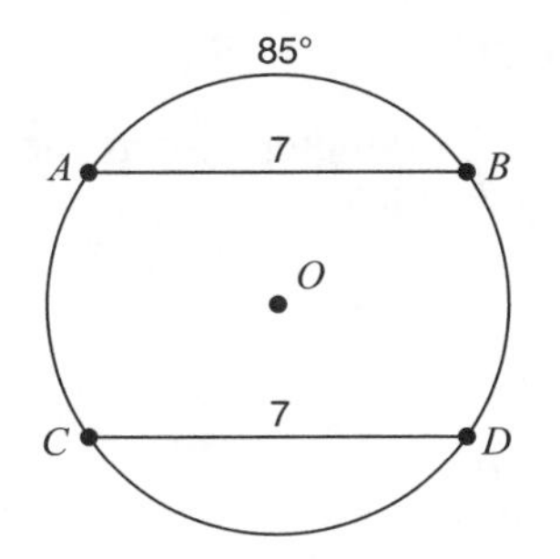

Figure 7.6 Congruent chords have congruent arcs.

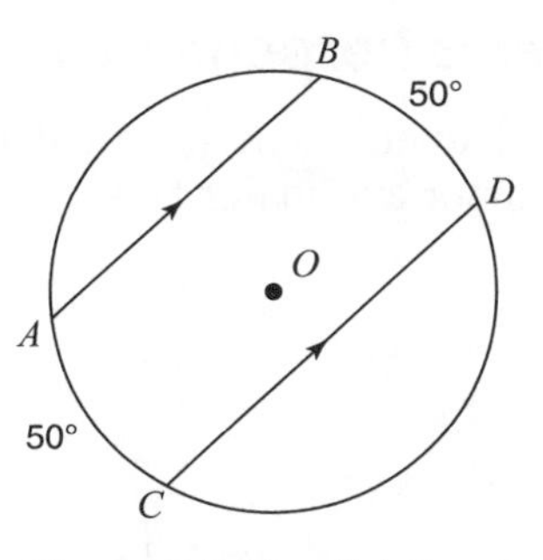

Figure 7.7 Parallel chords cut off congruent arcs.

In Figure 7.7, if chords $\overline{AB}$ and $\overline{CD}$ are parallel, then $\overparen{AC} \cong \overparen{BD}$.

Theorems: Arc–Chord Theorems
In the same or congruent circles:

- If two chords are congruent, then their arcs are congruent.
- If two arcs are congruent, then their chords are congruent.
- If two chords are parallel, then their intercepted arcs are congruent.

Example 1

In the accompanying diagram, $\overline{AB}$ is parallel to $\overline{CD}$. If m$\overparen{AB}$ = 110 and m$\overparen{CD}$ = 90, what is the degree measure of central angle *BOD*?

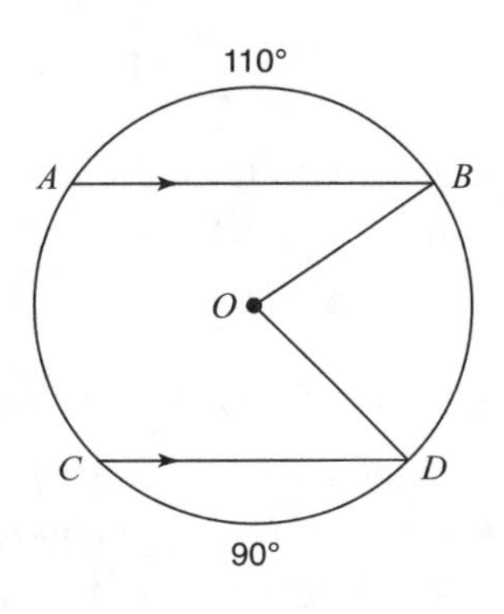

Solution: Since parallel chords intercept equal arcs, let $x = m\overparen{BD} = m\overparen{AC}$. The sum of the degree measures of the arcs that comprise a circle is 360. Thus

$$\begin{aligned} m\overparen{AB} + m\overparen{BD} + m\overparen{CD} + m\overparen{AC} &= 360 \\ 110 + x \quad + 90 \quad + x &= 360 \\ 2x + 200 &= 360 \\ x = \frac{160}{2} &= 80 \end{aligned}$$

Hence, $m\overparen{BD} = 80$. Since a central angle and its intercepted arc have the same degree measure, $m\angle BOD =$ **80**.

Diameter Perpendicular to a Chord

When a diameter is perpendicular to a chord, the diameter bisects the chord and its minor and major arcs, as shown in Figure 7.8.

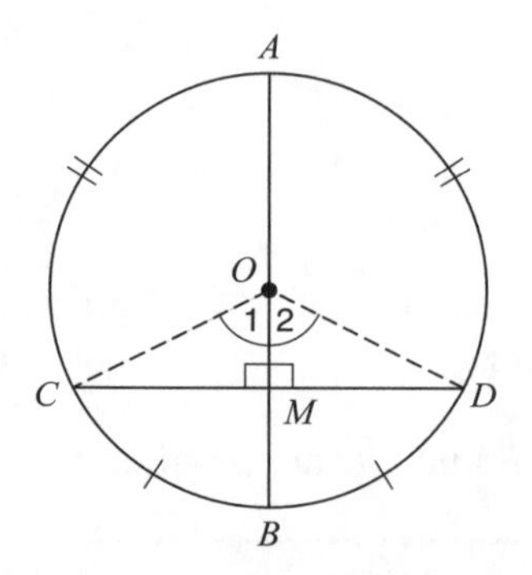

Figure 7.8 Diameter $\overline{AB} \perp$ chord $\overline{CD}$ makes $\overline{CM} \cong \overline{DM}$, $\overset{\frown}{CB} \cong \overset{\frown}{CD}$ and $\overset{\frown}{AC} \cong \overset{\frown}{AD}$.

> **Theorem: Diameter ⊥ Chord**
> If a diameter of a circle is perpendicular to a chord, then it bisects the chord and its arcs.

Paragraph Proof (Refer to Figure 7.8)

Given: Circle O, diameter $\overline{AB} \perp$ chord $\overline{CD}$.
Prove: a. $\overline{AB}$ bisects $\overline{CD}$.
b. $\overline{AB}$ bisects $\overset{\frown}{CD}$ and $\overset{\frown}{CAD}$.

a. Prove $\triangle COM \cong \triangle DOM$ to show that $\overline{CM} \cong \overline{DM}$.

- Draw radii $\overline{OC}$ and $\overline{OD}$ (two points determine a line).
- $\overline{OC} \cong \overline{OD}$ (Hyp) and $\overline{OM} \cong \overline{OM}$ (Leg). Right triangles COM and DOM are congruent by HL ≅ HL.
- $\overline{CM} \cong \overline{DM}$ (CPCTC) so $\overline{AB}$ bisects $\overline{CD}$.

b. Use congruent central angles to obtain the required pairs of congruent arcs.

- Since $\angle 1 \cong \angle 2$ (CPCTC), $\overset{\frown}{CB} \cong \overset{\frown}{DB}$. Hence, $\overline{AB}$ bisects $\overset{\frown}{CD}$.
- $\angle AOC \cong \angle AOD$ since supplements of congruent angles are congruent.
- $\therefore \overset{\frown}{AC} \cong \overset{\frown}{AD}$ so $\overline{AB}$ bisects $\overset{\frown}{CAD}$.

Example 2

A chord is 3 inches from the center of a circle whose radius is 5 inches. What is the length of the chord?

Solution: In the accompanying diagram of circle O, $\overline{OA}$ is a radius, and OE is the distance of chord $\overline{AB}$ from center O. It is given that $OA = 5$ in and $OE = 3$ in. Because $\triangle OEA$ is a 3–**4**–5 right triangle, $AE = 4$. Since $\overline{OE}$ is perpendicular to chord $\overline{AB}$ and passes through center O, it lies on a diameter, so it bisects $\overline{AB}$. Hence $AE = BE = 4$ in.

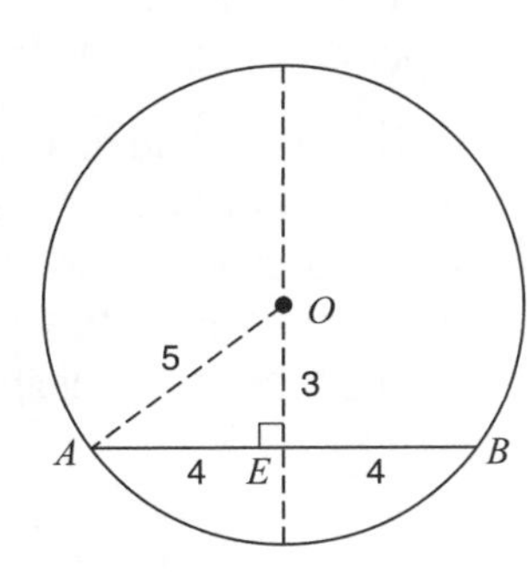

The length of the chord $\overline{AB}$ is, therefore, 4 + 4 or **8 in**.

Equidistant Chords

In Figure 7.9, $\overline{AB} \cong \overline{CD}$. The lengths of perpendicular segments $\overline{OP}$ and $\overline{OQ}$ represent the distances of the segments from the center O. When a line through the center of a circle is perpendicular to a chord, it bisects the chord. Hence, $\overline{AP} \cong \overline{CQ}$, as halves of congruent segments are congruent. Since $\overline{OA} \cong \overline{OB}$, $\triangle OPA \cong \triangle OQC$, which means $\overline{OP} \cong \overline{OQ}$.

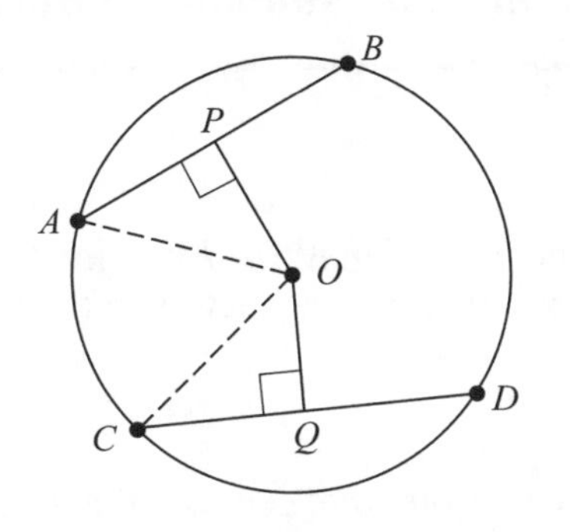

Figure 7.9 $\overline{AB} \cong \overline{CD} \Rightarrow OP = OQ$.

Conversely, if two chords are the same distance from the center of a circle, the chords are congruent.

> **Theorems: Equidistant Chord Theorems**
> In the same or congruent circles:
> - If two chords are congruent, they are the same distance from the center.
> - If two chords are the same distance from the center, they are congruent.

MATH FACTS

Deciding When Two Arcs Are Congruent

In the same circle or in congruent circles, two *arcs* are congruent when any one of the following statements is true:

- The central angles that intercept the arcs are congruent.
- The chords that the arcs determine are congruent.
- The arcs are between parallel chords.
- The arcs are formed by a diameter perpendicular to a chord.
- The arcs are semicircles.

Deciding When Two Chords Are Congruent

In the same circle or in congruent circles, two *chords* are congruent when

- The arcs that the chords intercept are congruent.
- The chords are the same distance from the center.

Check Your Understanding of Section 7.1

A. Multiple Choice

1. A chord 48 centimeters in length is 7 centimeters from the center of a circle. What is the number of centimeters in the length of a radius of this circle?
(1) 25 (2) 50 (3) 55 (4) $\sqrt{2,353}$

2. A chord is 5 inches from the center of a circle whose diameter is 26 inches. What is the number of centimeters in the length of the chord?
(1) 12 (2) 24 (3) 31 (4) 36

3. What is the distance, in inches, of a 30-inch chord from the center of a circle with a radius of 17 inches?
(1) 4 (2) 8 (3) 13 (4) 47

4. In the accompanying diagram of circle O, $\overline{BOC}$ is a diameter, radius $\overline{OA}$ is drawn, and $m\angle OAB = 30$. What is the measure of minor arc AC?
(1) 15 (3) 60
(2) 30 (4) 90

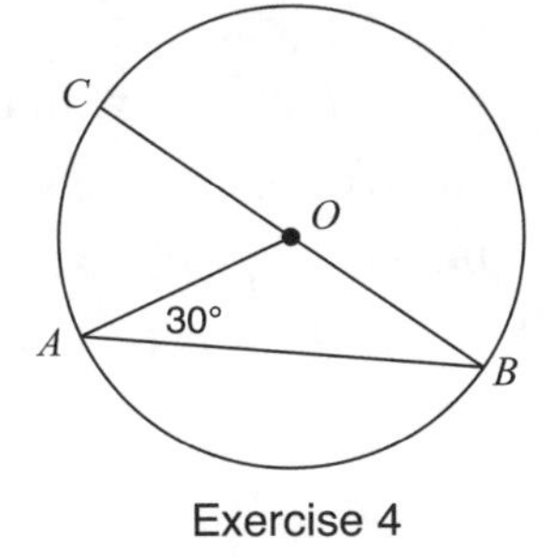

Exercise 4

B. *Show or explain how you arrived at your answer.*

5–7. In the accompanying diagram, $\overline{AB}$ is parallel to $\overline{CD}$ in circle O.

5. If $m\overparen{BD} = 75$ and $m\overparen{CD} = 90$, find $m\angle AOB$.

6. If $m\angle BAO = 40$ and $m\overparen{CD} = 70$, find $m\overparen{BD}$.

7. If $OC = CD$ and $m\overparen{AC} = 79$, find $m\angle AOB$.

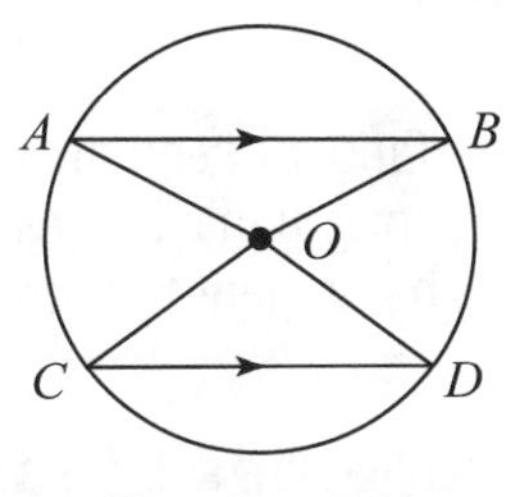

Exercises 5–7

8. In circle O, chord $\overline{AB}$ is 9 inches from the center. The diameter of the circle exceeds the length of $\overline{AB}$ by 2 inches. Find the length of $\overline{AB}$.

9. Given: In circle O, quadrilateral $OXEY$ is a square.
Prove: $\overparen{QP} \cong \overparen{JT}$.

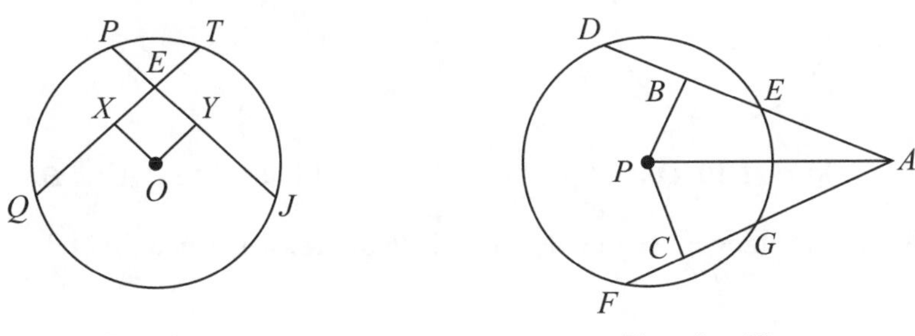

Exercise 9 Exercise 10

10. Given: In circle P, $\overline{PB} \perp \overline{DE}$, $\overline{PC} \perp \overline{FG}$, $\overline{DE} \cong \overline{FG}$.
Prove: $\overline{PA}$ bisects $\angle FAD$.

11. In the accompanying diagram of circle O, points E and F are midpoints of radii $\overline{OD}$ and $\overline{OC}$, respectively, $\widehat{AD} \cong \widehat{BC}$.

Prove: a. $\triangle AOE \cong \triangle BOF$.

b. $\angle DAE \cong \angle CBF$

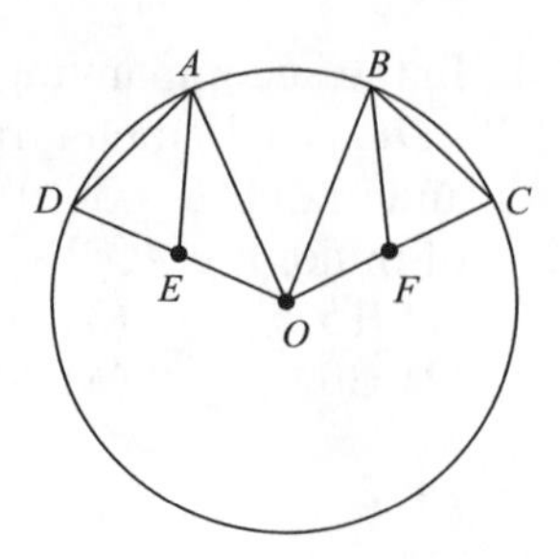

Exercise 11

7.2 TANGENTS AND CIRCLES

KEY IDEAS

Tangent segments drawn to a circle from the same exterior point are congruent. Circles may be tangent to each other. The same line may be tangent to more than one circle.

Properties of Tangents

In Figure 7.10, tangent line t intersects the circle at point A so radius $\overline{OA} \perp t$. Conversely, if line t is perpendicular to radius $\overline{OA}$ at A, then t is tangent to circle O at point A.

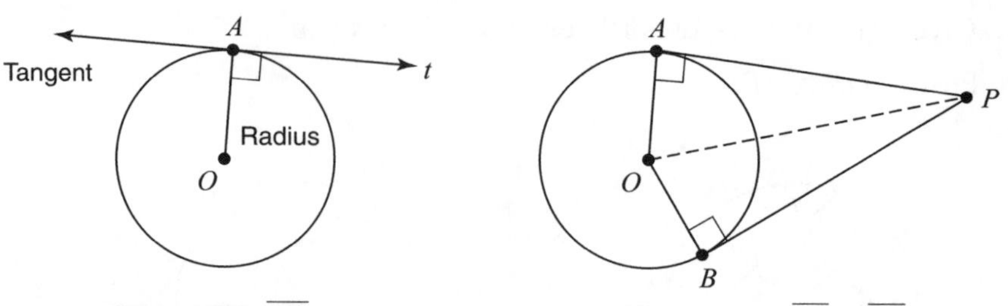

Figure 7.10 $\overline{OA} \perp t$.

Figure 7.11 $\overline{PA} \cong \overline{PB}$.

In Figure 7.11, tangent segments $\overline{PA}$ and $\overline{PB}$ are drawn. Since $\triangle OAP \cong \triangle OBP$ by Hy–Leg, $\overline{PA} \cong \overline{PB}$.

> **Theorems: Tangent Theorems**
> - **Theorem:** If a line is tangent to a circle, then it is perpendicular to the radius drawn to the point of contact.
> - **Theorem:** If a line is perpendicular to a radius at its endpoint on a circle, then the line is tangent to the circle at that point.
> - **Theorem:** If two tangent segments are drawn to a circle from the same exterior point, then they are congruent.

Example 1

From a point R that is 13 inches from the center of circle O, $\overline{RS}$ is drawn tangent to circle O at point S. If $RS = 12$ inches, what is the area of circle O expressed in terms of π?

Solution: Draw radius $\overline{OS}$. Since $\overline{OS} \perp \overline{RS}$, $\triangle OSR$ is a right triangle whose side lengths form a **5**–12–13 Pythagorean triple where $x = 5$. The area, A, of a circle is given by the formula $A = \pi r^2$. Because $r = 5$ inches, $A = \pi \times 5^2 = 25\pi \text{ in}^2$.

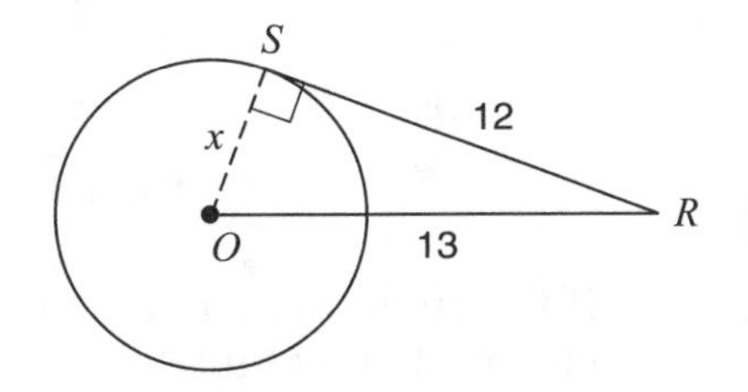

Example 2

In the accompanying figure, $\overrightarrow{PAB}$ and $\overrightarrow{PCD}$ are tangent to circle O at points A and B, respectively, $m\angle P = 48$, chord $\overline{AC}$ and diameter $\overline{AOK}$ are drawn. Find $m\angle KAC$.

Solution: Tangents $\overline{PA}$ and $\overline{PC}$ are congruent, so in $\triangle APC$, $m\angle PAC = m\angle PCA = x$:

$$x + x + 48 = 180$$
$$2x = 180 - 48$$
$$\frac{2x}{2} = \frac{132}{2}$$
$$x = 66$$

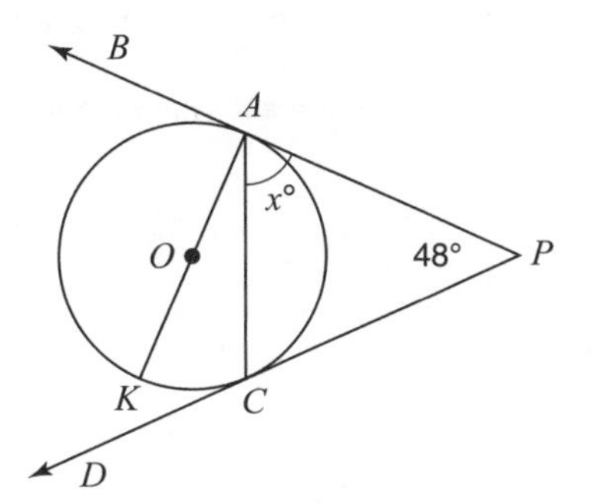

Angle KAP is a right angle so $m\angle KAP = 90$.

$$m\angle KAC + m\angle PAC = m\angle KAP$$
$$m\angle KAC + 66 = 90$$
$$m\angle KAC = 90 - 66$$
$$m\angle KAC = \mathbf{24}$$

Tangent Circles

Tangent circles are circles in the same plane that are tangent to the same line at the same point. Circles may be tangent internally or externally, as shown in Figure 7.12.

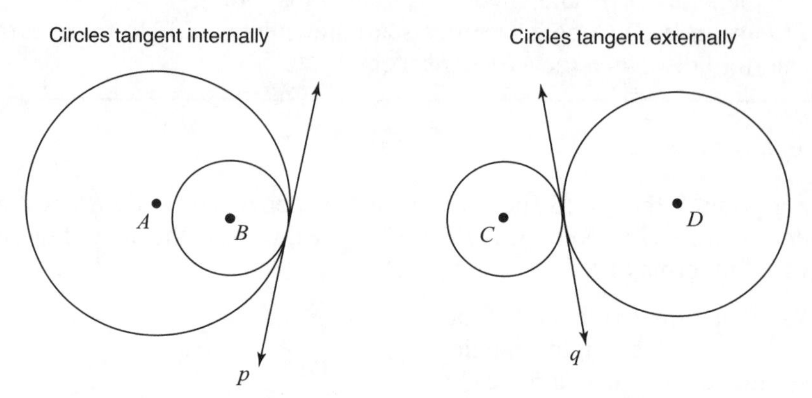

Figure 7.12 Tangent circles.

- If the circles lie on the same side of the common tangent line, the circles are **tangent internally.**
- If the circles lie on opposite sides of the common tangent line, the circles are **tangent externally**.

Common Tangents

A line that is tangent to two circles is a **common tangent**. Common tangents may be *external or internal*. A common *external* tangent does *not* intersect the line through the centers of the two circles at a point between the two circles. A common *internal* tangent intersects the line joining the centers of the two circles at a point between the two circles. Figure 7.13 shows the possible numbers of common tangents drawn to two circles.

One common external tangent

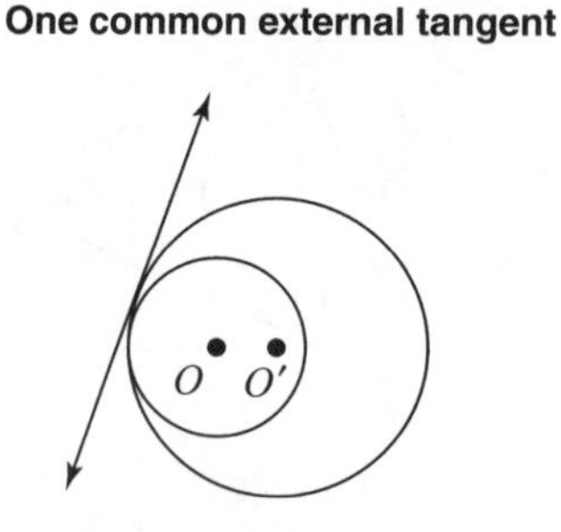

Internally tangent circles

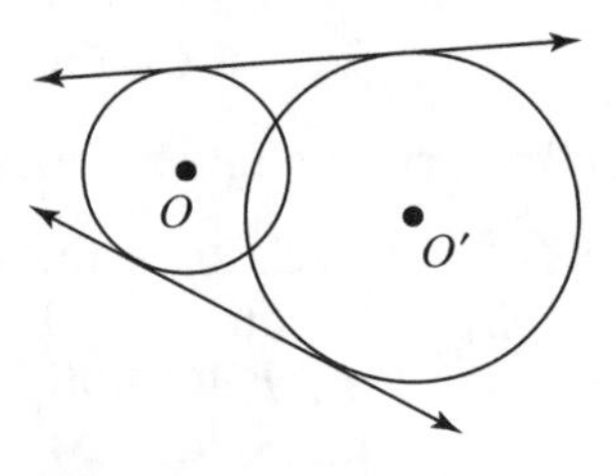

Circles intersecting at two points

Figure 7.14 Common tangents.

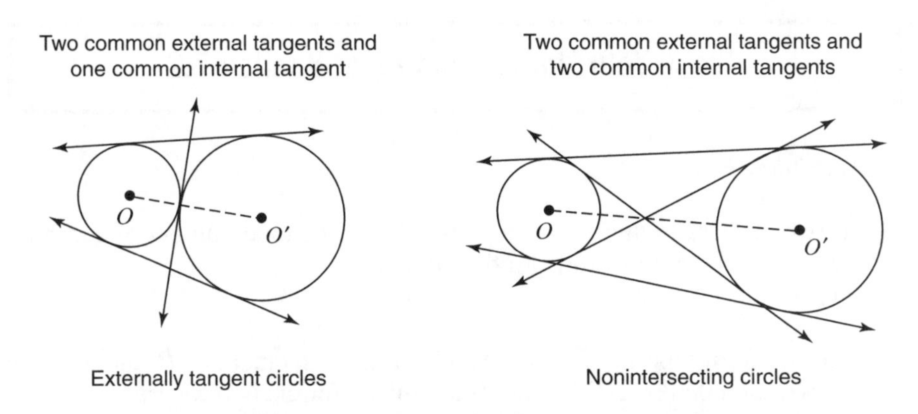

Figure 7.14 (continued) Common tangents.

The **length of a common tangent** is the length of the segment whose endpoints are the two points of contact.

Circumscribed and Inscribed Polygons

A polygon is **inscribed in a circle** if all the vertices of the polygon are points on the circle, as in Figure 7.14. The circle is said to be **circumscribed about the polygon**.

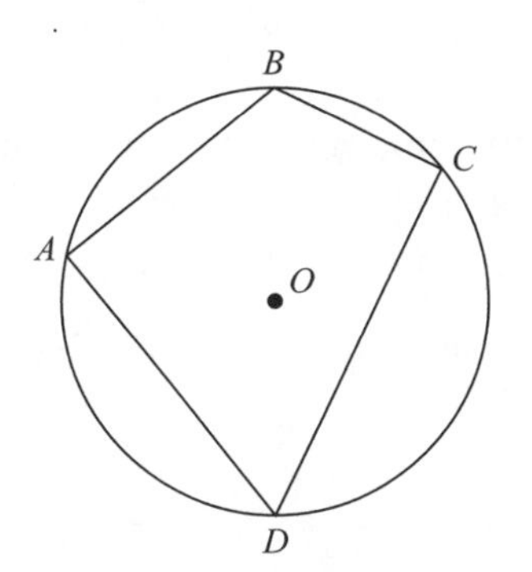

Figure 7.14 Quadrilateral *ABCD* is inscribed in circle *O*.

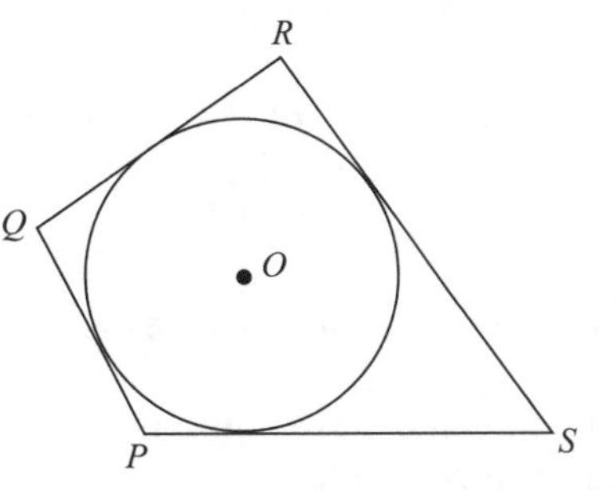

Figure 7.15 Quadrilateral *PQRS* is circumscribed about circle *O*.

A polygon is **circumscribed about a circle** if each side of the polygon is tangent to the circle, as in Figure 7.15. The circle is said to be **inscribed in the polygon**.

The **center** of a regular polygon is the point in the interior of the polygon that is equidistant from the vertices of the polygon. The centers of the inscribed and circumscribed circles of a regular polygon coincide with the center of the regular polygon.

Check Your Understanding of Section 7.2

A. Multiple Choice

1. If two circles are internally tangent, what is the total number of common tangents that can be drawn to the circles?
(1) 1 (2) 2 (3) 3 (4) 0

2. From exterior point L of circle O, tangent segments $\overline{LP}$ and $\overline{LQ}$ are drawn such that $m\angle PLQ = 60$. If a radius of circle O measures 6, what is the distance from the center of the circle to point L?
(1) $3\sqrt{3}$ (2) 2 (3) 3 (4) 12

3. If two circles with diameters of 6 and 2 are internally tangent, then the distance between their centers is
(1) 0 (2) 2 (3) 3 (4) 4

4. The radii of two circles are r and R, and the distance between their centers is d. If $d = r + R$, then the number of common tangents the two circles have is
(1) 1 (2) 2 (3) 3 (4) 4

5. If circles O and O' do not intersect, the maximum number of common tangents they may have is
(1) 1 (2) 2 (3) 3 (4) 4

6. If for two given circles exactly two common tangents are possible, the circles
(1) intersect at two points
(2) are concentric
(3) are tangent internally
(4) are tangent externally

7. In the accompanying diagram, equilateral $\triangle ABC$ is circumscribed about circle O, and equilateral $\triangle DEF$ is inscribed in the same circle. What is the ratio of AB to DE?

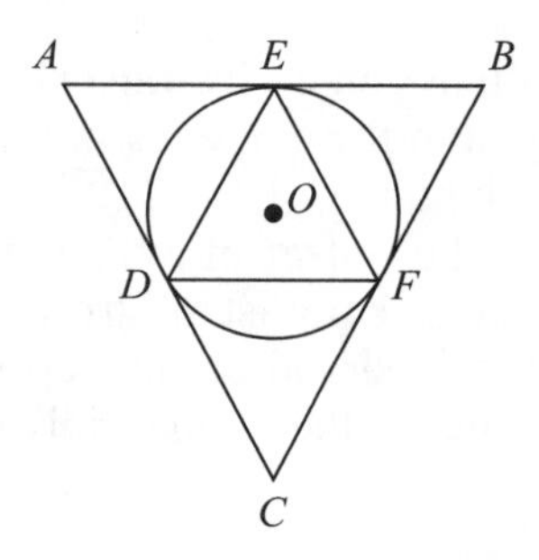

(1) 4:1 (3) $2\sqrt{3}:1$
(2) 2:1 (4) $\sqrt{3}:1$

8. Tangents $\overrightarrow{PA}$ and $\overrightarrow{PB}$ are tangent to circle O at points A and B, respectively, and radii $\overline{OA}$ and $\overline{OB}$ are drawn. Which statement is always true?
(1) $m\angle P = m\angle AOB$
(2) $m\angle P > m\angle AOB$
(3) $m\angle P + m\angle AOB = 90$
(4) $m\angle P + m\angle AOB = 180$

B. *Show or explain how you arrived at your answer.*

9. Externally tangent circles A and B have diameters of 8 inches and 4 inches, respectively. Point P lies on their common internal tangent at a distance of 3 inches from the point at which the two circles touch. Find the perimeter of $\triangle APB$ correct to the *nearest tenth of an inch*.

10. In the accompanying figure, $\overline{JK}$, $\overline{KL}$, and $\overline{JL}$ are tangent to circle O at points R, S, and T, respectively. What is the length of $\overline{JTL}$?

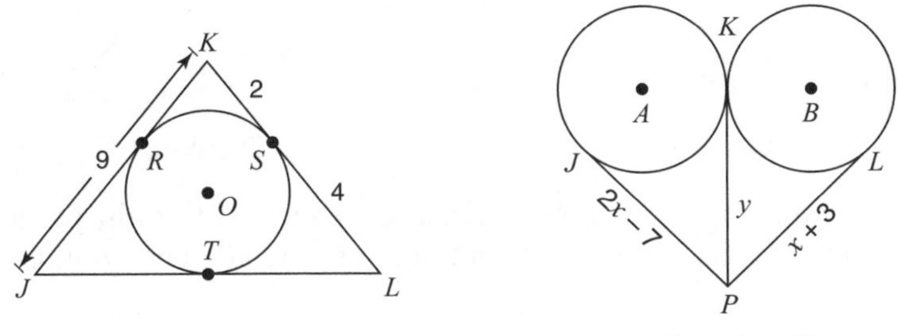

Exercise 10

Exercise 11

11. In the accompanying figure, $\overline{PJ}$ and $\overline{PL}$ are tangent segments drawn to circles A and B, respectively, and $\overline{PK}$ is tangent to both circles at K. Find the values of x and y.

12. From a point R that is 15 inches from the center of circle O, $\overline{RA}$ is drawn tangent to circle O at point A. If the area of circle O is 64π, what is the length of $\overline{RA}$?

13. In the accompanying diagram of circle O, the lengths of the radii of the two concentric circles are 4 and 5. Chord $\overline{AB}$ of the larger circle is tangent to the smaller circle. What is the length of $\overline{AB}$?

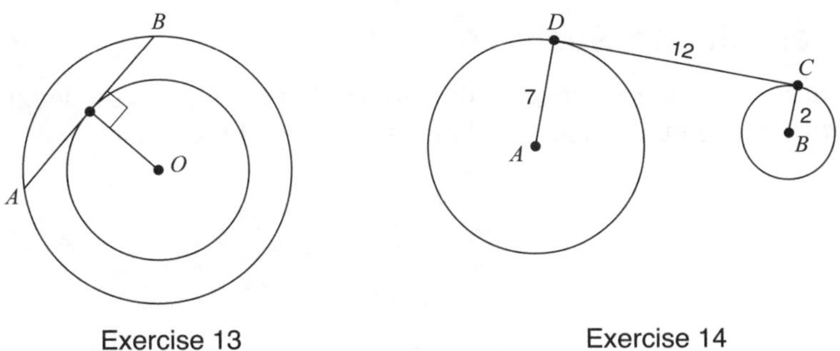

Exercise 13

Exercise 14

14. In the accompanying figure of circles A and B, $\overline{CD}$ is a common tangent segment. The lengths of radii $\overline{AD}$ and $\overline{BC}$ are 7 and 2, respectively. If the length of $\overline{CD}$ is 12, what is the length of $\overline{AB}$?

15. In the accompanying diagram of circle O, $\overline{PA}$ is drawn tangent to the circle at A. Assume B is any point on $\overline{PA}$ between P and A. Write an explanation or an informal proof that shows that $\overline{OB}$ is *not* perpendicular to $\overline{PA}$.

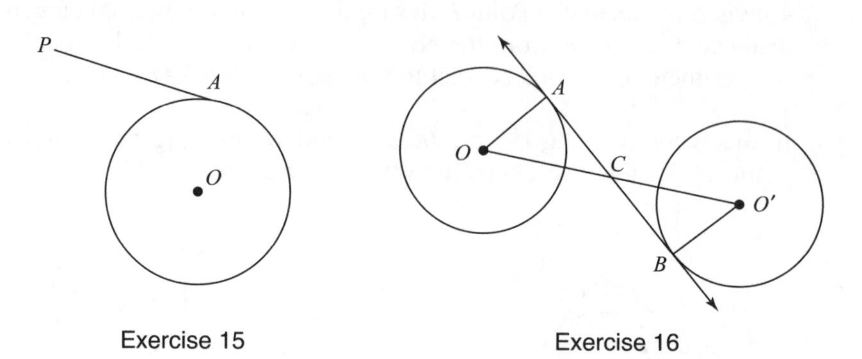

Exercise 15

Exercise 16

16. In the accompanying diagram, circle $O \cong$ circle O', radii $\overline{OA}$ and $\overline{O'B}$ are drawn, and $\overline{OO'}$ intersects common internal tangent $\overline{AB}$ at C.
Prove: $\overline{OB} \cong \overline{O'C}$.

7.3 ARC LENGTH AND AREA

KEY IDEAS

The **length of an arc** of a circle is the distance along the arc measured in *linear units* such as inches or centimeters. A **sector** of a circle is a pie-shaped slice of the circle bounded by two radii and their intercepted arc. The length of an arc and the area of a sector are each a fractional part of the circle containing them.

Arc Length and Sector Area

In the accompanying figure, r is the radius of circle O, L is the length of arc AB, $2\pi r$ is the circumference, and πr^2 is the area of the circle.

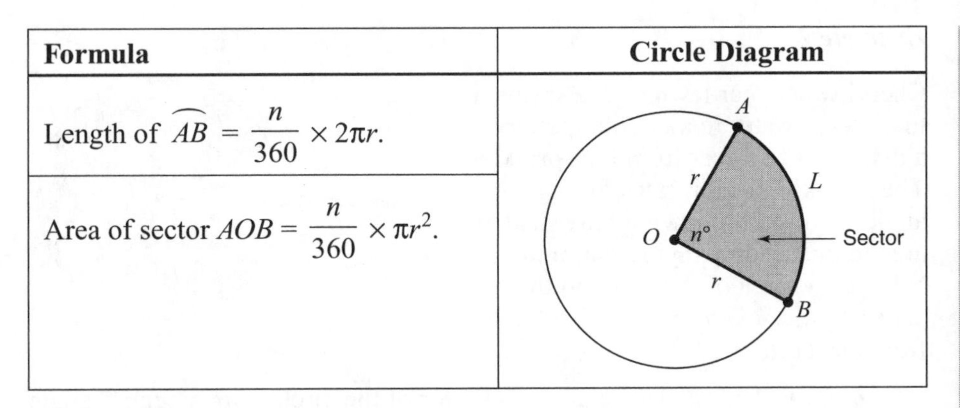

Formula	Circle Diagram
Length of $\overset{\frown}{AB} = \dfrac{n}{360} \times 2\pi r.$	
Area of sector $AOB = \dfrac{n}{360} \times \pi r^2.$	

Example 1

Cities H and K are located on the same line of longitude and the difference in the latitude of these cities is 9°, as shown in the accompanying diagram. If Earth's radius is 3,954 miles, how many miles north of city K is city H along arc HK? Round your answer to the *nearest tenth of a mile.*

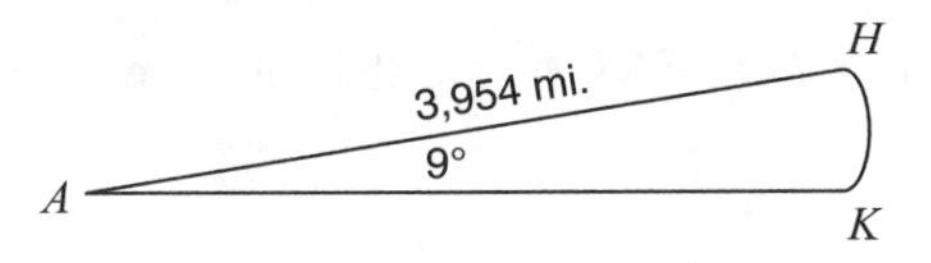

(Not drawn to scale)

Solution: Find the length of arc HK given that the central angle measures 9° with a radius length of 3,954 miles:

$$\text{Length of } \overset{\frown}{HK} = \frac{9}{360} \times \overbrace{2 \times \pi \times 3{,}954}^{\text{circumference}}$$

$$= \frac{1}{40} \times (2 \times \pi \times 3{,}954)$$

Use a calculator:

$$= (2 \times \pi \times 3{,}954) \div 40$$

$$= 621.0928676$$

To the *nearest tenth* of a mile, city H is **621.1 miles** north of city K along arc HK.

Example 2

Cheryl waters her lawn with a sprinker that sprays water in a circular pattern at a distance of 15 feet from the sprinkler. The sprinkler head rotates through an angle of 300°, as shown by the shaded area in the accompanying diagram. What is the area of the lawn, to the *nearest square foot*, that receives water from this sprinkler?

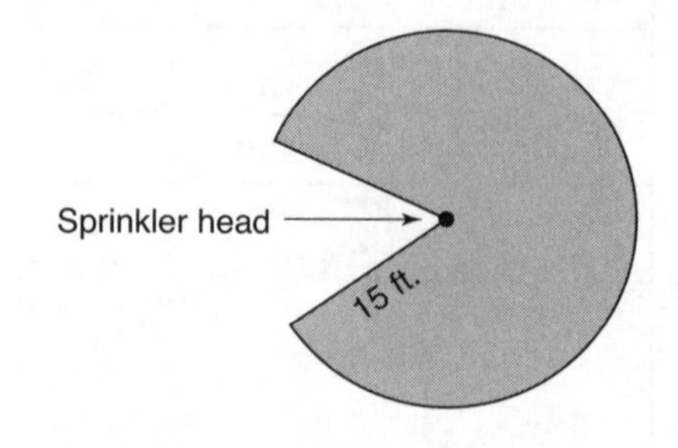

Solution: The shaded region is a sector of the circle with a central angle of 300°.

$$\text{Area of shaded region} = \frac{300}{360} \times \overbrace{(\pi \times 15^2)}^{\text{Area of circle}}$$

$$= (300 \times \pi \times 15 \times 15) \div 360$$

Use a calculator: $= 589.0486225 \text{ ft}^2$

To the *nearest square foot*, **589 ft²** of the lawn receives water from this sprinkler.

Exercise 3

Regular pentagon *ABCDE* is inscribed in circle *O*. If the area of circle *O* is $100\pi \text{ cm}^2$, find the length of $\overparen{ABC}$ in terms of π.

Solution: A pentagon has five sides, so the measure of each central angle is $\frac{360}{5} = 72$. Since area of circle $\pi r^2 = 100\pi \text{ cm}^2$, radius = 10 cm.

$$\text{Length of } \overparen{ABC} = \frac{72+72}{360} \times (2 \times \pi \times 10)$$

$$= \frac{144}{360} \times 20\pi$$

$$= 8\pi \text{ cm}$$

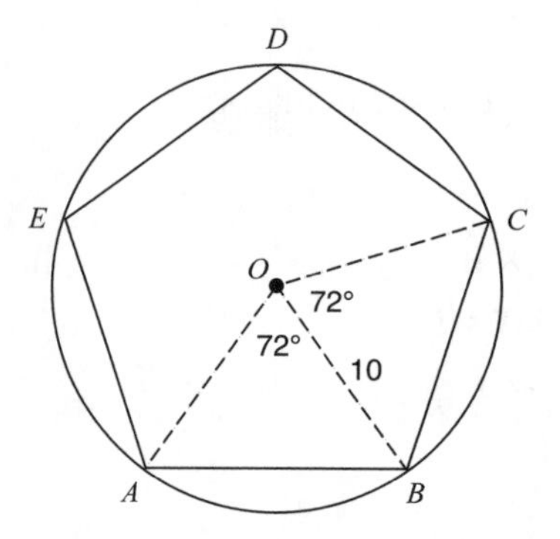

Area of a Segment

A **segment** of a circle is the interior region of the circle bound by a chord and its arc. The area of a segment can be calculated indirectly by subtracting areas.

Example 4

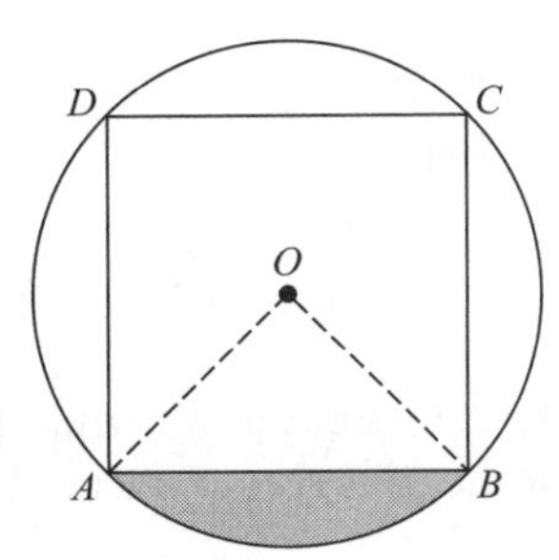

Square $ABCD$ is inscribed in circle O, as shown in the accompanying figure. If the radius is 8 inches, find the area of the shaded region in terms of π.

Solution: Draw radii $\overline{OA}$ and $\overline{OB}$. To find the area of segment AOB, subtract the area of $\triangle AOB$ from the area of sector AOB.

- The area of circle O is $\pi \times 8^2 = 64\pi$. Hence,

$$\text{Area of sector } AOB = \frac{90}{360} \times 64\pi = 16\pi \text{ in}^2$$

- Since a square is a regular polygon with 4 sides, the measure of central angle AOB is $\frac{360}{4} = 90$.

$$\text{Area of right triangle } AOB = \frac{1}{2} \times OA \times OB$$
$$= \frac{1}{2} \times 8 \times 8$$
$$= 32 \text{ in}^2$$

- The area of the shaded region (segment AOB) is the difference between the areas of sector AOB and right triangle AOB:

$$\text{The area of the shaded region} = \mathbf{16\pi - 32 \text{ in}^2}$$

Apothems and Central Angles

In Figure 7.16, O is the center of regular polygon $ABCDE$. $\overline{OA}$ is a radius R of the circumscribed circle, and $\overline{OY}$ is a radius of the inscribed circle. A **central angle** of a regular polygon is an angle whose vertex is the center of the polygon and whose sides are radii drawn to the consecutive vertices. The measure of a central angle of a regular n–gon is $\frac{360}{n}$. Since the regular polygon in Figure 7.16 has five sides,

$$m\angle AOB = \frac{360}{5} = 72$$

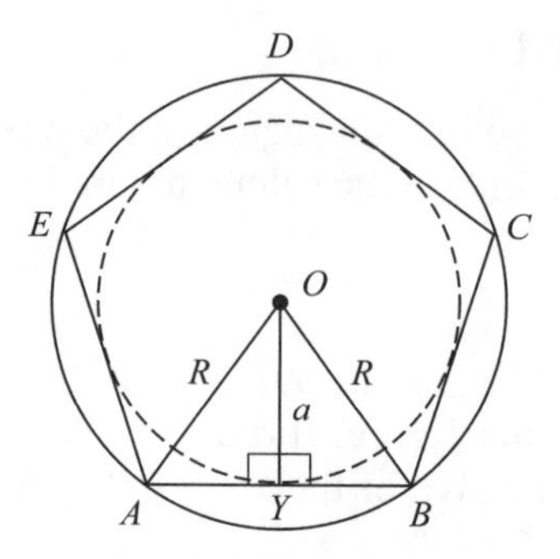

Figure 7.16 Central angle and apothem of a regular polygon.

An **apothem** of a regular polygon is the perpendicular distance from the center of the polygon to a side. In Figure 7.16, $\overline{OY}$ is the apothem drawn to side $\overline{AB}$.

- A regular polygon with n sides has n congruent apothems.
- Each apothem is the perpendicular bisector of the side to which it is drawn. Thus

$$\overline{OY} \perp \overline{AB} \text{ and } \overline{AY} \cong \overline{BY}$$

- An apothem coincides with a radius of the inscribed circle.

Area of a Regular Polygon: $A = \frac{1}{2}ap$

To find the area of any regular polygon with n sides each with length s:

- Separate the polygon into n triangles by drawing radii from its center to the endpoints of each of its sides, as shown in Figure 7.17. These triangles are congruent since $\overline{OA} \cong \overline{OB} \cong \overline{OC} \cong \cdots$, and the included central angles are congruent, which makes the triangles congruent by SAS $\cong$ SAS.
- Since the apothem, a, represents the height of each triangle, the area of representative triangle AOB is $\frac{1}{2} \times s \times a$.

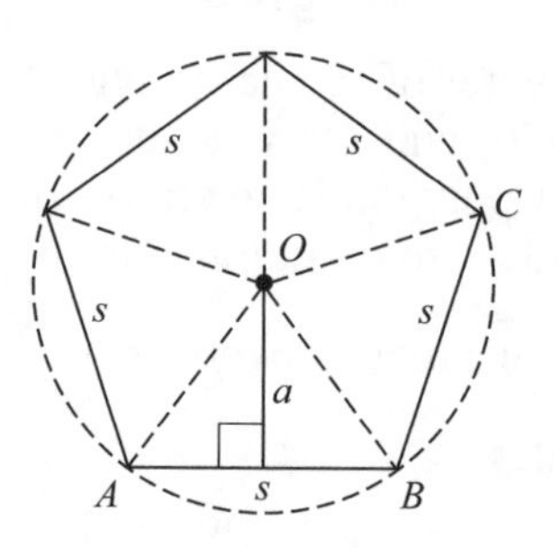

Figure 7.17 Finding the area of a regular polygon.

- Because the n triangles that comprise the regular polygon are congruent, multiplying the area of one such triangle by n gives the area of the entire regular polygon:

$$\begin{aligned}\text{Area of regular } n\text{-gon} &= \text{Total area of } n \text{ triangles}\\ &= n\left(\frac{1}{2}\times s\times a\right)\\ &= \frac{1}{2}a(n\times s) \quad \leftarrow n\times s = \text{perimeter } p\\ &= \frac{1}{2}ap\end{aligned}$$

Theorem: Area of a Regular Polygon

If the apothem of a regular polygon is represented by a and its perimeter represented by p, then its area, A, is given by the formula.

$$A = \frac{1}{2}ap$$

Example 5

A regular hexagon is inscribed in a circle whose area is 144π cm^2. What is the area of the hexagon?

Solution: The measure of a central angle of the regular hexagon is $\frac{360}{6} = 60$. Because the area of the circumscribed circle $= 144\pi$, $R^2 = 144$ so $R = OA = OB = \ldots = 12$ cm. Since $\triangle AOB$ is equiangular, it is also equilateral. Hence, $AB = 12$ cm. Apothem $\overline{OY}$ is the perpendicular bisector of $\overline{AB}$, so $AY = BY = 6$.

Method 1: Use the formula $A = \frac{1}{2}ap$.

- Using the 30–60 right triangle relationships,

  $$OY = AY\times\sqrt{3} = 6\sqrt{3} \text{ cm}$$

- Perimeter of $ABCDEF = 6\times 12$
  $= 72$ cm

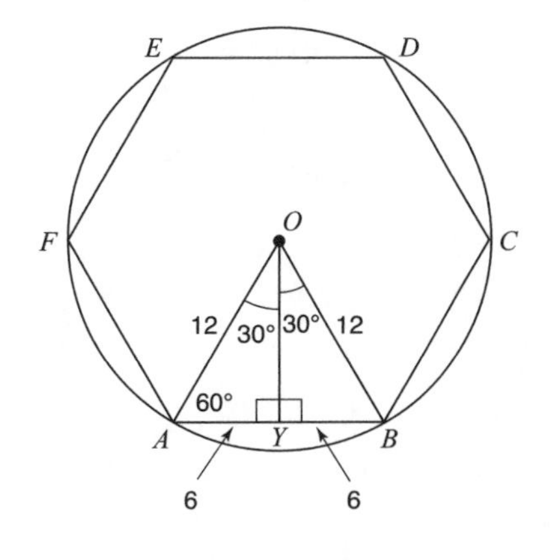

- Area $ABCDEF = \frac{1}{2}ap$

$$= \frac{1}{2} \times (6\sqrt{3}) \times (72)$$

$$= 216\sqrt{3}\ \text{cm}^2$$

Method 2: Find the area of a representative triangle and then multiply that area by 6.

- Since $\triangle AOB$ is equilateral, its area is given by the formula $A = \frac{s^2}{4}\sqrt{3}$

  where $s = 12$ cm:

$$A = \frac{12^2}{4}\sqrt{3} = \frac{144}{4}\sqrt{3} = 36\sqrt{3}\ \text{cm}^2$$

- Because the regular hexagon is comprised of six congruent triangles, find the total area by multiplying the area of $\triangle AOB$ by 6:

$$\text{Area of regular hexagon } 6 \times 36\sqrt{\mathbf{3}} = 216\sqrt{\mathbf{3}}\ \text{cm}^2.$$

The area of the regular hexagon is $\mathbf{216\sqrt{3}\ cm^2}$.

Example 6

Find the area of a regular decagon to the *nearest square inch* if each side measures 30 inches.

Solution: A decagon has ten sides. Draw representative triangle AOB where the measure of central angle AOB is $\frac{360}{10} = 36$.

- Apothem $\overline{OY}$ bisects $\angle AOB$ and $\overline{AB}$. Hence, $m\angle AOY = 18$ and $AY = 15$.
- In right triangle AOB, use the tangent ratio to find OY:

$$\tan 18° = \frac{AY}{OY}$$

$$\tan 18° = \frac{15}{OY}$$

$$OY = 15 \div \tan 18°$$

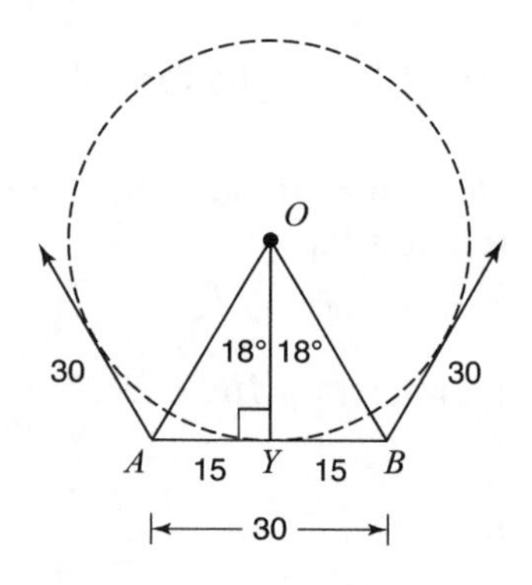

- Use the area formula for a regular polygon:

$$A = \frac{1}{2}ap$$

$$= \frac{1}{2} \times \overbrace{(15 \div \tan 18°)}^{\text{Apothem } OY} \times \overbrace{(30 \times 10)}^{\text{Perimeter}}$$

Use your calculator: $= 6{,}924.787959$

The area of the regular decagon, correct to the *nearest square inch*, is **6,925 in^2**.

Check Your Understanding of Section 7.3

A. Multiple Choice

1. In circle O, radii $\overline{OA}$ and $\overline{OB}$ are drawn such that $m\angle AOB = 72$. If the area of circle O is 20π in^2, what is the number of square inches in the area of sector AOB?
(1) 4π (2) 5π (3) 12π (4) 16π

2. In circle O, radii $\overline{OA}$ and $\overline{OB}$ are drawn such that $m\angle AOB = 45$. If the circumference of circle O is 40π cm, what is the number of centimeters in the circumference of minor arc AB?
(1) 4π (2) 5π (3) 8π (4) 10π

3. In circle O, radii $\overline{OA}$ and $\overline{OB}$ are drawn such that $m\angle AOB = 135$. If the area of circle O is 64π, what is the length of minor arc AB?
(1) 3π (2) 6π (3) 9π (4) 12π

4. In the accompanying figure, points A and B lie on circle O and equilateral triangle AOB is drawn. If the circumference of circle O is 12π, then what is the length of minor arc AB?
(1) π (3) 3π
(2) 2π (4) 4π

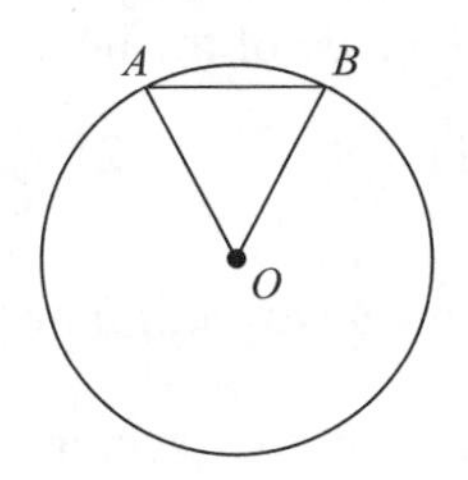

5. The area of a circle is 16π. What is the length of a side of the regular hexagon inscribed in the circle?
(1) 8 (2) 2 (3) 6 (4) 4

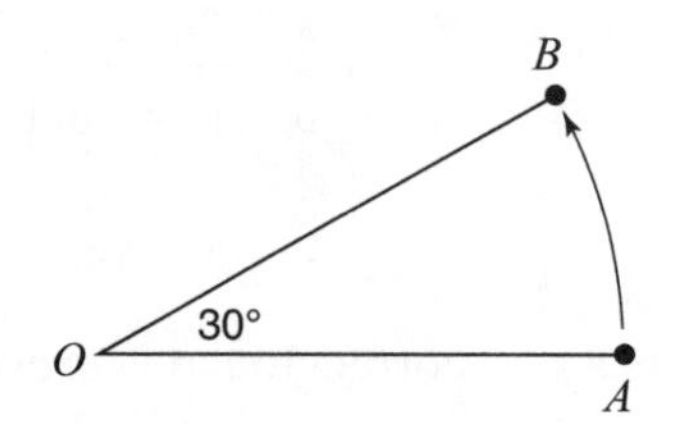

6. In the accompanying figure, a radar tracking beam is centered at O and sweeps in a circular pattern an angle of 30° as the end of the beam moves from point A to point B in the counterclockwise direction. If the length of $\overset{\frown}{AB}$ of circle O is 6π kilometers, what is the number of square kilometers in the area of the region that the radar beam covers when it moves from A to B?
(1) 36π (2) 54π (3) 72π (4) 108π

7. Kira buys a large circular pizza that is divided into eight equal slices. She measures along the outer edge of the crust of one piece and finds the distance to be $5\frac{1}{2}$ inches. What is the *diameter* of the pizza to the *nearest inch*?
(1) 14 (2) 8 (3) 7 (4) 15

8. In a circle, the minor sector formed by two perpendicular radii has an area of 16π. The length of the *major* intercepted arc is
(1) 12π (2) 8π (3) 6π (4) 4π

9. A regular hexagon is inscribed in a circle. What is the ratio of the length of a side of the hexagon to the minor arc that it intercepts?
(1) $\frac{\pi}{6}$ (2) $\frac{3}{6}$ (3) $\frac{3}{\pi}$ (4) $\frac{6}{\pi}$

B. *Show or explain how you arrived at your answer.*

10. In circle O, sector AOB has a central angle of 40°. If the area of the sector is 4π square inches, find in terms of π the number of inches in the length of minor arc AB.

11. A ball is rolling in a circular path that has a radius of 10 inches, as shown in the accompanying diagram. What distance has the ball rolled when the subtended arc measures 54°? Express your answer to the *nearest hundredth of an inch.*

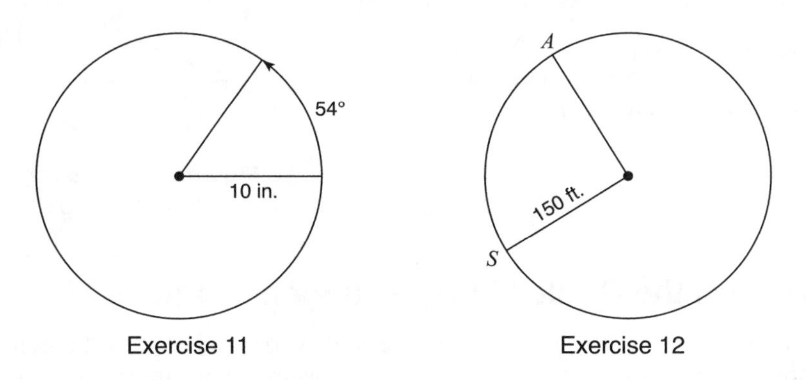

Exercise 11 Exercise 12

12. Kathy and Tami are at point A on a circular track that has a radius of 150 feet, as shown in the accompanying diagram. They run counterclockwise along the track from A to S, a distance of 247 feet. Find, to the *nearest degree,* the measure of minor arc AS.

13. A circle is circumscribed about a traffic sign that has the shape of a regular octagon. If the area of the circle is 256π in^2, find to the *nearest tenth* of an inch the distance along the circle, between two consecutive vertices of the octagon?

14. If each side of a regular polygon with 12 sides measures 54.0 cm, find
a. The length of the minor arc intercepted by a side of the polygon correct to the *nearest tenth* of a centimeter
b. The area of the polygon correct to the *nearest square centimeter*

7.4 CIRCLE ANGLE MEASUREMENT

KEY IDEAS

The vertex of an angle whose sides are chords, secants, or tangents may lie on the circle, inside the circle, or outside the circle. In each case, the location of the vertex determines the relationship between the measure of the angle and the measures of the intercepted arc or arcs.

Vertex on the Circle: Inscribed Angle

An **inscribed angle** of a circle is an angle whose vertex is on the circle and whose sides are chords. In Figure 7.18, inscribed angle ABC intercepts $\overset{\frown}{AC}$. An inscribed angle is measured by one-half of the measure of its intercepted arc:

$$m\angle ABC = \frac{1}{2}m\overset{\frown}{AC} = \frac{1}{2}(50) = 25$$

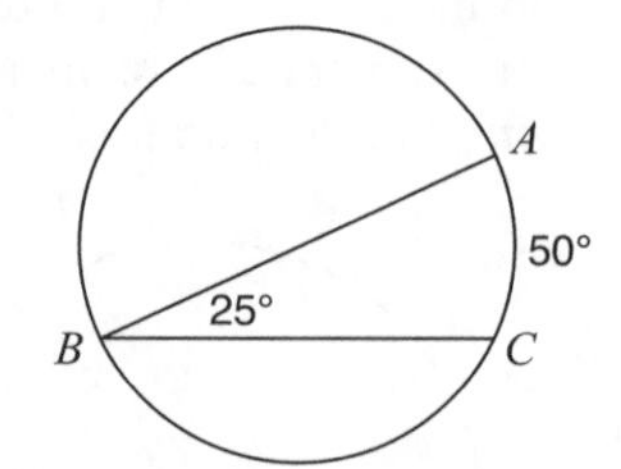

Figure 7.18 Measuring an inscribed angle.

Vertex on the Circle: Chord–Tangent Angle

An angle whose vertex is on the circle and whose sides are a tangent and a chord is measured by one-half of the measure of its intercepted arc. In Figure 7.19:

$$m\angle RST = \frac{1}{2}m\overset{\frown}{RS} = \frac{1}{2}(140) = 70$$

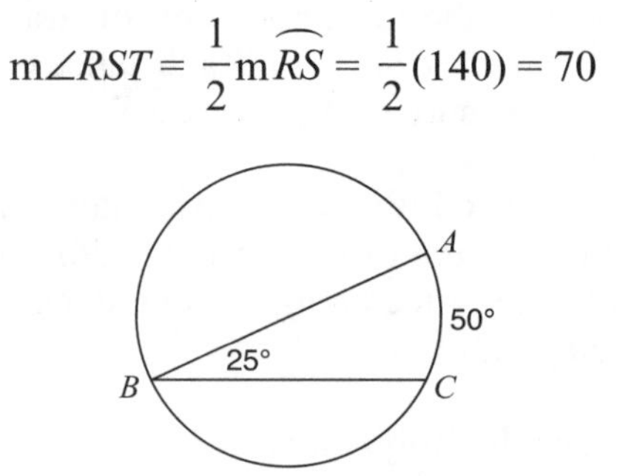

Figure 7.19 Measuring an angle formed by a tangent and a chord.

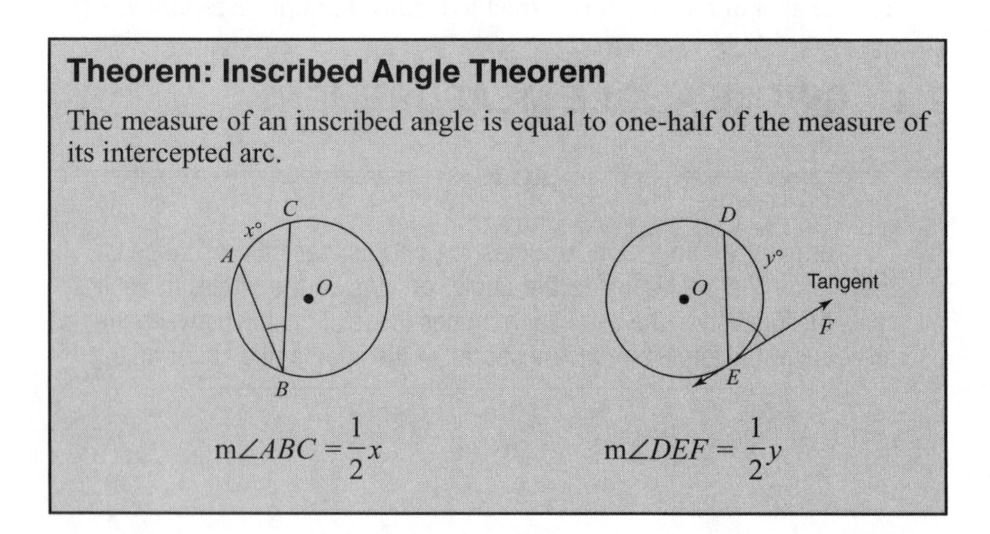

Theorem: Inscribed Angle Theorem

The measure of an inscribed angle is equal to one-half of the measure of its intercepted arc.

$$m\angle ABC = \frac{1}{2}x \qquad m\angle DEF = \frac{1}{2}y$$

Theorem: Chord–Tangent Theorem
The measure of an angle formed by a tangent and a chord drawn to the point of tangency is one-half of the measure of its intercepted arc.

Example 1

In each case, find the value of x.

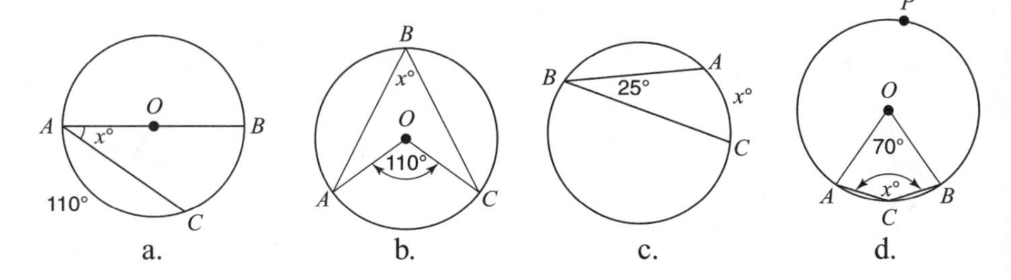

Solution:

a. $\overset{\frown}{ABC}$ is a semicircle, so $\text{m}\overset{\frown}{BC} = 180 - 110 = 70$.

$$\begin{aligned} x &= \frac{1}{2}\text{m}\overset{\frown}{BC} \\ &= \frac{1}{2}(70) \\ &= \mathbf{35} \end{aligned}$$

b. $\angle AOC$ is a central angle, so $\text{m}\overset{\frown}{AC} = 110$.

$$\begin{aligned} \text{m}\angle ABC &= \frac{1}{2}\text{m}\overset{\frown}{AC} \\ &= \frac{1}{2}(110) \\ &= \mathbf{55} \end{aligned}$$

c. The degree measure of an arc intercepted by an inscribed angle must be twice the degree measure of the inscribed angle. Hence

$$\begin{aligned} x &= 2(\text{m}\angle ABC) \\ &= 2(25) \\ &= \mathbf{50} \end{aligned}$$

d. $\angle AOB$ is a central angle, so $m\widehat{ACB} = 70$.

$$m\widehat{APB} = 360 - 70 = 290$$

$$x = \frac{1}{2}m\widehat{APB}$$

$$= \frac{1}{2}(290)$$

$$= \mathbf{145}$$

Example 2

In the accompanying diagram, $\overrightarrow{DE}$ is tangent to circle O at D, $\overline{AOB}$ is a diameter, and $\overline{CD}$ is parallel to $\overline{AOB}$. If $m\angle DAB = 21$, find $m\angle CDE$.

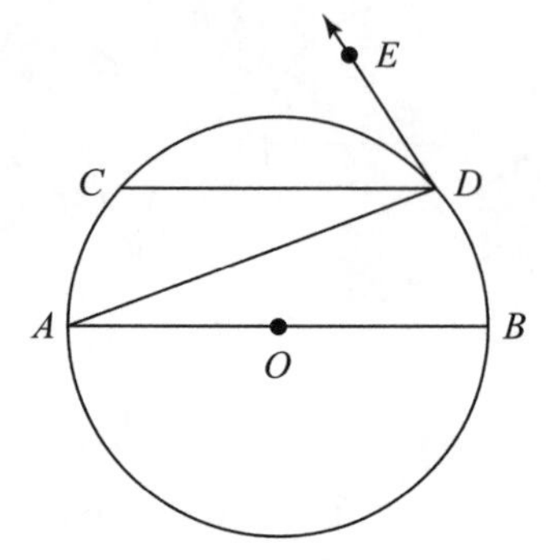

Solution:

- Angle CDE intercepts arc CD, so the degree measure of this arc is needed. Since the sum of the degree measures of the arcs of a semicircle is 180:

$$m\widehat{AC} + m\widehat{CD} + m\widehat{DB} = 180.$$

- The degree measure of inscribed angle DAB is 21. The degree measure of its intercepted arc, $\widehat{DB}$, must be twice as great, or 42. Since parallel chords intercept equal arcs, $m\widehat{AC} = m\widehat{DB} = 42$. Hence:

$$42 + m\widehat{CD} + 42 = 180$$

$$m\widehat{CD} = 180 - 84$$

$$= 96$$

- Since $\angle CDE$ is formed by a chord and a tangent:

$$m\angle CDE = \frac{1}{2}m\widehat{CD}$$

$$= \frac{1}{2}(96)$$

$$= \mathbf{48}$$

Corollaries of Inscribed Angle Theorem

Several important relationships follow directly from the Inscribed Angle Theorem.

- Corollary 1: If two inscribed angles intercept the same or congruent arcs, then the angles are congruent. In the accompanying figure, inscribed angles B and D are congruent because they intercept the same arc, $\widehat{AC}$.

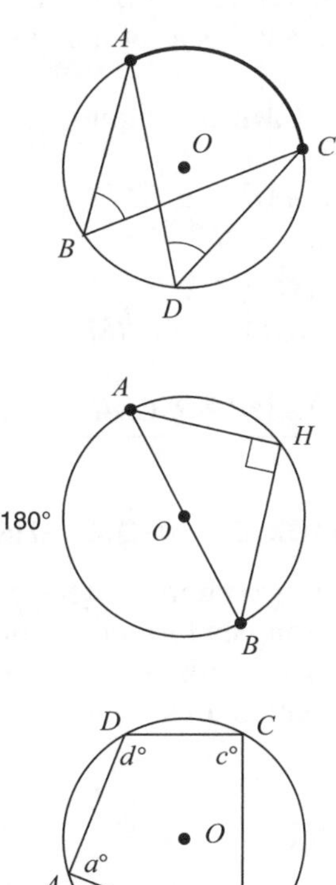

- Corollary 2: An angle inscribed in a semicircle is a right angle. In the accompanying figure, because $\angle AHB$ is inscribed in a semicircle, its intercepted arc is also a semicircle so

$$m\angle AHB = \frac{1}{2} m\widehat{AB} = \frac{1}{2}(180) = 90$$

- Corollary 3: The opposite angles of an inscribed quadrilateral are supplementary. In the accompanying figure,

$$a + c = 180 \quad \text{and} \quad b + d = 180$$

Circle Proofs

When the sides of triangles cut off arcs of a circle, circle angle–measurement relationships can sometimes be used to prove that pairs of angles in these triangles are congruent.

Example 3

Given: In circle O, $\overline{AB}$ is a diameter, $\widehat{AM} \cong \widehat{DM}$.
Prove: $\triangle AMB \cong \triangle CMB$.

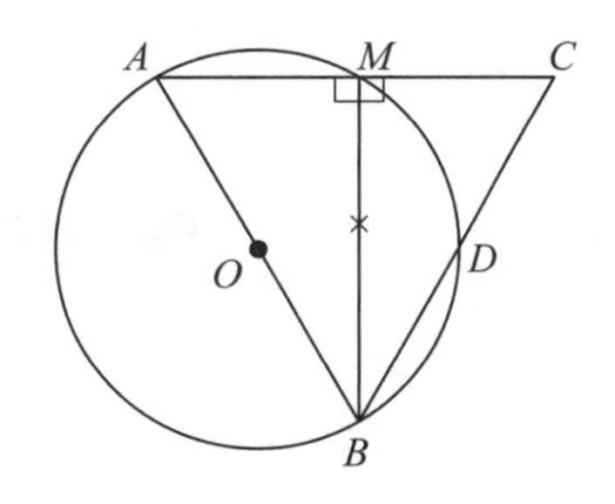

Solution: See the proof.

PLAN: The triangles can be proved congruent by using the ASA postulate.

Proof

Statement		Reason
1. In circle O, $\overline{AB}$ is a diameter.		1. Given.
2. $\angle AMB$ is a right angle.		2. An angle inscribed in a semicircle is a right angle.
3. $\angle CMB$ is a right angle.		3. The supplement of a right angle is a right angle.
4. $\angle AMB \cong \angle CMB$	Angle	4. All right angles are congruent.
5. $\overline{MB} \cong \overline{MB}$	Side	5. Reflexive property.
6. $\overset{\frown}{AM} \cong \overset{\frown}{DM}$		6. Given.
7. $\angle ABM \cong \angle CBM$	Angle	7. Inscribed angles that intercept congruent arcs are congruent.
8. $\triangle AMB \cong \triangle CMB$		8. ASA postulate.

Vertex of Angle: Inside the Circle

When two chords intersect in the interior of a circle, as in Figure 7.20, each of the angles formed are opposite two arcs of the circle. The angle formed by the two chords is measured by one-half the *sum* of the measures of the two intercepted arcs. Thus

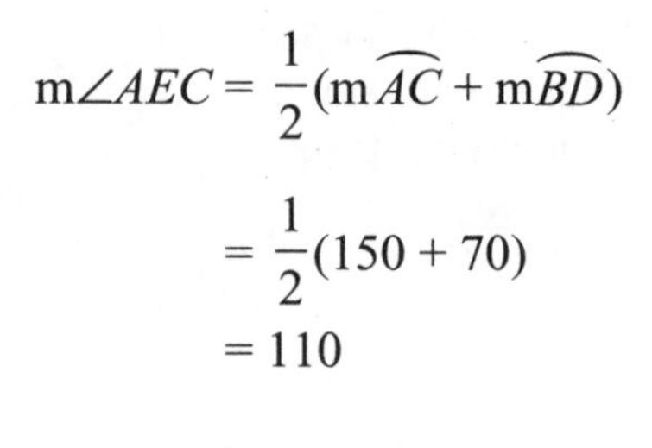

$$m\angle AEC = \frac{1}{2}(m\overset{\frown}{AC} + m\overset{\frown}{BD})$$

$$= \frac{1}{2}(150 + 70)$$

$$= 110$$

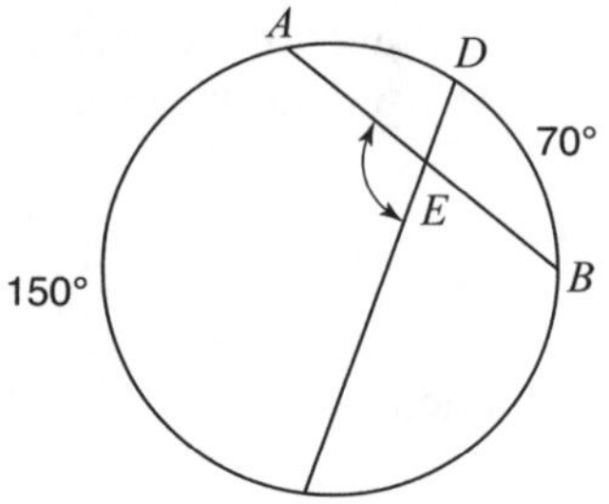

Figure 7.20 Vertex of angle inside circle.

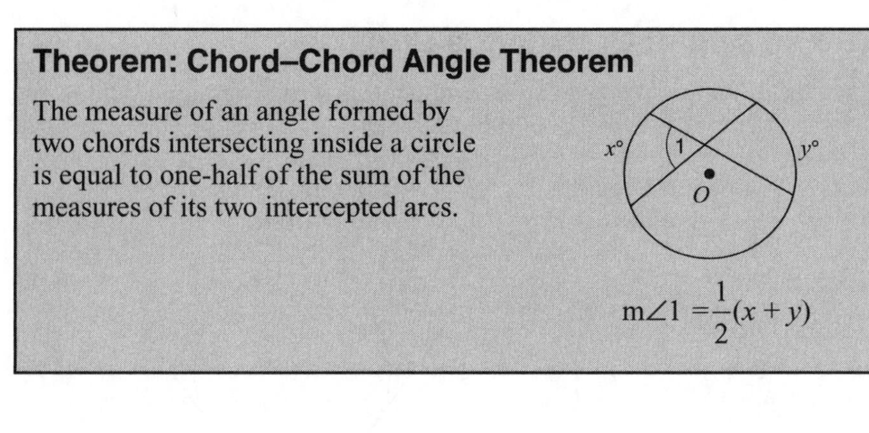

Theorem: Chord–Chord Angle Theorem

The measure of an angle formed by two chords intersecting inside a circle is equal to one-half of the sum of the measures of its two intercepted arcs.

$$m\angle 1 = \frac{1}{2}(x + y)$$

Example 4

In the accompanying diagram, regular pentagon $ABCDE$ is inscribed in circle O. Chords $\overline{AD}$ and $\overline{BE}$ intersect at F, and $\overleftrightarrow{BT}$ is tangent to circle O at B. Find:

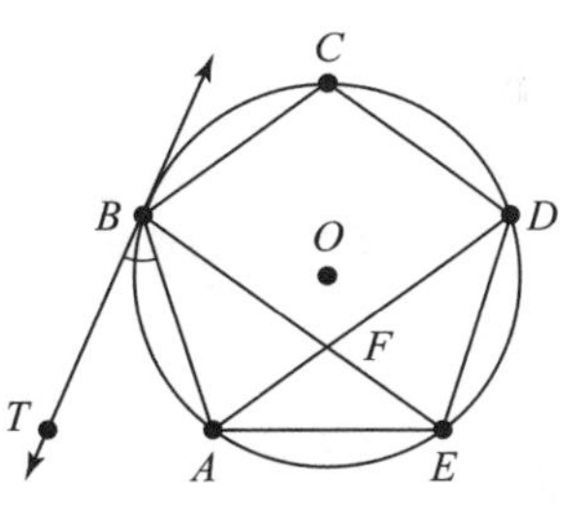

a. $m\angle ABT$ b. $m\angle AFE$

Solution:

a. In a regular pentagon all five sides have the same length; therefore, circle O is divided into five equal arcs. The degree measure of each arc is, therefore, $\frac{360}{5}$ or 72. Hence:

$$m\angle ABT = \frac{1}{2}m\widehat{AB} = \frac{1}{2}(72) = \mathbf{36}$$

b. Angle AFE intercepts arcs BCD and AE. Since $m\widehat{AE} = 72$ and $m\widehat{BCD} = m\widehat{BC} + m\widehat{CD} = 72 + 72 = 144$:

$$\begin{aligned} m\angle AFE &= \frac{1}{2}(\widehat{BCD} + m\widehat{AE}) \\ &= \frac{1}{2}(144 + 72) \\ &= \frac{1}{2}(216) \\ &= \mathbf{108} \end{aligned}$$

Vertex of Angle: Outside the Circle

The vertex of an angle whose sides intercept arcs of a circle may lie in the exterior of a circle, as illustrated in Figure 7.21.

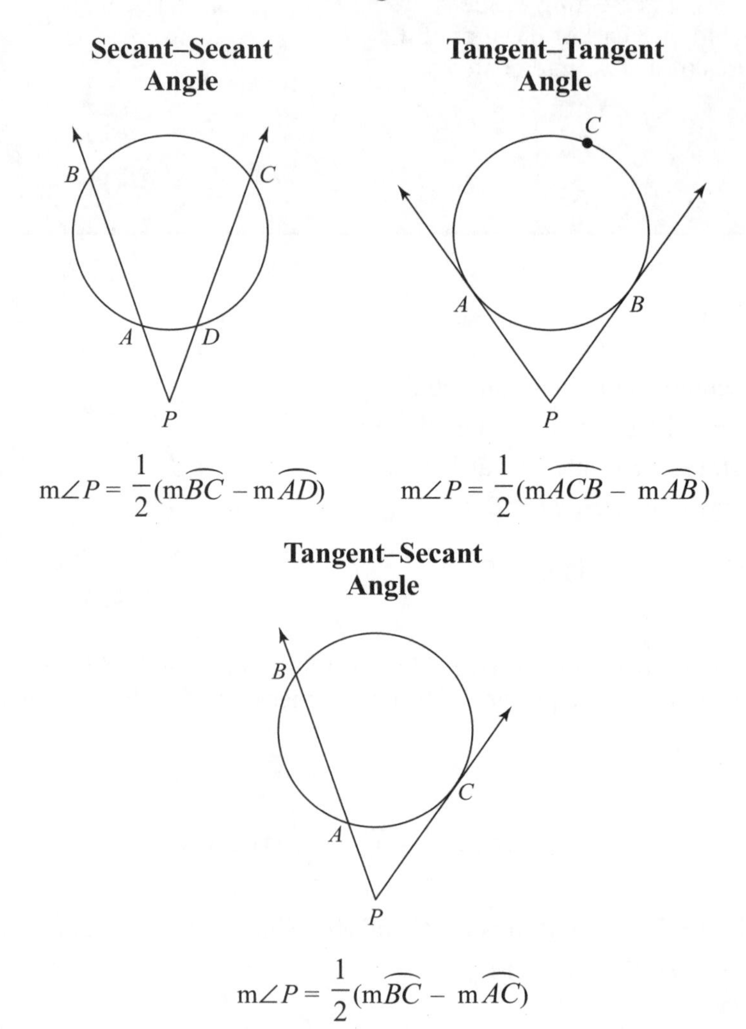

Figure 7.21 Angles whose vertices are in the exterior of the circle.

> **Theorems: Sec–Sec, Tan–Tan, Tan–Sec Angle Theorem**
> The measure of an angle formed by two secants, two tangents, or a tangent and a secant intersecting in the exterior of a circle is one-half of the difference in the measures of its two intercepted arcs.

Example 5

In each case, find the value of x.

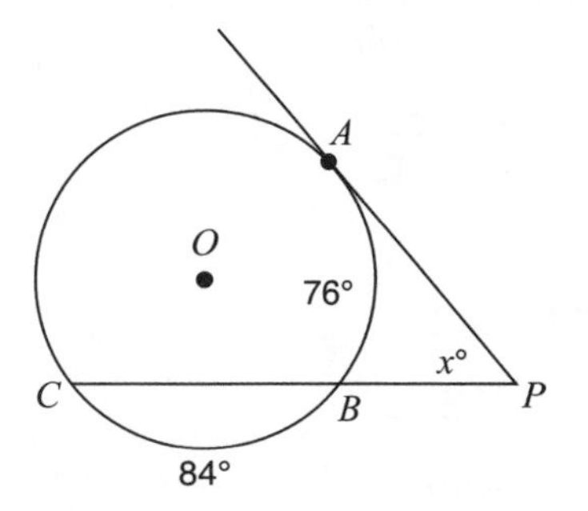

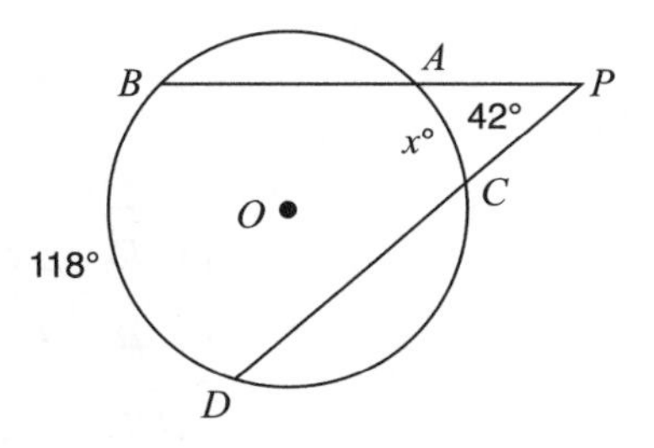

Solution:

a. $m\overset{\frown}{AC} + 76 + 84 = 360$

$$m\overset{\frown}{AC} = 360 - 160$$
$$= 200$$

Hence:

$$x = \frac{1}{2}(m\overset{\frown}{AC} - m\overset{\frown}{AB})$$
$$= \frac{1}{2}(200 - 76)$$
$$= \frac{1}{2}(124)$$
$$= \mathbf{62}$$

b.

$$m\angle P = \frac{1}{2}(m\overset{\frown}{BD} - m\overset{\frown}{AC})$$

$$42 = \frac{1}{2}(118 - x)$$
$$84 = 118 - x$$
$$x = 118 - 84$$
$$= \mathbf{34}$$

Example 6

In the accompanying diagram, $\overrightarrow{PA}$ is tangent to circle O at point A. Secant $\overline{PBC}$ is drawn. Chords $\overline{CA}$ and $\overline{BD}$ intersect at point E. If

$$m\overset{\frown}{AD} : m\overset{\frown}{AB} : m\overset{\frown}{DC} : m\overset{\frown}{BC} = 2:3:4:6$$

find the degree measure of each of the numbered angles.

Solution: First find the degree measures of the arcs of the circle. Let $2x = m\overset{\frown}{AD}$. Then

$$3x = m\overset{\frown}{AB}, \qquad 4x = m\overset{\frown}{CD}, \qquad 6x = m\overset{\frown}{BC}.$$

Since the sum of the degree measures of the arcs of a circle is 360:

$$\begin{aligned} m\overset{\frown}{AD} + m\overset{\frown}{AB} + m\overset{\frown}{DC} + m\overset{\frown}{BC} &= 360 \\ 2x + 3x + 4x + 6x &= 360 \\ 15x &= 360 \\ x = \frac{360}{15} &= 24 \end{aligned}$$

Hence

$$\begin{aligned} m\overset{\frown}{AD} &= 2x = 2(24) = 48 \\ m\overset{\frown}{AB} &= 3x = 3(24) = 72 \\ m\overset{\frown}{DC} &= 4x = 4(24) = 96 \\ m\overset{\frown}{BC} &= 6x = 6(24) = 144 \end{aligned}$$

- $\angle 1$ *is an inscribed angle:*

$$\begin{aligned} m\angle 1 &= \frac{1}{2} m\overset{\frown}{AB} \\ &= \frac{1}{2}(72) \\ &= \mathbf{36} \end{aligned}$$

- $\angle 2$ *is a chord–chord angle:*

$$\begin{aligned} m\angle 2 &= \frac{1}{2}(m\overset{\frown}{AD} + m\overset{\frown}{BC}) \\ &= \frac{1}{2}(48 + 144) \\ &= \mathbf{96} \end{aligned}$$

- $\angle 3$ *is a tangent–chord angle:*

$$\begin{aligned} m\angle 3 &= \frac{1}{2} m\overset{\frown}{AC} = \frac{1}{2}(m\overset{\frown}{AD} + m\overset{\frown}{DC}) \\ &= \frac{1}{2}(48 + 96) \\ &= \mathbf{72} \end{aligned}$$

- $\angle 4$ *is a secant–tangent angle:*

$$\begin{aligned} m\angle 4 &= \frac{1}{2}(m\overset{\frown}{AC} - m\overset{\frown}{AB}) \\ &= \frac{1}{2}(144 - 72) \\ &= \mathbf{36} \end{aligned}$$

- $\angle 5$: Angles 5 and *CBD* are supplementary. The measure of $\angle 5$ may be found indirectly by first finding m$\angle CBD$.

$$\begin{aligned} m\angle CBD &= \frac{1}{2} m\overset{\frown}{DC} = \frac{1}{2}(96) = 48 \\ m\angle 5 &= 180 - m\angle CBD \\ &= 180 - 48 \\ &= \mathbf{132} \end{aligned}$$

MATH FACTS

The measure of an angle whose sides cut off arcs of a circle depends on the location of the vertex of the angle.

Position of Vertex of Angle	Measure of Angle
• On the circle	• $\frac{1}{2}$ the degree measure of the intercepted arc.
• Interior of the circle	• $\frac{1}{2}$ the *sum* of the degree measures of the two intercepted (opposite) arcs.
• Exterior of the circle	• $\frac{1}{2}$ the *difference* of the degree measures of the two intercepted arcs.

Check Your Understanding of Section 7.4

A. Multiple Choice

1. Tangent $\overline{PA}$ and secant $\overline{PBC}$ are drawn to circle O from external point P, and chord $\overline{AB}$ is drawn. If $m\overset{\frown}{AC} = 2\ m\overset{\frown}{AB}$, what is the ratio of $m\angle PAB$ to $m\angle ABC$?
(1) 1:1 (2) 1:2 (3) 3:2 (4) 1:4

2. In the accompanying diagram, $\triangle ABC$ is inscribed in circle R. If $m\overset{\frown}{AB} = 2x$, $m\overset{\frown}{BC} = 5x$, and $m\overset{\frown}{AC} = 150$, which type of triangle is $\triangle ABC$?
(1) right
(2) isosceles
(3) equilateral
(4) obtuse

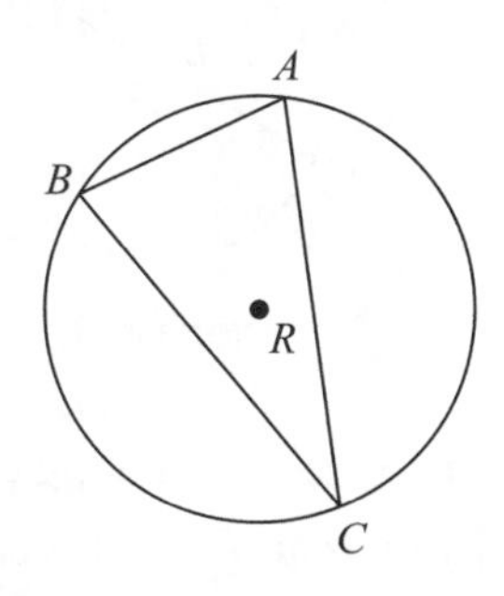

3. In circle O, $\overline{PA}$ and $\overline{PB}$ are tangent to the circle from point P. If the ratio of the measure of major arc AB to the measure of minor arc AB is 5:1, then $m\angle P$ is
(1) 60 (2) 90 (3) 120 (4) 180

4. In the accompanying diagram, quadrilateral $ABCD$ is inscribed in circle O, $m\angle BAD = 80$ and $\overrightarrow{BCE}$ is drawn. What is $m\angle DCE$?
(1) 60 (2) 80 (3) 100 (4) 120

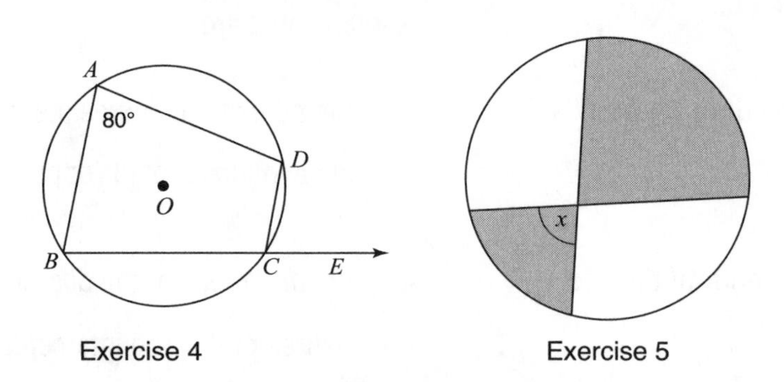

Exercise 4 Exercise 5

5. The accompanying diagram shows a child's spin toy that is constructed from two chords intersecting in a circle. The curved edge of the larger shaded section is one-quarter of the circumference of the circle, and the curved edge of the smaller shaded section is one-fifth of the circumference of the circle. What is the measure of angle x?
(1) 40 (2) 72 (3) 81 (4) 108

6. In the accompanying diagram, chords $\overline{AB}$ and $\overline{CD}$ intersect at E. If $m\angle DEB = 55$ and $m\overset{\frown}{DB} = 70$, what is $m\overset{\frown}{AC}$?
(1) 15 (2) 40 (3) 55 (4) 70

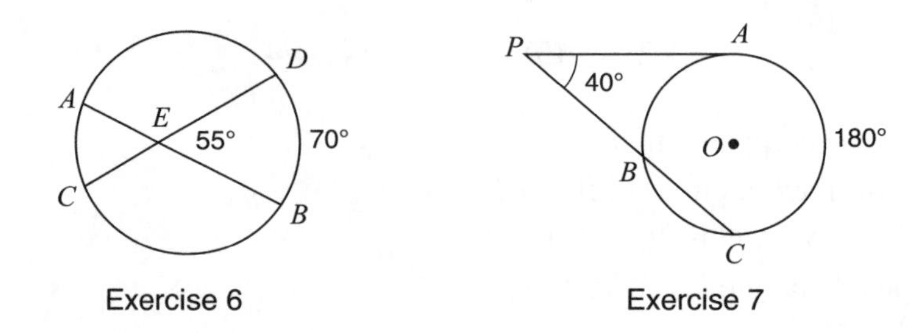

Exercise 6 Exercise 7

7. In the accompanying diagram of circle O, the measure of angle P formed by tangent $\overline{PA}$ and secant $\overline{PBC}$ is 40. If $m\overset{\frown}{AC} = 180$, find $m\overset{\frown}{AB}$.
(1) 50 (2) 70 (3) 100 (4) 140

8. Two tangents to a circle from the same external point intercept a major arc whose measure is 210. What is the measure of the angle formed by the tangents?
(1) 30 (2) 75 (3) 90 (4) 150

9. In the accompanying figure, $\overline{AB}$ is tangent to circle O at E, $\overline{DE}$, $\overline{EF}$, and $\overline{FD}$ are chords, and $m\angle AEF = 136$. What is $m\angle D$?
(1) 68 (2) 22 (3) 88 (4) 44

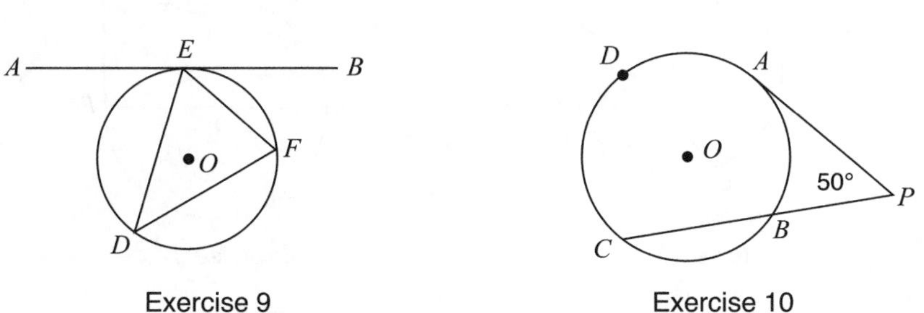

Exercise 9 Exercise 10

10. In the accompanying diagram, tangent $\overline{PA}$ and secant $\overline{PBC}$ are drawn to circle O. If $m\overset{\frown}{ADC}$ is twice $m\overset{\frown}{AB}$ and $m\angle P$ is 50, what is $m\overset{\frown}{AB}$?
(1) 25 (2) 50 (3) 100 (4) 200

B. *Show or explain how you arrived at your answer.*

11–14. Find the value of x.

11.

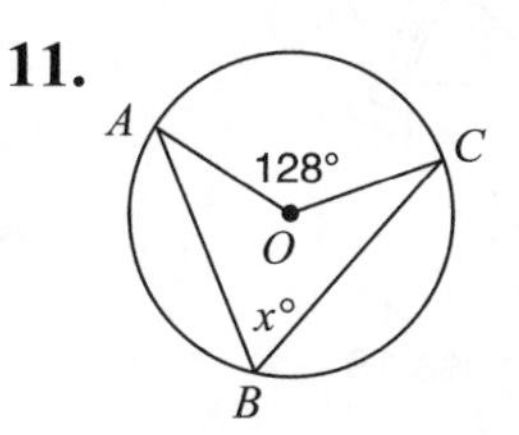

13.

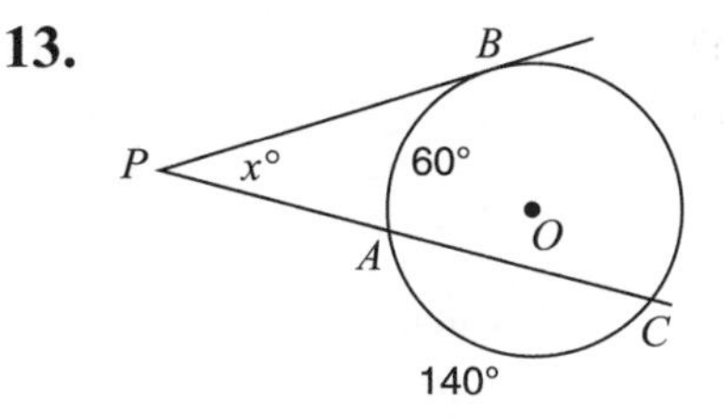

12.

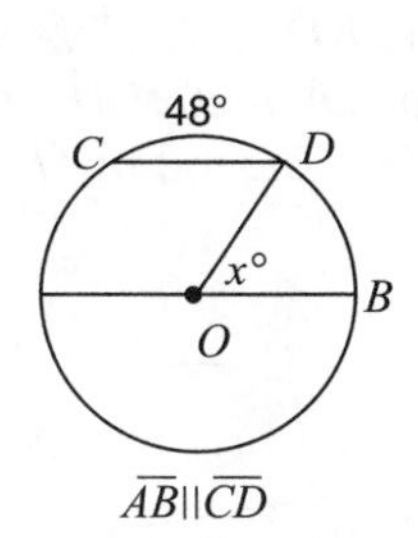

14.

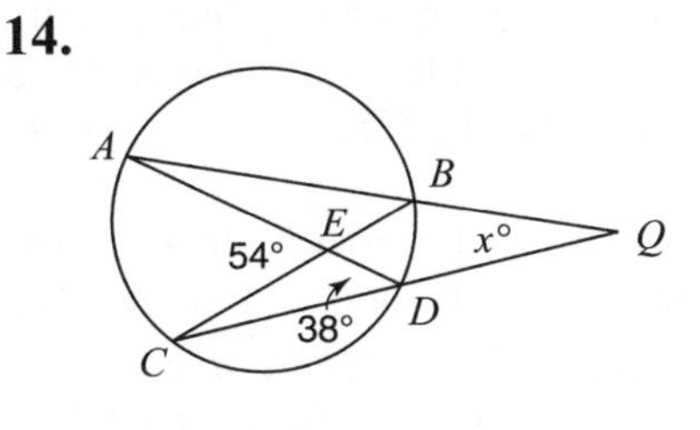

15. Point P lies outside circle O, which has a diameter of $\overline{AOC}$. The angle formed by tangent $\overline{PA}$ and secant $\overline{PBC}$ measures 30°. Find the number of degrees in the measure of minor arc CB.

16. In the accompanying diagram, chords $\overline{AB}$ and $\overline{CD}$ intersect at E. If $m\angle AEC = 4x$, $m\widehat{AC} = 120$, and $m\widehat{DB} = 2x$, what is the value of x?

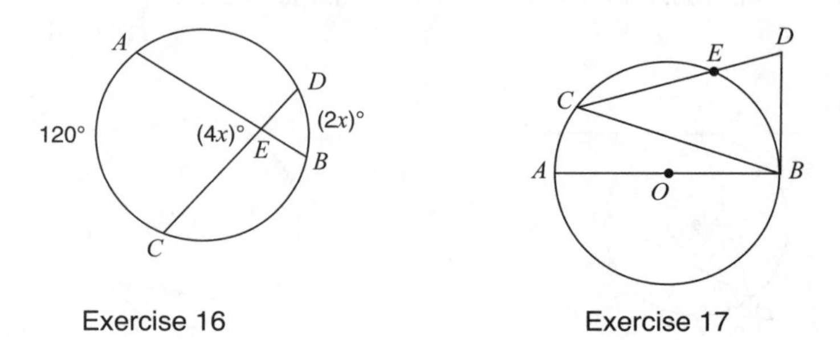

Exercise 16 Exercise 17

17. In the accompanying diagram, $\overline{CED}$ is a secant, $\overline{BD}$ is tangent to circle O at B, $\overline{BC}$ is a chord, and $\overline{BOA}$ is a diameter. If $m\widehat{AC} : m\widehat{CB} = 1:4$ and $m\widehat{CE} = 68$, find $m\angle BDC$.

18. In the accompanying diagram, chord $\overline{BE}$ bisects $\angle ABC$. If $m\angle ABC = 70$ and $m\,\widehat{BAE} = 200$, find $m\angle AFE$.

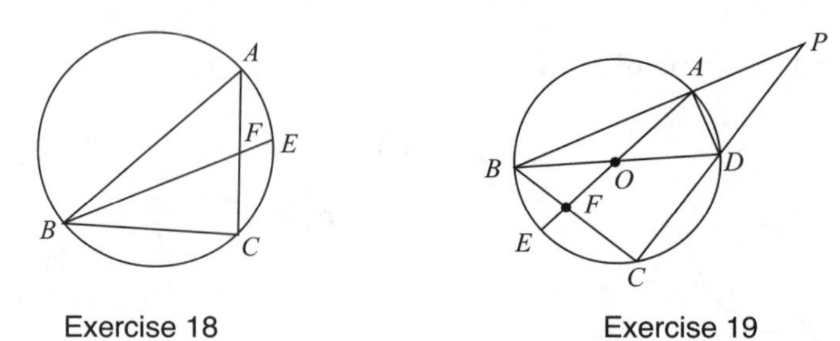

Exercise 18 Exercise 19

19. In the accompanying diagram of circle O, diameters $\overline{BD}$ and $\overline{AE}$, secants $\overline{PAB}$ and $\overline{PDC}$, and chords $\overline{AD}$ and $\overline{BC}$ are drawn. Diameter $\overline{AE}$ intersects chord $\overline{BC}$ at F. If $m\angle ABD = 20$ and $m\angle BFE = 75$, find $m\angle P$.

20. In the accompanying diagram of circle O, chords $\overline{AB}$ and $\overline{CD}$ intersect at E and $m\widehat{AC} : m\widehat{CB} : m\widehat{BD} : m\widehat{DA} = 4:2:6:8$. What is $m\angle DEB$?

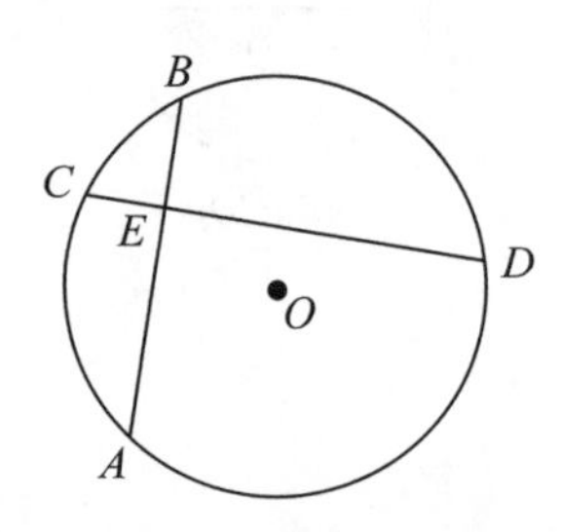

21. A machine part consists of a circular wheel with an inscribed triangular plate, as shown in the accompanying diagram. If $\overline{SE} \cong \overline{EA}$, $SE = 10$ inches, and m$\overset{\frown}{SE} = 140$, find the length of $\overline{SA}$ to the *nearest tenth of an inch*.

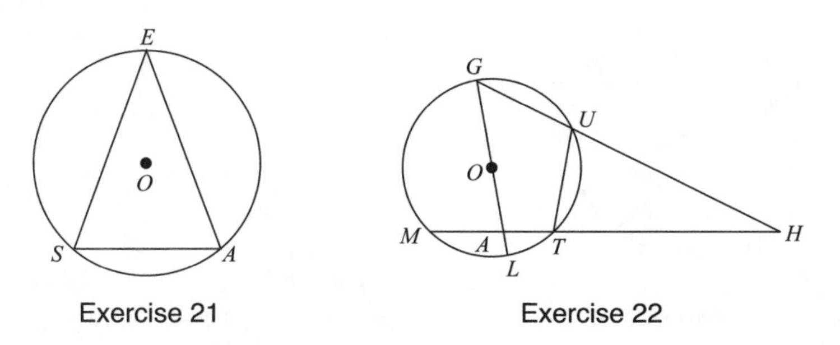

Exercise 21   Exercise 22

22. Given circle O with diameter $\overline{GOAL}$; secants $\overline{HUG}$ and $\overline{HTAM}$ intersect at point H; m$\overset{\frown}{GM}$: m$\overset{\frown}{ML}$:m$\overset{\frown}{LT} = 7 : 3 : 2$; and $\overline{GU} \cong \overline{UT}$. Find the ratio of m$\angle UGL$ to m$\angle H$.

23. In circle O, tangent $\overrightarrow{PW}$ and $\overline{PST}$ are drawn. Chord $\overline{WA}$ is parallel to chord $\overline{ST}$. Chords $\overline{AS}$ and $\overline{WT}$ intersect at point B. If m$\overset{\frown}{WA}$: m$\overset{\frown}{AT}$: m$\overset{\frown}{ST}$ = $1:3:5$, find

a. m$\angle TBS$
b. m$\angle TWP$
c. m$\angle WPT$
d. m$\angle ASP$

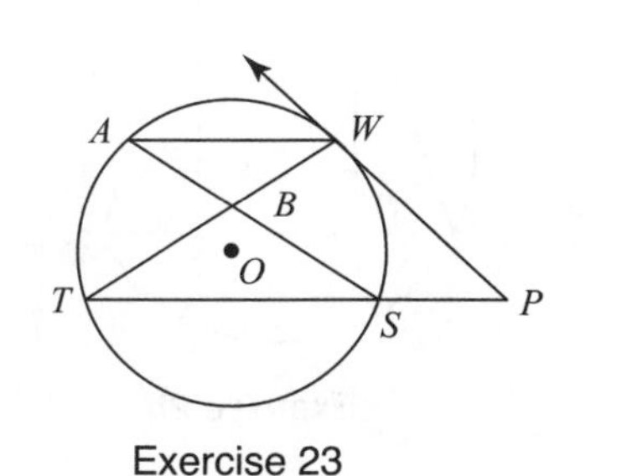

Exercise 23

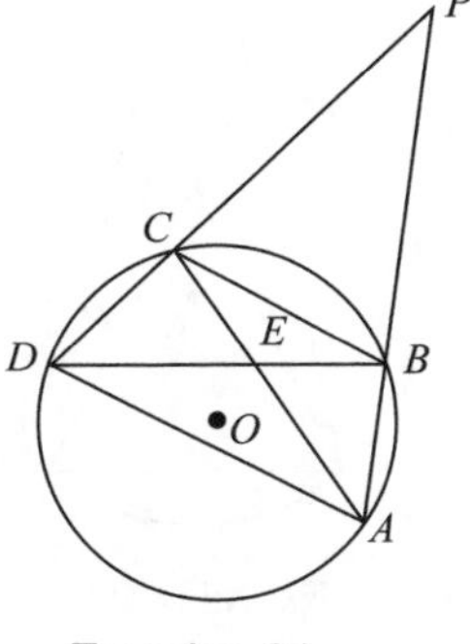

Exercise 24

24. In the accompanying diagram, $\overline{PCD}$ and $\overline{PBA}$ are secants from external point P to circle O. Chords $\overline{DA}$, $\overline{DEB}$, $\overline{CEA}$, and $\overline{CB}$ are drawn. $\overline{AB} \cong \overline{DC}$, m$\overset{\frown}{BC}$ is twice m$\overset{\frown}{AB}$, and m$\overset{\frown}{AD}$ is 60 more than m$\overset{\frown}{BC}$. Find:

a. m$\angle P$
b. m$\angle PCB$

25. In the accompanying diagram of circle O, $\overline{AB} \cong \overline{CD}$, $m\angle DFB = 70$, $m\overset{\frown}{AB} = 115$, and $m\overset{\frown}{DE} = 65$. Find the measure of $\angle P$.

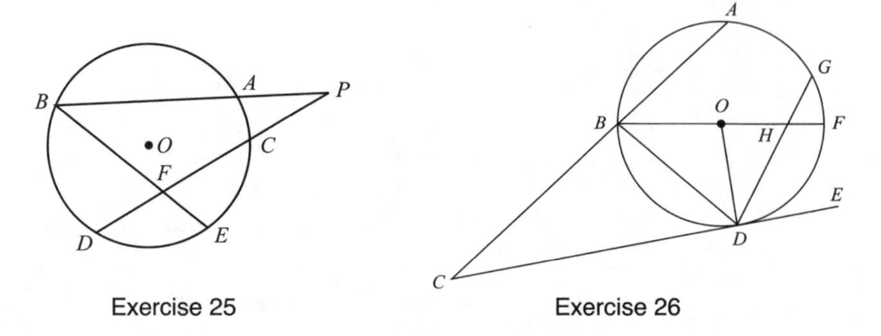

Exercise 25

Exercise 26

26. In the accompanying diagram, circle O has radius $\overline{OD}$, diameter $\overline{BOHF}$, secant $\overline{CBA}$, and chords $\overline{DHG}$ and $\overline{BD}$; $\overline{CE}$ is tangent to circle O at D; $m\overset{\frown}{DF} = 80$; and $m\overset{\frown}{BA} : m\overset{\frown}{AG} : m\overset{\frown}{GF} = 3 : 2 : 1$.
Find $m\overset{\frown}{GF}$, $m\angle BHD$, $m\angle BDG$, $m\angle GDE$, $m\angle C$, and $m\angle BOD$.

27. Given: Secants $\overline{PDA}$ and $\overline{PCB}$ are drawn to circle O, $\overline{PDA} \cong \overline{PCB}$, chords $\overline{AC}$ and $\overline{BD}$ intersect at F.
Prove: $\overline{AF} \cong \overline{BF}$.

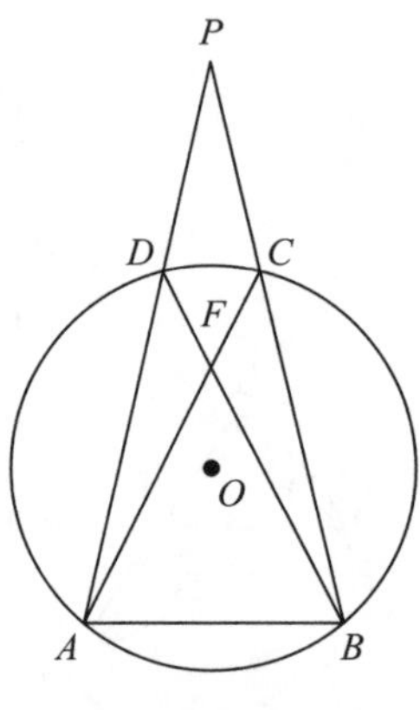

Exercise 27

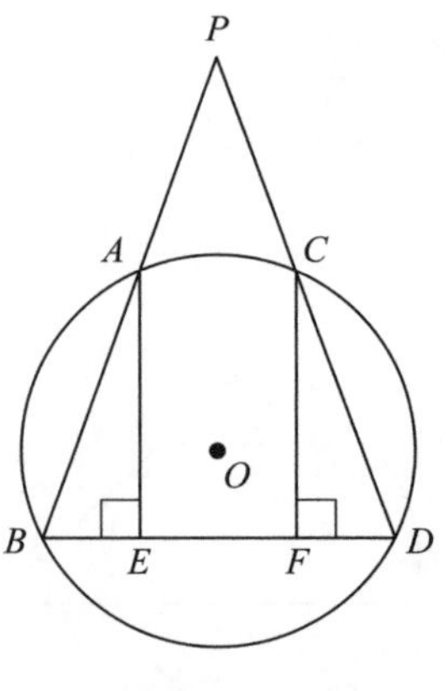

Exercise 28

28. Given: Secants $\overline{PAB}$ and $\overline{PCD}$ are drawn to circle O, $\overline{PAB} \cong \overline{PCD}$, chords $\overline{AE} \perp \overline{BD}$, and $\overline{CF} \perp \overline{BD}$.
Prove: $\overline{AE} \cong \overline{CF}$.

7.5 SIMILAR TRIANGLES AND CIRCLES

Proving that triangles with chord, tangent, or secant segments as sides are similar can lead to some useful relationships involving the lengths of these segments.

Similar Triangle Proofs with Circles

Circle angle–measurement relationships can be used to help prove two triangles are similar.

Example 1

Given: $\overline{DB}$ is tangent to circle P at B, $\overline{AB}$ is a diameter, and $\overline{CD} \perp \overline{DB}$.

Prove: $\frac{AB}{BC} = \frac{BC}{CD}$.

Solution: See the proof.

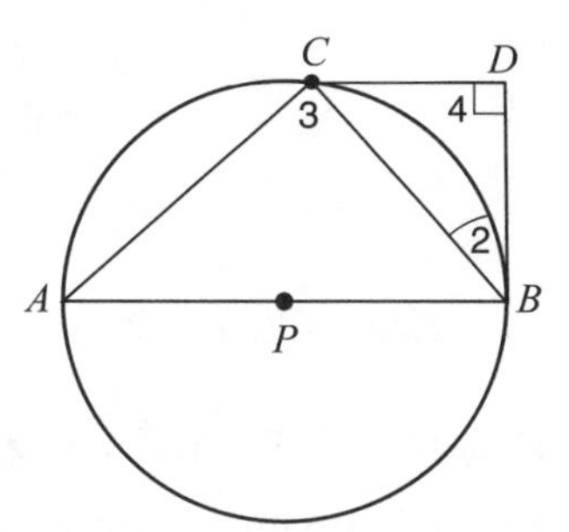

PLAN: Prove that triangles ABC and CBD are similar.

Proof

- Angle 1 is an inscribed angle, and $\angle 2$ is formed by a tangent and a chord. Since the degree measure of each of these angles is one-half of the degree measure of the same arc, $\overset{\frown}{BC}$, $\angle 1 \cong \angle 2$.
- Angle 3 is inscribed in a semicircle, and $\angle 4$ is formed by perpendicular lines. Since each angle is a right angle, $\angle 3 \cong \angle 4$.
- Hence, $\triangle ABC \sim \triangle CBD$ by the AA Theorem of Similarity. Since the lengths of corresponding sides of similar triangles are in proportion:

$$\frac{AB}{BC} = \frac{BC}{CD}$$

Segments of Intersecting Chords

When two chords intersect inside a circle, similar triangles can be used to prove that the product of the lengths of the segments of one chord is equal to the product of the segments of the other chord. In Figure 7.22, $\triangle AED \sim \triangle CEB$ from which it follows:

$$AE \times EB = CE \times ED$$

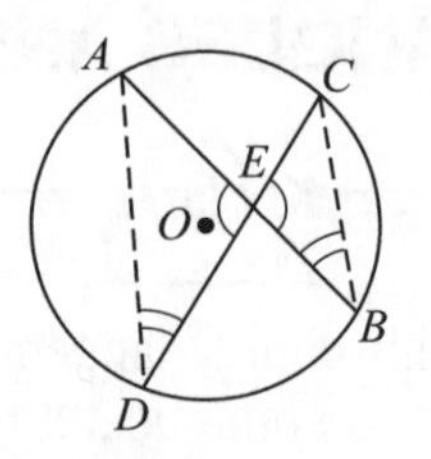

Figure 7.22 $AE \times EB = CE \times ED$

Example 2

In the accompanying diagram of circle O, chord $\overline{AB}$ bisects chord $\overline{CD}$ at E. If $AE = 4$ and $EB = 9$, what is the length of $\overline{CD}$?

Solution: If $x = CE = ED$, then

$$(CE)(ED) = (EB)(AE)$$
$$(x)(x) = (9)(4)$$
$$x^2 = 36$$
$$x = \sqrt{36} = 6$$

Hence, $CD = CE + ED = 6 + 6 =$ **12**.

Tangent–Secant Segments

In the accompanying figure of circle O, the length of tangent segment PT is denoted by t, and the length of secant $\overline{PAB}$ is denoted by s. Segment $\overline{PA}$ falls outside the circle and is called an **external secant segment**, denoted by e. Drawing $\overline{AT}$ and $\overline{BT}$ forms similar triangles:

$$\triangle PAT \sim \triangle PTB$$

It follows that

$$\frac{s}{t} = \frac{t}{e} \quad \text{or} \quad t^2 = s \times e$$

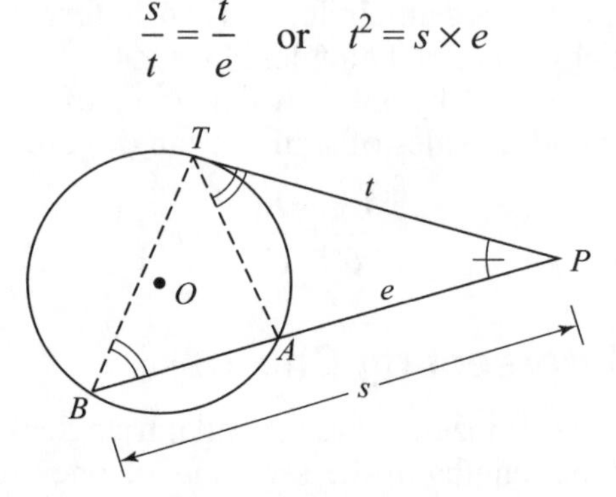

Figure 7.23 Tangent–secant segments.

Example 3

In the accompanying diagram, $\overline{PT}$ is tangent to circle O at T and $\overline{PAB}$ is a secant. If $PA = 4$ and $AB = 12$, find PT.

Solution: Since $PA = 4$ and $AB = 12$, $PAB = 4 + 12 = 16$. Hence:

$$\begin{aligned}(PT)^2 &= (PAB)(PA)\\ &= (16)(4)\\ &= 64\\ PT &= \sqrt{64} = \mathbf{8}\end{aligned}$$

Secant–Secant Segments

In Figure 7.24, e_1 and e_2 represent the external segments of secants $\overline{PCD}$ and $\overline{PAB}$, respectively. Drawing $\overline{AD}$ and $\overline{BC}$ forms similar triangles:

$$\triangle PDA \sim \triangle PBC$$

It follows that

$$\frac{s_1}{s_2} = \frac{e_2}{e_1} \quad \text{or} \quad s_1 \times e_1 = s_2 \times e_2$$

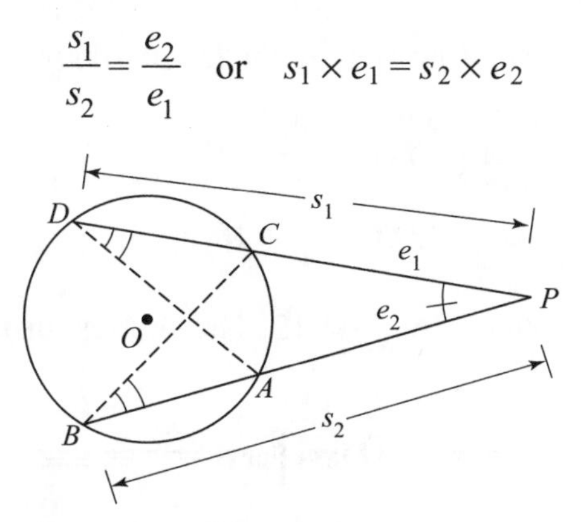

Figure 7.24 Secant–secant segments.

Example 4

In the accompanying diagram, $\overline{NEW}$ and $\overline{NTA}$ are secants to circle O. If $NE = 5$, $NT = 4$, and $TA = 6$, find WE.

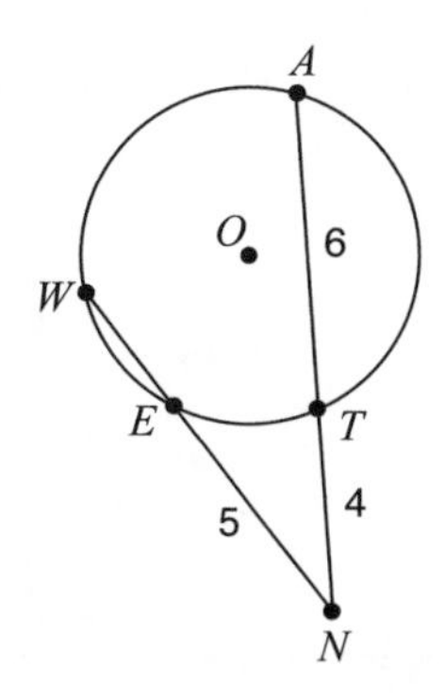

Solution: Find WE by first finding NEW:

$$(NE) \times (NEW) = (NT) \times (NTA)$$
$$5 \times (NEW) = 4 \times (4 + 6)$$
$$NEW = \frac{40}{5} = 8$$

Hence, $WE = NEW - NE = 8 - 5 = \mathbf{3}$.

Example 5

In the accompanying diagram, cabins B and G are located on the shore of a circular lake, and cabin L is located near the lake. Point D is a dock on the lake shore, and is collinear with cabins B and L. The road between cabins G and L is 8 miles long and is tangent to the lake. The path between cabin L and dock D is 4 miles long. What is the shortest distance, in miles, from cabin B to dock D?

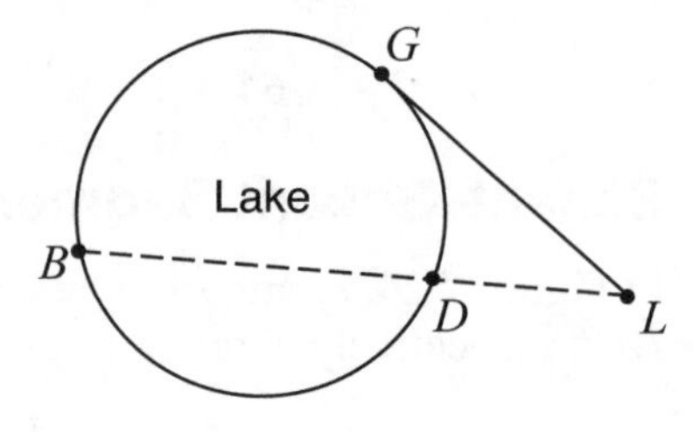

Solution: Find BD by first finding the length of secant $\overline{LDB}$:

$$(LD)(LDB) = (LG)^2$$
$$(4)(LDB) = (8)^2$$
$$LDB = \frac{64}{4} = 16$$

Hence, $BD = LDB - LD = 16 - 4 = 12$. The shortest distance from cabin to dock is **12 mi**.

MATH FACTS

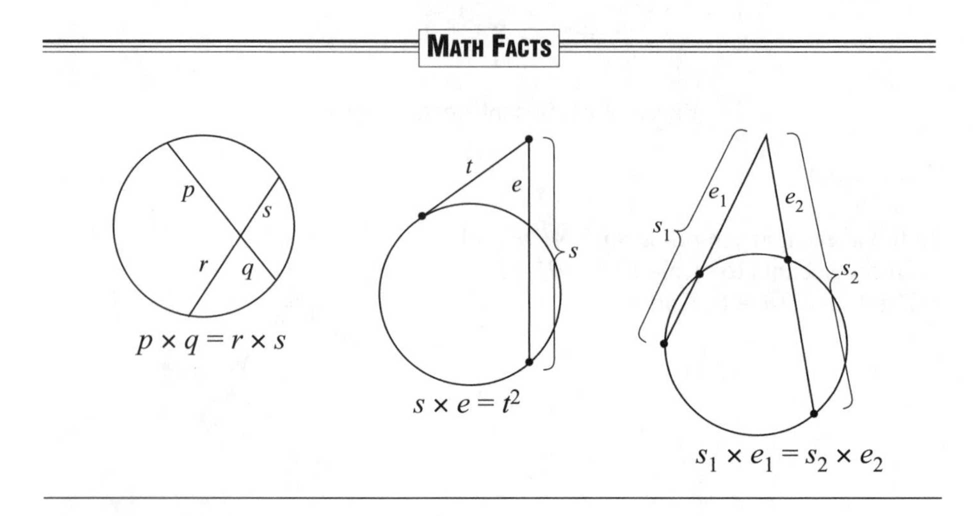

Check Your Understanding of Section 7.5

A. Multiple Choice

1. In the accompanying diagram, $\overline{PAB}$ and $\overline{PCD}$ are secants drawn to circle O, $\overline{PA} = 8$, $\overline{PB} = 20$, and $\overline{PD} = 16$. What is the length of $\overline{PC}$?
(1) 6.4 (2) 10 (3) 12 (4) 40

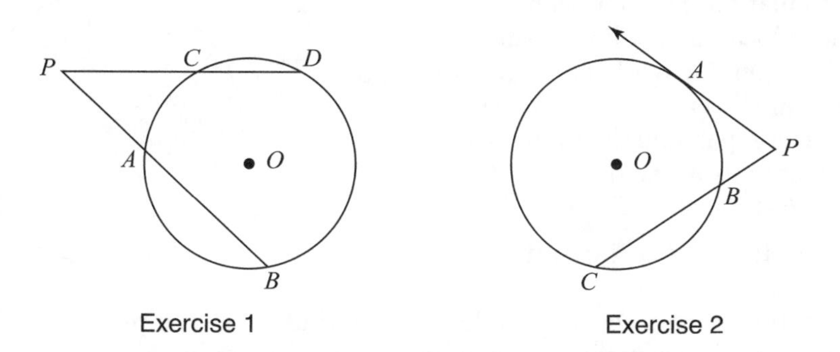

Exercise 1 Exercise 2

2. In the accompanying diagram, $\overrightarrow{PA}$ is tangent to circle O at A. If $CB = 12$ and $PB = 4$, what is the length of $\overline{PA}$?
(1) $4\sqrt{3}$ (2) 48 (3) $16\sqrt{3}$ (4) 8

3. In circle O, chords $\overline{AB}$ and $\overline{CD}$ intersect at E. If $AE = 4$, $EB = 12$, and $ED = 16$, then CE equals
(1) 19 (2) 16 (3) 3 (4) 48

4. In the accompanying diagram, $\overrightarrow{BA}$ is tangent to circle O at point A and $\overrightarrow{BCD}$ is a secant. If $AB = 12$ and $BC = 9$, what is CD?
(1) 7 (2) 9 (3) 16 (4) 108

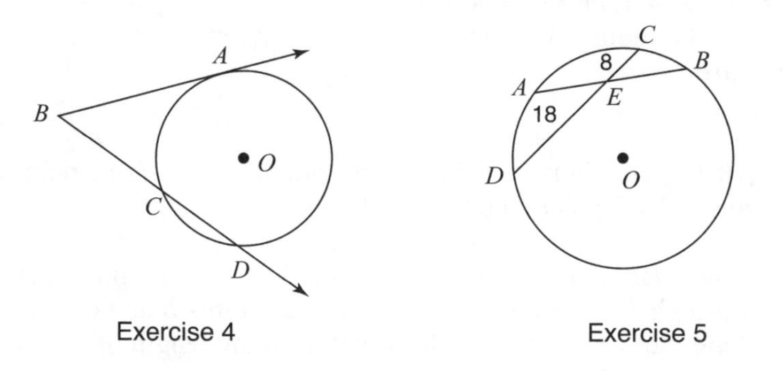

Exercise 4 Exercise 5

5. In the accompanying diagram, chords $\overline{AB}$ and $\overline{CD}$ of circle O intersect at point E, the midpoint of $\overline{AB}$. If $CE = 8$ and $ED = 18$, what is the length of $\overline{AB}$?
(1) 16 (2) 24 (3) 12 (4) 10

6. In circle O, chords $\overline{AB}$ and $\overline{CD}$ intersect at point E. If $AE = 9$, $EB = 16$, $CE = x$, and $ED = y$, then
(1) $xy = 144$ (2) $9x = 16y$ (3) $16x = 9y$ (4) $x + y = 25$

7. Kim wants to determine the radius of a circular pool without getting wet. She is located at point K, which is 4 feet from the pool and 12 feet from the point of tangency, as shown in the accompanying diagram. What is the radius of the pool?
(1) 16 ft (3) 32 ft
(2) 20 ft (4) $4\sqrt{10}$ ft

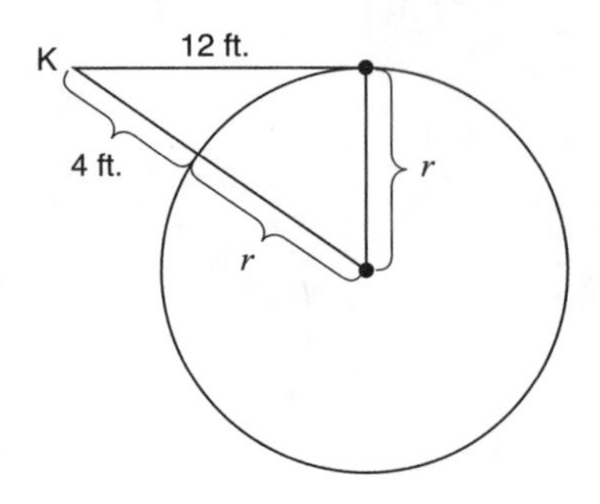

8. Two chords intersect in a circle whose radius is 8. The product of the lengths of the segments of one of the chords can not be
(1) 80 (2) 63 (3) 48 (4) 32

B. *Show or explain how you arrived at your answer.*

9. In circle O, chords $\overline{AB}$ and $\overline{CD}$ intersect at point E. If $AE = 2$, $CD = 9$, and $CE = 4$, find AB.

10. In circle O, diameter $\overline{AB}$ is perpendicular to chord $\overline{CD}$ at point E, $CD = 8$, and $EB = 2$. What is the length of diameter $\overline{AB}$?

11. In the accompanying diagram, secants $\overline{PAB}$ and $\overline{PCD}$ are drawn to circle O. If $PC = 5$, $CD = 7$, and $PA = 4$, find the length of a radius of circle O to the *nearest tenth.*

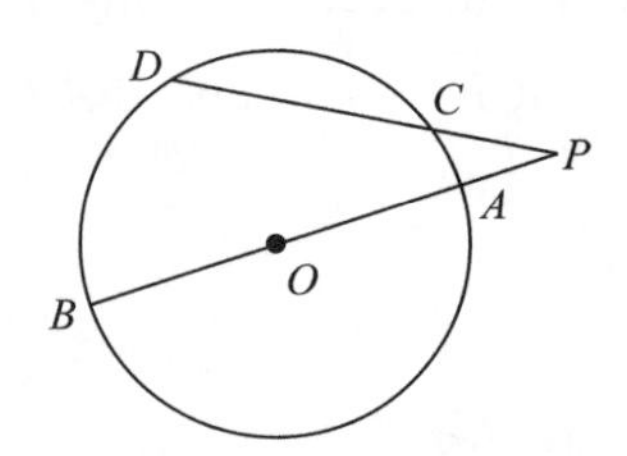

12. Chord $\overline{AB}$ bisects chord $\overline{CD}$ in the interior of circle O at point E. If $AE = 4$ and $AB = 20$, find the length of $\overline{CD}$.

13. From point P, $\overline{PA}$ is drawn tangent to circle O at A. A secant drawn from point P through point O intersects the circle at points B and C. If $PA = 12$ and the length of a radius of circle O is 9, find the length of the external segment of secant $\overline{PBC}$.

14. Secants $\overline{PAB}$ and $\overline{PCD}$ are drawn to circle O. If A is the midpoint of $\overline{PAB}$, chord $CD = 14$, and $PC = 4$, what is the length of $\overline{PAB}$?

15. Tangent $\overrightarrow{PC}$ and secant $\overline{PAB}$ are drawn to circle O, as shown in the accompanying diagram. If PA is 6 less than PC and AB is 1 more than two times PC, find the length of $\overline{PC}$.

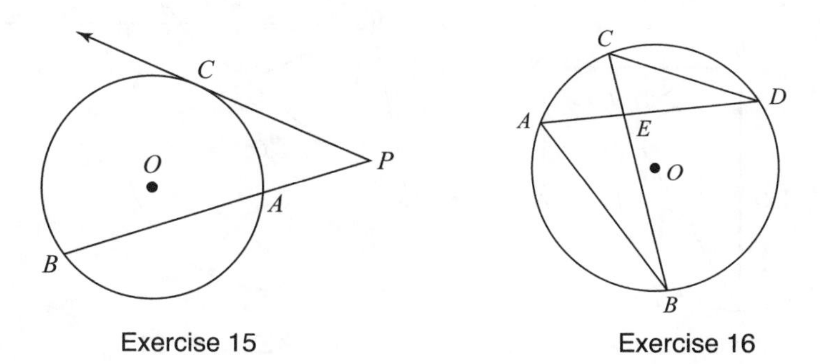

Exercise 15 Exercise 16

16. In the accompanying diagram of circle O, chords $\overline{AD}$ and $\overline{BC}$ intersect at E, chords $\overline{AB}$ and $\overline{CD}$ are drawn. Write an explanation or an informal proof that shows $\triangle AEB$ and $\triangle CED$ are similar.

17. Given: $\overline{MJ}$ is tangent to circle O at point J, $\overline{KJ}$ is a diameter, N is the midpoint of $\widehat{LNJ}$.

Prove: $\dfrac{KL}{KJ} = \dfrac{KP}{KM}$.

18. Given: $\overline{MJ}$ is tangent to circle O at point J, $\overline{KJ}$ is a diameter, $\overline{JP} \cong \overline{JM}$.
Prove: $KL \times KM = JK \times LP$.

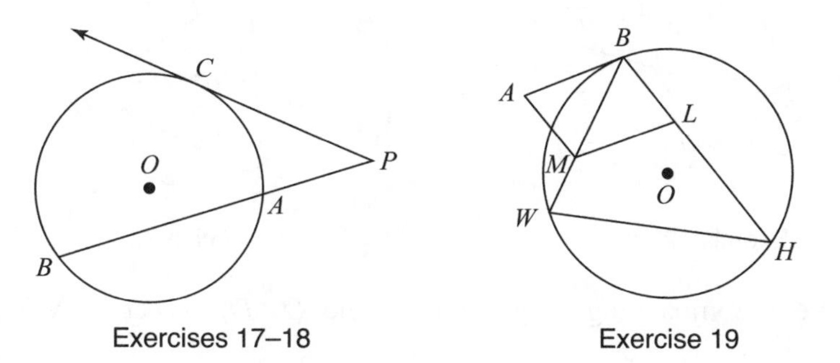

Exercises 17–18 Exercise 19

19. Given: $\overline{AB}$ is tangent to circle O at point B, chords $\overline{WB}$, $\overline{BH}$, and $\overline{HW}$ are drawn; quadrilateral $ABLM$ is a parallelogram.

Prove: $\dfrac{BL}{BW} = \dfrac{BM}{BH}$.

20. Given: $\overline{CEB}$ is tangent to circle O at B, diameter $\overline{AB} \parallel \overline{CD}$. Secants $\overline{DFB}$ and $\overline{EFA}$ intersect circle O at F.

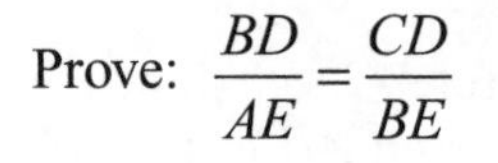

Prove: $\dfrac{BD}{AE} = \dfrac{CD}{BE}$

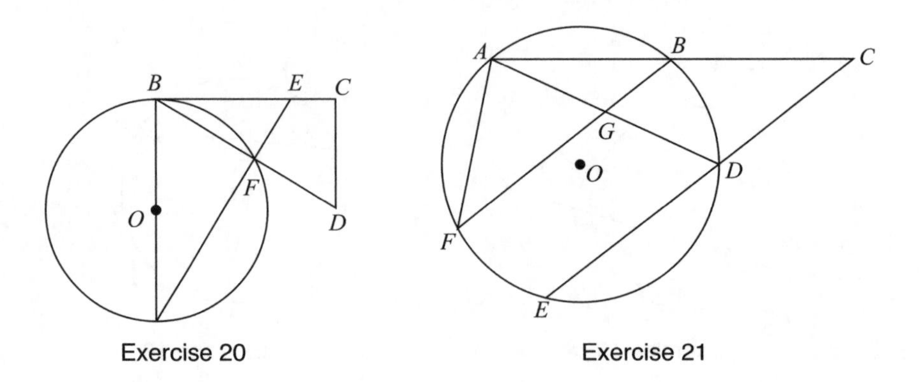

Exercise 20

Exercise 21

21. Given: Circle O, chord $\overline{BGF}$ is parallel to secant $\overline{CDE}$, B is the midpoint of $\overset{\frown}{AD}$, secant $\overline{CBA}$, chord $\overline{AGD}$, and chord $\overline{AF}$ are drawn.
Prove: $CD \times BF = CA \times BA$.

22. Given in the accompanying diagram, $\overleftrightarrow{TP}$ is a common tangent to circles A and B at P.
Prove: $MN \times PR = PN \times QR$.

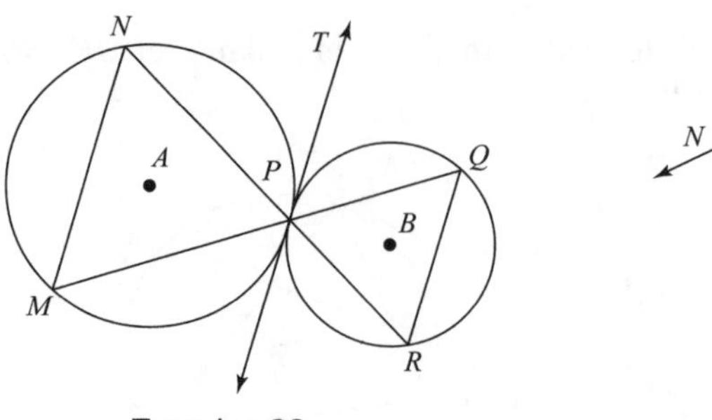

Exercise 22

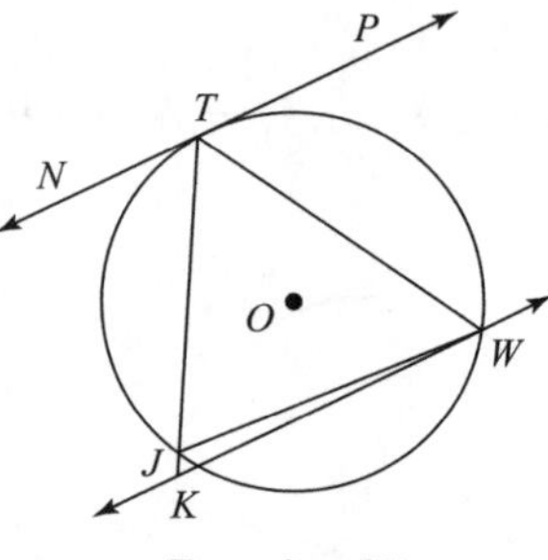

Exercise 23

23. In the accompanying diagram of circle O, $\overline{TK}$ bisects $\angle NTW$, and $\overline{WK} \cong \overline{WT}$.
Prove: $(TW)^2\ JT \times TK$.

24. In the accompanying figure, $\overrightarrow{PA}$ is a tangent to circle O at point A. Secant $\overline{PDC}$ passes through the center O and is perpendicular to chord $\overline{AB}$ at E. Radii $\overline{OA}$ and $\overline{OE}$ are drawn.

Prove: $\dfrac{OP}{OB} = \dfrac{PA}{BE}$

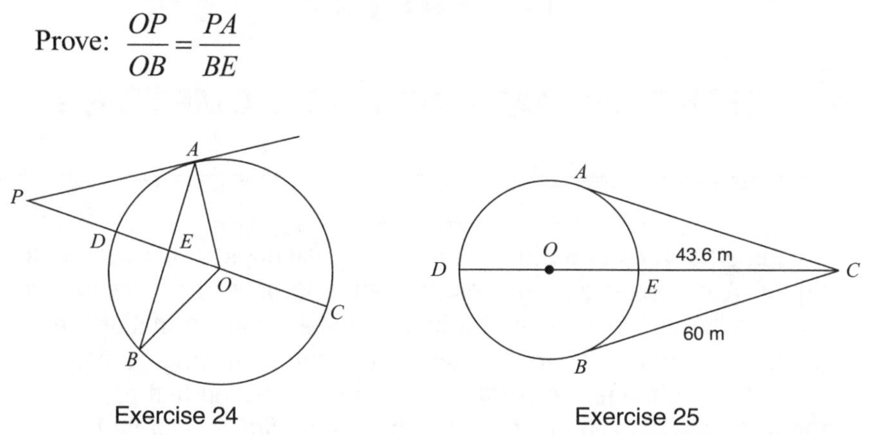

Exercise 24

Exercise 25

25. An architect is designing a park with an entrance represented by point C and a circular garden with center O, as shown in the accompanying diagram. The architect plans to connect three points on the circumference of the garden, A, B, and D, to the park entrance at C with walkways. Walkways $\overline{CA}$ and $\overline{CB}$ are tangent to the garden, and walkway $\overline{DOEC}$ is a path through the center of the garden. If $m\overset{\frown}{ADB} : m\overset{\frown}{AEB} = 3 : 2$, $BC = 60$ meters, and $EC = 43.6$ meters, find

a. The measure of the angle between walkways $\overline{CA}$ and $\overline{CB}$, correct to the *nearest degree*.

b. The diameter of the circular garden, correct to the *nearest meter*.

26. Given: Isosceles triangle ABC is inscribed in circle O with $\overline{AC} \cong \overline{BC}$. Chords $\overline{CD}$ and $\overline{AB}$ intersect at E. Chord $\overline{AD}$ is drawn.
Prove: $(AC)^2 = CE \times CD$.

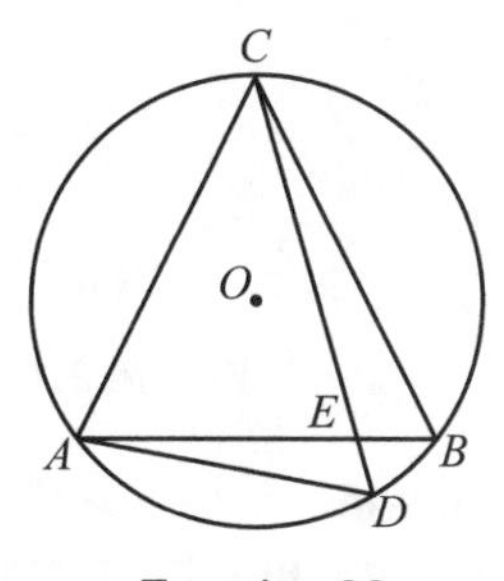

Exercise 26

CHAPTER 8
TRANSFORMATION GEOMETRY

8.1 TRANSFORMATIONS AND ISOMETRIES

KEY IDEAS

A geometric **transformation** changes the position, size, or shape of a figure. A particular type of transformation is defined by describing where it "takes" each point of a figure. Under a transformation, each point of the original figure is taken to its **image**. The original point is called the **preimage**. An **isometry** is a transformation that preserves the distance between points. If two geometric figures can be related by an isometry, then they are congruent.

Transformation Notation

Figure 8.1 shows one possible transformation of $\triangle ABC$ onto $\triangle A'B'C'$. The set of points that comprise $\triangle ABC$ belong to the *preimage* set and the set of points that form $\triangle A'B'C'$ represent the *image* set.

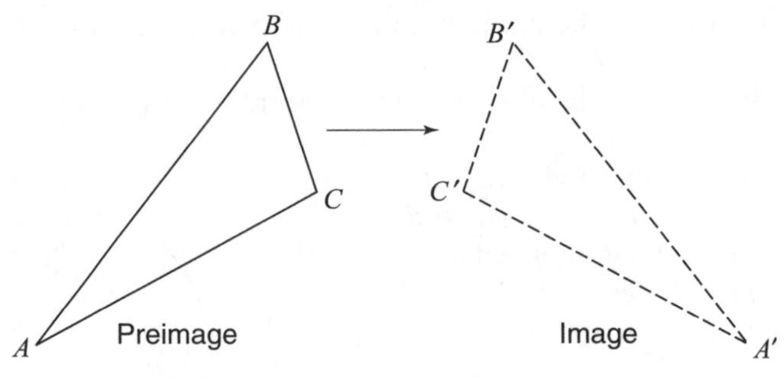

Figure 8.1 Transformation of $\triangle ABC$ onto $\triangle A'B'C'$.

It is customary to use the same capital letter to represent a preimage point and its matching image point. To distinguish between the two points, the letter representing the image point is followed by a prime mark ($'$), as when A' is the image of A. The pairing of a preimage image point with its corresponding image point can also be indicated using "arrow" notation. Referring to Figure 8.1, $A \to A'$ indicates that point A' is the image of point A under the given transformation. Similarly, the notation $\triangle ABC \to \triangle A'B'C'$ means that $\triangle A'B'C'$ is the image of $\triangle ABC$ under the given transformation.

Definition of Transformation

Because each point of $\triangle ABC$ in Figure 8.1 corresponds to exactly one point of $\triangle A'B'C'$ and each point of $\triangle A'B'C'$ corresponds to exactly one point of $\triangle ABC$, there exists a one-to-one mapping or pairing of the points of the two figures.

MATH FACTS

A **transformation** is a one–to–one mapping of the points of one set, called the **preimage** set, onto the points of a second set, called the **image** set. When a transformation is applied to a geometric figure, the transformed figure is the image and the original figure is the preimage.

Congruence Transformations

A transformation that maintains the distance between any two points of a figure is called an **isometry**. Under any isometry:

- Collinearity and betweenness of points are preserved. In Figure 8.2, A' and B' are the image points of A and B under some isometry. If C is between A and B, then C' is between A' and B'.
- The image of a line segment is a congruent line segment. In Figure 8.2, if $AB = 6$, then $A'B' = 6$.
- The image of an angle is a congruent angle, as illustrated in Figure 8.2.

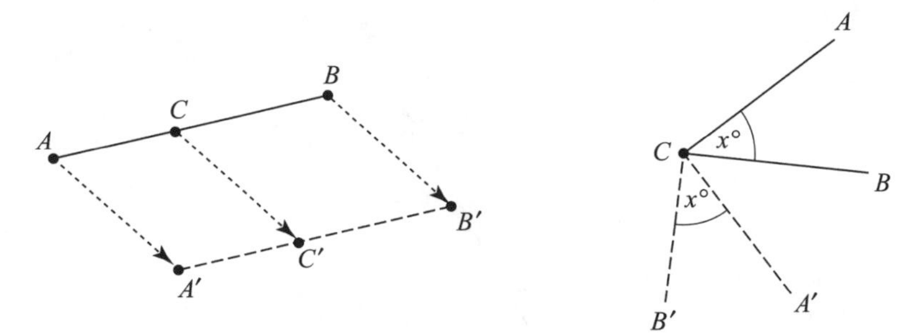

Figure 8.2 Properties of Isometries.

Some Basic Isometries

Figure 8.3 illustrates three different types of transformations that are isometries: reflection, translation, and rotation. Because an isometry always produces an image congruent to the original figure, it is sometimes referred to as a **congruence transformation**.

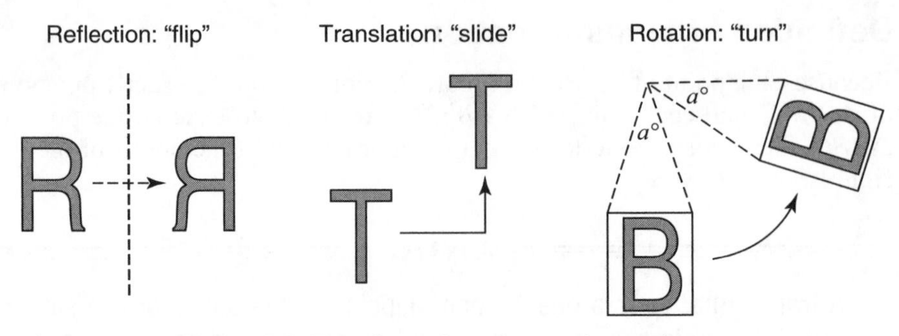

Figure 8.3 Transformations that are isometries.

Reflections

A line *reflection* "flips" an object over the line so that the image appears "backwards" much like how the reflected image of that object would appear in a mirror. The transformation represented in Figure 8.1 is actually the reflection of $\triangle ABC$ over a vertical line (not drawn) midway between $\triangle ABC$ and $\triangle A'B'C'$.

Figure 8.4 shows how to determine the reflected image of point A over line ℓ:

- Draw a line segment from A perpendicular to line ℓ. Extend that segment its own length to A'.
- The shorthand notation $r_{\ell}(A) = A'$ indicates the reflection of point A over line ℓ is A'.

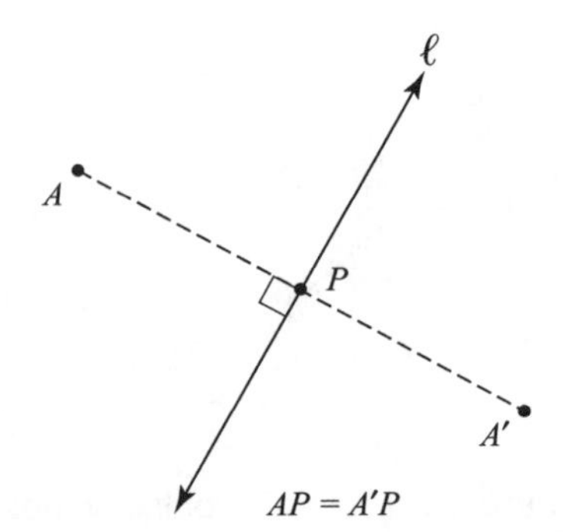

Figure 8.4 Reflecting point A over line ℓ.

To reflect a polygon over a line, reflect each of its vertices over the line. Then connect the reflected image points with line segments as shown in Figure 8.5.

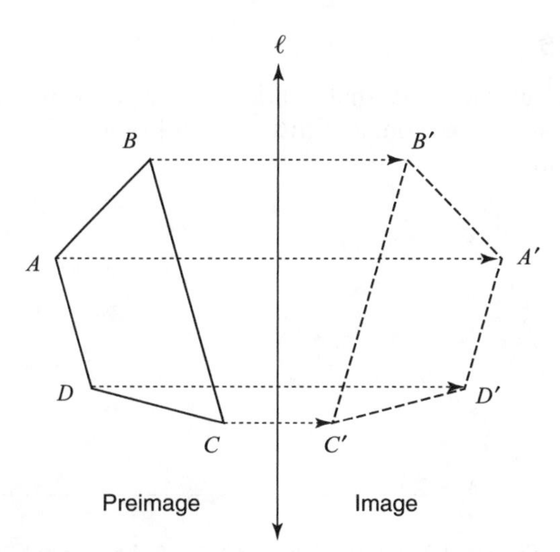

Figure 8.5 Reflecting Trapezoid *ABCD* over line ℓ.

A **reflection** over a line is an isometry that maps all points of a figure such that each image point is on the opposite side of the reflecting line and the same distance from it as its preimage. If a point of the figure is on the reflecting line, then its image is the point itself.

Orientation

Every convex polygon has two orientations: clockwise and counterclockwise. The orientation assigned to a polygon depends on the direction of the path traced when moving along consecutive vertices. In Figure 8.6, $\triangle ABC$ has clockwise orientation while $\triangle A'B'C'$, its reflected image in line p, has counterclockwise orientation. A reflection, therefore, *reverses* orientation. It is this property of a reflection that makes the reflected image appear "backwards."

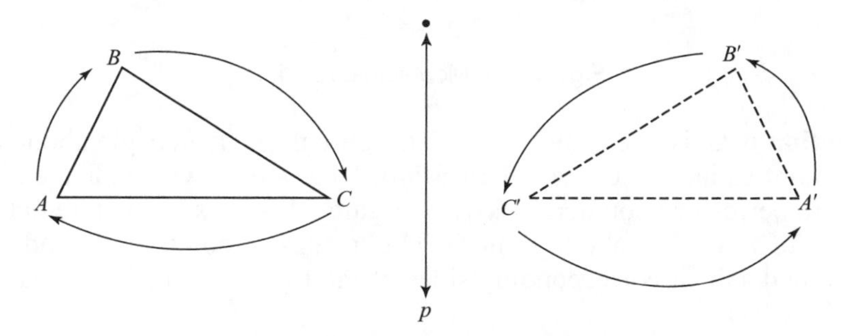

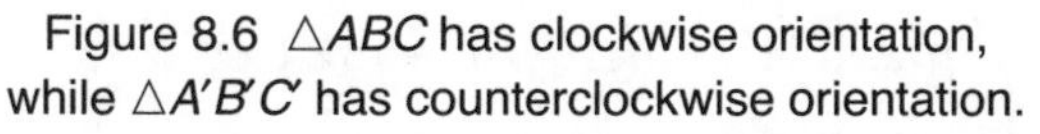
Figure 8.6 $\triangle ABC$ has clockwise orientation, while $\triangle A'B'C'$ has counterclockwise orientation.

Translations

A **translation** is an isometry that "slides" each point of a figure the same distance in the same direction, as illustrated in Figure 8.7. A translation preserves orientation.

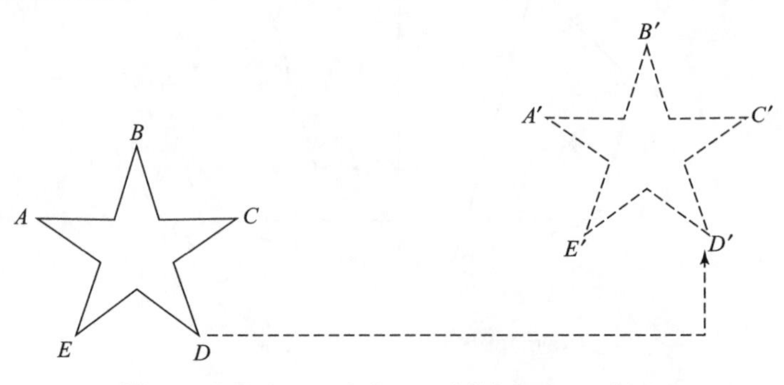

Figure 8.7 A translation or "slide" of an object.

Rotations

Suppose two identical pieces of paper have the same smiley face drawn in the same location. A pin is pushed through the papers when their edges are aligned. A rotation of the smiley face can be modeled by holding one paper fixed and turning the other paper, as illustrated in Figure 8.8. The pin represents the *center of rotation*.

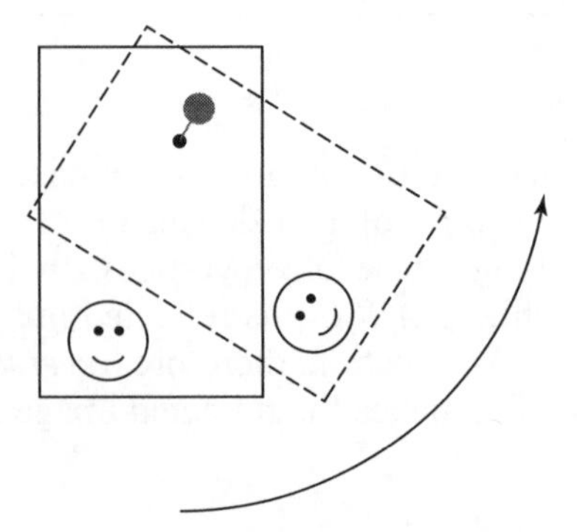

Figure 8.8 Modeling a rotation.

A **rotation** is an isometry that "turns" a figure through an angle about some fixed point called the **center of rotation**. Unless otherwise indicated, rotations are performed counterclockwise. Figure 8.9 shows a counterclockwise rotation of $\triangle ABC$ $x°$ about point O. The images of points A, B, and C are determined so that corresponding sides of the figure and its image have the same lengths and

$$m\angle AOA' = m\angle BOB' = m\angle COC' = x$$

The shorthand notation $R_{x^\circ}(A) = A'$ indicates that the rotated image of point A after a counterclockwise rotation of x° is point A'. You should verify that the rotation preserves orientation.

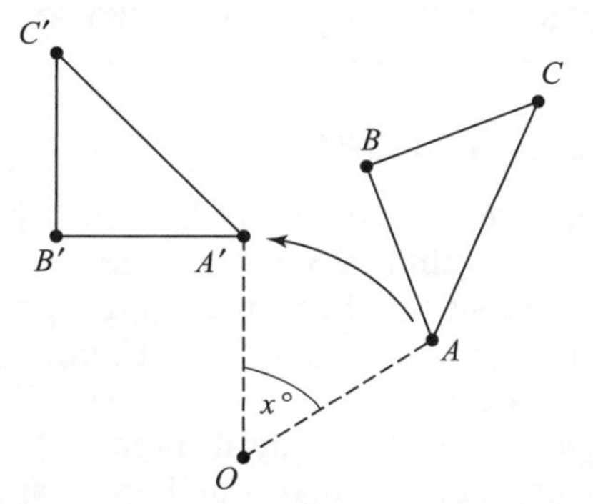

Figure 8.9 Counterclockwise rotation of $\triangle ABC$ x° about point O.

Glide Reflection

There are only four types of isometries: reflections, translations, rotations, and *glide reflections*. A **glide reflection** is an isometry that combines a reflection over a line with a translation or "glide" in the direction parallel to the reflecting line, as in Figure 8.10. The line reflection and translation may be performed in either order. Reflections and glide reflections reverse orientation; translations and rotations have the same orientation.

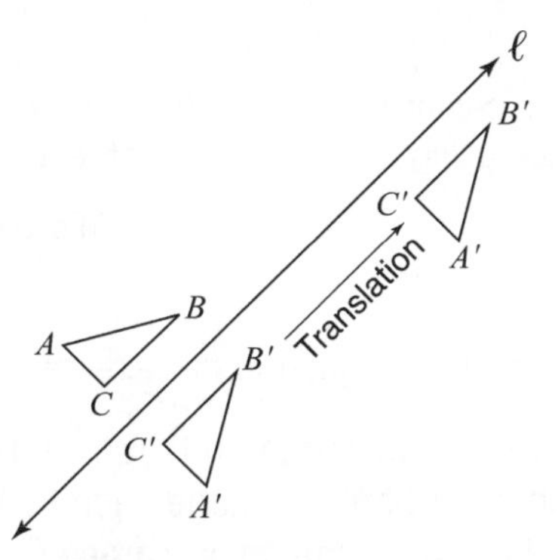

Figure 8.10 Glide reflection.

Classifying Isometries

Isometries are given special names according to whether they maintain or reverse orientation.

- A **direct isometry** is an isometry that preserves orientation. Translations and rotations are direct isometries.
- An **opposite isometry** is an isometry that reverses orientation. Line and glide reflections are opposite isometries.

MATH FACTS

If two geometric figures are related such that either one is the image of the other under some isometry, then the two figures are congruent.

Similarity Transformations

Not all transformations produce congruent images. When the image size setting of an office copying machine is set to a value other than 100 percent, the copy machine changes the size of the figure being reproduced without affecting its shape. This process is an example of a *dilation* in which the reproduced copy is the *dilated* image of the original. The image size setting represents the *scale factor* of the dilation. Figures 8.11 and 8.12 illustrate dilations with scale factors greater than 1 and less than 1.

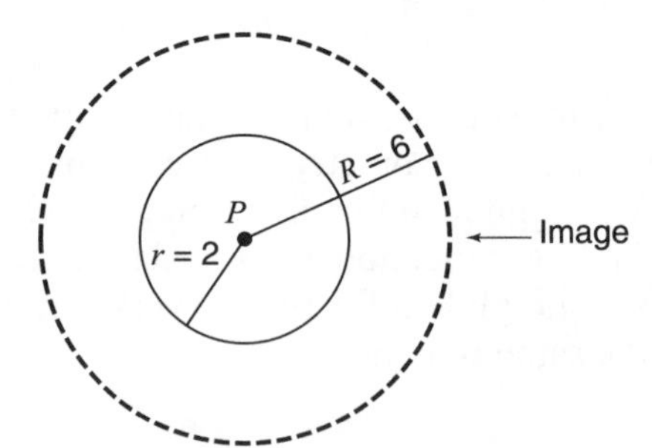

Figure 8.11 Dilation with center P of a circle with radius 2 using a scale factor of 3.

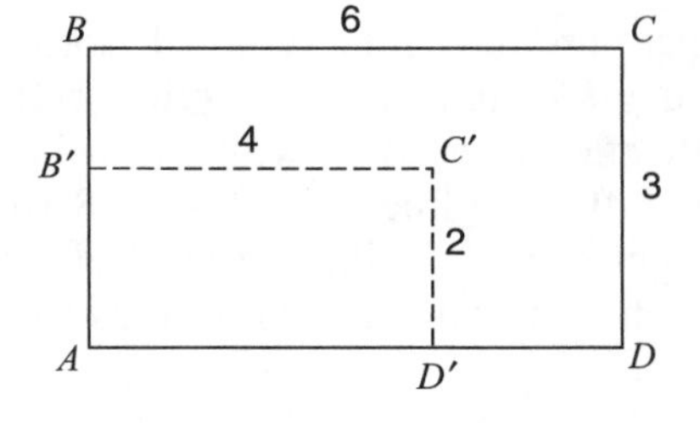

Figure 8.12 Dilation with center A of rectangle $ABCD$ using a scale factor of $\frac{2}{3}$.

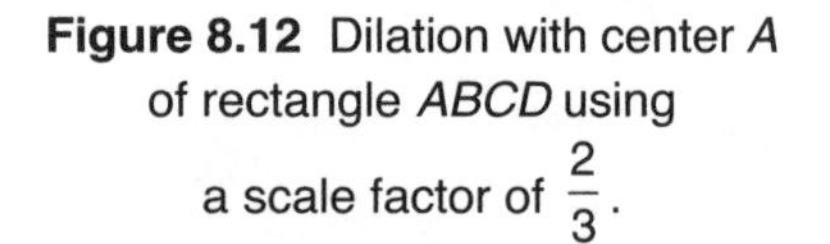

MATH FACTS

A **dilation** is a similarity transformation that changes the size of a figure by mapping each point onto its image such that the distance from the center of the dilation to the image is c times the distance from the center to the preimage. The multiplying factor c is called the **scale factor** or **constant of dilation**.

- If $c > 1$, the dilation enlarges the figure, as in Figure 8.11.
- If $0 < c < 1$, the dilatation shrinks the figure, as in Figure 8.12.

Example 1

Which transformation appears to represent an isometry?

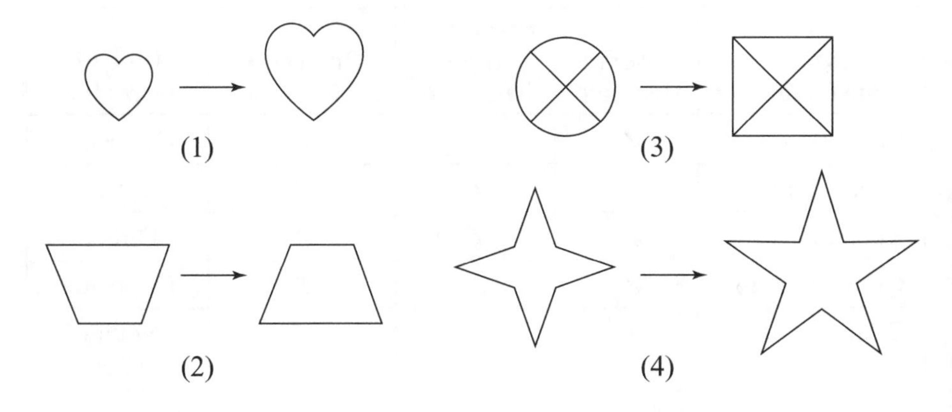

Solution: A transformation is an isometry if the preimage (original figure) and image are congruent. In choice (2), the pair of figures appear congruent, although the image is turned upside-down.

Example 2

Under what type of transformation, shown in the accompanying figure, is $\triangle AB'C'$ the image of $\triangle ABC$?

(1) dilation
(2) translation
(3) rotation about point A
(4) reflection in line ℓ

Solution: Consider each choice in turn:

- Choice (1): A dilation changes the size of the original figure. Since $\triangle AB'C'$ and $\triangle ABC$ are the same size, the figure does *not* represent a dilation.
- Choice (2): Since $\triangle AB'C'$ cannot be obtained by sliding $\triangle ABC$ in the horizontal (sideways) or vertical (up and down) direction, or in both directions, the figure does *not* represent a translation.
- Choice (3): A rotation about a fixed point "turns" a figure about that point. Since angles BAB' and CAC' are straight angles, $\triangle AB'C'$ is the image of $\triangle ABC$ after a rotation of 180° about point A.
- Choice (4): Since line ℓ is *not* the perpendicular bisector of $\overline{BB'}$ and $\overline{CC'}$, points B' and C' are not the reflected images of points B and C, respectively. Hence, the figure does *not* represent a reflection.

The correct choice is **(3)**.

Table 8.1 summarizes some key properties of transformations.

Table 8.1 Properties of Transformations

Type of Transformation	Preserves Distance	Preserves Angle Measure	Preserves Orientation	Type of Isometry
Reflection	✓	✓	✗	Opposite
Translation	✓	✓	✓	Direct
Rotation	✓	✓	✓	Direct
Glide reflection	✓	✓	✗	Opposite
Dilation	✗	✓	✓	Not an Isometry

Check Your Understanding of Section 8.1

A. Multiple Choice

1. Which figure best represents a line reflection?

(1) **P q** (2) **P ᓂ** (3) **P** $_{\mathbf{P}}$ (4) $_{\mathbf{P}}$**P**

2. One function of a movie projector is to enlarge the image on the film. This procedure is an example of a
(1) dilation (2) reflection (3) rotation (4) translation

3. A reflection does *not* preserve
(1) collinearity
(2) segment measure
(3) orientation
(4) angle measure

4. In the accompanying diagram, $\triangle A'B'C'$ is the image of $\triangle ABC$ under a transformation in which $\triangle ABC \cong \triangle A'B'C'$. This transformation is an example of a
(1) line reflection
(2) rotation
(3) translation
(4) dilation

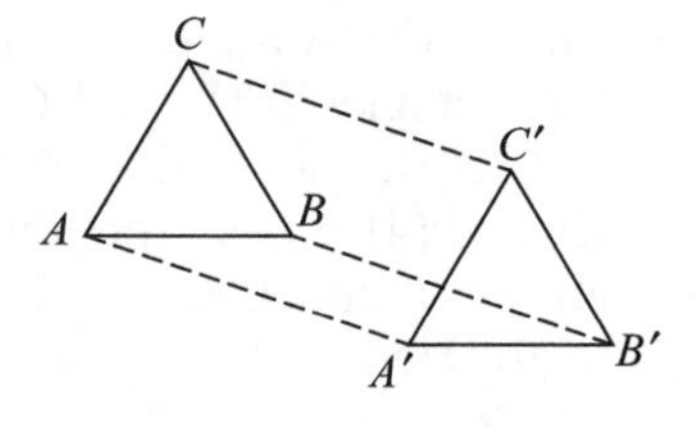

5. Ms. Brewer's art class is drawing reflected images. She wants her students to draw images reflected over a line. Which diagram represents a correctly drawn image?

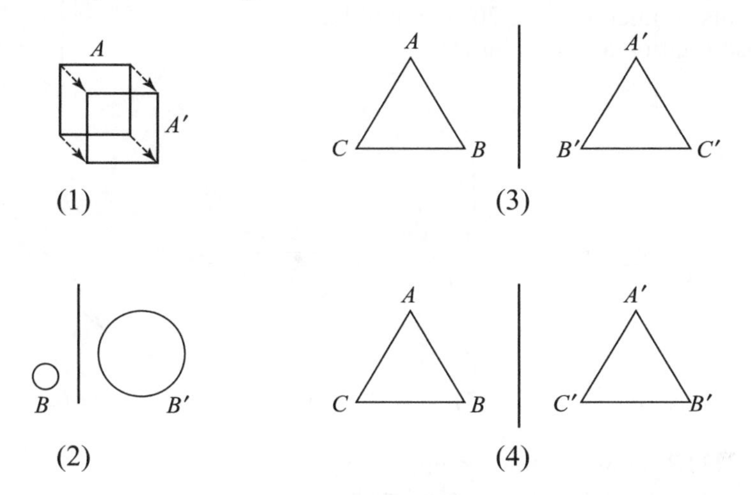

6. In the accompanying diagram of right triangle ACB with the right angle at C, line ℓ is drawn through C and is parallel to $\overline{AB}$. If $\triangle ABC$ is reflected in line ℓ, forming the image $\triangle A'B'C'$, which statement is *not* true?

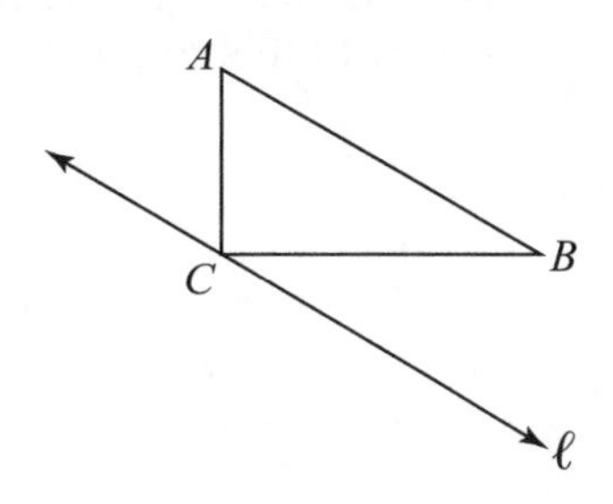

(1) C and C' are the same point.
(2) $m\angle ABC = m\angle A'B'C'$.
(3) Line ℓ is equidistant from A and A'.
(4) Point C is the midpoint of $\overline{AA'}$ area of $\triangle A'B'C'$.

7. In the diagram, figure B is the image of figure A under which transformation?

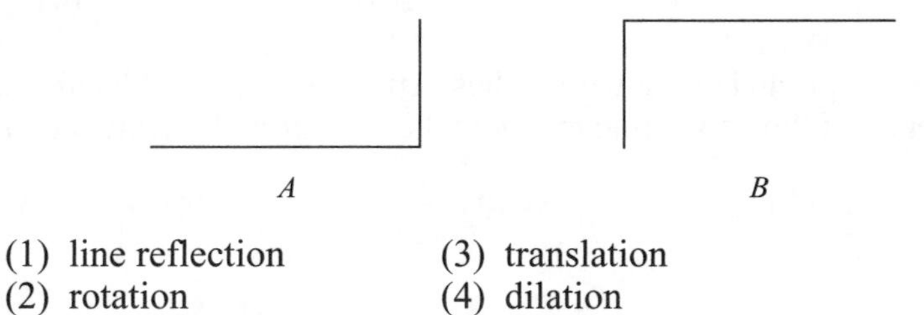

(1) line reflection
(2) rotation
(3) translation
(4) dilation

8. The accompanying diagram shows the starting position of the spinner on a board game. Which figure represents the image of this spinner after a 270° counterclockwise rotation about point P?

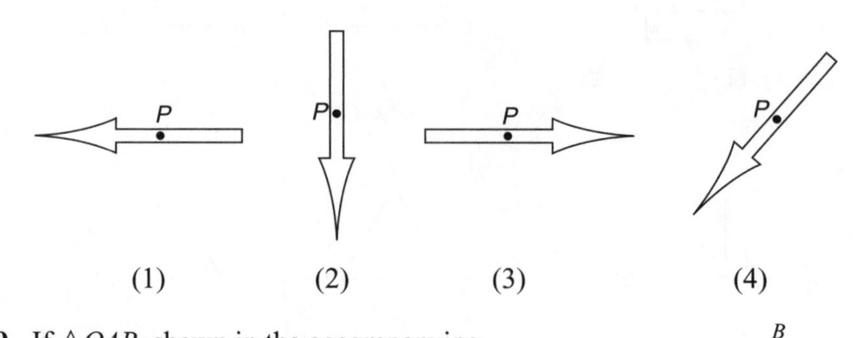

9. If $\triangle OAB$, shown in the accompanying diagram, is rotated 90° *clockwise* about point O, which figure represents the image of this rotation?

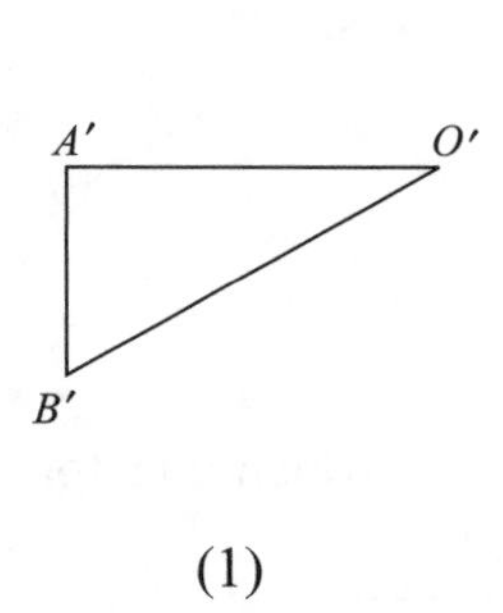

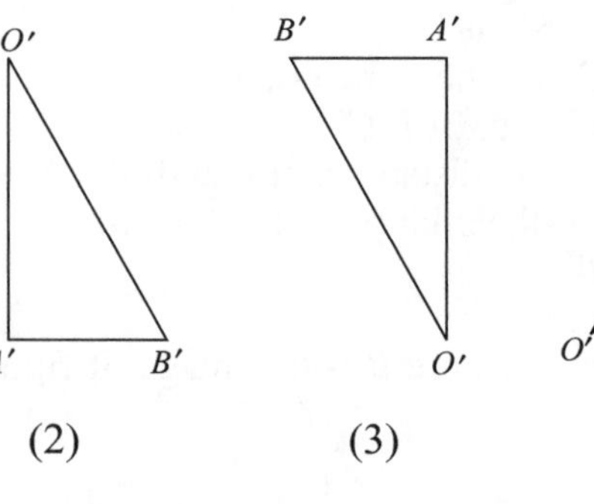

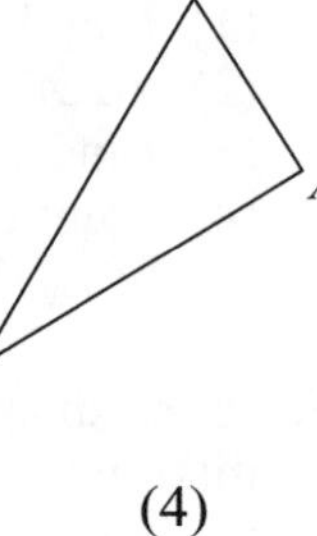

10. The area of a circle is 16 square inches. After the circle is dilated, the circumference of the new circle is 16π inches. What is the scale factor?

(1) 1 (2) 2 (3) $\frac{1}{2}$ (4) $\frac{1}{4}$

11. In the accompanying diagram, K is the image of A after a translation. Under the same translation, which point is the image of J?
(1) B
(2) C
(3) E
(4) F

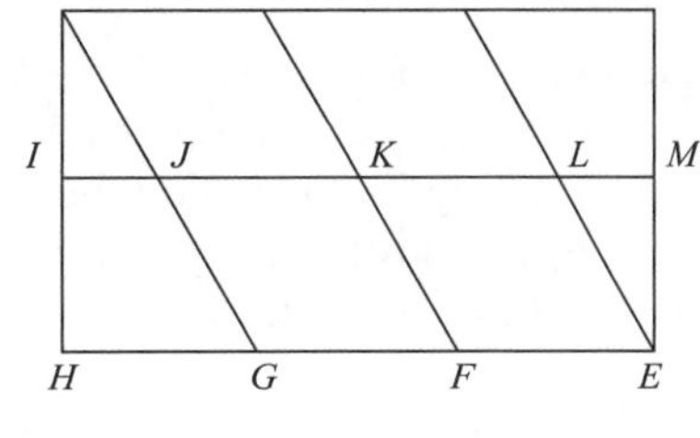

B. *Show or explain how you arrived at your answer.*

12. Write the converse of "If a transformation is a line reflection, then it is an isometry." Give a counterexample to show that the converse is false.

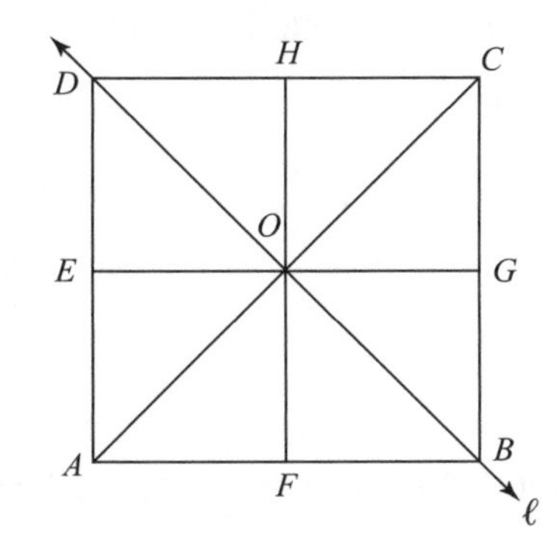

13. In the accompanying diagram of square $ABCD$, F is the midpoint of $\overline{AB}$, G is the midpoint of $\overline{BC}$, H is the midpoint of $\overline{CD}$, and E is the midpoint of $\overline{DA}$.
a. Find the image of $\triangle EOA$ after it is reflected in line ℓ.
b. Is this isometry direct or opposite? Give a reason for your answer.

8.2 TYPES OF SYMMETRY

KEY IDEAS

There are many examples of *symmetry* in nature. People's faces, leaves, and butterflies have real or imaginary "lines of symmetry" that divide the figures into two parts that are mirror images. If a geometric shape has line symmetry, it can be "folded" along the line of symmetry so that the two parts coincide. A figure can also have *rotational* symmetry or *point* symmetry.

Line Symmetry

The objects in Figure 8.13 have *line symmetry*. A figure has line symmetry if it can be reflected in a line such that the image coincides with the preimage. The reflecting line is called the **line of symmetry** and divides the figure into two congruent reflected parts. The line of symmetry may be a horizontal line, a vertical line, or neither.

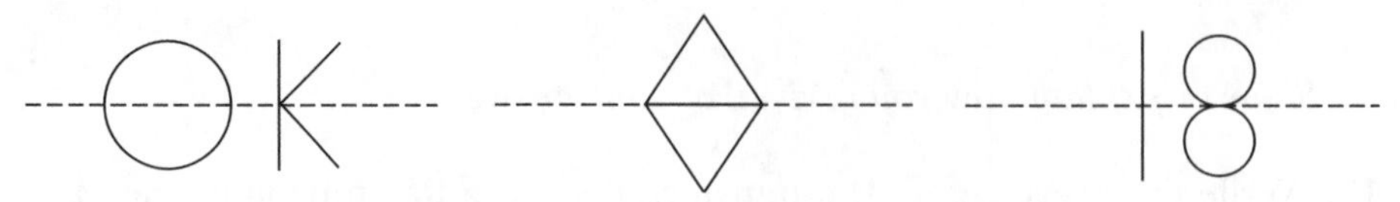

Figure 8.13 Horizontal line symmetry.

The objects in Figure 8.14 have vertical line symmetry.

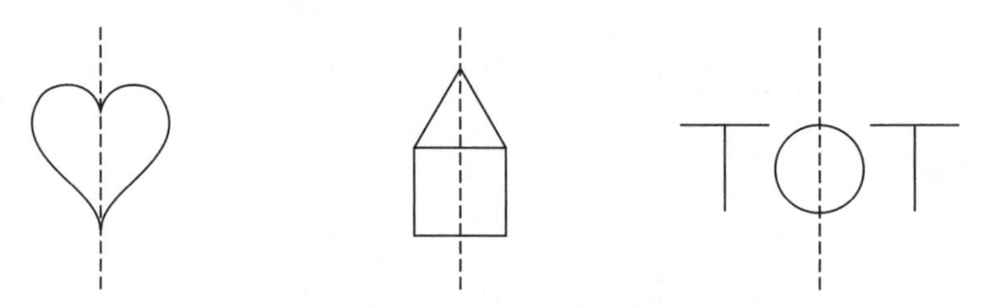

Figure 8.14 Vertical line symmetry.

Figure 8.15 illustrates that a figure may have both a horizontal and a vertical line of symmetry.

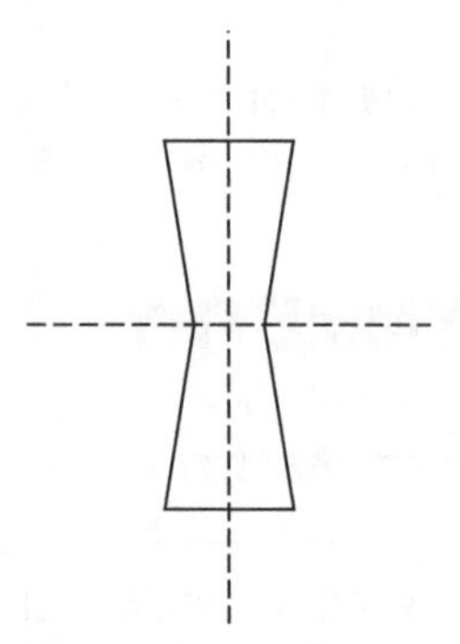

Figure 8.15 Both Horizontal and Vertical Line Symmetry.

As shown in Figure 8.16, a figure may have many lines of symmetry or may have no line of symmetry.

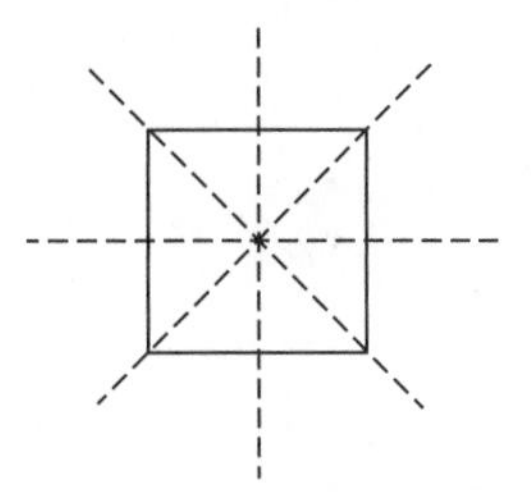

Four lines of symmetry

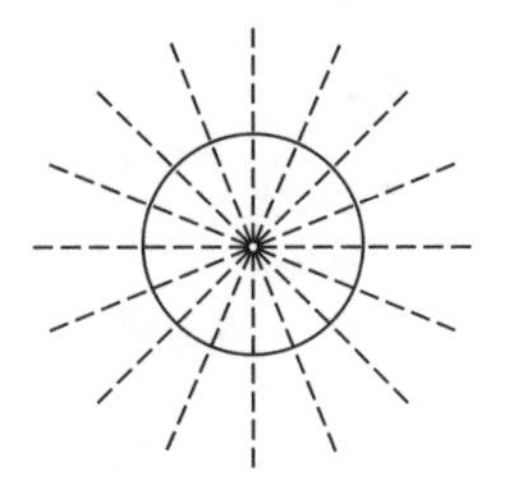

An infinite number of lines of symmetry

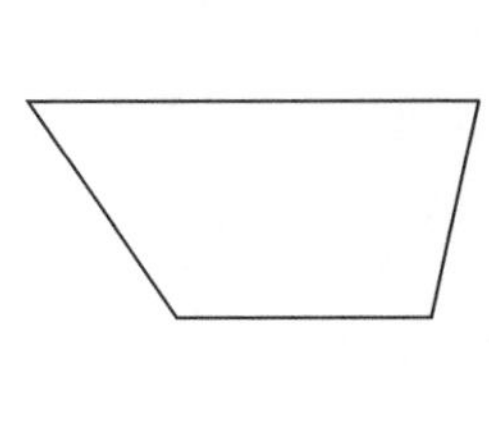

No line of symmetry

Figure 8.16 Figures with many or no lines of symmetry.

Rotational Symmetry

After a clockwise rotation of 60° about its center *O*, a regular hexagon will coincide with itself, as indicated in Figure 8.17. Regular hexagon *ABCDEF* has 60° rotational symmetry.

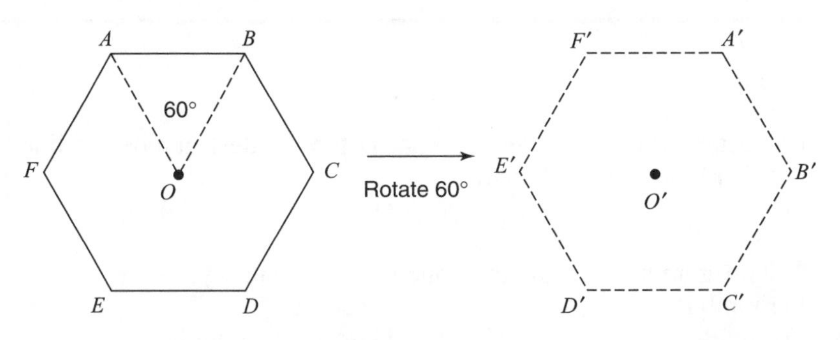

Figure 8.17 Rotational symmetry with a 60° angle of rotation.

A figure has **rotational symmetry** if it coincides with its image after a rotation through some positive angle less than 360°. Every regular polygon enjoys rotational symmetry about its center for an angle of rotation of $\frac{360°}{n}$, where n is the number of sides of the polygon. For an equilateral triangle, square, regular pentagon, and regular octagon, the angles of rotation are 120°, 90°, 72°, and 45°, respectively.

Point Symmetry

A figure has **point symmetry** with respect to a point when it has 180° rotational symmetry about that point, as in Figure 8.18.

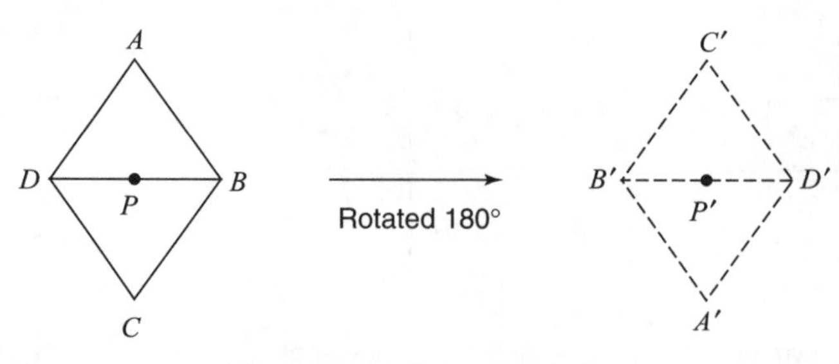

Figure 8.18 Point symmetry.

If you are not sure whether a figure has point symmetry, turn the page on which the figure is drawn upside down. Now compare the rotated figure with the original. If they look exactly the same, the figure has point symmetry.

Check Your Understanding of Section 8.2

A. Multiple Choice

1. If a rectangle is not a square, what is the greatest number of lines of symmetry that can be drawn?
(1) 1 (2) 2 (3) 3 (4) 4

2. Which figure has one and only one line of symmetry?
(1) rhombus
(2) circle
(3) square
(4) isosceles triangle

3. Which type of symmetry, if any, does a square have?
(1) line symmetry, only
(2) point symmetry, only
(3) both line and point symmetry
(4) no symmetry

4. Which letter has both point and line symmetry?
(1) **Z** (2) **T** (3) **C** (4) **H**

5. What is the total number of lines of symmetry for an equilateral triangle?
(1) 1 (2) 2 (3) 3 (4) 4

6. Which letter has point symmetry but no line symmetry?
(1) **E** (2) **S** (3) **W** (4) I

7. Which number has horizontal and vertical line symmetry?
(1) **818** (2) **383** (3) **414** (4) **100**

8. Which letter has line symmetry but no point symmetry?
(1) **O** (2) **X** (3) **N** (4) **M**

9. Which geometric shape does *not* have any lines of symmetry?

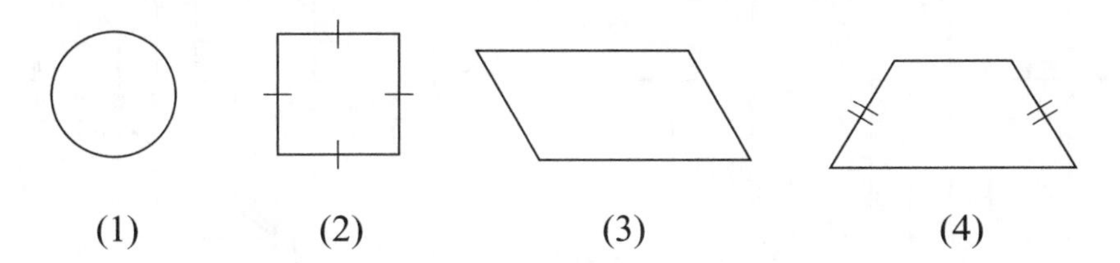

10. Which figures, if any, have *both* point symmetry and line symmetry?

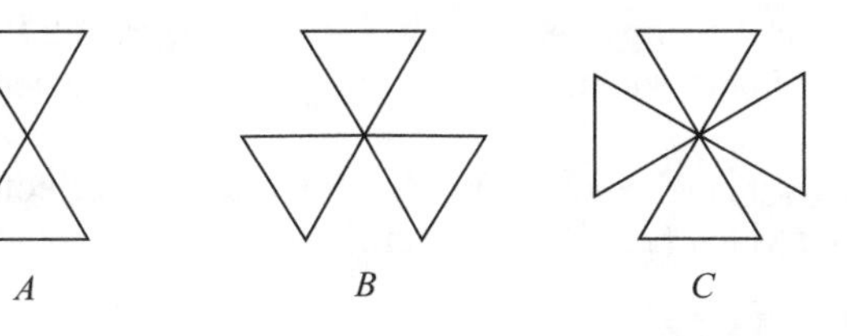

(1) A and C only
(2) B and C only
(3) none of the figures
(4) all of the figures

8.3 TRANSFORMATIONS USING COORDINATES

KEY IDEAS

Transformations can be performed in the coordinate plane.

Reflections Using Coordinates

To reflect a *point* over a coordinate axis, flip it over the axis so that the image is the same distance from the reflecting line as the original point. In Figure 8.19:

- A' is the reflection of A over the x-axis. Point A' has the same x-coordinate as point A but the opposite y-coordinate.
- A'' is the reflection of A over the y-axis. Point A'' has the same y-coordinate as point A but the opposite x-coordinate.
- A''' is the reflection of A in the origin. The coordinates of point A''' are opposite those of point A.

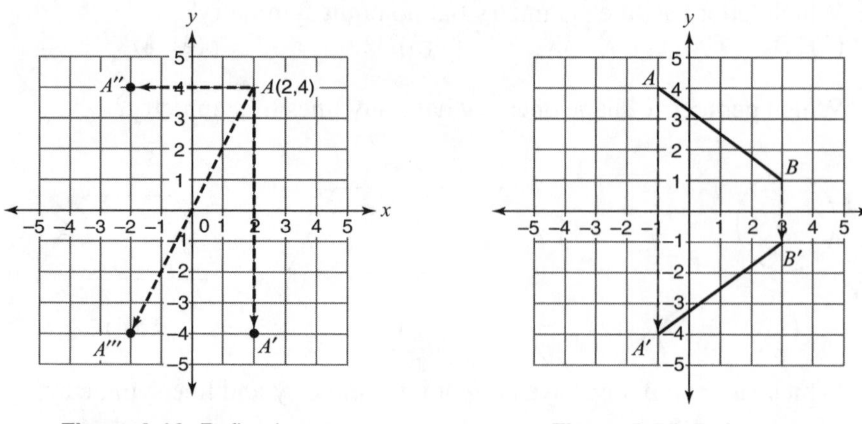

Figure 8.19 Reflecting over an axis and in the origin.

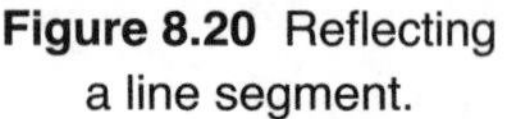

Figure 8.20 Reflecting a line segment.

The notation $r_{x\text{-axis}}\,(2, 4) = (2, -4)$ indicates that the reflected image of point (2, 4) over the x-axis is (2,–4). In general,

- $r_{x\text{-axis}}\,(x, y) = (x, -y)$
- $r_{y\text{-axis}}\,(x, y) = (-x, y)$
- $r_{\text{origin}}\,(x, y) = (-x, -y)$

To reflect a *line segment* over a coordinate axis, flip it over the axis by reflecting each endpoint of that segment. Then connect the images of the endpoints. If the endpoints of $\overline{AB}$ are $A(-1, 4)$ and $B(3, 1)$, then, after a reflection of $\overline{AB}$ over the x-axis, the image is $\overline{A'B'}$ with endpoints $A'(-1,-4)$ and $B'(3,-1)$. See Figure 8.20. To reflect a point over the line $y = x$ or over the line $y = -x$, as illustrated in Figure 8.21, use these rules:

- $r_{y=x}\,A(x, y) = A'(y, x)$
- $r_{y=-x}\,A(x, y) = A''(-y, -x)$

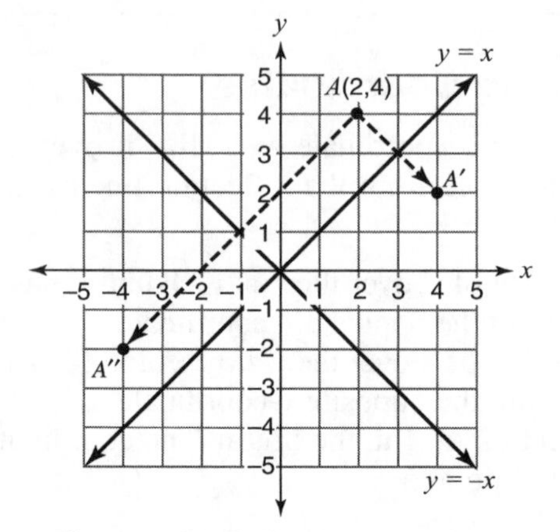

Figure 8.21 Reflecting over $y = \pm x$.

Example 1

Graph $\triangle ABC$ with coordinates $A(1, 3)$, $B(5, 7)$, and $C(8, -3)$. On the same set of axes graph $\triangle A'B'C'$, the reflection of $\triangle ABC$ over the y-axis.

Solution: After graphing $\triangle ABC$, reflect points A, B, and C over the y-axis, as shown in the accompanying figure. Then connect the image points A', B', and C' with line segments.

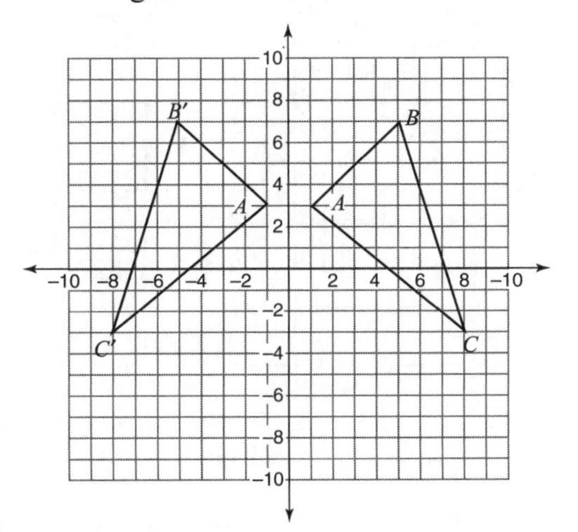

Reflection in a Point

A point may be reflected in the coordinate plane in a point other than the origin. Under a reflection in point P, the image of point A is point A' provided P is the midpoint of $\overline{AA'}$, as illustrated in Figure 8.22.

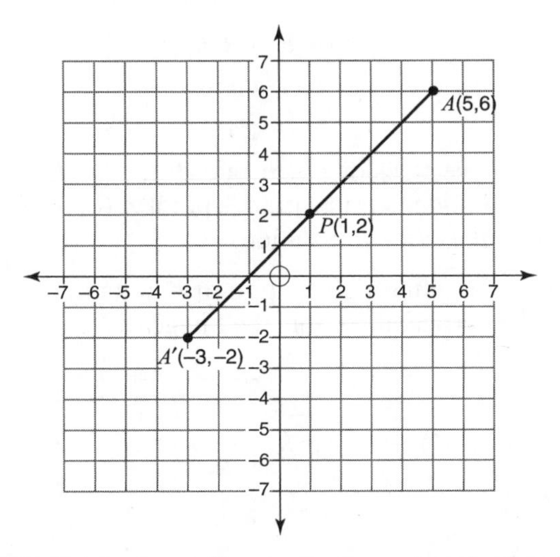

Figure 8.22 Points A and A' are reflections of each other in point P.

Translations Using Coordinates

In Figure 8.23, $\triangle A'B'C'$ is the image of $\triangle ABC$ under a translation that shifts each point of $\triangle ABC$ 7 units to the left and 2 units up. If (x, y) is any point of $\triangle ABC$, then $(x - 7, y + 2)$ represents the coordinates of the translated image point.

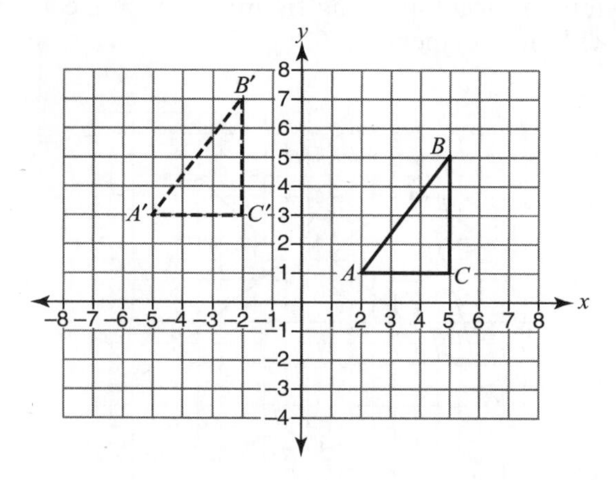

Figure 8.23 Translation of $\triangle ABC$.

The shorthand notation $T_{h,k}$ is sometimes used to represent a translation of a figure h units horizontally and k units vertically. The signs of h and k indicate direction: $h > 0$ means right, $h < 0$ is left; $k > 0$ represents up, and $k < 0$ is down. Referring again to Figure 8.23,

$$T_{-7,2}\, A(2, 1) = A'(2 + (-7), 1 + 2) = A'(-5, 3)$$

In general,

$$T_{h,k}\, P(x, y) = P'(x + h, y + k)$$

Example 2

The coordinates of the vertices of $\triangle ABC$ are $A(2, -3)$, $B(0, 4)$, and $C(-1, 5)$. If the image of point A under a translation is point $A'(0, 0)$, find the images of points B and C under the same translation.

Solution: In general, after a translation of h units in the horizontal direction and k units in the vertical direction, the image of $P(x, y)$ is $P'(x + h, y + k)$. Since

$$A(2, -3) \rightarrow A'(\overbrace{2 + h}, \underbrace{-3 + k}) = A'(0, 0)$$

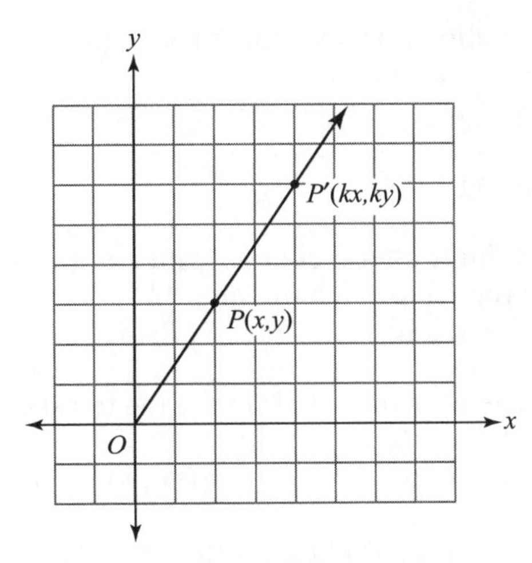

Figure 8.25 Dilation of point P.

Example 4

After a dilation with respect to the origin, the image of $A(2, 3)$ is $A'(4, 6)$. What are the coordinates of the point that is the image of $B(1, 5)$ after the same dilation?

Solution: Determine the constant of dilation. The constant of dilation is 2 since

$$A(2, 3) \rightarrow A'(\underline{2} \times 2, \underline{2} \times 3) = A'(4, 6)$$

Under the same dilation, the x– and y–coordinates of point B are also multiplied by 2:

$$D_2(1, 5) \rightarrow B'(\underline{2} \times 1, \underline{2} \times 5) = B'(2, 10)$$

Check Your Understanding of Section 8.3

A. *Multiple Choice*

1. Which transformation represents a dilation?
(1) $(8, 4) \rightarrow (11, 7)$
(2) $(8, 4) \rightarrow (-8, 4)$
(3) $(8, 4) \rightarrow (-4, -8)$
(4) $(8, 4) \rightarrow (4, 2)$

2. Which transformation is an example of an opposite isometry?
(1) $(x, y) \to (x + 3, y - 6)$
(2) $(x, y) \to (3x, 3y)$
(3) $(x, y) \to (y, x)$
(4) $(x, y) \to (y, -x)$

3. The image of point A after a dilation with a scale factor of 3 is (6, 15). What was the original location of point A?
(1) (2, 5) (2) (3, 12) (3) (9, 18) (4) (18, 45)

4. What is the image of point (–3, 4) under the translation that shifts (x, y) to $(x - 3, y + 2)$?
(1) (0, 6) (2) (6, 6) (3) (–6, 8) (4) (–6, 6)

5. The image of point (–2, 3) after a certain translation is (3, –1). What is the image of point (4, 2) after the same translation?
(1) (–1, 6) (2) (0, 7) (3) (5, 4) (4) (9, –2)

6. What is image of point (–3, –1) after a rotation of 90° about the origin?
(1) (3, 1) (2) (1, –3) (3) (3, –1) (4) (1, 3)

7. The three vertices of $\triangle ABC$ are located in Quadrant II. The image of $\triangle ABC$ after a reflection in the x-axis is $\triangle A'B'C'$. In which quadrant is the image of $\triangle A'B'C'$ located after a reflection in the y-axis?
(1) I (2) II (3) III (4) IV

8. A function, f, is defined by the set {(2, 3), (4, 7), (–1, 5)}. If f is reflected in the line $y = x$, which point will be in the reflection?
(1) (–5, 1) (2) (5, –1) (3) (1, –5) (4) (–1, 5)

9. Which mapping rule does *not* represent an isometry in the coordinate plane?
(1) $(x, y) \to (2x, 2x)$
(2) $(x, y) \to (x + 2, y + 2)$
(3) $(x, y) \to (-x, y)$
(4) $(x, y) \to (x, -y)$

10. Point P' is the image of point $P(-3, 4)$ after a translation defined by $T_{(7, -1)}$. Which other transformation on P would also produce P' as its image?
(1) $r_{y=-x}$ (2) $r_{y\text{-axis}}$ (3) $R_{90°}$ (4) $R_{-90°}$

B. *Show or explain how you arrived at your answer.*

11–14. In the accompanying figure, each grid box is 1 unit. Identify each of the given transformations as either a reflection, translation, rotation, dilation, or glide reflection. State the reflection line, translation rule, center and angle of rotation, or the reflecting line and translation for a glide reflection.

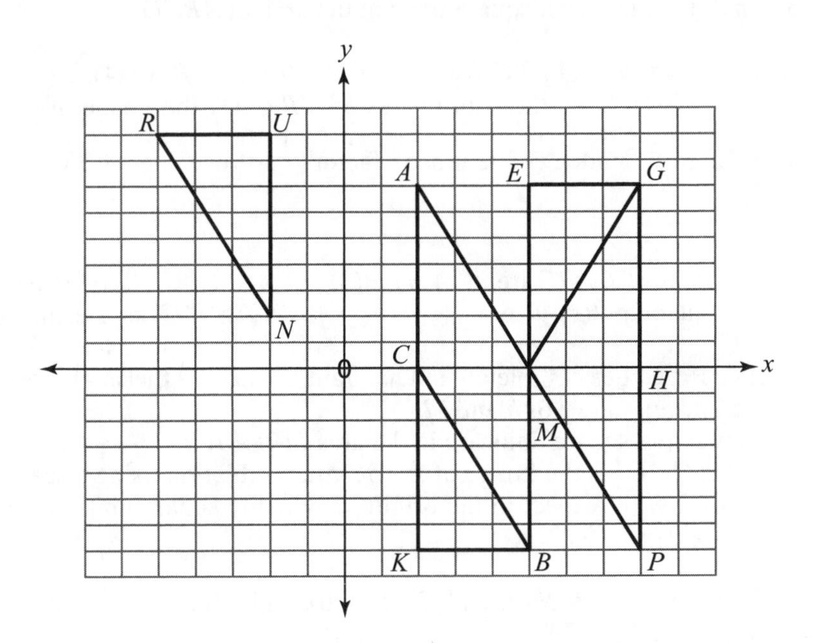

11. $\triangle GHM \rightarrow \triangle ACM$

12. $\triangle BMC \rightarrow \triangle NUR$

13. $\triangle BKC \rightarrow \triangle GEM$

14. $\triangle ACM \rightarrow \triangle PHM$

15. If $T_{h,k}(2, -1) = (-2, 1)$, what are the coordinates of the image of $(-3, 4)$ under the same translation?

16. If $T_{-2,3}(x, y) = (2, -1)$, what are the coordinates of the preimage point?

17. Under a reflection in point $X(1, 2)$, the image of point $P(3, -1)$ is P'. Determine the coordinates of point P'.

18. Carson is a decorator. He often sketches his room designs on the coordinate plane. He has graphed a square table on his grid so that its corners are at the coordinates $A(2, 6)$, $B(7, 8)$, $C(9, 3)$, and $D(4, 1)$. To graph a second identical table, he reflects $ABCD$ over the y-axis.

a. Using graph paper, sketch and label $ABCD$ and its image $A'B'C'D'$, which show the locations of the two tables.

b. Find the number of square units in the area of $ABCD$.

19. Given: Quadrilateral $ABCD$ with vertices $A(-2, 2)$, $B(8, -4)$, $C(6, -10)$, and $D(-4, -4)$. State the coordinates of $A'B'C'D'$, the image of quadrilateral $ABCD$ under a dilation of factor $\frac{1}{2}$. Prove that $A'B'C'D'$ is a parallelogram.

20. The vertices of $\triangle ABC$ are $A(-4, 7)$, $B(3, -2)$, and $C(8, -2)$. After a translation that maps (x, y) onto $(x + h, y + k)$, $\triangle A'B'C'$ is the image of $\triangle ABC$.

a. If $\triangle A'B'C'$ lies completely in Quadrant I, what are the smallest possible integer values of h and k?

b. How many square units are in the area of $\triangle ABC$?

c. If $\triangle A''B''C''$ is the image of $\triangle ABC$ after a dilation using a scale factor of 2 with respect to the origin, how many square units are in the area of $\triangle A''B''C''$?

21. The vertices of $\triangle PEN$ are $P(1, 2)$, $E(3, 0)$, and $N(6, 4)$. On graph paper, draw and label $\triangle PEN$.

a. Graph and state the coordinates of $\triangle P'E'N'$, the image of $\triangle PEN$ after a reflection in the y-axis.

b. Graph and state the coordinates of $\triangle P''E''N''$, the image of $\triangle PEN$ under the translation $(x, y) \rightarrow (x + 4, y - 3)$.

22. Triangle SAM has coordinates $S(3, 4)$, $A(3, -5)$, and $M(-4, -2)$. On graph paper, graph and label $\triangle SAM$.

a. Graph and label $\triangle S'A'M'$, the image of $\triangle SAM$ after a reflection in the line $y = x$.

b. Graph and label $\triangle S''A''M''$, the image of $\triangle SAM$ after a dilation of 2. Express in simplest form the ratio of the area of $\triangle SAM$ to the area of $\triangle S''A''M''$.

8.4 COMPOSING TRANSFORMATIONS

KEY IDEAS

A glide reflection is an example of a *composite transformation* as it combines two other transformations to form a new transformation. When evaluating a composite transformation, the order in which the transformations are performed matters.

Composite Transformations

A **composite transformation** is a series of transformations, one followed by the other, in which the image of one transformation is used as the preimage of the next transformation.

Example 1

In the accompanying figure, p and q are lines of symmetry for regular hexagon $ABCDEF$ intersecting at point O, the center of the hexagon. Determine the final image of the composite transformation of the reflection of $\overline{AB}$ in line q followed by a reflection of its image in line p.

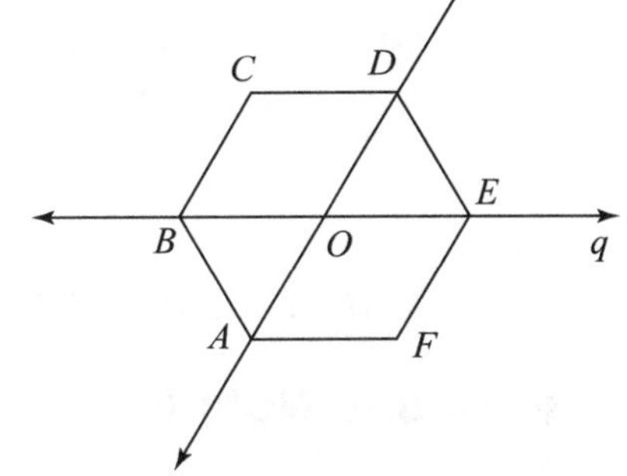

Solution: Since $r_{\text{line } q}(\overline{AB}) = \overline{CB}$ and $r_{\text{line } p}(\overline{CB}) = \overline{EF}$, the final image is $\overline{EF}$.

Example 2

The coordinates of the vertices of $\triangle ABC$ are $\triangle(2, 0)$, $B(1, 7)$, and $C(5, 1)$.

a. Graph $\triangle A'B'C'$ the reflection of $\triangle ABC$ over the y-axis and graph $\triangle A''B''C''$ the reflection of $\triangle A'B'C'$ over the x-axis.
b. What single type of transformation maps $\triangle ABC$ onto $\triangle A''B''C''$?

Solution:

a. Under the given composite transformation

$$A\,(2, 0) \to A'\,(-2, 0) \to A''\,(-2, 0)$$
$$B\,(1, 7) \to B'\,(-1, 7) \to B''\,(-1, 7)$$
$$C\,(5, 1) \to C'\,(-5, 1) \to C''\,(-5, 1)$$

See the accompanying figure.

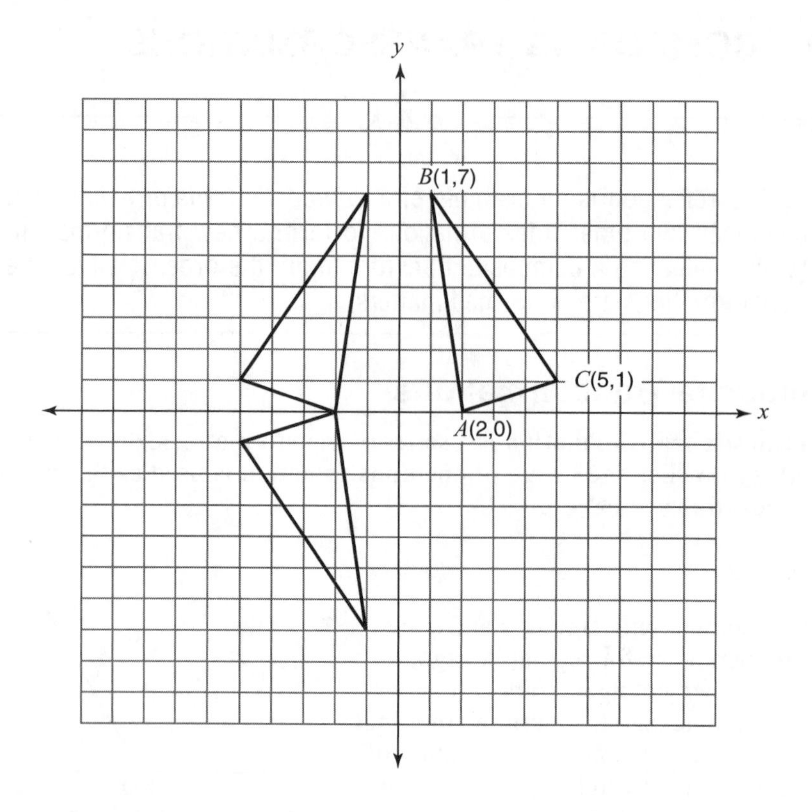

b. $R_{180^\circ}(\triangle ABC) = \triangle A''B''C''$.

Composite Notation

A composite transformation is indicated by placing a centered circle between two transformations, as in $r_{x\text{-axis}} \circ T_{1,2}(-1, 4)$. The transformation on the right side of the centered circle is always evaluated first. Thus, $r_{x\text{-axis}} \circ T_{1,2}(-1, 4)$ represents a composite transformation consisting of a translation followed by a reflection. To find the final image point,

- First perform the translation:

$$T_{1,2}(-1, 4) = (-1 + 1, 4 + 2) = (0, 6)$$

- Then reflect the translated image point:

$$r_{x\text{-}axis}(0, 6) = (0, -6)$$

Thus, $r_{x\text{-axis}} \circ T_{1,2}(-1, 4) = (0, -6)$. The composite transformation $r_{x\text{-axis}} \circ T_{1,2}(-1, 4)$ can also be written as $r_{x\text{-axis}}(T_{1,2}(-1, 4))$ where the transformation enclosed by the parentheses is performed first.

Example 3

Using the same figure as in Example 1, identify the image of the composite transformation $r_q \circ R_{60^\circ}(E)$.

Solution: The notation $r_q \circ R_{60^\circ}(E)$ represents the composite transformation of a 60° counterclockwise rotation of point E followed by a reflection of the rotated image point in line q. Because, $R_{60^\circ}(E) = D$,

$$r_q \circ R_{60^\circ}(E) = r_q(D) = F$$

The final image point is ***F***.

Composing Reflections over Two Lines

The composition of two rotations with the same center is a *rotation*. The composition of two translations is a *translation*. The composition of two reflections, however, is *not* a reflection. There are two possibilities to consider:

- Composing two reflections over parallel lines *translates* the original figure, as shown in Figure 8.26.
- Composing two reflections over intersecting lines *rotates* the original figure, as shown in Figure 8.27.

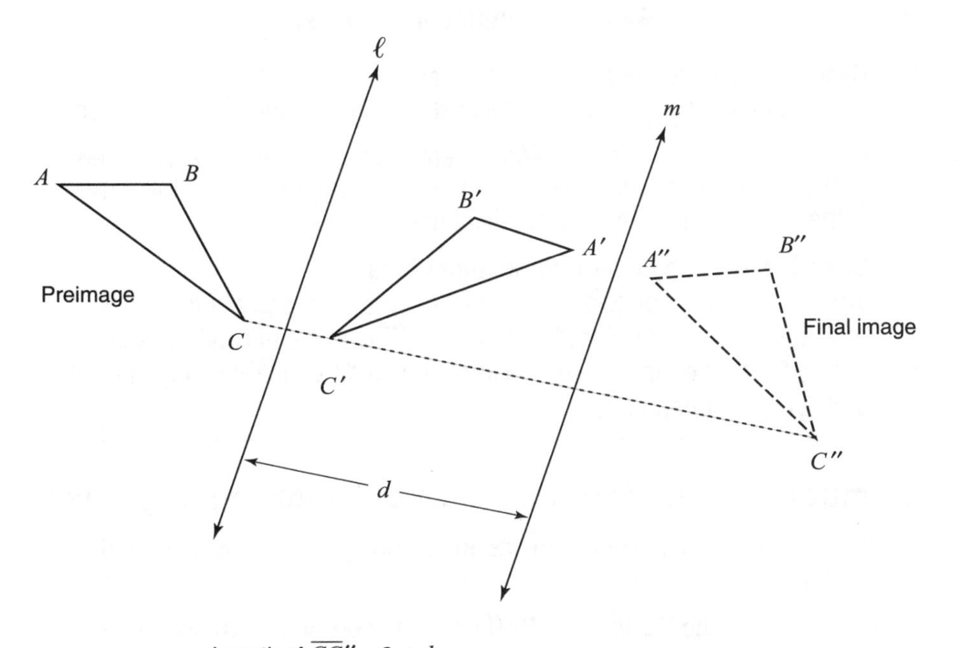

Figure 8.26 Composing reflections over two parallel lines.

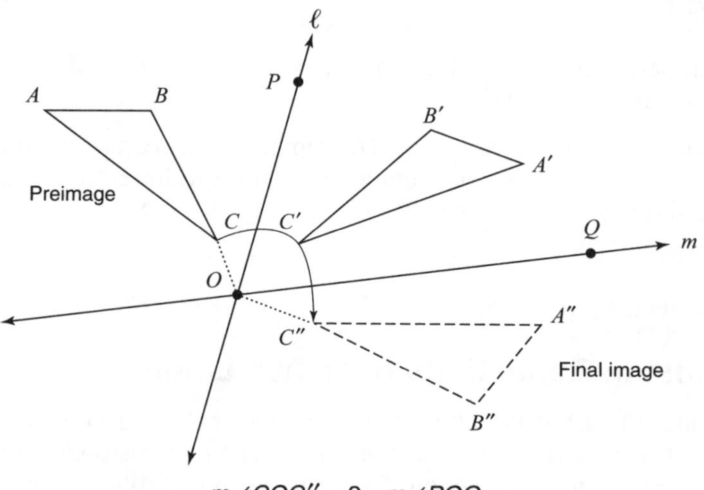

$$m\angle COC'' = 2 \times m\angle POQ$$

Figure 8.27 Composing reflections over two lines intersecting at point *O*.

MATH FACTS

Reflection–Reflection Theorem

Case 1: Reflecting over Parallel Lines
The composition of two reflections over two parallel lines is a translation.

- The direction of the translation is perpendicular to the reflecting lines.
- The distance between the final image and the preimage is two times the distance between the parallel lines.

Case 2: Reflecting over Intersecting Lines
The composition of two reflections over two intersecting lines is a rotation about their point of intersection. The angle of rotation is equal to two times the measure of the angle formed by the reflecting lines at their point of intersection.

Composite Transformations and Congruent Figures

Here are some other observations about composite transformations that you should know:

- Because of the Reflection–Reflection Theorem, any translation or rotation can be expressed as the composition of two reflections.
- If two figures are congruent, there exists a transformation that maps one figure onto the other.
- In a plane, one of two congruent figures can be mapped onto the other by a composition of at most three reflections.

Check Your Understanding of Section 8.4

A. Multiple Choice

1. The composition of two equal glide reflections is equivalent to
(1) a translation that is twice the distance of a single glide reflection
(2) a dilation with a scale factor of 2
(3) a rotation
(4) a reflection in a line perpendicular to the direction of the translation.

2–5. In the accompanying diagram of regular octagon *ABCDEFGH*, lines ℓ and m are lines of symmetry.

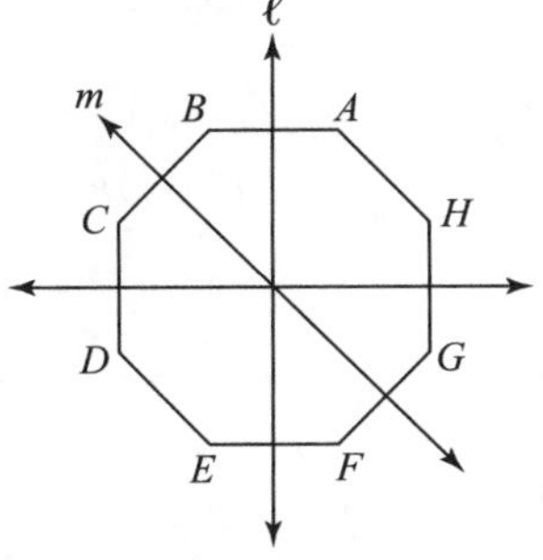

Exercises 2–5

2. What is the image of the reflection of $\overline{AB}$ over line m followed by a reflection of the image over line ℓ?
(1) $\overline{CD}$ (2) $\overline{AH}$ (3) $\overline{HG}$ (4) $\overline{FG}$

3. What is the image of a 135° counterclockwise rotation of point H followed by a reflection of the rotated image over line m?
(1) B (2) D (3) E (4) G

4. What is $r_m \circ r_\ell\,(\overline{FG})$?
(1) $\overline{CD}$ (2) $\overline{AH}$ (3) $\overline{HG}$ (4) $\overline{BC}$

5. What is $r_m \circ r_\ell \circ r_m\,(H)$?
(1) A (2) E (3) F (4) G

6. What are the coordinates of $r_{y=x} \circ R_{x\text{-axis}}\,(3, 1)$?
(1) (1, –3) (2) (–3, 1) (3) (–1, 3) (4) (3, –1)

7. What are the coordinates of $r_{x\text{-axis}} \circ R_{90°}(4, -2)$?
(1) (2, 4) (2) (2, −4) (3) (4, 2) (4) (−4, 2)

8. The coordinates of $\triangle JRB$ are $J(1, -2)$, $R(-3, 6)$, and $B(4, 5)$. What are the coordinates of the vertices of its image after the transformation $T_{2,-1} \circ r_{y\text{-axis}}$?
(1) (3, 1), (−1, −7), (6, −6)
(2) (3, −3), (−1, 5), (6, 4)
(3) (1, −3), (5, 5), (−2, 4)
(4) (−1, −2), (3, 6), (−4, 5)

9–10. In the accompanying diagram, p and q are lines of symmetry for figure $ABCDEF$.

9. What is $r_p \circ r_q \circ r_p(A)$?
(1) B (2) D (3) E (4) F

10. What is $r_q \circ r_p \circ r_q(\overline{BC})$?
(1) $\overline{AB}$ (2) $\overline{BC}$ (3) $\overline{DE}$ (4) $\overline{EF}$

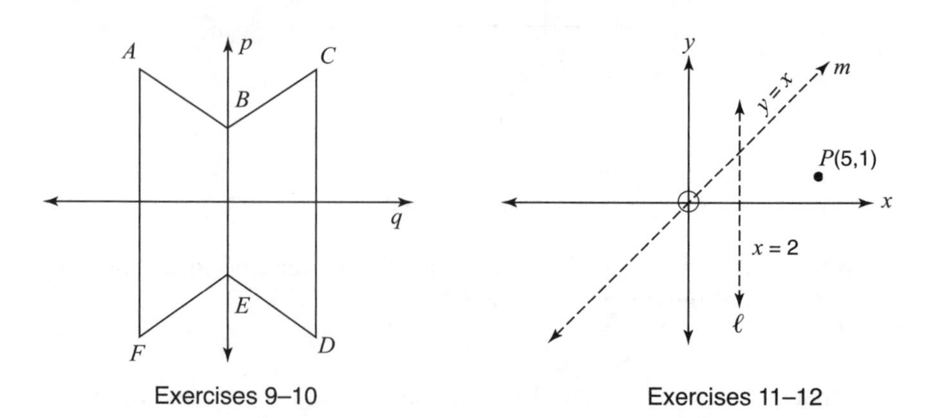

Exercises 9–10

Exercises 11–12

11–12. In the accompanying diagram, the equation of line ℓ is $x = 2$ and the equation of line m is $y = x$.

11. What are the coordinates of the image of $r_\ell \circ r_m(P)$?
(1) (3, 5) (2) (−1, 1) (3) (3, −5) (4) (1, −1)

12. What are the coordinates of the image of $r_{\text{origin}} \circ r_\ell \circ r_{x\text{-axis}}(P)$?
(1) (−1, 1) (2) (1, 1) (3) (3, 5) (4) (−5, −3)

13. What are the coordinates of the image of point $A(-4, 1)$ after the composite transformation $R_{90^\circ} \circ r_{y=x}$ where the origin is the center of rotation?
(1) $(-1, -4)$ (2) $(-4, -1)$ (3) $(1, 4)$ (4) $(4, 1)$

14. If the coordinates of point A are $(-2, 3)$, what is the image of A under the composite transformation $r_{y\text{-axis}} \circ D_3$?
(1) $(-6, -9)$ (2) $(9, -6)$ (3) $(5, 6)$ (4) $(6, 9)$

15. The composite transformation that reflects point P through the origin, the x-axis, and the line $y = x$, in the order given, is equivalent to which rotation of point P about the origin?
(1) R_{90° (2) R_{180° (3) R_{270° (4) R_{360°

16. In the accompanying diagram, $s \parallel t$. Which is equivalent to the composition of line reflections $r_s \circ r_t(\triangle ABC)$?
(1) a rotation
(2) a line reflection
(3) a translation
(4) a glide reflection

17. Given the transformations: $R(x, y) \to (-x, y)$ and $S(x, y) \to (y, x)$. What is $(R \circ S)(5, -1)$?
(1) $(1, 5)$ (2) $(1, -5)$ (3) $(-1, 5)$ (4) $(-1, -5)$

18. Which transformation is equivalent to the composite line reflections $r_{y\text{-axis}} \circ r_{y=x}(\overline{AB})$?
(1) a rotation (3) a translation
(2) a dilation (4) a glide reflection

B. *Show or explain how you arrived at your answer.*

19. Given point $A(-2, 3)$. State the coordinates of the image of A under the composite transformation $T_{-3,4} \circ r_{x\text{-axis}}$.

20. Draw the image of the figure shown in the accompanying grid after the composite transformation $r_{y\text{-axis}} \circ R_{90^\circ}$. State the coordinates of P', the image of point P under this transformation.

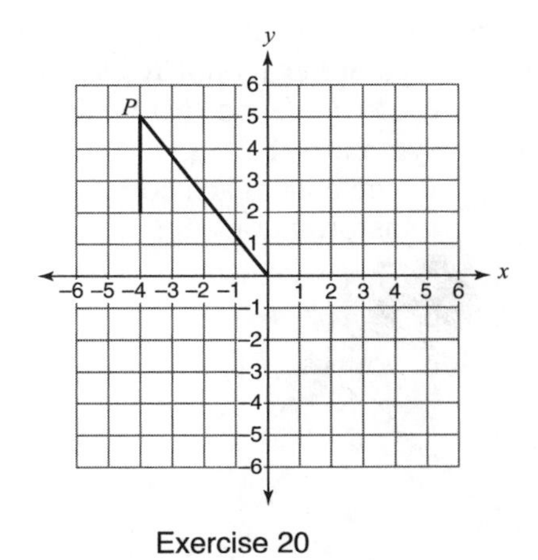

Exercise 20

21. Using the same figure as for Exercises 11–14 in Section 8.3, state the composite transformation rule that maps $\triangle RUN$ onto $\triangle GEM$.

22. If (–4, 8) is the image under the composite transformation $T_{h,3} \circ T_{-2,k}(-3, 0)$, what are the coordinates of the image of (2, –1) under the same composite transformation?

23. Given $A(0, 5)$ and $B(2, 0)$, graph and label $\overline{AB}$. Under the transformation $r_{x\text{-axis}} \circ r_{y\text{-axis}}(\overline{AB})$, A maps to A'', and B maps to B''. Graph and label $\overline{A''B''}$. What single transformation would map $\overline{AB}$ to $\overline{A''B''}$?

24. The coordinates of the vertices of $\triangle ABC$ are $A(-1, 2)$, $B(6, 2)$, and $C(3, 4)$. On graph paper, draw and label $\triangle ABC$.
a. On graph paper, graph and state the coordinates of $\triangle A'B'C'$, the image of $\triangle ABC$ after the composition $R_{90°} \circ r_{x\text{-axis}}$.
b. Write a transformation equivalent to $R_{90°} \circ r_{x\text{-axis}}$.

25. a. On graph paper, graph and label the triangle whose vertices are $A(0, 0)$, $B(8, 1)$, and $C(8, 4)$. Then graph, label, and state the coordinates of $\triangle A'B'C'$, the image of $\triangle ABC$ under the composite transformation $r_{x=0} \circ r_{y=x}$ ($\triangle ABC$). Which single transformation maps $\triangle ABC$ onto $\triangle A'B'C'$
(1) rotation
(2) dilation
(3) glide reflection
(4) translation

b. On the same set of axes, graph, label, and state the coordinates of $\triangle A''B''C''$, the image of $\triangle ABC$ under the composite transformation $r_{y=2} \circ r_{y=0}$ ($\triangle ABC$). Which single transformation maps $\triangle ABC$ onto $\triangle A''B''C''$?
(1) rotation
(2) dilation
(3) glide reflection
(4) translation

26. In the accompanying diagram of regular hexagon $ABCDEF$ with center O, L and P are lines of symmetry. State the final image under each composite transformation.
a. $r_P \circ R_{-120°}(C)$
b. $r_L \circ r_P(\overline{AB})$
c. $R_{60°} \circ r_O(A)$

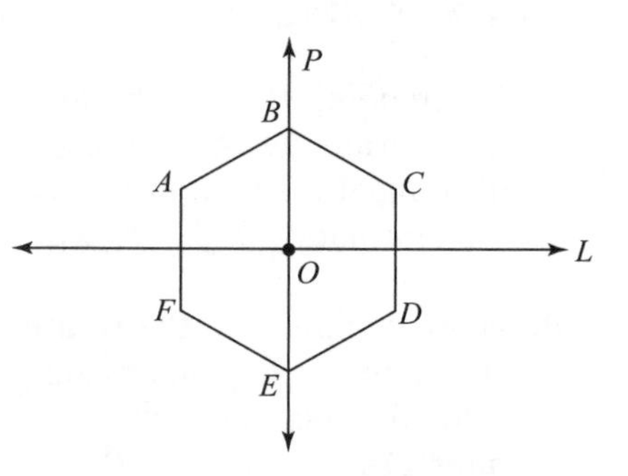

CHAPTER 9
LOCUS AND COORDINATES

9.1 SIMPLE LOCUS

KEY IDEAS

A **locus** may be thought of as a path consisting of the set of all points, and only those points, that satisfy a particular condition. A locus that satisfies only one condition, called a **simple locus**, takes the form of a line, a pair of lines, or a curve such as a circle. The plural of locus is *loci.*

Finding a Simple Locus: An Example

To find the locus of points that are 3 units from point K:

- Make a diagram. Keep drawing points 3 units from point K until you discover a pattern.
- Connect the points with a broken curve as shown in the accompanying figure.
- Write a sentence that describes what you have discovered: "The locus of points that are 3 units from point K is a circle that has point K as its center and a radius of 3 units."

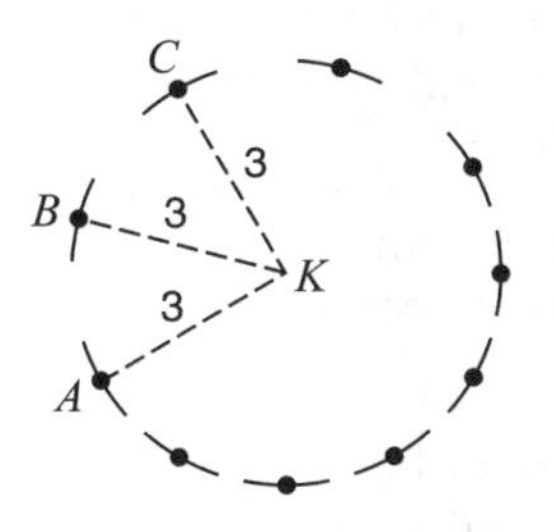

Five Basic Loci

The accompanying table summarizes the five basic loci that you need to know.

The Five Basic Loci

Locus	Diagram
1. **Condition:** Locus at a fixed distance d from a given point P. **Locus:** A circle with radius d and center P.	1. 
2. **Condition:** Locus at a fixed distance d from a line ℓ. **Locus:** Two lines parallel to line ℓ and each at distance d from line ℓ.	2. 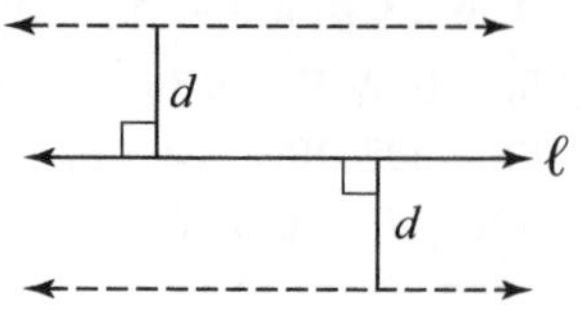
3. **Condition:** Locus equidistant from two points A and B. **Locus:** The perpendicular bisector of the segments whose endpoints are A and B.	3. 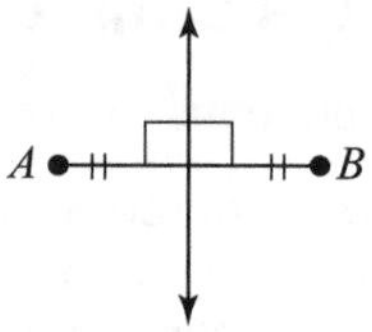
4. **Condition:** Locus equidistant from two parallel lines. **Locus:** A line parallel to the pair of lines and midway between them.	4. 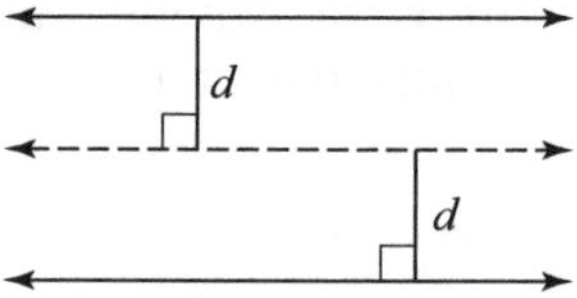
5. **Condition:** Locus equidistant from two intersecting lines. **Locus:** The bisectors of each pair of vertical angles formed by the lines.	5. 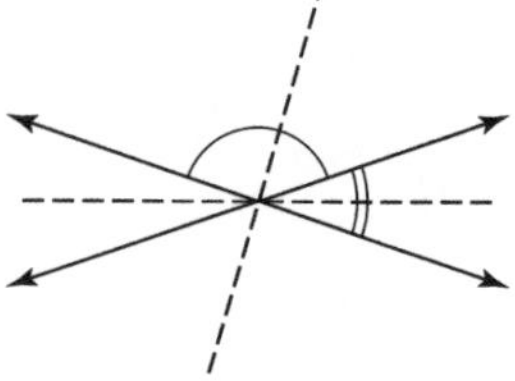

Example 1

Find the locus of points:

a. Equidistant from two concentric circles having radii of lengths 3 and 7 centimeters, respectively.

b. Equidistant from two parallel lines that are 8 inches apart.

Solution:

a. The locus of points equidistant from two concentric circles is another circle having the same center as the original circles and a radius length of 5 cm, which is the average of the lengths of the radii of the two given circles.

$$\frac{3\text{ cm} + 7\text{ cm}}{2} = 5\text{ cm}$$

b. The locus is a line that is parallel to the given lines and is at a distance of 4 inches from each of them:

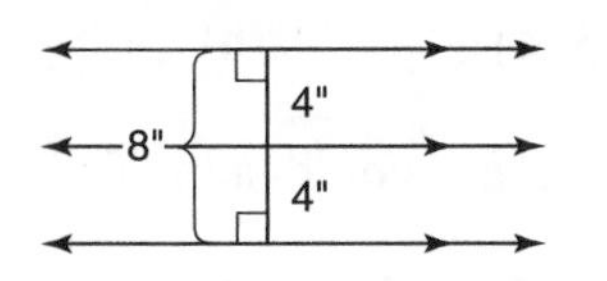

Locus and Coordinates

Locus can be applied to points in the coordinate plane where the x-coordinate is sometimes referred to as the **abscissa** and the y-coordinate as the **ordinate**. For example, the locus of all points whose ordinates exceed their abscissas by 1 is the line whose equation is $y = x + 1$.

Example 2

What is an equation of the locus of points equidistant from points $A(-2, 1)$ and $B(4, 1)$?

Solution: The required locus is the perpendicular bisector of $\overline{AB}$. Since the coordinates of A and B determine a horizontal segment whose midpoint is (1,2), the required locus is a vertical line through (1,2).

An equation of the required locus is $\mathbf{x = 1}$.

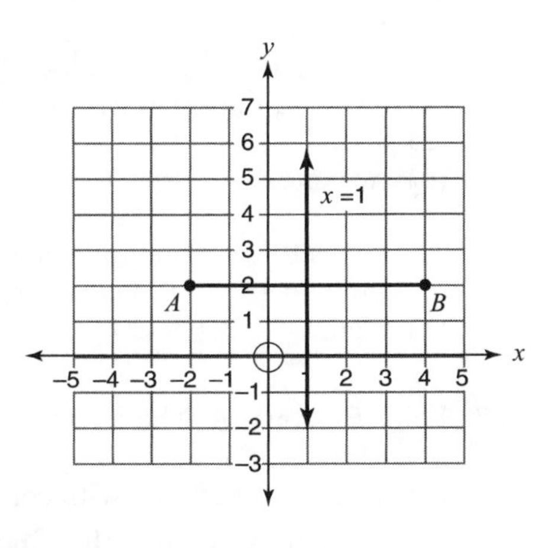

MATH FACTS

Let point $(\bar{x}, \bar{y})$ represent the midpoint of $\overline{AB}$.

- If $\overline{AB}$ is a horizontal segment, then the locus of points equidistant from the endpoints of $\overline{AB}$ is the vertical line $x = \bar{x}$.
- If $\overline{AB}$ is a vertical segment, then the locus of points equidistant from the endpoints of $\overline{AB}$ is the horizontal line $y = \bar{y}$.

Check Your Understanding of Section 9.1

A. Multiple Choice

1. The locus of the midpoints of the radii of a circle is
(1) a point (2) two lines (3) one line (4) a circle

2. Which point lies on the locus of points equidistant from points $A(2, 0)$ and $B(6, 0)$?
(1) (0, 4) (2) (2, 5) (3) (5, 4) (4) (4, 5)

3. Which is an equation of the locus of points equidistant from points $A(2, 1)$ and $B(2, 5)$?
(1) $x = 3$ (2) $y = 3$ (3) $x = 2$ (4) $y = 2$

4. Which is an equation of the locus of points whose ordinates exceed twice their abscissas by 3?
(1) $x = 2y + 3$
(2) $x + 3 = 2y$
(3) $y = 2x + 3$
(4) $y + 3 = 2x$

5. The locus of the center of a coin that rolls along a flat table top in a straight line is
(1) a line parallel to the table top
(2) a pair of lines parallel to the table top
(3) a circle
(4) two circles

6. The locus of points equidistant from two concentric circle having radii of 7 cm and 11 cm is a concentric circle whose radius is
(1) 8 cm (2) 9 cm (3) 18 cm (4) 4 cm

7. Point P is on $\overleftrightarrow{AB}$. The locus of the centers of all circles of radius 5 that pass through P is
(1) a circle of radius 5 with center at P
(2) a line passing through P and perpendicular to $\overleftrightarrow{AB}$
(3) two lines, both perpendicular to $\overleftrightarrow{AB}$ and 5 units on either side of P
(4) two lines parallel to $\overleftrightarrow{AB}$, one 5 units above $\overleftrightarrow{AB}$ and the other 5 units below $\overleftrightarrow{AB}$

B. Show or explain how you arrived at your answer.

8–14. Write an equation or equations that completely describe the given locus.

8. The locus of points equidistant from the endpoints of the segment determined by (–3, 5) and (7, 5).

9. The locus of points equidistant from the endpoints of the segment determined by (–1,–2) and (–1,6).

10. The locus of points equidistant from the lines $y = -2$ and $y = 8$.

11. The locus of points equidistant from the lines $x = -7$ and $x = -1$.

12. The locus of points 2 units from the y-axis.

13. The locus of points 5 units from the line $x = 3$.

14. The locus of points 4 units from the line $y = -1$.

9.2 SATISFYING MORE THAN ONE LOCUS CONDITION

KEY IDEAS

Sometimes it is necessary to find the set of points satisfying more than one locus condition. To find such a **compound locus**, use the same diagram to describe each locus condition. Then locate the points, if any, at which all of the loci intersect.

Finding a Compound Locus

To find the number of points that are 2 inches from point P on line ℓ *and*, at the same time, 2 inches from point P:

- Draw line ℓ and label any convenient point, P, on it.
- Represent the first locus as a pair of parallel lines on either side of line ℓ, and at a distance of 2 inches from it.
- Using the same diagram, represent the second locus as a circle that has P as its center and a radius of 2 inches.

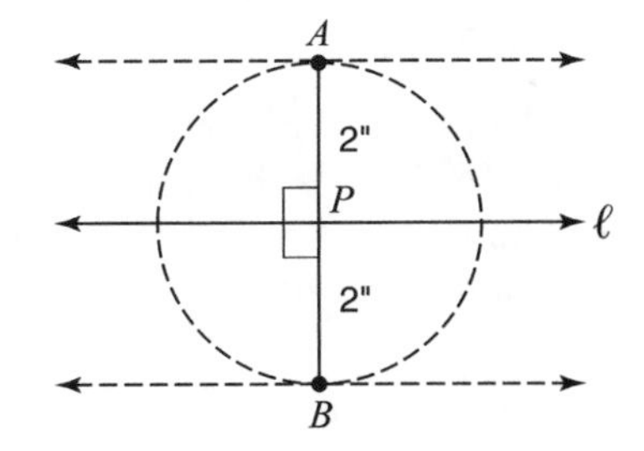

Since the two loci intersect at points A and B, there are **two points** that satisfy both of the given locus conditions.

Example 1

Point P is located on one of two parallel lines that are 8 inches apart. What is the number of points that are equidistant from the parallel lines and also 4 inches from P?

Solution:

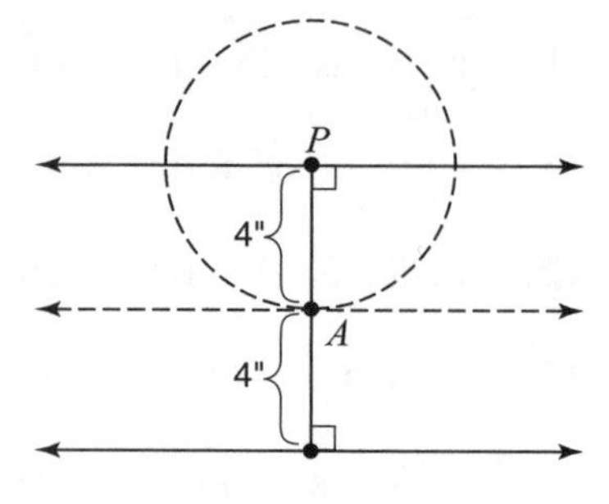

- The locus of points equidistant from two parallel lines is another parallel line midway between them.
- The locus of points 4 inches from P is a circle that has P as its center and a radius of 4 inches.
- Because the circle is tangent to the line that is midway between the original pair of parallel lines, the loci intersect only at point A.

There is exactly **one point** that satisfies both conditions.

Example 2

In Example 1 what is the number of points that are equidistant from the parallel lines and also 3 inches from P?

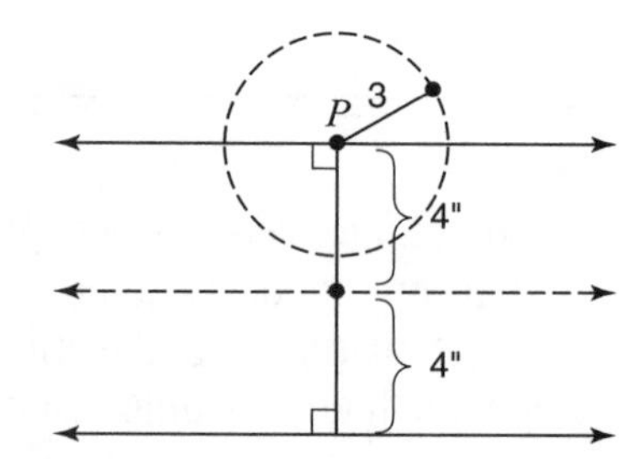

Solution: If the circle has a radius of 3 inches, it does *not* intersect the broken line. Since the loci do *not* intersect, **no point** satisfies both conditions.

Example 3

Points A and B are 5 inches apart. What is the number of points equidistant from points A and B, and 3 inches from A?

Solution:

- The locus of points equidistant from A and B is the perpendicular bisector of $\overline{AB}$. If M is the midpoint of $\overline{AB}$, then $AM = MB = 2.5$.
- The locus of points 3 inches from A is a circle that has A as its center and a radius of 3 inches.

- Since the radius of circle A is greater than 2.5 inches, the circle will intersect the perpendicular bisector at two points, labeled C and D in the accompanying figure.

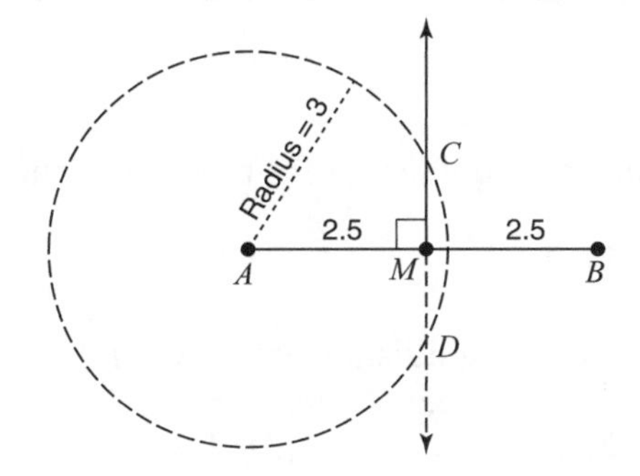

There are exactly **two points** that satisfy both locus conditions.

Example 4

How many points are 3 units from the origin and 2 units from the y-axis?

Solution:

- The locus of points 3 units from the origin is a circle that has (0,0) as its center and a radius of 3. The circle intersects each coordinate axis at 3 and –3.
- The locus of points 2 units from the y-axis is a pair of vertical lines 2 units on either side of it.
- Since the distance of the vertical lines from the y-axis is less than the radius of the circle ($2 < 3$), the lines intersect the circle in four points labeled A, B, C, and D.

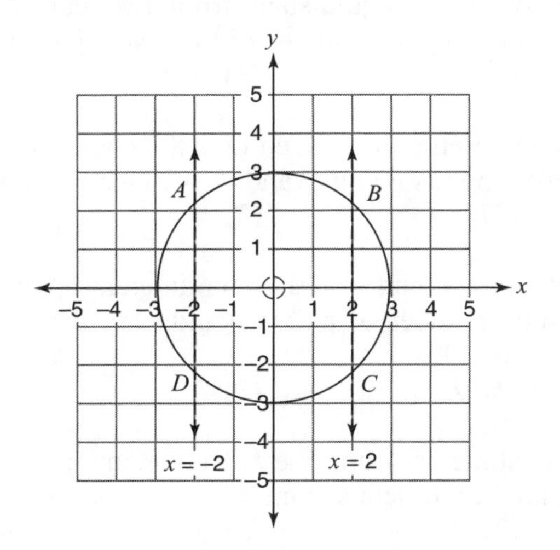

There are **four points** that satisfy both locus conditions.

Check Your Understanding of Section 9.2

A. Multiple Choice

1. How many points are equidistant from points A and B and also 4 inches from $\overline{AB}$?
(1) 1 (2) 2 (3) 3 (4) 4

2. How many points are equidistant from two intersecting lines and also 3 cm from their point of intersection?
(1) 1 (2) 2 (3) 3 (4) 4

3. How many points are 4 units from the origin and also 4 units from the x-axis?
(1) 1 (2) 2 (3) 3 (4) 4

4. How many points are equidistant from points $P(2, 1)$ and $Q(2, 5)$ and also 3 units from the origin?
(1) 1 (2) 2 (3) 3 (4) 4

5. Lines ℓ and m are parallel lines 8 centimeters apart, and point P is on line m. What is the total number of points that are equidistant from lines ℓ and m and 4 centimeters from P?
(1) 1 (2) 2 (3) 0 (4) 4

6. The number of points equidistant from two parallel lines and also equidistant from two points on one of the given lines is *exactly*
(1) 1 (2) 2 (3) 3 (4) 4

7. The distance between points P and Q is 8 inches. How many points are equidistant from points P and Q and also 3 inches from P?
(1) 1 (2) 2 (3) 0 (4) 4

8. Two parallel lines, r and s, are 5 centimeters apart. Point A lies on line r. The total number of points equidistant from lines r and s and 5 centimeters from A is
(1) 1 (2) 2 (3) 3 (4) 4

9. Point A is 4 centimeters from line k. How many points are 1 centimeter from line k and 5 centimeters from A?
(1) 1 (2) 2 (3) 3 (4) 4

10. The total number of points 3 centimeters from $\overleftrightarrow{JK}$ and 6 centimeters from point K is:
(1) 1 (2) 2 (3) 3 (4) 4

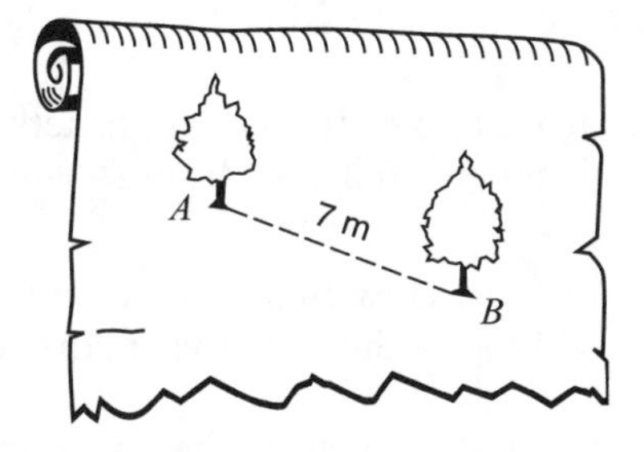

11. The accompanying diagram shows a map. A treasure is buried at a distance of 3 meters from the foot of tree A and 5 meters from the foot of tree B. If $AB = 7$ meters, what is the total number of possible places in which the treasure might be buried?
(1) 1 (2) 2 (3) 3 (4) 4

12. Point P is 2 inches from line ℓ. The total number of points 5 inches from line ℓ and 7 inches from point P is
(1) 1 (2) 2 (3) 3 (4) 4

13. The total number of points equidistant from (3, 0) and (–11, 0) and 4 units from the origin is
(1) 1 (2) 2 (3) 0 (4) 4

14. Kevin draws the locus of points d units from the line $x = -2$. He then draws on the same set of axes the locus of points 3 units from the origin. If Kevin determines that the two loci intersect at a total of 3 points, what is the value of d?
(1) 1 (2) 2 (3) 3 (4) 4

15. The length of the line segment joining points A and B is 10. There will be only one point equidistant from both A and B and also at a distance q from A if
(1) $q = 10$ (2) $q > 10$ (3) $q = 5$ (4) $5 < q < 10$

B. *Show or explain how you arrived at your answer.*

16. The distance between points P and Q is 9 inches. Find the number of points that are equidistant from points P and Q and also x inches from point P when
a. $x = 4$ inches
b. $x = 5$ inches

17. A tree is located 30 feet east of a fence that runs north to south. Kelly tells her brother Billy that their dog buried Billy's hat a distance of 15 feet from the fence and also 20 feet from the tree. Draw a sketch to show where Billy should dig to find his hat. Based on your sketch, how many locations for the hat are possible?

18. Point P is x inches from line ℓ. If there are exactly three points that are 2 inches from line ℓ and also 6 inches from P, what is the value of x?

19. Point P is located on $\overleftrightarrow{AB}$. How many points are 3 units from $\overleftrightarrow{AB}$ and 5 units from point P? Draw a diagram to support your answer.

20. Lines AB and CD are parallel to each other and 6 inches apart. Point P is located between the two parallel lines and 1 inch from $\overleftrightarrow{AB}$. How many points are equidistant from $\overleftrightarrow{AB}$ and $\overleftrightarrow{CD}$ and, at the same time, 2 inches from point P? Draw a diagram to support your answer.

21. a. Describe completely the locus of points 2 units from the line whose equation is $x = 3$.
b. Describe completely the locus of points n units from the point $P(3,2)$.
c. Determine the total number of points that satisfy the locus conditions in parts a and b simultaneously for $n = 2$?

22. Point M is the midpoint of $\overline{AB}$.
a. Describe fully the locus of all points in a plane that are
(1) equidistant from A and B
(2) 6 units from $\overleftrightarrow{AB}$
(3) d units from M
b. For what value of d will there be exactly two points that simultaneously satisfy all three conditions in part **a**.

23. Point P is on $\overleftrightarrow{AB}$.
a. Describe fully the locus of points:
(1) d units from $\overleftrightarrow{AB}$
(2) D units from P.
b. Find the number of points that satisfy the two locus conditions when
(1) $D = d$
(2) $D < d$
(3) $D > d$

9.3 CONCURRENCY THEOREMS

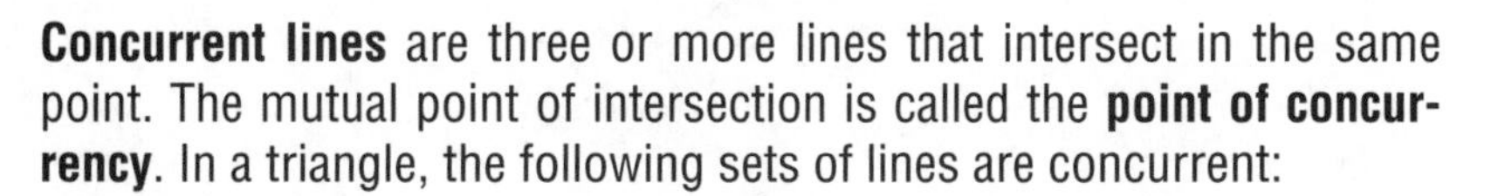

Concurrent lines are three or more lines that intersect in the same point. The mutual point of intersection is called the **point of concurrency**. In a triangle, the following sets of lines are concurrent:

- The three medians
- The three altitudes
- The perpendicular bisectors of the three sides
- The three angle bisectors

Concurrency of Medians

The three medians of a triangle are concurrent in a point called the **centroid** of the triangle. The distance from each vertex to the centroid is two-thirds of the length of the entire median drawn from that vertex, as shown in Figure 9.1. Thus, if $AX = 9$, then $AP = \frac{2}{3} \times 9 = 6$ and $PX = 3$.

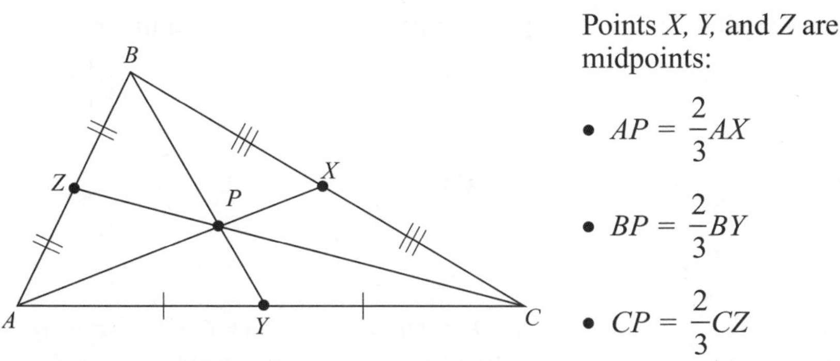

Figure 9.1 Point *P* is the *centroid* of the triangle.

The centroid of a triangle is its *center of gravity*. A metal triangular plate can be made to balance horizontally in space by placing a single support directly underneath the centroid of the triangle.

Concurrency of Perpendicular Bisectors

The perpendicular bisectors of the three sides of a triangle are concurrent in a point equidistant from the vertices of the triangle. In Figure 9.2, the perpendicular bisectors of the sides of $\triangle ABC$ are concurrent in point O. Since point O is equidistant from the vertices of the triangle, $OA = OB = OC$.

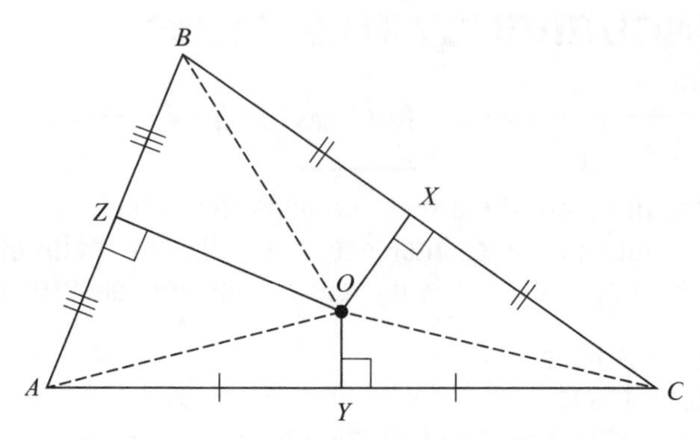

Figure 9.2 Point O is the circumcenter of $\triangle ABC$.

Using OA as a radius and point O as a center, a circle can be *circumscribed* about $\triangle ABC$ so that each of its vertices are points on the circle. Point O is called the **circumcenter** of the triangle.

Location of the Circumcenter of a Triangle

The circumcenter of a triangle can fall in the interior of the triangle, on a side of the triangle, or in the exterior of the triangle, as illustrated in Figure 9.3.

Acute triangle

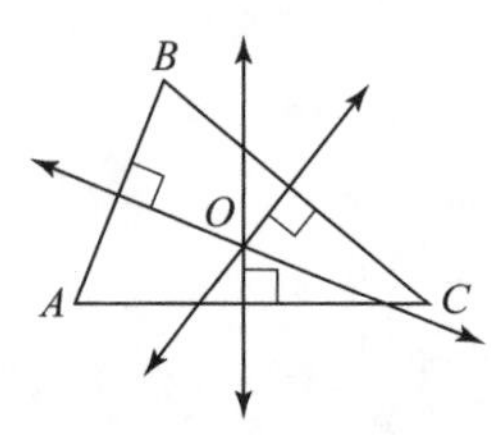

Point O in the interior of the acute triangle.

Right triangle

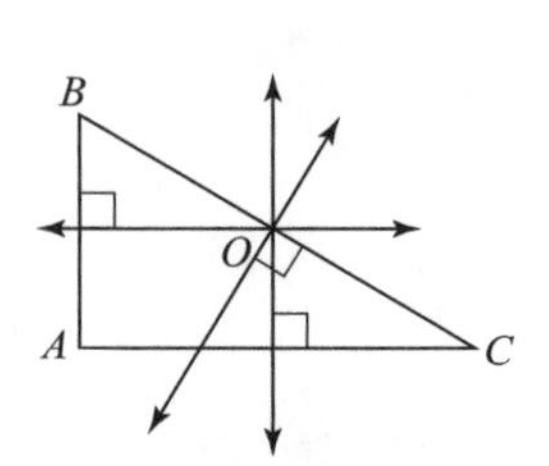

Point O on the hypotenuse of a right triangle.

Obtuse triangle

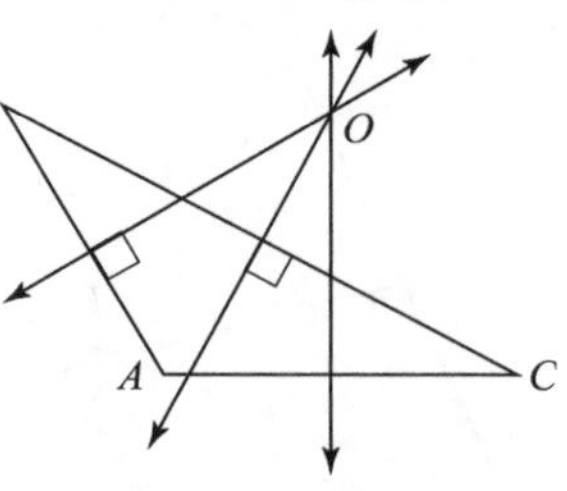

Point O in the exterior of the obtuse triangle.

Figure 9.3 Locating the circumcenter O of a triangle.

MATH FACTS

The center of a circle can be located by finding the point of intersection of the perpendicular bisectors of any two non-parallel chords of the circle.

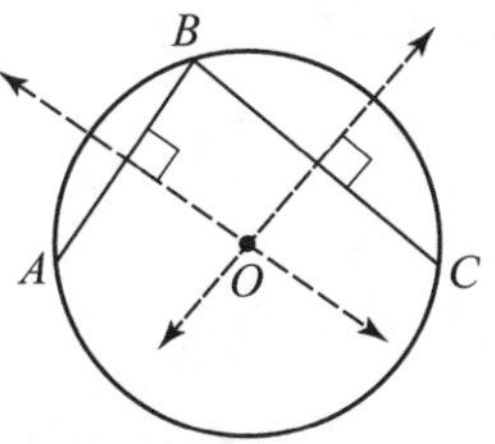

Concurrency of Angle Bisectors

The three angle bisectors of a triangle are concurrent in a point equidistant from the sides of the triangle. In Figure 9.4, the bisectors of the angles of $\triangle ABC$ are concurrent in point Q. Since point Q is equidistant from the sides of the triangle, $QX = QY = QZ$.

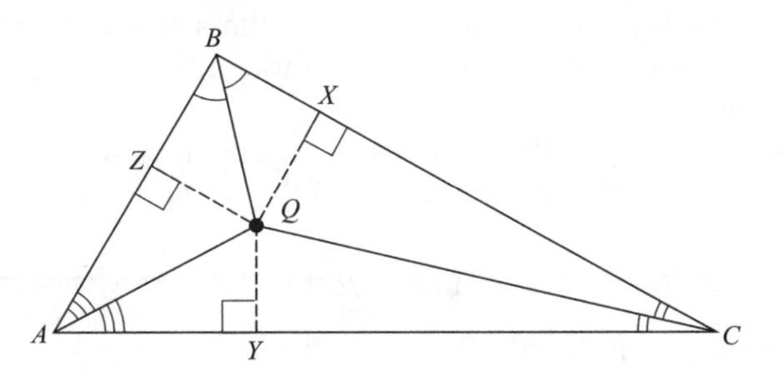

Figure 9.4 The bisectors of the angles are concurrent at a point Q equidistant from the sides of $\triangle ABC$.

Using QX as the radius and point Q as the center, a circle can be *inscribed* in $\triangle ABC$ so that points X, Y, and Z are on the circle, as shown in Figure 9.5. Point Q is called the **incenter** of the triangle. Unlike the circumcenter of a triangle, the incenter of a triangle always lies in the interior of the triangle.

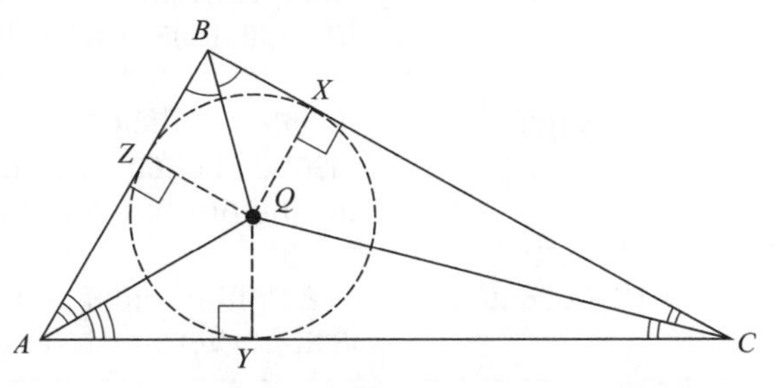

Figure 9.5 Point Q is the incenter of $\triangle ABC$.

Concurrency of Altitudes

The altitudes of a triangle, extended if necessary, are concurrent in a point called the **orthocenter** of the triangle. The orthocenter of a triangle can fall in the interior of the triangle, on a side of the triangle, or in the exterior of the triangle, as illustrated in Figure 9.6.

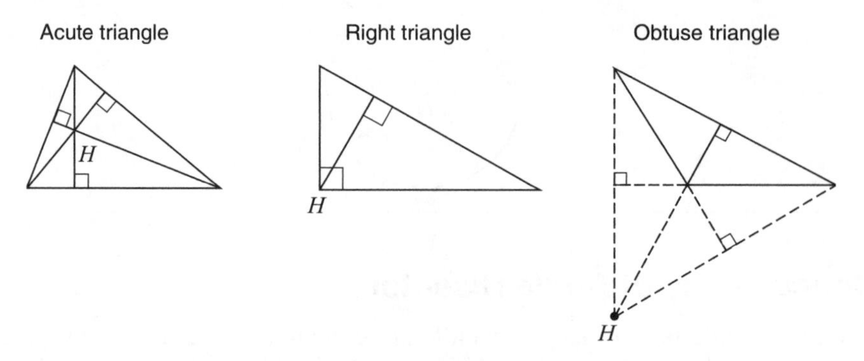

Figure 9.6 Point H is the *orthocenter* of each triangle.

- In an acute triangle, the orthocenter lies in the interior of the triangle.
- In a right triangle, either leg serves as the altitude drawn to the other leg. The point of concurrency of the three altitudes is the vertex of the right angle.
- In an obtuse triangle, the altitudes and a side need to be extended so that the orthocenter falls in the exterior of the triangle.

MATH FACTS

Set of Lines in a Triangle	Point of Concurrency	Related Facts
Medians	Centroid	Centroid divides each median into segment lengths with the ratio 2:1 as measured from each vertex.
Perpendicular bisectors of sides	Circumcenter	Circumcenter is the center of the circumscribed circle. In a right triangle, it lies on the hypotenuse.
Angle bisectors	Incenter	Incenter is the center of the inscribed circle and always lies in the interior of the triangle.
Altitudes	Orthocenter	In a right triangle, the orthocenter is at the vertex of the right angle.

Check Your Understanding of Section 9.3

A. *Multiple Choice.*

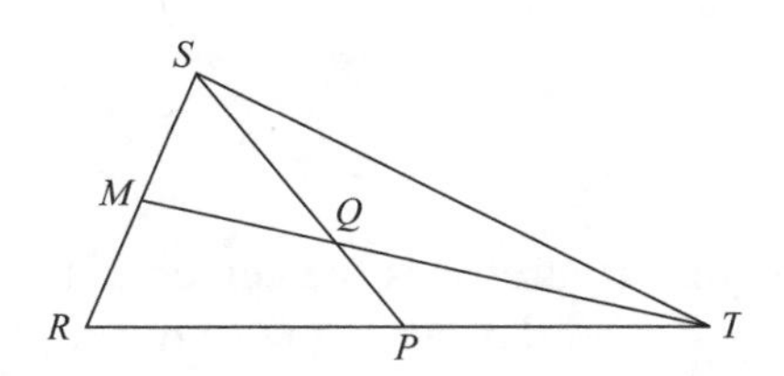

1. In $\triangle RST$, medians $\overline{TM}$ and $\overline{SP}$ are concurrent at point Q. If $TQ = 3x - 1$ and $QM = x + 1$, what is the length of median $\overline{TM}$?
(1) 3 (2) 8 (3) 12 (4) 11

2. If circle O is circumscribed about $\triangle ABC$, then point O always lies on
(1) side $\overline{AC}$
(2) the median to side $\overline{AC}$
(3) the bisector of $\angle ABC$
(4) the perpendicular bisector of $\overline{AC}$

3. In $\triangle ABC$, points J, K, and L are the midpoints of sides $\overline{AB}$, $\overline{BC}$, and $\overline{AC}$, respectively. If the three medians of the triangle intersect at point P and the length of $\overline{LP}$ is 6, what is the length of median $\overline{BL}$?
(1) 18 (2) 12 (3) 9 (4) 4

4. If circle O is inscribed in $\triangle ABC$, then point O is always on
(1) side $\overline{AC}$
(2) the median to side $\overline{AC}$
(3) the bisector of $\angle ABC$
(4) the perpendicular bisector of $\overline{AB}$

5. Given right triangle ABC with right angle at C. The locus of points equidistant from points A and B intersects $\overline{AB}$ at the
(1) center of the inscribed circle
(2) center of the circumscribed circle
(3) orthocenter of the triangle
(4) centroid of the triangle

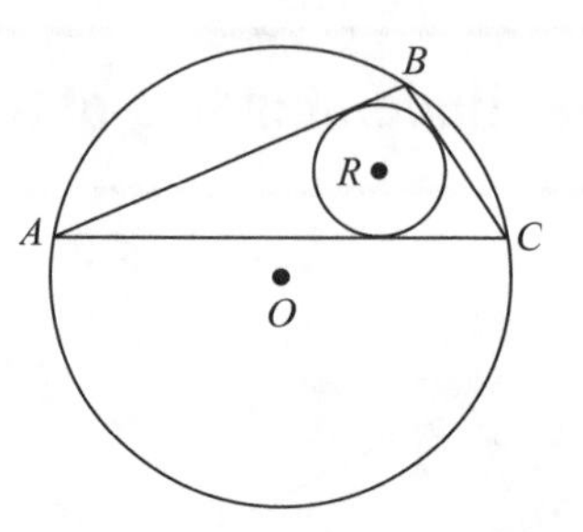

6. In the accompanying figure, O is the center of the circle circumscribed about scalene $\triangle ABC$ and R is the center of the circle inscribed in scalene $\triangle ABC$.
   Which statement is *false*?
   (1) Point R is equidistant from $\overline{AC}$ and $\overline{BC}$.
   (2) Point O is equidistant from points B and C.
   (3) Point R lies on the bisectors of angles A and C.
   (4) Point O is the point at which the altitudes drawn to sides $\overline{AC}$, $\overline{AB}$, and $\overline{BC}$ intersect.

B. *Show or explain how you arrived at your answer.*

7. Given right triangle ABC with right angle at C.
   a. Describe the locus of points equidistant from sides $\overline{AC}$ and $\overline{AB}$.
   b. In how many points, if any, will the locus described in part a intersect the locus of points equidistant from sides $\overline{AB}$ and $\overline{BC}$.
   c. The locus described in parts a and b can be used to determine the
      (1) center of the inscribed circle
      (2) center of the circumscribed circle
      (3) orthocenter of the triangle
      (4) centroid of the triangle

9.4 WRITING AN EQUATION OF A LINE

KEY IDEAS

The set of all ordered pairs (x, y) that satisfy a two-variable linear equation is represented by a line in the coordinate plane. To be able to write an equation of an oblique (slanted) line, you need to know two facts about the line:

- The slope and the coordinates of a point on the line or
- The coordinates of two points on the line.

Slope–Intercept Equation: $y = mx + b$

When a two-variable linear equation is solved for y, it has the general form $\boldsymbol{y = mx + b}$, where

- m, the coefficient of x, is the slope of the line
- b, the y-intercept, indicates where the line crosses the y-axis

To read the slope and y-intercept of a line from its equation, it may be necessary to rewrite it in $y = mx + b$ form. If an equation of line ℓ is $2y - 6x = 10$, then $\frac{2y}{2} = \frac{6x}{2} + \frac{10}{2}$ and $y = 3x + 5$ so the slope of line ℓ is 3 and its y-intercept is 5.

Example 1

An equation of line ℓ is $2y = 3x + 6$. Which equation represents a line that is perpendicular to line ℓ?

(1) $y = \frac{3}{2}x + 2$ (3) $y = \frac{2}{3}x - 3$

(2) $y = -\frac{3}{2}x + 2$ (4) $y = -\frac{2}{3}x - 3$

Solution: Perpendicular lines have slopes that are negative reciprocals. Compare the slope of line ℓ with the slopes of the lines in the answer choices.

- Find the slope of line ℓ. If $2y + 3x = 6$, then solving for y gives $y = -\frac{3}{2}x + 3$.

  Hence, the slope of line ℓ is $-\frac{3}{2}$.

- The slope of the required line must be $\frac{2}{3}$; the negative reciprocal, $-\frac{3}{2}$.

- Because the slope of the line in answer choice (3) is $\frac{2}{3}$, this line is perpendicular to line ℓ.

The correct choice is **(3)**.

Example 2

The equation of line r is $3y + 6x = 12$ and the equation of line s is $8y - 12 = 4x$. Which statement is true about lines r and s?

(1) $r \parallel s$ (3) lines r and s coincide

(2) $r \perp s$ (4) lines r and s have the same y-intercept

Solution: Rewrite each equation in slope-intercept form.

- Line r: $3y + 6x = 12$, so $3y = -6x + 12$ and $y = -2x + 4$.
- Line s: $8y - 12 = 4x$, so $8y = 4x + 12$ and $y = \frac{1}{2}x + \frac{3}{2}$.
- The slope of line r is -2, and the slope of line s is $\frac{1}{2}$. Since -2 and $\frac{1}{2}$ are negative reciprocals, $r \perp s$.

The correct choice is **(2)**.

Example 3

Write an equation of the line that passes through the point $(-1, 3)$ and is parallel to the line whose equation is $y - 2x = 3$.

Solution: If $y - 2x = 3$, then $y = 2x + 3$ so the slope of this line is 2.

- Because parallel lines have the same slope, the slope of the required line is also 2, so its equation has the form, $y = 2x + b$.
- It is also given that $(-1, 3)$ is a point on the line. Find b by substituting $x = -1$ and $y = 3$ in $y = 2x + b$, which gives $3 = 2(-1) + b$, so $b = 5$.
- Since $m = 2$ and $b = 5$, an equation of the required line is $\boldsymbol{y = 2x + 5}$.

Point–Slope Equation: $y - k = m(x - h)$

In point–slope form, an equation of a line is written as $y - k = m(x - h)$, where (h,k) is any point on the line and m is the slope of the line. If the line that has a slope of -2 passes through the point $(4, 3)$, then its equation can be written in point–slope form by letting $h = 4$, $k = 3$, and $m = -2$:

$$y - k = m(x - h)$$

$$\downarrow \quad \downarrow \quad \downarrow$$

$$y - 3 = -2(x - 4)$$

If necessary, the equation $y - 3 = -2(x - 4)$ can be written in $y = mx + b$ form by removing the parentheses on the right side of the equation and isolating y:

$$\begin{aligned} y - 3 &= -2(x - 4) \\ y - 3 &= -2x + 8 \\ y &= -2x + 11 \end{aligned}$$

Example 4

Write an equation of the line ℓ that passes through $(6, -7)$ and is parallel to the line $y - 3x = -4$. What is the y-intercept of line ℓ?

Solution: Let m represent the slope of line ℓ and (h, k) represent the coordinates of any point on line ℓ.

- Rewrite $y - 3x = -4$ as $y = 3x - 4$. Because line ℓ is parallel to the given line, the slope of line ℓ is also 3.
- As it is also given that line ℓ passes through $(6, -7)$, $(h, k) = (6, -7)$.
- Write an equation of line ℓ using the point–slope form where $h = 6$, $k = -7$, and $m = 3$:

$$\begin{aligned} y - k &= m(x - h) \\ y - (-7) &= 3(x - 6) \\ y + 7 &= 3x - 18 \\ y &= 3x - 25 \end{aligned}$$

The y-intercept of line ℓ is **–25**.

Example 5

Find an equation of the line that is the perpendicular bisector of the line segment whose endpoints are $R(-8, 7)$ and $S(4, 3)$.

Solution:

- Find the midpoint $(\bar{x}, \bar{y})$ of $\overline{RS}$:

$$\bar{x} = \frac{-8+4}{2} = \frac{-4}{2} = -2 \quad \text{and} \quad \bar{y} = \frac{7+3}{2} = \frac{10}{2} = 5$$

The midpoint of $\overline{RS}$ is $(-2, 5)$.
- Find the slope of $\overline{RS}$:

$$m = \frac{\Delta y}{\Delta x} = \frac{3-7}{4-(-8)} = \frac{-4}{12} = -\frac{1}{3}$$

The slope of the perpendicular bisector of $\overline{RS}$ is the negative reciprocal of $-\frac{1}{3}$ or 3.
- Use the point–slope equation form where $(h, k) = (\bar{x}, \bar{y}) = (-2, 5)$ and $m = 3$:

$$\begin{aligned} y - k &= m(x - h) \\ y - 5 &= 3(x - (-2)) \\ \mathbf{y - 5} &\mathbf{= 3(x + 2)} \end{aligned}$$

If required, the equation can also be written in $y = mx + b$ form:

$$\begin{aligned} y - 5 &= 3(x + 2) \\ y - 5 &= 3x + 6 \\ \mathbf{y} &\mathbf{= 3x + 11} \end{aligned}$$

Example 6

The vertices of $\triangle ABC$ are $A(-3, -1)$, $B(1,7)$, and $C(6, -3)$.
a. Write an equation of the line that contains the median from C to $\overline{AB}$.
b. Write an equation of the line that contains the altitude from A to $\overline{BC}$.
c. Prove that the altitude determined in part b, when extended, passes through the point (7, 4).

Solution:

a. Label the point at which the median from C intersects side $\overline{AB}$ as point M.

- Find the midpoint, $M(\bar{x}, \bar{y})$, of $\overline{AB}$:

$$\bar{x} = \frac{-3+1}{2} = \frac{-2}{2} = -1 \qquad \text{and} \qquad \bar{y} = \frac{-1+7}{2} = \frac{6}{2} = 3$$

The midpoint is $M(-1, 3)$.

- Find the slope, m, of $\overline{CM}$:

$$m = \frac{\Delta y}{\Delta x} = \frac{3-(-3)}{-1-6} = -\frac{6}{7}$$

- Use the point–slope equation form where $(h, k) = M(-1, 3)$ and $m = -\frac{6}{7}$:

$$\text{Equation of } \overline{CM}\text{: } \mathbf{y - 3 = -\frac{6}{7}(x - 3)}$$

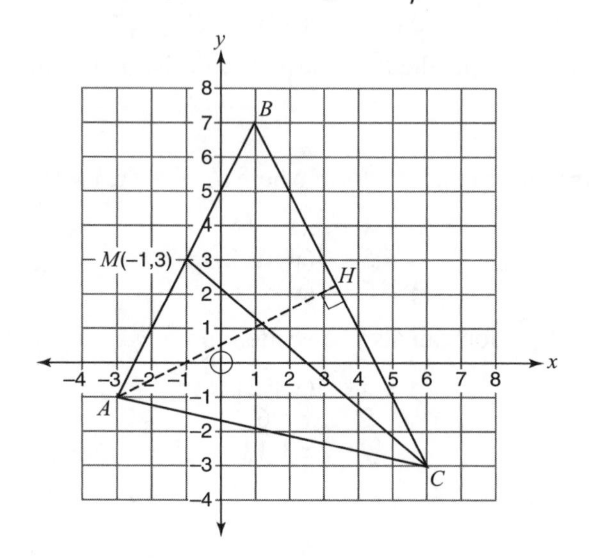

b. Label the point at which the altitude intersects $\overline{BC}$ as point H.

- Find the slope, m, of $\overline{BC}$:

$$m = \frac{\Delta y}{\Delta x} = \frac{-3-7}{6-1} = \frac{-10}{5} = -2$$

- Find the slope of $\overline{AH}$. Since $\overline{AH} \perp \overline{BC}$, slope of $\overline{AH} = \frac{1}{2}$.

- Use the point–slope equation form where $(h, k) = A(-3, -1)$ and $m = \frac{1}{2}$:

$$y-(-1) = \frac{1}{2}(x-(-3))$$

Equation of $\overline{AH}$: $\mathbf{y + 1 = \frac{1}{2}(x + 3)}$

If required, the equation can also be written in $y = mx + b$ form:

$$y + 1 = \frac{1}{2}(x + 3)$$

$$y + 1 = \frac{1}{2}x + \frac{3}{2}$$

Slope–intercept form of $\overline{AH}$: $y = \frac{1}{2}x + \frac{1}{2}$

c. Show that the point (7, 4) satisfies the equation of $\overline{AH}$:

$$y = \frac{1}{2}x + \frac{1}{2}$$

$$4 \;\Big|\; \frac{1}{2}(7) + \frac{1}{2}$$

$$\frac{7}{2} + \frac{1}{2}$$

$$\frac{8}{2}$$

$$4 = 4 \checkmark$$

Check Your Understanding of Section 9.4

A. Multiple Choice

1. Which line is perpendicular to the line whose equation is $5y + 6 = -3x$?

(1) $y = -\frac{5}{3}x + 7$ (3) $y = -\frac{3}{5}x + 7$

(2) $y = \frac{5}{3}x + 7$ (4) $y = \frac{3}{5}x + 7$

2. The graph of $x - 3y = 6$ is parallel to the graph of

(1) $y = -3x + 7$ (3) $y = 3x - 8$

(2) $y = -\frac{1}{3}x + 5$ (4) $y = \frac{1}{3}x + 8$

3. The graph of which equation is perpendicular to the graph of $y - 3 = \frac{1}{2}x$?

(1) $y = -\frac{1}{2}x + 5$ (3) $y = 2x + 5$

(2) $2y = x + 3$ (4) $y + 2x = 3$

4. Which is an equation of the line that is parallel to $y = 2x - 8$ and passes through the point $(0,-3)$?

(1) $y = 2x + 3$ (3) $y = -\frac{1}{2}x + 3$

(2) $y = 2x - 3$ (4) $y = -\frac{1}{2}x - 3$

5. Which is an equation of the line that is parallel to $y - 3x + 5 = 0$ and has the same y-intercept as $y = -2x + 7$?

(1) $y = 3x - 2$ (3) $y = 3x + 7$

(2) $y = -2x - 5$ (4) $y = -2x - 7$

B. *Show or explain how you arrived at your answer.*

6. a. Determine an equation of the line that is the perpendicular bisector of the line segment whose endpoints are $A(-1, 8)$ and $B(-5, 2)$.
b. Determine the coordinates of the point at which the perpendicular bisector of $\overline{AB}$ intersects the y-axis.

7. Write an equation that describes the locus of points equidistant from the lines $y = 3x - 1$ and $y = 3x + 9$.

8. Given points $A(6, 3)$ and $B(2, 2)$.
a. If A' is the image of point A after a reflection over the line $y = x$, find an equation of $\overleftrightarrow{A'B}$.
b. Determine the number of square units in the area of $\triangle ABA'$.

9. Kim graphed the line represented by the equation $3y + 2x = 4$. Write an equation of a line that is
a. Parallel to the line Kim graphed and that contains the point $(1, -6)$.
b. Perpendicular to the line Kim graphed and that passes through the origin.
c. Neither parallel nor perpendicular to the line Kim graphed and that has the same y-intercept as the line Kim graphed.

10. The vertices of $\triangle RST$ are $R(2, 7)$, $S(8, 9)$, and $T(6, 3)$.
a. Write an equation of the perpendicular bisector of $\overline{RT}$.
b. Prove that the perpendicular bisector passes through vertex S.

11. The vertices of $\triangle ABC$ are $A(-4, 1)$, $B(2, 13)$, and $C(10, 9)$.
a. Find the slope of $\overline{AB}$.
b. Write an equation of the line that passes through the midpoint, M, of $\overline{BC}$ and is parallel to $\overline{AB}$.
c. If the line determined in part b intersects side $\overline{AC}$ at point D, the ratio of the length of $\overline{DM}$ to the length of $\overline{AB}$ is
(1) 1:1 (2) 1:2 (3) 1:3 (4) 1:4

12. The vertices of right triangle ABC are $A(3, 3)$, $B(7, 7)$, and $C(7, -1)$.
a. Write an equation of the line which passes through B and is parallel to $\overline{AC}$.
b. If circle O is circumscribed about $\triangle ABC$, find the coordinates of O.

13. Given points $A(2, 2)$ and $B(6, 3)$.
a. Find the coordinates of A', the image of A after a dilation of constant 4 with respect to the origin. Write an equation of $\overleftrightarrow{AA'}$.
b. Find the coordinates of B', the image of B after a reflection in $\overleftrightarrow{AA'}$.
c. Show that $ABA'B'$ is *not* a parallelogram.

14. The vertices of $\triangle PQR$ are $P(8, 6)$, $Q(-1, 13)$, and $R(5, -5)$. The median drawn from P intersects $\overline{QR}$ at point M.
a. Write an equation of $\overline{PM}$.
b. Using the methods of coordinate geometry, prove that $\overline{PM}$ is perpendicular to $\overline{QR}$.
c. Explain why the median and the altitude to side $\overline{QR}$ of $\triangle PQR$ coincide.

15. Given $\triangle ABC$ with vertices $A(3, -1)$, $B(7, 3)$, and $C(-1, 7)$, and $\overline{CD}$ is the altitude to $\overline{AB}$.
a. Write an equation of the line that contains altitude $\overline{CD}$.
b. Find the coordinates of the midpoint of $\overline{AB}$. Show that altitude $\overline{CD}$ intersects $\overline{AB}$ at its midpoint.

16. The coordinates of the vertices of $\triangle ABC$ are $A(-6, -8)$, $B(6, 4)$, and $C(-6,10)$.
a. Write an equation of the altitude from C to $\overline{AB}$.
b. Write an equation of the altitude from B to $\overline{AC}$.
c. Find the x-coordinate of the point of intersection of the two altitudes in parts a and b.

17. The coordinates of the vertices of $\triangle NYC$ are $N(-2, 9)$, $Y(6, 3)$, and $C(4, -7)$.
a. Write an equation of the line that joins the midpoints of sides $\overline{NY}$ and $\overline{NC}$.
b. Write an equation of $\overleftrightarrow{YC}$.
c. Show by means of coordinate geometry that the lines determined in parts a and b are parallel.

9.5 GENERAL EQUATION OF A CIRCLE

KEY IDEAS

In the coordinate plane, the locus of points at a fixed distance of r units from point $O(h, k)$ is a circle centered at (h, k) with radius r. Applying the distance formula:

$$\sqrt{(x-h)^2+(y-k)^2} = r$$

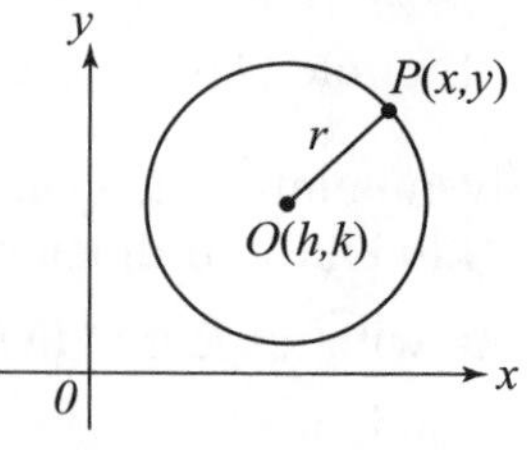

Squaring both sides gives a general equation of a circle:

$$(x-h)^2 + (y-k)^2 = r^2$$

Equation of a Circle: Center–Radius Form

The equation $(x - h)^2 + (y - k)^2 = r^2$ describes a circle whose center is at (h, k) with radius r. If the circle is centered at the origin, then $h = k = 0$ and the equation of the circle simplifies to $x^2 + y^2 = r^2$.

- If the center of a circle is at (2,–1) and its radius is 5, then an equation of the circle is $(x - h)^2 + (y - k)^2 = r^2$, where $h = 2$, $k = -1$, and $r = 5$. Making the substitutions gives $(x - 2)^2 + (y - [-1])^2 = 5^2$ or, equivalently, $(x - 2)^2 + (y + 1)^2 = 25$.
- The center and radius of a circle can be read from its equation. By rewriting the equation $(x - 3)^2 + (y + 4)^2 = 36$ in center–radius form, you can determine that $h = 3$, $k = -4$, and $r = 6$:

$$(x-3)^2 + (y+4)^2 = 36$$

$$(x-3)^2 + (y-(-4))^2 = 6^2$$

$$\downarrow \qquad\qquad \downarrow \qquad\qquad \downarrow$$

$$(x-h)^2 + (y-k)^2 = r^2$$

Thus, the center of this circle is (3, –4) and its radius is 6.

Example 1

The coordinates of the endpoints of diameter $\overline{AB}$ of circle O are $A(1, 2)$ and $B(-7, -4)$. Find an equation of circle O.

Solution:

- To find the coordinates (h, k) of the center of the circle, use the midpoint formula:

$$h = \frac{1+(-7)}{2} = \frac{-6}{2} = -3 \qquad \text{and} \qquad k = \frac{2+(-4)}{2} = \frac{-2}{2} = -1$$

  The coordinates of the center of circle O are, therefore, (–3, –1).
- To find the radius length, use the distance formula to find the distance from the center of the circle to any point on the circle, such as point A. If $(x_1, y_1) = A(1, 2)$ and $(x_2, y_2) = O(-3, -1)$, then

$$OA = \sqrt{(x_2 - x_1)^2 + (y_2 - y_1)^2}$$

$$= \sqrt{(-3-1)^2 + (-1-2)^2}$$

$$= (-4)^2 + (-3)^2$$

$$= \sqrt{16+9}$$

$$= 5$$

- Substitute $h = -3$, $k = -1$, and $r = 5$ into the center–radius form of the equation of a circle:

$$(x - (-3))^2 + (y - (-1))^2 = 5^2$$
$$(x + 3)^2 + (y + 1)^2 = 25$$

Example 2

Given the equation of circle O is $(x - 3)^2 + (y + 4)^2 = 169$.
a. Show that $P(-2, 8)$ is a point on the circle.
b. Find an equation of the line that is tangent to circle O at point P.

Solution:

a. Verify that $P(-2, 8)$ satisfies the equation of circle O:

$$(x - 3)^2 + (y + 4)^2 = 169$$
$$(-2 - 3)^2 + (8 + 4)^2$$
$$(-5)^2 + (12)^2$$
$$25 + 144$$
$$169 = 169 \checkmark$$

b. The center of the circle is $O(3, -4)$.
- Find the slope of radius $\overline{OP}$:

$$\frac{\Delta y}{\Delta x} = \frac{-4 - 8}{3 - (-2)} = -\frac{12}{5}$$

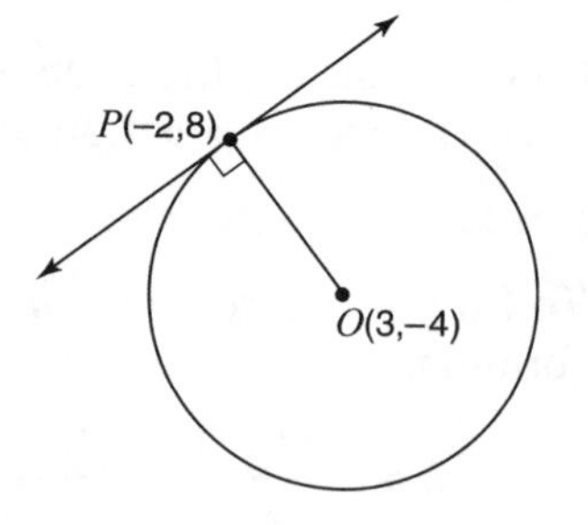

- Since radius $\overline{OP}$ is perpendicular to the tangent line, the slope of the tangent line is $\frac{5}{12}$.

- Use the point–slope equation form where $(h, k) = (-2, 8)$ and $m = \frac{5}{12}$:

$$y - k = m(x - h)$$
$$y - 8 = \frac{5}{12}(x - (-2))$$
$$\mathbf{y - 8 = \frac{5}{12}(x + 2)}$$

Dilations of Circles

If two circles have the same center, then either circle may be considered to be the **dilation** of the other circle. The **center of dilation** is the common center of the two circles, and the **constant of dilation** is the ratio of their radii.

- The circle $x^2 + y^2 = 64$ is a dilation of the circle $x^2 + y^2 = 16$ since each circle has the origin as its center. As the radius of the larger circle is 8 and the radius of the smaller circle is 4, the constant of dilation is $\frac{8}{4} = 2$.
- The dilation of the circle $x^2 + y^2 = 9$, using a constant of dilation of 4, is the circle $x^2 + y^2 = 144$. You should verify that since the radius of the smaller circle is 3, the radius of the larger circle is $3 \times \underline{\mathbf{4}} = 12$.

Translations of Circles

The circles $(x - h)^2 + (y - k)^2 = r^2$ and $x^2 + y^2 = r^2$ have exactly the same size and shape, but they differ only in their locations in the coordinate plane. Compared to the center of the circle $x^2 + y^2 = r^2$, the center of the circle $(x - h)^2 + (y - k)^2 = r^2$ is shifted h units horizontally and k units vertically. For example, since the center of the circle $(x - 1)^2 + (y + 3)^2 = 100$ is $(1, -3)$, this circle is the image of the circle $x^2 + y^2 = 100$ after it has been translated 1 unit horizontally to the right and 3 units vertically down.

Solving A Linear–Quadratic System

To solve a system of equations consisting of a linear equation and either a circle or a parabola equation, graph the two equations on the same set of axes. The solution consists of the set of all points, if any, at which the two graphs intersect. A linear–quadratic system may have one solution, two solutions, or, if the graphs do not intersect, no solution.

Example 3

Solve graphically:

$$y - x = 6$$
$$(x - 1)^2 + (y - 6)^2 = 25$$

Solution: Graph the two equations on the same set of axes.

- Graph the line $y - x = 6$ by first finding two convenient points that satisfy the equation, such as $(0, 6)$ and $(-6, 0)$. After drawing a line through these two points, label the graph with its equation, as shown in the accompanying figure.

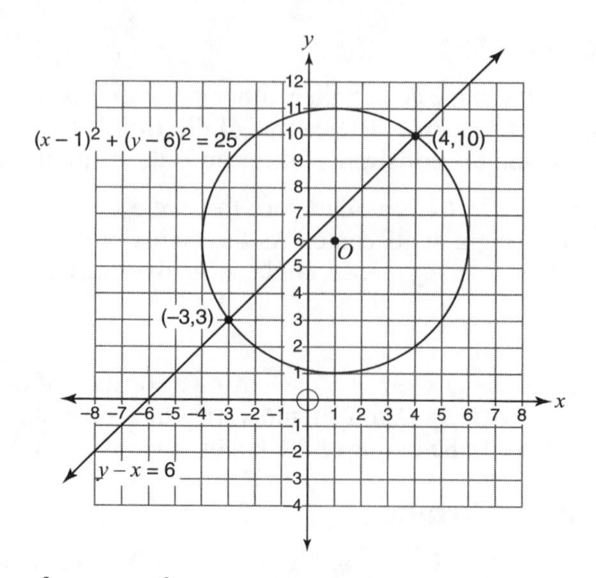

- Graph $(x - 1)^2 + (y - 6)^2 = 25$ by locating its center, (1, 6), and then using a compass to draw a circle with a radius 5 units. Label the graph with its equation.
- Read the coordinates of the points of intersection from the graph. The solution consists of the points **(–3, 3)** and **(4, 10)**.

Example 4

Solve graphically:

$$y = -x^2 + 4x - 3$$
$$x + y = 1$$

Solution: Graph the two equations on the same set of axes.

- Graph $x + y = 1$ using two points that satisfy the equation such as (0,1) and (3, –2).
- Graph $y = -x^2 + 4x - 3$ using a table of values created by your graphing calculator. Make sure to include the vertex of the parabola and three points on either side of it. The x-coordinate of the vertex can be obtained using the formula $x = \frac{-b}{2a}$ where $x = \frac{-b}{2a}$ a is the coefficient of the x^2 term in the parabola equation and b is the coefficient of the x term:

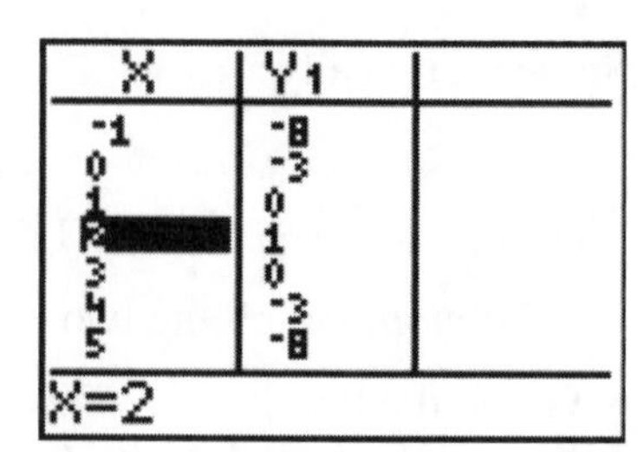

X	Y1
-1	-8
0	-3
1	0
2	1
3	0
4	-3
5	-8

X=2

$$x = -\frac{-4}{2(1)} = 2$$

- Label each graph with its equation, and label the points of intersection with their coordinates, as shown in the accompanying figure.

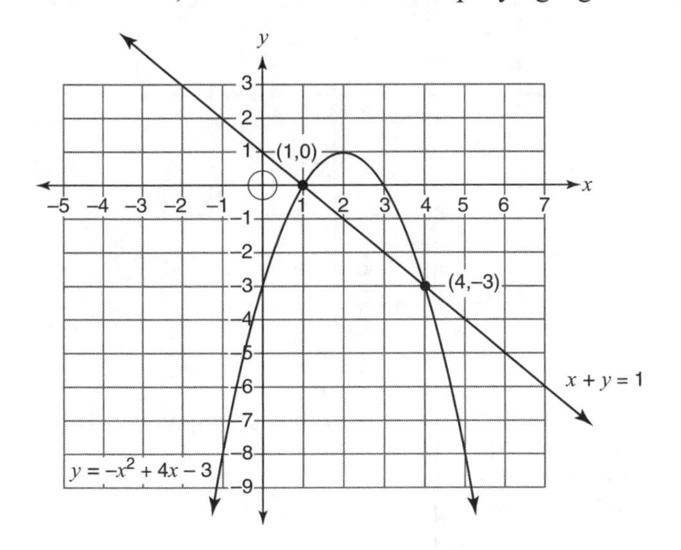

The solution consists of the points **(1,0)** and **(4,–3)**.

Check Your Understanding of Section 9.5

A. *Multiple Choice*

1. Which is an equation of the circle with center at (4, –2) and radius length of 3?

(1) $(x + 4)^2 + (y - 2)^2 = 9$ (3) $(x + 4)^2 + (y - 2)^2 = 3$
(2) $(x - 4)^2 + (y + 2)^2 = 9$ (4) $(x - 4)^2 + (y + 2)^2 = 3$

2. When a circle whose equation is $x^2 + y^2 = 25$ is shifted 3 units horizontally to the right and 2 units vertically down, it coincides with the circle whose equation is

(1) $(x + 3)^2 + (y - 2)^2 = 25$ (3) $(x - 3)^2 + (y - 2)^2 = 25$
(2) $(x - 3)^2 + (y + 2)^2 = 25$ (4) $(x + 3)^2 + (y + 2)^2 = 25$

3. Which could be an equation of a circle that touches the x-axis at one point and whose center is at (–6, 8)?

(1) $(x - 6)^2 + (y - 8)^2 = 36$ (3) $(x + 6)^2 + (y - 8)^2 = 36$
(2) $(x - 6)^2 + (y + 8)^2 = 64$ (4) $(x + 6)^2 + (y - 8)^2 = 64$

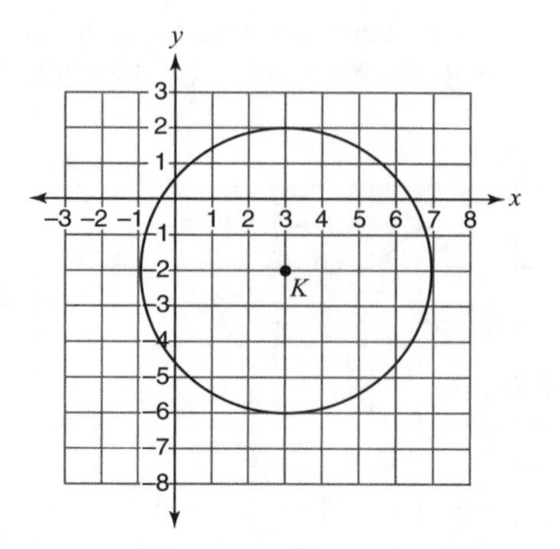

4. Which is an equation of circle K in the accompanying figure?
(1) $(x + 3)^2 + (y - 2)^2 = 16$
(2) $(x - 3)^2 + (y + 2)^2 = 16$
(3) $(x + 3)^2 + (y - 2)^2 = 4$
(4) $(x - 3)^2 + (y + 2)^2 = 4$

5. What is an equation of the locus of points 7 units from point (−2, 3)?
(1) $(x + 2)^2 + (y - 3)^2 = 49$
(2) $(x + 2)^2 + (y - 3)^2 = 7$
(3) $(x - 2)^2 + (y + 3)^2 = 49$
(4) $(x - 2)^2 + (y + 3)^2 = 7$

6. Which pair of graphs intersect at exactly one point?
(1) $x^2 + y^2 = 4$ and $x = -3$
(2) $x^2 + y^2 = 4$ and $y = -2$
(3) $x^2 + y^2 = 4$ and $y = x$
(4) $x^2 + y^2 = 4$ and $x = 4$

7. A circle whose equation is $(x + 1)^2 + (y - 2)^2 = 36$ is translated 1 unit horizontally to the left and 3 units vertically down. What is an equation of the image of this circle?
(1) $(x + 2)^2 + (y + 1)^2 = 36$
(2) $(x + 2)^2 + (y - 5)^2 = 36$
(3) $x^2 + (y + 1)^2 = 36$
(4) $x^2 + (y - 5)^2 = 36$

8. Circle O has its center at the origin and passes through point (−3, 4). What is an equation of the image of circle O after a dilation with a scale factor of 2?
(1) $x^2 + y^2 = 100$
(2) $(x + 3)^2 + (y - 4)^2 = 100$
(3) $x^2 + y^2 = 25$
(4) $x^2 + y^2 = 50$

9. What is an equation of the circle with a diameter that has endpoints at (2, 5) and (−6, 5)?
(1) $(x - 2)^2 + (y + 5)^2 = 64$
(2) $(x - 2)^2 + (y + 5)^2 = 16$
(3) $(x + 2)^2 + (y - 5)^2 = 64$
(4) $(x + 2)^2 + (y - 5)^2 = 16$

10. The center of a circle that has a radius of 6 is (–1, –2). What is an equation of the image of this circle after it is shifted 2 units horizontally to the right and 1 unit vertically down?

(1) $(x-3)^2+(y-1)^2=36$ (3) $(x-1)^2+(y+3)^2=36$

(2) $(x+1)^2+(y+3)^2=36$ (4) $(x-3)^2+(y-3)^2=36$

B. *Show or explain how you arrived at your answer.*

11. What is an equation of the circle that has its center at $O(2, -3)$ and passes through $P(-2, 0)$?

12. The coordinates of the vertices of quadrilateral $ABCD$ are $A(1, 2)$, $B(9, 8)$, $C(15, 0)$, and $D(7, -6)$.

a. Prove that $ABCD$ is a rhombus.

b. Write an equation of the circle whose center is at the point at which the diagonals of rhombus $ABCD$ intersect and whose radius is equal in length to a side of rhombus $ABCD$.

13. In the accompanying figure, $\overleftrightarrow{ASB}$ is tangent to circle O at point $S(-3,4)$.

a. Write an equation of circle O.

b. Write an equation of $\overleftrightarrow{ASB}$ and state the y-intercept.

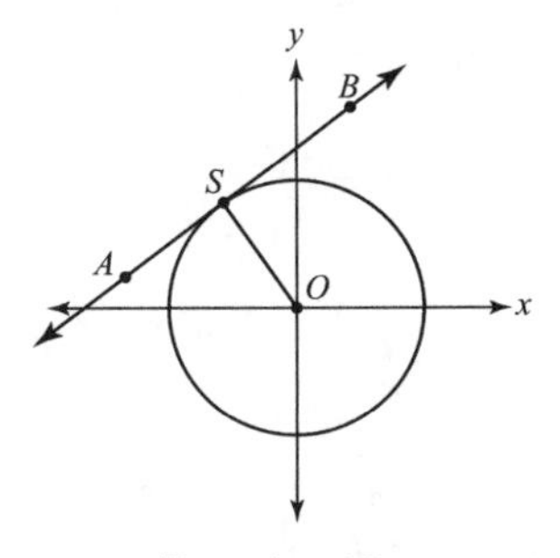

Exercise 13

14. The vertices of $\triangle ABC$ are $A(-4, -2)$, $B(2, 6)$, and $C(2, -2)$.

a. Write an equation of the locus of points equidistant from vertex B and vertex C.

b. Write an equation of the locus of points 4 units from vertex C. What is the total number of points that satisfy this locus and the locus described in part a?

15. Given the equation of circle O is $(x+2)^2+(y^2-5)^2=100$.

a. Show that $P(6, -1)$ is a point on the circle.

b. Find an equation of the line that is tangent to circle O at point P.

16. a. Write an equation of the locus of points n units from the point $P(3, 2)$.

b. Write the equation or equations that represent the locus of points 2 units from the line whose equation is $x = 3$.

c. What is the total number of points that satisfy the conditions in parts a and b simultaneously for the following values of n?

(1) $n < 2$ (2) $n = 2$

17. On graph paper, draw the graph of the circle $x^2 + y^2 = 9$ and label it A.
a. On the same set of axes, draw the image of circle A after the translation $(x, y) \rightarrow (x + 5, y - 3)$ and label it B.
b. On the same set of axes, draw the image of circle B after a reflection in the x-axis and label it C. What is the area of the triangle formed by connecting the centers of circles A, B, and C?

18 a. On graph paper, draw the graph of circle O, which is represented by the equation $(x - 1)^2 + (y + 3)^2 = 16$.
b. On the same set of axes, draw the image of circle O after the translation $(x, y) \rightarrow (x - 2, y + 4)$ and label it O'. Write an equation of circle O'.
c. Write an equation of $\overleftrightarrow{OO'}$.

19–22. Solve each of the following line–circle systems of equations graphically.

19. $x^2 + y^2 = 25$
$x + 2y = 10$

20. $x^2 + (y - 2)^2 = 9$
$x - y = 1$

21. $(x - 2)^2 + (y - 3)^2 = 100$
$y + 13 = x$

22. $(x + 4)^2 + (y - 3)^2 = 25$
$2y + x = 7$

23–25. Solve each of the following line–parabola systems of equations graphically.

23. $y = x^2 + 4x - 1$
$y - x = 3$

24. $y = -x^2 - 2x + 8$
$y = x + 4$

25. $y = x^2 - 6x + 5$
$y + 7 = 2x$

26. Graph on the same set of axes the two circles whose equations are $(x - 3)^2 + (y - 5)^2 = 25$ and $(x - 7)^2 + (y - 5)^2 = 9$. Write an equation of the line passing through the two points at which the circles intersect.

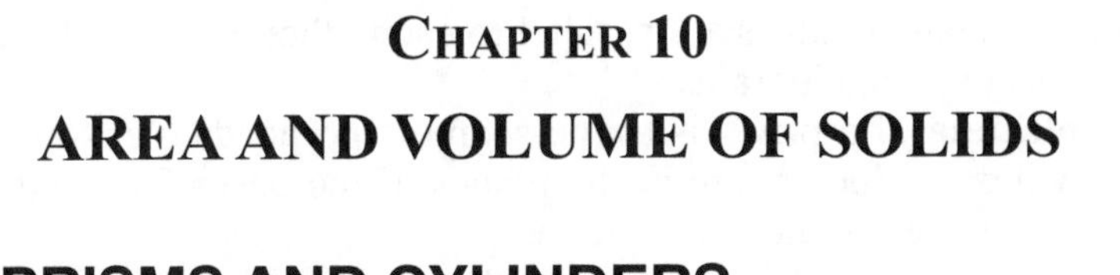

CHAPTER 10
AREA AND VOLUME OF SOLIDS

10.1 PRISMS AND CYLINDERS

KEY IDEAS

The solid figure counterpart of a polygon is called a *polyhedron*. A **polyhedron** is a solid figure whose sides or **faces** are polygons. The line segment where two faces intersect is called an **edge**. A **vertex** is a point at which three or more edges intersect.

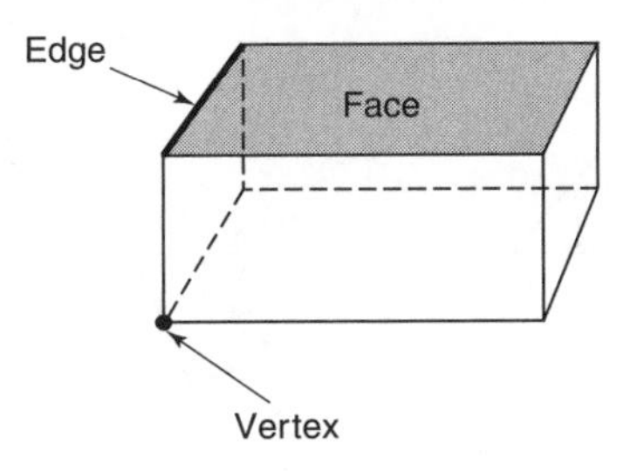

Prisms

A **prism** is a polyhedron formed by connecting the corresponding vertices of two congruent polygons that lie in parallel planes. The two congruent polygons are called **bases**. Figure 10.1 shows a prism with pentagon bases. The lines joining the corresponding vertices of the bases are the **lateral edges** and are parallel to each other. The nonbase faces are called **lateral faces**.

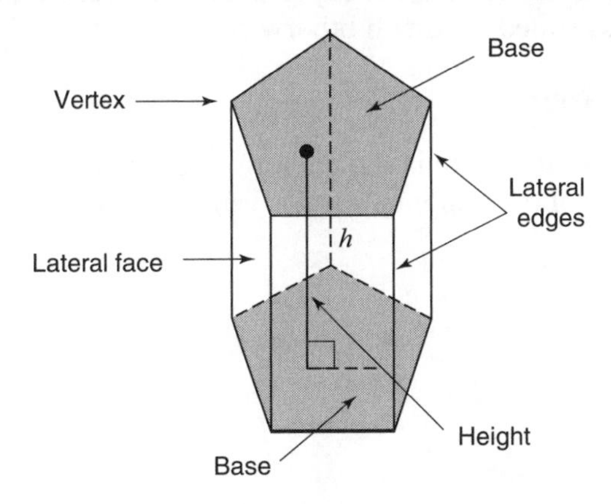

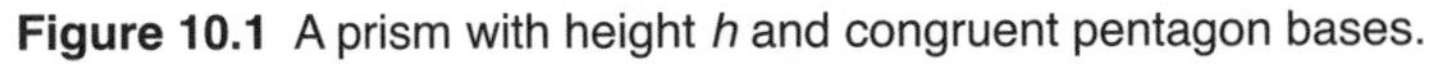

Figure 10.1 A prism with height h and congruent pentagon bases.

- Since the lateral edges are parallel to each other, the lateral faces of all prisms are parallelograms.
- An **altitude** of a prism is a line segment perpendicular to both bases and whose endpoints are in the planes of the bases. The length of the altitude is the **height** of the prism.
- A prism is named by the type of polygons that form their bases. A prism with triangles as bases, for example, is a *triangular prism*.

Right Prisms

A prism may be *right* or *oblique*, as in Figure 10.2, where h is the height of the prism. A **right prism** is one whose lateral edges are perpendicular to its bases. The bases of a right prism are aligned directly above the other. In a right prism, the length of a lateral edge represents the height of the prism, and the lateral faces are all rectangles.

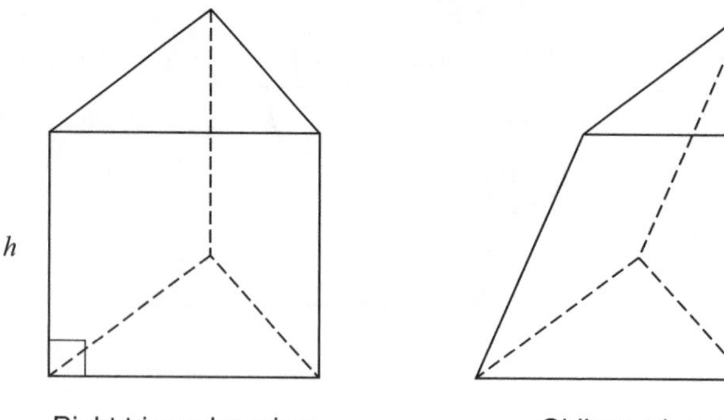

Figure 10.2 Right and oblique prisms.

If the lateral edges of a prism are slanted because they are *not* perpendicular to a base, the prism is called an **oblique prism**. You can assume a prism is a right prism unless stated or drawn otherwise.

Volume of a Prism

The **volume** of a solid figure is the amount of space it occupies as measured by the number of nonoverlapping $1 \times 1 \times 1$ unit cubes that can exactly fill the interior of the solid.

Theorem: Volume of a Prism

- The volume, V, of a right rectangular prism (box) is the product of its three dimensions:

$$V_{\text{Box}} = \underbrace{\text{length} \times \text{width}}_{\text{Area of base}} \times \text{height} \quad \text{and} \quad V_{\text{Cube}} = (\text{edge})^3$$

- The volume, V, of any prism is equal to the area of a base times the height:

$$V = B \times h$$

where B is the area of a base and h is the height. Thus, two prisms have equal volumes if their bases have equal areas and their heights are equal.

Example 1

Find the volume of the right triangular prism in the accompanying figure.

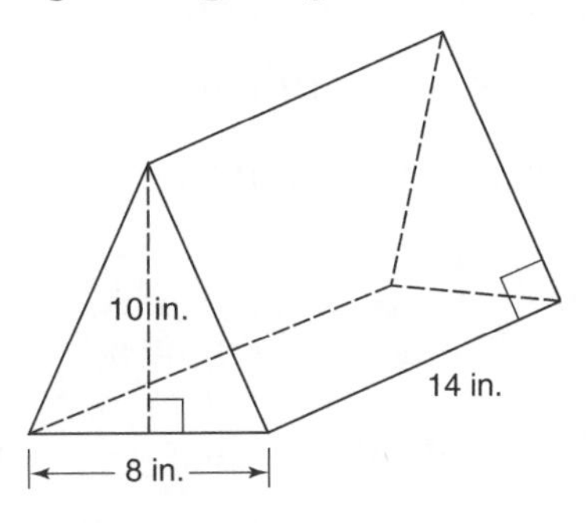

Solution:

- If B represents the area of a triangular base, then

$$B = \frac{1}{2} \times \text{base of triangle} \times \text{height of a triangle}$$
$$= \frac{1}{2} \times 8 \text{ in} \times 10 \text{ in}$$
$$= 40 \text{ in}^2$$

- Since the height of the prism is the perpendicular distance between the two triangular bases, $h = 14$ in.
- Use the volume formula for a prism where $B = 40$ in^2 and $h = 14$ in:

$$V = B \times h$$
$$= 40 \text{ in}^2 \times 14 \text{ in}$$
$$= \mathbf{560 \text{ in}^3}$$

Lateral Area of a Right Prism

The **lateral area** of a prism is the sum of the areas of only its nonbase (lateral) faces. To find the lateral area of the right prism in Figure 10.3, add the areas of its three lateral sides: *ABCD*, *ABEF*, and *ECDF*. Since $BC = AD = 12$ in and $AB = CD = 5$ in:

$$\begin{aligned} \text{Area rectangle } ABCD &= 5 \times 12 = 60 \text{ in}^2 \\ &+ \\ \text{Area rectangle } ABEF &= 5 \times 10 = 50 \text{ in}^2 \\ &+ \\ \text{Area rectangle } ECDF &= 5 \times 10 = 50 \text{ in}^2 \\ \hline \text{Lateral Area} &= 160 \text{ in}^2 \end{aligned}$$

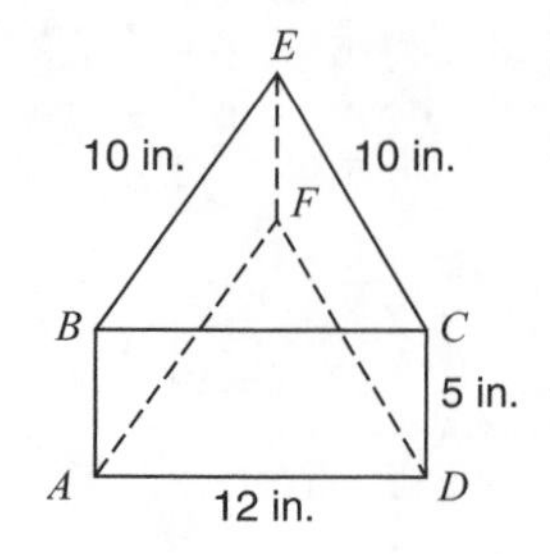

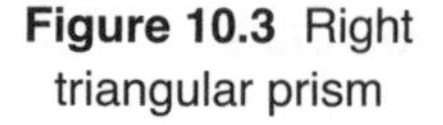

Figure 10.3 Right triangular prism

In a right prism, there is a relationship between the lateral area, height, and base perimeter:

$$\begin{aligned} \text{Lateral Area } &= 5 \times 12 + 5 \times 10 + 5 \times 10 \\ &= \underbrace{5} \times \underbrace{(12 + 10 + 10)} \\ &= \text{Height} \times \text{Perimeter of base} \end{aligned}$$

> **Theorem: Lateral Area of a Right Prism**
>
> If in a right prism, h is the height and p is the perimeter of a base, then
>
> $$\text{Lateral Area (L.A.)} = h \times p\,.$$
>
> The **surface area** (S.A.) of a prism is the total area of all of its faces including the two congruent bases.

Example 2

Find the volume of the prism in Figure 10.3.

Solution: The accompanying figure shows a two–dimensional view of the lower triangular base in which $\overline{FP}$ is the altitude drawn to the base of isosceles triangle *AFD*. The lengths of the sides of right triangle *APF* form a 6–***8***–10 Pythagorean triple in which $FP = 8$.

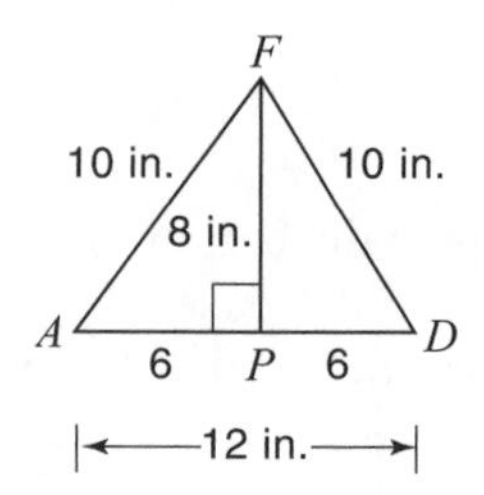

- Determine B, the area of the base of the prism:

$$\text{Area of } \triangle AFD = \frac{1}{2} \times (AD) \times (FP)$$
$$= \frac{1}{2} \times 12 \text{ in} \times 8 \text{ in}$$
$$= 48 \text{ in}^2$$

Thus, $B = 48 \text{ in}^2$.

- Use the volume formula for a prism where $B = 48 \text{ in}^2$ and $h = CD = 5$ in:

$$V = B \times h$$
$$= 48 \text{ in}^2 \times 5 \text{ in}$$
$$= \mathbf{240 \text{ in}^3}$$

Example 3

In a right equilateral triangular prism, the length of a side of a base is 20 cm and a lateral edge measures 25 cm. Find the volume and lateral area of the prism.

Solution: In a right prism, the lateral edge represents an altitude, so $h = 25$ cm.

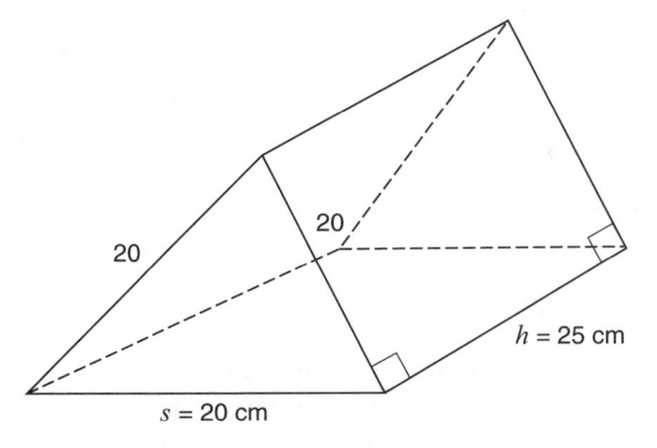

- Find the area of the equilateral triangle base using the formula $B = \frac{s^2}{4}\sqrt{3}$ where s is the length of a side of the triangle. Since it is given that $s = 20$ cm:

$$B = \frac{(20)^2}{4}\sqrt{3}$$
$$= \frac{400}{4}\sqrt{3}$$
$$= 100\sqrt{3} \text{ cm}^2$$

- Use the volume formula for a prism where $B = 100\sqrt{3}$ and $h = 25$ cm:

$$V = B \times h$$
$$= (100\sqrt{3})(25)$$
$$= \mathbf{2500\sqrt{3}\ cm^3}.$$

- Use the formula for the lateral area of a prism where $h = 25$ cm and $p = 20 + 20 + 20 = 60$ cm:

$$\text{L.A.} = h \times p$$
$$= 25\text{ cm} \times 60\text{ cm}$$
$$= \mathbf{1{,}500\ cm^2}$$

Cylinders

A cylinder, like a prism, has two congruent bases in parallel planes. The bases of a cylinder, however, are circles rather than polygons. The radius, r, of a base is the **radius** of the cylinder. A cylinder may be right or oblique (slanted), as illustrated in Figure 10.4.

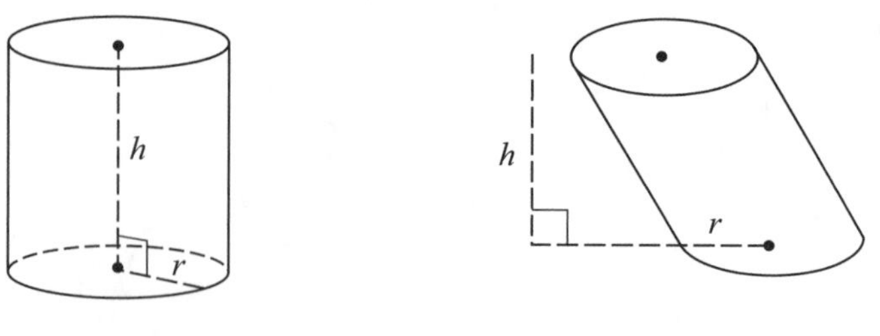

Figure 10.4 Right and oblique cylinders.

In a **right cylinder,** the line joining the centers of the two bases is perpendicular to the bases so that the centers of the bases are directly aligned. A cylinder is **oblique** if the line joining the centers is *not* perpendicular to the bases. The height, h, of a cylinder is the perpendicular distance between the two bases.

Volume and Area Formulas for a Cylinder

The volume and area formulas for a cylinder and right cylinder correspond to those of a prism and right prism except that for a cylinder πr^2 is substituted for the area of the base and $2\pi r$ is used as the perimeter (circumference) of the base.

Theorem: Cylinder Formulas

- The volume, V, of any cylinder is equal to the area of a base times the height:

$$V = B \times h$$

where $B = \pi r^2$ is the area of a circular base and h is the height.

- If in a right cylinder, h is the height and p is the perimeter of a base, then

$$\text{Lateral Area (L.A.)} = p \times h = 2\pi rh.$$

Example 4

A closed cylindrical can has a diameter of 1.0 ft and a height of 1.8 feet. Find

a. The volume of the can correct to the *nearest tenth* of a cubic foot.
b. The lateral area of the can.
c. The surface area of the can correct to the *nearest tenth* of a square foot.

Solution: Let $r = \frac{1}{2} \times 1.0 \text{ ft} = 0.5 \text{ ft}$ and $h = 1.8$ feet.

a. Determine the volume of the cylinder using the formula $V = \pi r^2 h$:

$$V = \pi \times (0.5)^2 \times 1.8$$
$$= \pi \times 0.25 \times 1.8$$
$$= 0.45\pi \text{ ft}^3$$

Multiply using the stored calculator value for π: $= 1.413716694 \text{ ft}^3$

Round: $= \mathbf{1.41 \text{ ft}^3}$

b. Use the formula for lateral area of a right cylinder:

$$\text{L.A.} = 2\pi rh$$
$$= 2\pi \times 0.5 \times 1.8$$
$$= \mathbf{1.8\pi \text{ ft}^2}$$

c. The surface area (S.A.) of the can is the sum of the areas of the two circular bases and the lateral area:

$$\text{S.A.} = \text{L.A.} + 2\pi r^2$$
$$= 1.8\pi + 2\pi(0.5)^2$$
$$= 1.8\pi + 0.5\pi$$
$$= 2.3\pi \text{ ft}^2$$
$$= \mathbf{7.2 \text{ ft}^2}$$

Check Your Understanding of Section 10.1

A. Multiple Choice

1. What is the number of inches in the radius of a cylinder 2 inches in height if its volume is 288π in^3?
(1) 3 (2) 6 (3) 12 (4) 24

2. In the accompanying figure of a right prism, the bases are right triangles. Which statement about its volume (V) and lateral area (L.A.) is true?
(1) $V = 288$ in^3 and L.A. $= 288$ in^2
(2) $V = 288$ in^3 and L.A. $= 216$ in^2
(3) $V = 720$ in^3 and L.A. $= 288$ in^2
(4) $V = 720$ in^3 and L.A. $= 216$ in^2

10 in.
12 in.
6 in.

3. If in a right square prism, the length of each side of the two bases is doubled and the height is tripled, then the volume of the prism is
(1) multiplied by 12
(2) multiplied by 8
(3) multiplied by 6
(4) multiplied by 4

4. The amount of light produced by a cylindrical-shaped fluorescent light bulb depends on its lateral area. A certain cylindrical-shaped fluorescent light bulb is 36 inches in length, has a 1 inch diameter, and is manufactured to produce 0.283 watts of light per square inch. What is the best estimate for the total amount of light that it is able to produce?
(1) 32 watts
(2) 34 watts
(3) 48 watts
(4) 64 watts

5. The bases of a right prism are right triangles whose legs measure 5 inches and 12 inches. If the lateral edge of the prism measures 15 inches, what is the number of square inches in the lateral area of the prism?
(1) 120 (2) 255 (3) 360 (4) 450

6. The lateral area of a right cylinder whose height is two times the length of the radius is 100π in^2. What is the number of inches in the height of the cylinder?
(1) 5 (2) 10 (3) 15 (4) 20

B. *Show or explain how you arrived at your answer.*

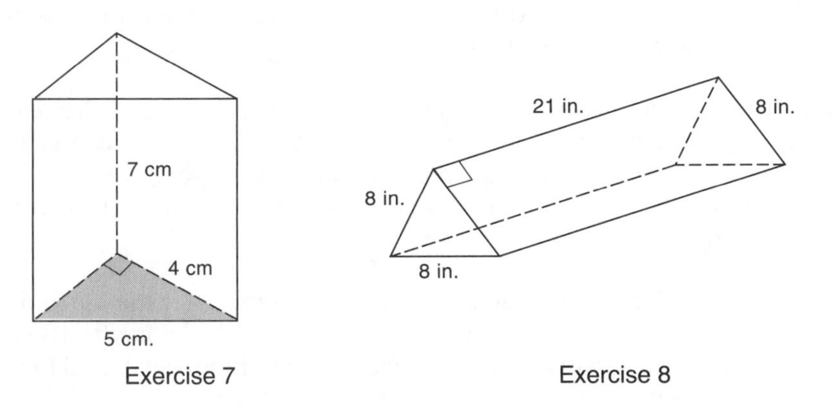

Exercise 7

Exercise 8

7–8. For each right prism above, find the:

a. Lateral area.

b. Volume.

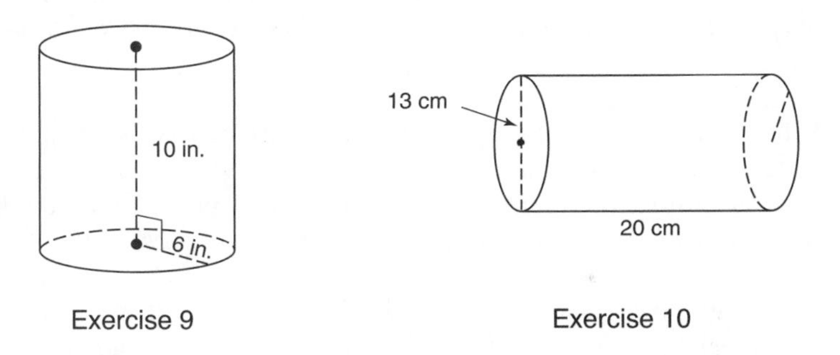

Exercise 9

Exercise 10

9–10. For each right cylinder above, find

a. The lateral area in terms of π.

b. The volume correct to the *nearest tenth* of a cubic unit.

11. The length of a side of the base of a right square prism is 12 cm. If the length of a lateral edge is 9 cm, find the surface area of the prism.

12. A prism has isosceles triangle bases whose legs measure 10 inches and base measures 16 inches. If the height of the prism is equal to the height of the triangular base, what is the volume of the prism?

13. A right circular cylinder has a lateral area of 306π ft^2. If the height of the cylinder is 17 ft, express in terms of π the number of cubic feet in the volume of the cylinder.

14. The base of a right prism is a right triangle whose legs measure 18 and 24 inches, respectively. If the length of a lateral edge is 16 inches, what is the surface area of the prism?

15. A piece of ice that has the shape of a right cylinder has a diameter of 6.0 cm and a height of 5.0 cm. If the ice cube melts at a constant rate of 13.0 cm^2 per minute, how many minutes elapse before the ice cube is completely melted? Round your answer to the *nearest hundredth* of a minute.

16. A cylindrical can is manufactured using aluminum for the top and bottom bases and tin for the remainder of the can. If the can is 10 cm in height and 6 cm in diameter, what percentage of the metal used to manufacture the can is tin? Round your answer to the *nearest tenth.*

17. The diameter of a cylindrical water storage tank is 12 feet and its height is 15 feet. A leak develops in the tank and water leaks out of it at a constant rate of 18.0 in^3/sec. If the tank is initially full, how many *hours* elapse before the tank is empty? Round off your answer to the *nearest hour*.

18. A hot water tank with a capacity of 85.0 gallons of water is being designed to have the shape of a right circular cylinder with a diameter 1.8 feet. Assuming that there are 7.48 gallons in 1 cubic foot, how high will the tank have to be? Round your answer to the *nearest tenth* of a foot.

19. A rectangle 6 inches in width and 8 inches in length is rotated in space about its longer side. Find in terms of π the volume and surface area of the solid generated.

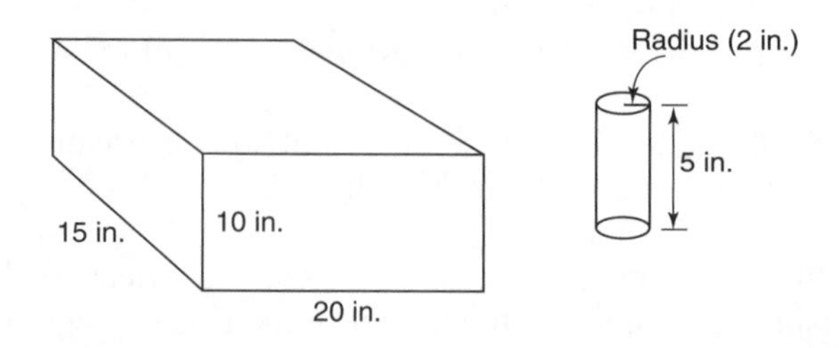

20. In the accompanying diagram, a rectangular container with the dimensions 10 inches by 15 inches by 20 inches is to be filled with water, using a cylindrical cup whose radius is 2 inches and whose height is 5 inches. What is the maximum number of full cups of water that can be placed into the container without the water overflowing the container?

10.2 PYRAMIDS AND CONES

KEY IDEAS

Unlike prisms and cylinders, pyramids and cones have only one base.

Pyramids

A **pyramid** is a polyhedron formed by connecting all of the vertices of a polygon to a point in a different plane than the polygon, as in Figure 10.5. The polygon is called the **base**, and the point to which the vertices are connected is called the **vertex** of the pyramid. The line segments joining the vertices of the base to the vertex are the **lateral edges**. The lateral (nonbase) faces of a pyramid are always triangles. The **height**, h, of a pyramid is the perpendicular distance from the vertex to the plane of the base.

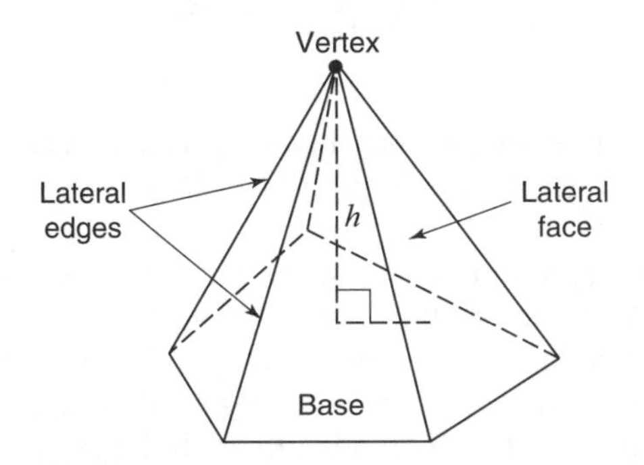

Figure 10.5 Pyramid with a pentagon base.

Volume of a Pyramid

Like a prism, the volume of a pyramid depends on the area of its base and the height.

Theorem: Volume of a Pyramid
The volume, V, of a pyramid is equal to one-third of the area of the base times the height:

$$V = \frac{1}{3}B \times h$$

where B is the area of the base and h is the height.

Regular Pyramid

The base of a pyramid can be any polygon. If the base is a regular polygon and its lateral edges are congruent, then the pyramid is a **regular pyramid**. Because the base of the pyramid in Figure 10.6 is a regular hexagon, the pyramid is classified as a regular hexagonal pyramid. You can assume a pyramid is regular unless stated or drawn otherwise.

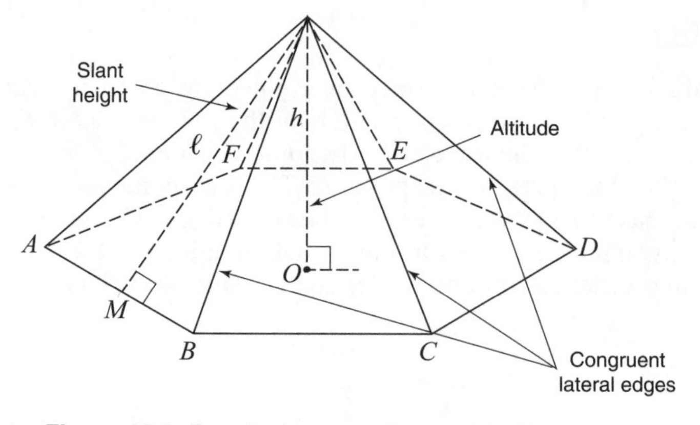

Figure 10.6 Regular hexagonal pyramid with center O, slant height ℓ ($\overline{VM}$), and $\overline{AM} \cong \overline{BM}$.

In Figure 10.6, VM represents the slant height. The **slant height** of a regular pyramid is the perpendicular distance from the vertex of the pyramid to a side of the base. The slant height is usually denoted by ℓ. Because the lateral faces of a regular pyramid are congruent isosceles triangles, the slant height is the perpendicular bisector of the side to which it is drawn. In any regular pyramid,

- The lateral faces are congruent isosceles triangles.
- The altitude drawn from the vertex intersects the base at its center.
- The altitude from the vertex and the slant height determine a right triangle, as illustrated in the regular square pyramid in Figure 10.7. Because O is the center of the square base and M is the midpoint of $\overline{AB}$, OM is one-half the length of a side of the square base.

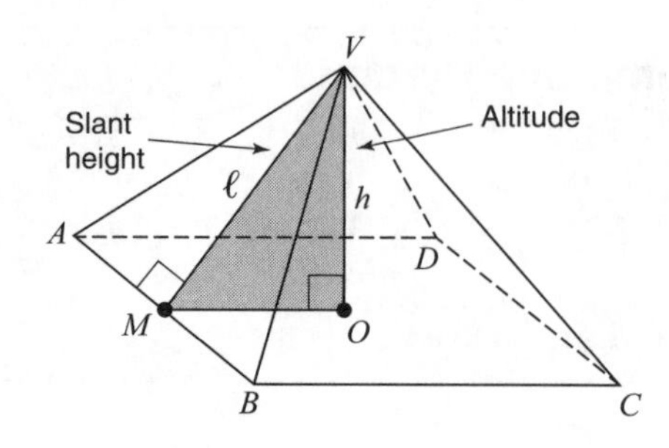

Figure 10.7 Right triangle determined by slant height $\overline{VM}$ and altitude $\overline{VO}$.

Lateral Area of a Regular Pyramid

The lateral area of a regular pyramid depends on its slant height and perimeter.

> **Theorem: Lateral Area Formula for a Regular Pyramid**
> If in a regular pyramid ℓ is the slant height and p is the perimeter of the base, then:
> $$\text{L.A.} = \frac{1}{2}p\ell$$

Example 1

The perimeter of the base of a regular square pyramid is 40 cm and a lateral edge measures 13 cm. Find

a. The lateral area of the pyramid.
b. The height of the pyramid in radical form.
c. The volume of the pyramid correct to the *nearest tenth* of a cubic centimeter.

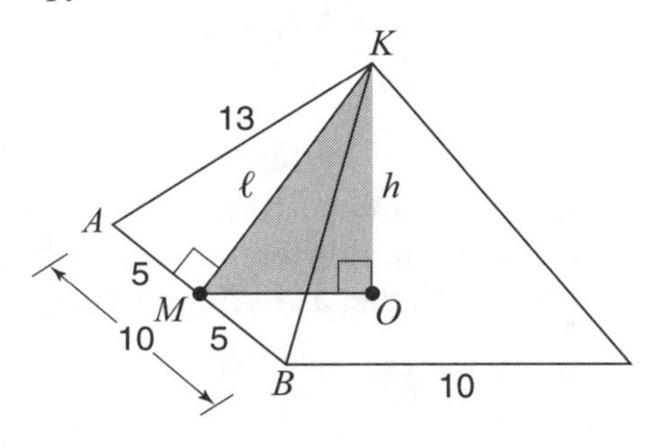

Solution:

a. First find the slant height.

- Each side of the base measures $\frac{40 \text{ cm}}{4} = 10$ cm. Because it is given that a lateral edge measures 13 cm, $KA = 13$ cm.
- Lateral face AKB is an isosceles triangle in which altitude $\overline{KM}$ bisects base $\overline{AB}$. Thus, $\triangle AMK$ is a 5–***12***–13 right triangle in which KM, the slant height, is 12 cm.
- Use the formula for lateral area of a pyramid where $p = 40$ cm and $\ell = 12$ cm:

$$\begin{aligned}\text{L.A.} &= \frac{1}{2}p\ell \\ &= \frac{1}{2}(40 \text{ cm})(12 \text{ cm}) \\ &= \mathbf{240\ cm^2}\end{aligned}$$

b. To find the height, h, of the pyramid, apply the Pythagorean Theorem in right triangle KOM where the length of hypotenuse $\overline{KM}$ is 12. Since OM is one-half of the side length of the square, $OM = \frac{1}{2} \times 10 = 5$. Thus,

$$h^2 + (OM)^2 = (KM)^2$$
$$h^2 + 5^2 = 12^2$$
$$h^2 = 144 - 25$$
$$\boldsymbol{h = \sqrt{119}\ \textbf{cm}}$$

c. Use the formula for the volume of a pyramid where $B = 10\text{ cm} \times 10\text{ cm} = 100\text{ cm}^2$ and $h = \sqrt{119}$ cm:

$$V = \frac{1}{3}Bh$$
$$= \frac{1}{3} \times 100\text{ cm}^2 \times \sqrt{119}\text{ cm}$$

Use your calculator: $= \mathbf{363.6\ cm^3}$

Cones

A **cone**, unlike a pyramid, has a circular base with each point on its circumference joined to a single point in a different plane called the **vertex** of the cone. A cone may be right or oblique, as in Figure 10.8.

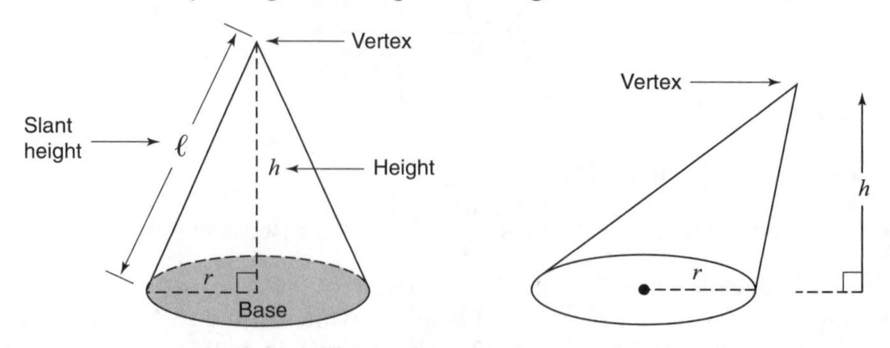

Figure 10.8 Right and oblique cones.

A **right cone** is a cone in which the line segment joining the vertex and the center of the base is perpendicular to the plane of the base, as in Figure 10.8. You can assume a cone is a right circular cone unless stated or drawn otherwise. The **slant height**, ℓ, of a right circular cone is the distance from the vertex to any point on the circumference of the circular base. The volume and area formulas for a cone correspond to those of a pyramid except that πr^2 and $2\pi r$ are substituted for the area and the perimeter of the base, respectively.

Theorem: Cone Formulas

- The volume, V, of any cone is equal to one-third of the area of the base times the height:

$$V = \frac{1}{3}B \times h = \frac{1}{3}\pi r^2 h$$

where B is the area of the base, h is the height, and r is the radius of the base.

- If in a right circular cone ℓ is the slant height and r is the radius, then

$$\text{L.A.} = \frac{1}{2} \times \underbrace{\text{Circumference}}_{\text{"base perimeter"}} \times \ell = \frac{1}{2}(2\pi r)\ell = \pi r \ell$$

Example 2

Find the volume and surface area of the cone in the accompanying figure. Answers may be left in terms of π.

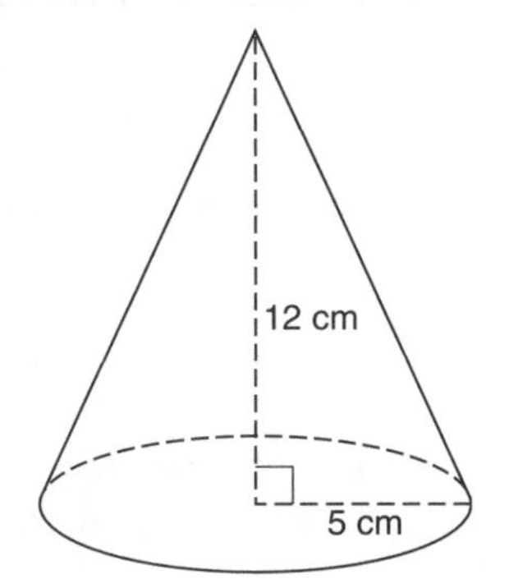

Solution: To find the volume of the cone, use the volume formula where $r = 5$ cm and $h = 12$ cm:

$$\begin{aligned} V &= \frac{1}{3}\pi r^2 h \\ &= \frac{1}{3}\pi \times 5^2 \times 12 \\ &= \mathbf{100\,\pi\,cm^3} \end{aligned}$$

To find the surface area, first find the lateral area.

- Since the slant height is the hypotenuse of a 5–12–***13*** right triangle, $\ell = 13$ cm. Hence,

$$\text{L.A.} = \pi r \ell = \pi \times 5 \text{ cm} \times 13 \text{ cm} = 65\pi \text{ cm}^2.$$

- The surface area includes both the lateral area and the area of the base:

$$\text{S.A.} = \text{L.A.} + \pi r^2 = 65\pi\mathbf{cm^2} + \pi(5 \text{ cm})^2 = \mathbf{90\pi cm^2}.$$

Table 10.1 summarizes volume and lateral area formulas where B = area of base, h = height, p = base perimeter, r = base radius, and ℓ = slant height. The formulas for lateral area only apply to right prisms, right cylinders, regular pyramids, or right cones.

Table 10.1 Formulas for Volume and Lateral Area.

Prism	Cylinder	Pyramid	Cone
• $V = Bh$	• $V = Bh$ where $B = \pi r^2$	• $V = \frac{1}{3}Bh$	• $V = \frac{1}{3}Bh$ where $B = \pi r^2$
• L.A. = hp	• L.A. = $2\pi rh$	• L.A. = $\frac{1}{2}p\ell$	• L.A. = $\pi r\ell$

Check Your Understanding of Section 10.2

A. Multiple Choice

1. For any regular pyramid with height h and slant height ℓ, which statement is always true?
(1) $h > \ell$
(2) $h < \ell$
(3) $h = \ell$
(4) $\ell >$ lateral edge

2. If a regular square pyramid with side length s and a right cone with radius r have congruent altitudes and equal volumes, then
(1) $s = \sqrt{\pi r}$
(2) $s = \frac{\sqrt{r}}{\pi}$
(3) $s = \pi\sqrt{r}$
(4) $s = r\sqrt{\pi}$

3. For any right cone, which statement is always true?
(1) lateral area > area of base
(2) area of base > lateral area
(3) area of base = lateral area
(4) surface area = 2 × area of base

4. In the accompanying figure of a regular square pyramid, the length of a side of the base is 40 meters and the slant height measures 25 meters. Which statement about its volume (V) and lateral area (L.A.) is true?

(1) $V = 8{,}000 \text{ m}^3$ and L.A. $= 2{,}000 \text{ m}^2$
(2) $V = 8{,}000 \text{ m}^3$ and L.A. $= 1{,}200 \text{ m}^2$
(3) $V = 2{,}000 \text{ m}^3$ and L.A. $= 2{,}000 \text{ m}^2$
(4) $V = 2{,}000 \text{ m}^3$ and L.A. $= 1{,}200 \text{ m}^2$

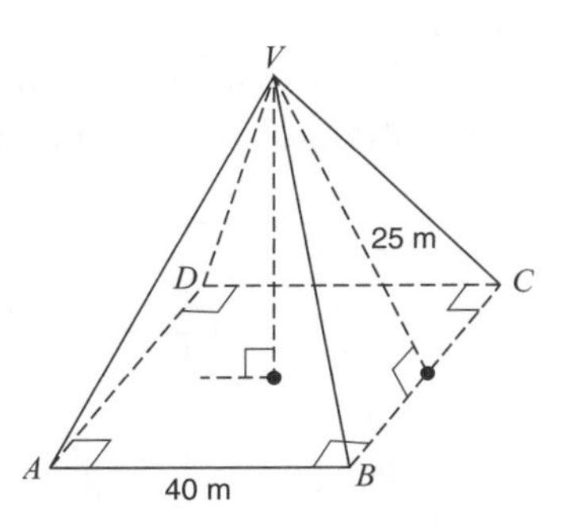

5. A right cone has a diameter of 18 feet and a slant height of 15 feet. Which statement about its volume (V) and lateral area (L.A.) is true?

(1) $V = 405\pi \text{ ft}^3$ and L.A. $= 135\pi \text{ ft}^2$
(2) $V = 405\pi \text{ ft}^3$ and L.A. $= 324\pi \text{ ft}^2$
(3) $V = 324\pi \text{ ft}^3$ and L.A. $= 135\pi \text{ ft}^2$
(4) $V = 324\pi \text{ ft}^3$ and L.A. $= 324\pi \text{ ft}^2$

B. *Show or explain how you arrived at your answer.*

6. In the accompanying diagram of a square pyramid, the slant height is 15 cm, and the length of a side of the base is 24 cm. Find the lateral area and volume of the prism.

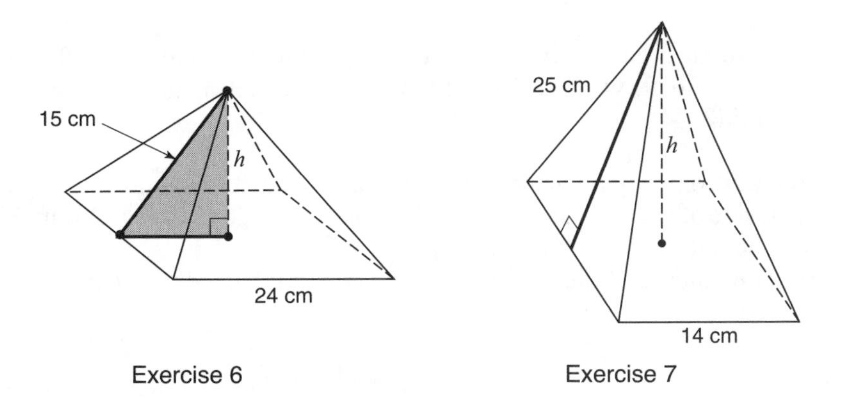

Exercise 6

Exercise 7

7. In the accompanying diagram of a square pyramid, the length of a lateral edge is 25 cm and the length of a side of the base is 14 cm. Find

a. The surface area of the pyramid
b. The volume of the pyramid correct to the *nearest tenth* of a cubic centimeter

8. A right circular cone has a lateral surface area of 80π square inches. If the slant height of the cone is 10 inches, what is the volume of the cone?

9. In the accompanying figure of a right cone, the vertex angle measures 60° and the slant height is 15 cm. Find, to the *nearest tenth* of a cubic centimeter, the volume of the cone.

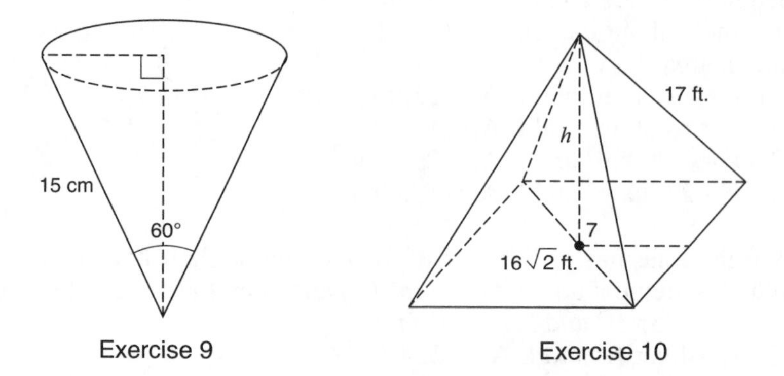

Exercise 9

Exercise 10

10. In the accompanying figure of a regular pyramid, the length of a diagonal of the square base is $16\sqrt{2}$ ft. If a lateral edge measures 17 ft, find the number of cubic feet in the volume of the pyramid. Round your answer to the *nearest tenth* of a cubic foot.

11. The altitude of a right circular cone makes an angle of 45° with the slant height. If the radius of the base is 8 inches, what is the volume of the cone in terms of π?

12. The lateral area of a regular square pyramid is 1,040 ft^2. If the slant height is 26 ft, what is the number of cubic feet in the volume of the pyramid?

13. In the accompanying diagram of a regular triangular pyramid, the length of each side of the base is 20 cm and the length of a lateral edge is 26 cm. Find

a. The lateral area of the pyramid

b. The volume of the pyramid to the *nearest cubic centimeter*

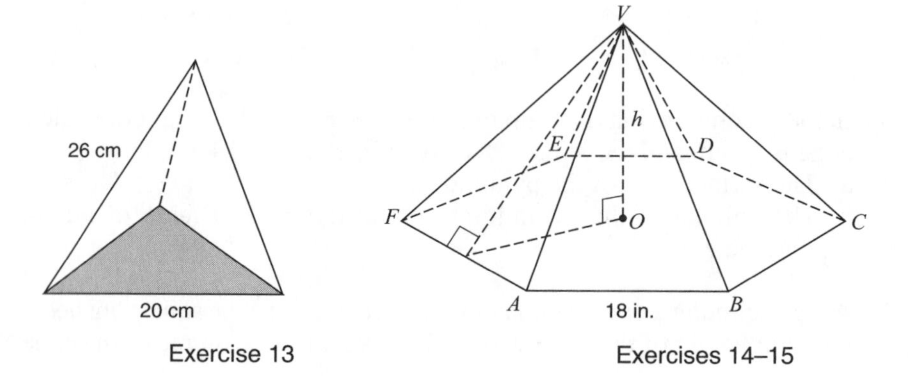

Exercise 13

Exercises 14–15

14–15. In the accompanying diagram of a regular hexagonal pyramid, the length of each side of the base is 18 cm, and the length of a lateral edge is 41 cm.

14. Find the lateral area of the pyramid.

15. Find the volume of the pyramid correct to the *nearest cubic centimeter*.

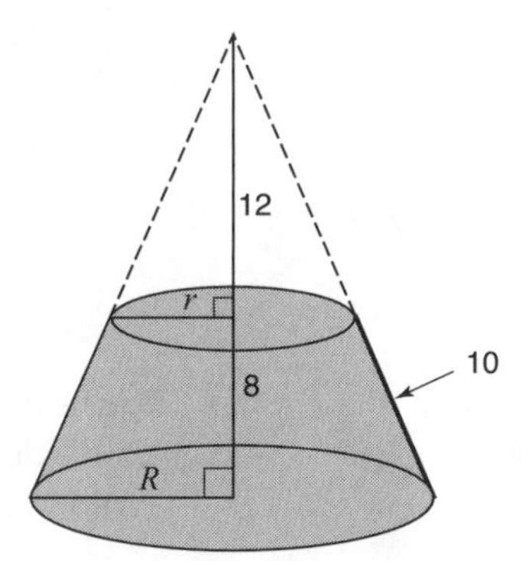

16. A lamp shade with a circular base has the shape of a solid called a *frustrum*. If a plane parallel to the base of a cone intersects the cone below its vertex, the truncated part of the cone between the slicing plane and the base is called the **frustrum** of the cone. In the accompanying figure, the shaded region represents a frustrum of a right cone in which the portion of the original cone that lies 12 inches below its vertex has been cut off. Find

a. The radius, R, of the original cone

b. The volume of the frustrum expressed in terms of π

17. A cone has a volume of 135 in^3. A plane parallel to the base divides the cone into two parts such that one part is a cone whose height is two-thirds that of the height of the original cone. Find the volumes of this smaller cone and the remaining frustrum.

10.3 SPHERES AND SIMILAR SOLIDS

KEY IDEAS

The solid figure counterpart of a circle is a *sphere*. Two solids are **similar** if they have the same shape and their corresponding dimensions are in proportion. The ratio of corresponding linear dimensions of two similar solids is called the **similarity ratio**. Any pair of cubes or pair of spheres are similar.

Spheres and Great Circles

A **sphere** is the set of all points in space at a fixed distance from a given point called the **center**. The **radius** of a sphere is the length of a line segment joining the center of the sphere to any point on the sphere. See Figure 10.9.

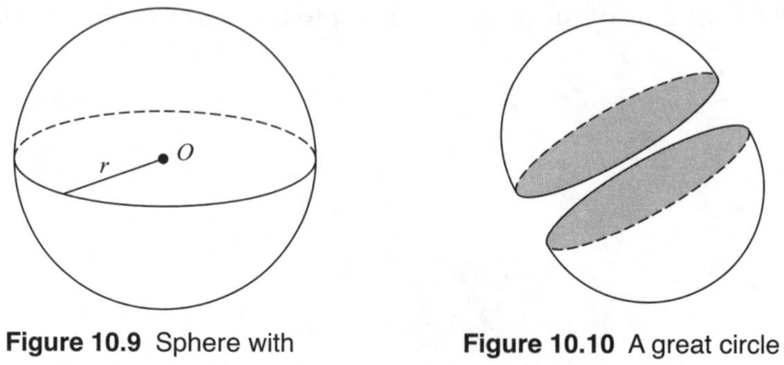

Figure 10.9 Sphere with center *O* and radius *r*.

Figure 10.10 A great circle divides the sphere into two hemispheres.

A **great circle** is the largest circle that can be drawn on a sphere. It has the same center as the sphere and divides the sphere into two equal parts called **hemispheres**, as in Figure 10.10.

- The intersection of a plane and a sphere is a circle. A great circle is the intersection of a plane passing through the center of a sphere. The circumference of a great circle may be drawn through any two points on the surface of a sphere as these two points and the center of the sphere determine a plane. See Figure 10.11.

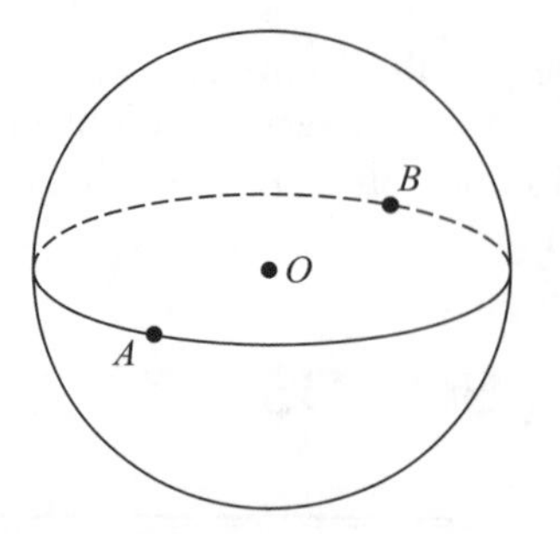

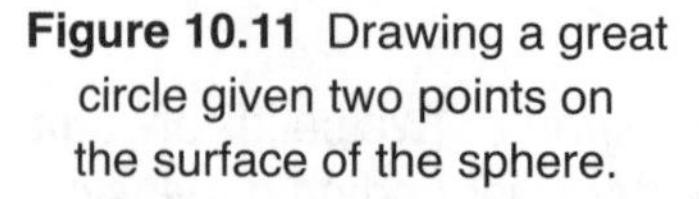

Figure 10.11 Drawing a great circle given two points on the surface of the sphere.

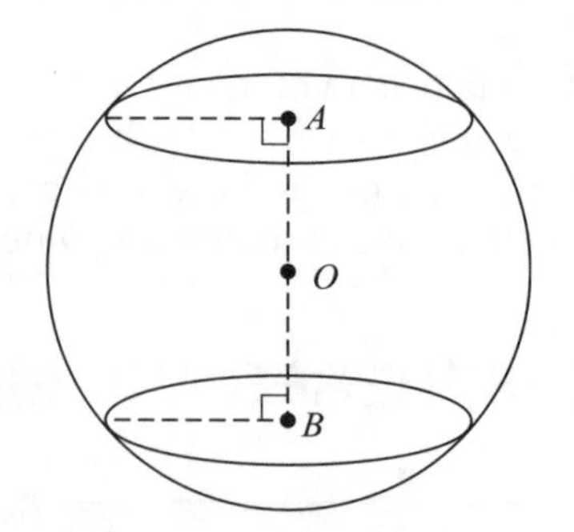

Figure 10.12 Planes intersecting sphere *O* in circles *A* and *B*.

- Two planes equidistant from the center of the sphere and intersecting the sphere do so in congruent circles. If in sphere *O* shown in Figure 10.12, $OA = OB$, then circles *A* and *B* are congruent.
- The surface area and volume of a sphere depends on its radius.

Theorem: Sphere Formulas
If a sphere has radius r, then

$$\text{Surface Area} = 4\pi r^2 \quad \text{and} \quad \text{Volume} = \frac{4}{3}\pi r^3$$

Example 1

Find the volume of a sphere whose surface area is 144π in^2.

Solution: Use the formula for surface area to find the radius of the sphere:

$$\begin{aligned} \text{S.A.} &= 4\pi r^2 = 144\pi \text{ in}^2 \\ r^2 &= \frac{144\pi}{4\pi} = 36 \text{ in}^2 \\ r &= 6 \text{ in} \end{aligned}$$

Now use the formula for volume of a sphere where $r = 6$ in

$$\begin{aligned} V &= \frac{4}{3}\pi r^3 \\ &= \frac{4}{3}\pi\,(6^3) \\ &= \frac{4\pi(\cancel{216})}{\cancel{3}} \\ &= 288\pi \end{aligned}$$

The volume of the sphere is **288π in^3**.

Comparing Areas and Volumes of Similar Solids

If the ratio of corresponding linear dimensions of two similar solids is $\frac{a}{b}$, then

- The ratio of their corresponding areas is $\left(\frac{a}{b}\right)^2$.
- The ratio of their corresponding volumes is $\left(\frac{a}{b}\right)^3$.

For example, if the ratio of the heights of two similar cylinders is $\frac{2}{3}$, then the ratio of their lateral areas is $\left(\frac{2}{3}\right)^2$ or $\frac{4}{9}$, and the ratio of their volumes is $\left(\frac{2}{3}\right)^3$ or $\frac{8}{27}$.

Example 2

Two spheres have radii of 8 cm and 20 cm. If the volume of the smaller sphere is 104π cm^3, what is the volume of the larger sphere?

Solution: Since the similarity ratio is $\frac{8}{20}$ or, equivalently, $\frac{2}{5}$:

$$\frac{\text{Volume of smaller sphere}}{\text{Volume of larger sphere}} = \left(\frac{2}{3}\right)^3$$

If x represents the volume of the larger sphere, then

$$\frac{104\pi}{x} = \frac{8}{125}$$

$$8x = (104\pi)(125)$$

$$\frac{8x}{8} = \frac{13,000\pi}{8}$$

$$x = 1,625\pi$$

The volume of the larger sphere is **$1,625\pi$ cm^3**

Example 3

The lateral areas of two similar prisms are 324 in^2 and 729 in^2. If the volume of the larger prism is 2,916 in^3, what is the volume of the smaller prism?

Solution: First find the similarity ratio. The lateral areas of the two similar prisms have the same ratio as the *square* of the similarity ratio. If $\frac{a}{b}$ represents the similarity ratio, then

$$\left(\frac{a}{b}\right)^2 = \frac{324}{729}$$

$$\frac{a}{b} = \frac{\sqrt{324}}{\sqrt{729}}$$

$$= \frac{18}{27}$$

$$= \frac{2}{3}$$

Let x represent the volume of the smaller prism. Since the volumes of the two similar prisms have the same ratio as the *cube* of the similarity ratio:

$$\frac{\text{Volume of smaller prism}}{\text{Volume of larger prism}} = \left(\frac{2}{3}\right)^3$$

$$\frac{x}{2916} = \frac{8}{27}$$

$$27x = 23{,}328$$

$$\frac{27x}{27} = \frac{23{,}328}{27}$$

$$x = 864$$

The volume of the smaller prism is **864 in^3**.

Check Your Understanding of Section 10.3

A. Multiple Choice

1. If the diameter of a sphere is tripled, the volume of the sphere is multiplied by
(1) 4 (2) 9 (3) 16 (4) 27

2. The density of lead is approximately $0.41 \frac{\text{pounds}}{\text{in}^3}$. What is the approximate weight, in pounds, of a lead ball that has a 5 inch diameter?
(1) 26.8 (2) 78.5 (3) 80.4 (4) 214.7

3. Which diagram represents the figure with the greatest volume?

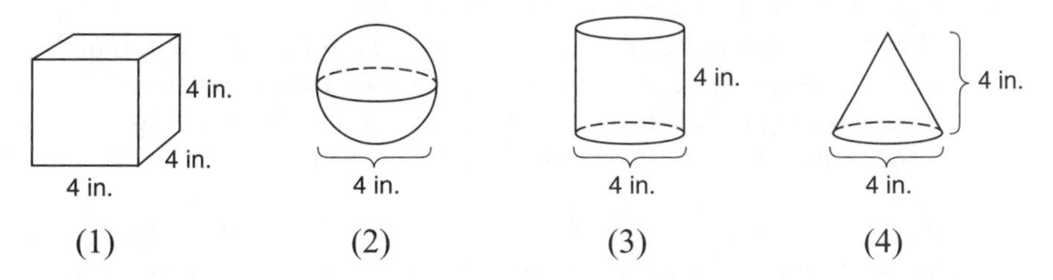

4. Triangle ABC represents a metal flag on pole $\overline{ABD}$, as shown in the accompanying diagram. On a windy day the triangle spins around the pole so fast that it looks like a three-dimensional shape. Which shape would the spinning flag create?
(1) sphere
(2) cone
(3) right cylinder
(4) pyramid

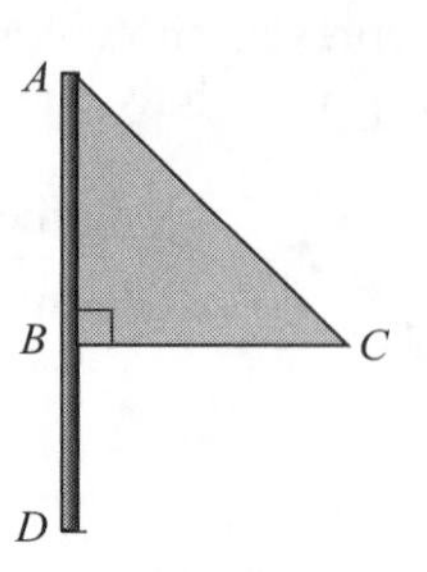

5. What is the best approximation of the number of square inches in the surface area of a spherical ball that fits tightly in a cube-shaped box with edges 8 inches in length?
(1) 67 (2) 201 (3) 268 (4) 804

6. If the volume of a sphere is $7{,}776\pi$ cm^3, what is the number of square centimeters in the surface area of the sphere?
(1) 324π (2) 648π (3) $1{,}296\pi$ (4) $2{,}592\pi$

7. If the ratio of the volumes of two similar solids is 27 to 64, what is the ratio of their areas?
(1) $\sqrt{27}$ to 8 (2) 3 to 8 (3) 9 to 16 (4) 3 to 4

8. The surface areas of two similar prisms are 112 in^2 and 63 in^2. The ratio of the volume of the larger prism to the volume of the smaller prism is
(1) 4 to 3
(2) 16 to 9
(3) 64 to 27
(4) 256 to 81

B. *Show or explain how you arrived at your answer.*

9. Tamika has a hard rubber ball whose circumference measures 13 inches. She wants to box it for a gift but can only find cube-shaped boxes of sides of 3 inches, 4 inches, 5 inches, or 6 inches.
a. What is the *smallest* box that the ball will fit into with the top on?
b. Tamika has a square sheet of wrapping paper that measures 1 ft by 1 ft. Assuming no waste, does Tamika have enough wrapping paper to completely cover the box? Explain your answer.

10. A solid sphere with a radius of 8 inches weighs 32 pounds. What is the weight, in pounds, of a sphere made of the same material whose radius is 6 inches?

11. A ball is placed inside a cube-shaped boxed that measures 6 inches on each side. When the box is closed, each of its sides touches the ball. What is the approximate volume of the enclosed space between the sphere and the sides of the box correct *to the nearest cubic inch*?

12. A bookend is shaped like a pyramid and weighs 0.24 pounds. How many pounds does a similarly shaped bookend weigh if it is made of the same material and each corresponding dimension is $2\frac{1}{2}$ times as large?

13. The lateral areas of two similar cylindrical cans are 567 in^2 and 1,008 in^2. If the volume of the smaller can is 1,161 in^3, what is the volume of the larger can?

14. The volumes of two similar pyramids are 2,744 cm^3 and 9,261 cm^3. If the lateral area of the smaller pyramid is 1,624 cm^2, what is the lateral area of the larger pyramid?

15. In the accompanying figure, a cone with a solid hemisphere at its top has a slant height of 13 cm. If the radii of the cone and hemisphere are 5 cm, find the volume of the figure correct to the *nearest cubic centimeter*.

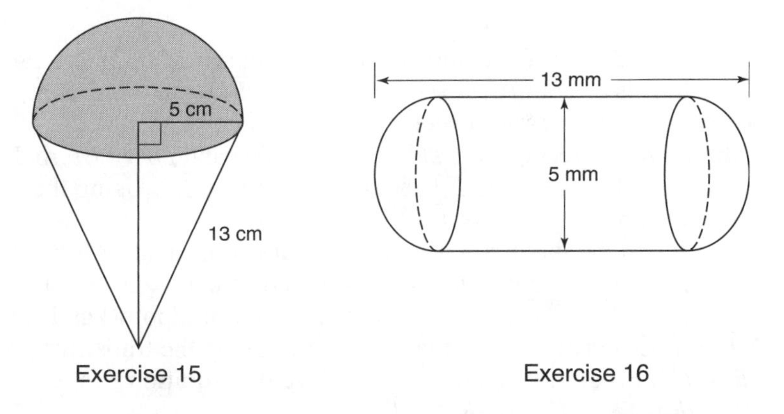

Exercise 15 Exercise 16

16. In the accompanying figure, a vitamin tablet is 13 mm long and 5 mm in diameter. Find the volume of the tablet to the *nearest cubic millimeter*.

17. A sealed cylindrical can holds three tennis balls each with a diameter of 2.5 inches. If the can is designed to have the smallest possible volume, find the number of cubic inches of unoccupied space inside the can correct to the *nearest tenth of a cubic inch*.

Answers and Solution Hints to Practice Exercises

Chapter 1

Lesson 1.1

1. (3)	**4.** (4)	**7.** (1)	**10.** 37	**13.** 134
2. (4)	**5.** (3)	**8.** (3)	**11.** 48	**14.** 20
3. (2)	**6.** (4)	**9.** 7	**12.** 3	**15.** 142

Lesson 1.2

1. (2)	**4.** 21, 21, 31; 21, 26, 26	**7.** 119
2. (2)	**5.** $\frac{16}{3}$	
3. (2)	**6.** 10	

Lesson 1.3

1. (3)	**3.** (3)	**5.** 110	**7.** 70	**9.** 45	**11.** 143
2. (1)	**4.** (3)	**6.** 77	**8.** 63	**10.** 150	**12.** 17

Lesson 1.4

1. (2) **2.** (2) **3.** (1) **4.** (2)

5. Subtract $m\angle MOW$ from angles 1 and 2 to obtain $\angle TOM \cong \angle BOW$.

6. Since halves of congruent segments ($\overline{AC} \cong \overline{BD}$) are congruent, $\overline{AE} \cong \overline{BE}$ so $\triangle AED$ is isosceles.

7. Subtracting corresponding sides of $AB = CD$ and $DE = BF$ makes $\overline{CE} \cong \overline{AF}$. Because $\triangle ADF$ is equilateral, $\overline{AF} \cong \overline{AD}$. Using the transitive property, $\overline{CE} \cong \overline{AD}$.

8. Since supplements of congruent angles are congruent, $\angle BAC \cong \angle BCA$. Because halves of congruent angles are congruent, $\angle 3 \cong \angle 4$.

9. First show $\overline{JR} \cong \overline{KT}$: $\overline{MR} \cong \overline{MT}$ (definition of midpoint) and $\overline{KT} \cong \overline{MT}$, $\overline{MR} \cong \overline{KT}$. It is also given that $\overline{JR} \cong \overline{MR}$. Using the transitive property, $\overline{JR} \cong \overline{KT}$. From the Given, $\overline{SJ} \cong \overline{SK}$. Use the Addition Property: $SJ + JR = SK + KT$ so $\overline{SR} \cong \overline{ST}$.

Lesson 1.5

1. (2)	**5.** (2)	**b.** T	**9a.** T	**e.** F
2. (3)	**6.** (3)	**c.** F	**b.** T	**10a.** (1) $\sim(p \wedge q)$
3. (4)	**7.** $\sim x \wedge y$	**d.** F	**c.** F	(2) $\sim p \vee \sim q$
4. (4)	**8a.** F	**e.** T	**d.** T	

10b.

p	q	$\sim(p \wedge q)$	$\sim p \vee \sim q$
T	T	F	F
T	F	T	T
F	T	T	T
F	F	T	T

c. Yes, since they all have the same truth values.

Lesson 1.6

1. (3) **2.** (4) **3.** (4) **4.** (3) **5.** (3) **6.** (3) **7.** (4) **8.** (4) **9.** (3) **10.** (2) **11.** (3) **12.** (2) **13.** (1) **14.** (3) **15.** (4) **16.** (1) **17.** (3) **18.** (3)

19. $\sim r \leftrightarrow \sim s$

20. **a.** $q \rightarrow \sim p$ **b.** $p \rightarrow \sim q$

21. **a.** $q \rightarrow \sim p$ **b.** $\sim p \rightarrow q$ **c.** $\sim(p \rightarrow \sim q)$

22. **a.** $p \rightarrow \sim q$ **b.** (2)

23. **a.** If a triangle is isosceles, then it has two congruent sides.
b. If a triangle does not have two congruent sides, then it is not isosceles.
c. A triangle is isosceles if and only if it has two congruent sides.

24. **a.** $\sim p \rightarrow \sim q$ **b.** $p \rightarrow \sim q$ **c.** $q \rightarrow p$

Chapter 2

Lesson 2.1

1. (3) **2.** (4) **3.** (3) **4.** (3) **5.** (4) **6.** 140 **7.** 29 **8.** 50 **9.** 30

10. **a.** Because $p \parallel q$, corresponding angles 1 and 2 are congruent so $m\angle 2 = 90$ and $m \perp q$.
b. If a line is perpendicular to one of two parallel lines, then it is also perpendicular to the other line.

11. $\angle 1 \cong \angle 3$ (alternate interior angles formed by parallel lines); $\angle 2 \cong \angle 4$ (corresponding angles formed by parallel lines); $\angle 1 \cong \angle 2$ (definition of angle bisector) so $\angle 3 \cong \angle 4$ (transitive property).

Lesson 2.2

1. (3)

2. (1)

3. No, since $x = 31$ so $(3x + 5)° = 98° \neq 82°$.

4. $\angle 1 \cong \angle 3$ (given); $\angle 2 \cong \angle 3$ (definition of angle bisector); $\angle 1 \cong \angle 3$ (transitive property). Because corresponding angles are congruent, $\overleftrightarrow{AD} \parallel \overleftrightarrow{BC}$.

5. $\overline{TC} \parallel \overline{HK}$ (lines perpendicular to the same line are parallel); $\angle 1 \cong \angle 2$ (congruent alternate interior angles formed by parallel lines); $\angle 2 \cong \angle 3$ (definition of angle bisector); $\angle 1 \cong \angle 3$ (transitive property); $\angle 3 \cong \angle 4$ (congruent corresponding angles formed by parallel lines); $\angle 1 \cong \angle 4$ (transitive property).

Lesson 2.3

1. (2)	**4.** (1)	**7.** (1)	**10.** (1)	**13.** 270	**16.** 20
2. (2)	**5.** (2)	**8.** (3)	**11.** (3)	**14.** 25	**17.** 97
3. (2)	**6.** (3)	**9.** (4)	**12.** (4)	**15.** 53	

Lesson 2.4

1. (4)	**3.** (2)	**5.** (3)	**7.** (3)	**9.** 1,980	**11.** 15
2. (1)	**4.** (1)	**6.** (3)	**8.** (3)	**10.** 8	

CHAPTER 3

Lesson 3.1

1. (4) **2.** (4) **3.** (1) **4.** (2)

5. $\angle ABD \cong \angle CBD$ (definition of angle bisector), $\overline{BD} \cong \overline{BD}$, $\angle BAD \cong \angle BCD$ (given) so $\triangle ABD \cong \triangle CBD$ by AAS $\cong$ AAS.
6. $\angle B \cong \angle E$ (right angles are congruent), $\overline{AC} \cong \overline{DF}$ (addition property), $\angle ACB \cong \angle EFD$ (congruent alternate interior angles) so $\triangle ABC \cong \triangle DEF$ by AAS $\cong$ AAS.
7. $\overline{AM} \parallel \overline{TH}$ so $\angle AMR \cong \angle HTS$ (congruent alternate interior angles); $\overline{RA} \cong \overline{SH}$ (given); $\overline{RA} \parallel \overline{SH}$ so $\angle ARM \cong \angle HST$ (congruent alternate exterior angles). Thus, $\triangle ARM \cong \triangle HST$ by AAS $\cong$ AAS.
8. $\angle 3 \cong \angle 4$ (complements of congruent angles are congruent), $\overline{AD} \cong \overline{AD}$, $\angle EAD \cong \angle FAD$ (definition of angle bisector) so $\triangle AED \cong \triangle AFD$ by ASA $\cong$ ASA.
9. $\overline{AB} \cong \overline{AB}$ (Hyp) and $\overline{BD} \cong \overline{AE}$ (Leg) so right $\triangle AEB \cong$ right $\triangle BDA$ by HL $\cong$ HL.
10. $\angle AEC \cong \angle BDC$ (all right angles are congruent), $\angle C \cong \angle C$, and $\overline{CE} \cong \overline{CD}$ (given) so $\triangle AEC \cong \triangle BDC$ by ASA $\cong$ ASA.
11. $\angle A \cong \angle C$ (given), $\overline{AB} \cong \overline{BC}$ (given), and $\angle B \cong \angle B$ so $\triangle AEB \cong \triangle CDB$ by ASA $\cong$ ASA.
12. $\angle B \cong \angle B$, $\overline{BD} \cong \overline{BE}$ (given), and $\angle AEB \cong \angle CDB$ (all right angles are congruent) so $\triangle ABE \cong \triangle CBD$ by ASA $\cong$ ASA.
13. Using the addition property, $\overline{DEB} \cong \overline{AEC}$; $\angle 1 \cong \angle 2$; $\overline{BC} \cong \overline{BC}$ so $\triangle DBC \cong \triangle ACB$ by SAS $\cong$ SAS.
14. $\overline{AB} \cong \overline{DC}$ and $m\angle 1 = m\angle 2$ (given); $\angle ABE \cong \angle DCE$ (all right angles are congruent); $m\angle ABE + m\angle 1 = m\angle DCE + m\angle 2$ so $\angle ABC \cong \angle DCB$ (addition property); $\overline{BC} \cong \overline{BC}$; $\triangle ABC \cong \triangle DCB$ by SAS $\cong$ SAS.

Lesson 3.2

1. Draw $\overline{AD}$ (two points determine a line). $\triangle ADB \cong \triangle ADC$ by SSS $\cong$ SSS so $\angle B \cong \angle D$ by CPCTC.
2.

Statement		Reason
1. $\overline{MP} \cong \overline{ST}$	Side	1. Given.
2. $\overline{MP} \parallel \overline{ST}$		2. Given.
3. $\angle MPL \cong \angle STR$	Angle	3. If two lines are parallel, then their corresponding angles are congruent.
4. $\overline{PL} \cong \overline{RT}$	Side	4. Given.
5. $\triangle RST \cong \triangle LMP$		5. SAS postulate.
6. $\angle SRT \cong \angle MLP$		6. CPCTC.
7. $\overline{RS} \parallel \overline{LM}$		7. Two lines are parallel if a pair of corresponding angles are congruent.

3. $\angle H \cong \angle K$ and $\angle IJH \cong \angle LJK$ (vertical angles are congruent); $\overline{HK}$ bisects $\overline{IL}$ (given) so $\overline{IJ} \cong \overline{LJ}$; $\triangle IJH \cong \triangle LJK$ by AAS $\cong$ AAS so $\overline{HJ} \cong \overline{KJ}$ by CPCTC. Thus, $\overline{IL}$ bisects $\overline{HK}$ (reverse of the definition of segment bisector).
4. $\overline{BD} \cong \overline{BD}$, $\angle ADB \cong \angle CDB$ (given), $\overline{AD} \cong \overline{CD}$ (sides of an equilateral triangle are congruent); $\triangle ADB \cong \triangle CDB$ by SAS $\cong$ SAS so $\angle ABD \cong \angle CBD$ by CPCTC. Thus, $\overline{BD}$ bisects $\angle ABC$ (reverse of the definition of angle bisector).
5. $\triangle AEB \cong \triangle CFD$ by HL$\cong$ HL so $\angle EAB \cong \angle FCD$ by CPCTC. Since alternate interior angles are congruent, $\overline{AB} \parallel \overline{CD}$.
6. $\overline{DF} \cong \overline{BE}$ (given); $\overline{DF} \parallel \overline{BE}$ so $\angle DFA \cong \angle BEC$ (congruent alternate interior angles); $\overline{AE} \cong \overline{CF}$ so $\overline{AF} \cong \overline{CE}$ (addition property). $\triangle DFA \cong \triangle CEB$ by SAS $\cong$ SAS so $\angle DAF \cong \angle BCE$ by CPCTC. Since alternate interior angles are congruent, $\overline{AD} \parallel \overline{BC}$.
7. $\triangle RSP \cong \triangle TSW$ by SAS $\cong$ SAS since $\overline{RS} \cong \overline{TS}$, $\angle PRS \cong \angle WST$, and $\overline{RP} \cong \overline{SW}$. By CPCTC, corresponding angles *RSP* and *WTS* are congruent, making $\overline{SP} \parallel \overline{TW}$.
8. $\triangle TLS \cong \triangle SWT$ by Hy-Leg $\cong$ Hy-Leg, since $\overline{ST} \cong \overline{ST}$ (hypotenuse) and, from the given, $\overline{TL} \cong \overline{SW}$ (leg). Hence, $\overline{SL} \cong \overline{TW}$ by CPCTC.
9. $\overline{TL}$ and $\overline{SW}$ are altitudes (given); $\angle SWR \cong \angle TLR$ (altitudes form right angles, and all right angles are congruent); $\angle R \cong \angle R$ (reflexive property); $\overline{RS} \cong \overline{RT}$ (given); $\triangle SWR \cong \triangle TLR$ (AAS); $\overline{RW} \cong \overline{RL}$ (CPCTC).
10. Since $\overline{CM} \cong \overline{MP}$ (given), $\angle BMC \cong \angle AMP$ (vertical angles are congruent), and $\overline{BM} \cong \overline{AM}$ (median divides a segment into two congruent segments, $\triangle AMP \cong \triangle BMC$ by SAS $\cong$ SAS. By CPCTC, $\angle P \cong \angle BCM$ so $\overline{AP} \parallel \overline{CB}$ (if alternate interior angles are congruent, the lines are parallel).

Lesson 3.3

1. (1)	**3.** (3)	**5.** (1)	**7.** (2)	**9.** 32
2. (2)	**4.** (3)	**6.** (3)	**8.** (2)	**10.** 6

11. Isosceles since $\angle ABD \cong \angle ADB$.

12. Since $\angle 2 \cong \angle 4$, $\overline{AD} \cong \overline{CD}$; $\angle BDA \cong \angle BDC$ (given); $\overline{BD} \cong \overline{BD}$. Thus, $\overline{AB} \cong \overline{CB}$ by CPCTC so $\triangle ADC$ is isosceles.

13. Since $\overline{AB} \cong \overline{BC}$ (given), $\angle BAC \cong \angle BCA$; $\angle 1 \cong \angle 3$ (given) so, by the subtraction property, $\angle 2 \cong \angle 4$. Using the converse of the Base Angles Theorem, $\overline{AD} \cong \overline{CD}$ so $\triangle ADC$ is isosceles.

14. $\overline{BD} \cong \overline{BE}$ (given) so $\angle BDE \cong \angle BED$ and $\angle ADB \cong \angle CEB$ (supplements of congruent angles are congruent). Using the subtraction property, $\overline{AD} \cong \overline{CE}$. $\triangle ADB \cong \triangle CEB$ (SAS $\cong$ SAS). By CPCTC, $\overline{AB} \cong \overline{BC}$ so $\triangle ABC$ is isosceles.

15. $\overline{DG} \cong \overline{CG}$, $\overline{AD} \cong \overline{FC}$, and $\overline{BC} \cong \overline{BED}$ (given); $\angle ACB \cong \angle FDE$ (Base Angles Theorem); $\overline{AC} \cong \overline{FD}$ (addition property); $\triangle ABC \cong \triangle FED$ (SAS $\cong$ SAS); $\angle B \cong \angle E$ (CPCTC).

16. Since $\overline{GE} \cong \overline{GF}$, $\angle GEF \cong \angle GFE$ and $\angle AED \cong \angle BFC$ (vertical angles are congruent); $\overline{DE} \cong \overline{CF}$ (given); $\angle ADE \cong \angle BCF$ (supplements of congruent angles are congruent) so $\triangle DAE \cong \triangle CBF$ (ASA $\cong$ ASA).

17. $m\angle 1 = m\angle 2 + m\angle C$. Because $\overline{DB} \cong \overline{BC}$, $m\angle 2 = m\angle C$ so $m\angle 1 = 2m\angle C$. Since $\overline{AD} \cong \overline{DB}$, $m\angle 1 = m\angle A$. Using the substitution property, $m\angle A = 2m\angle C$.

18. $\angle BRT \cong \angle STR$ (congruent alternate interior angles); $\angle BRT \cong \angle SRT$ (definition of angle bisector); $\angle SRT \cong \angle STR$ so $\overline{RT} \cong \overline{TS}$; $\overline{MR} \cong \overline{MT}$ (definition of midpoint); $\overline{SM} \cong \overline{SM}$ so $\triangle TSM \cong \triangle RSM$ (SSS $\cong$ SSS). By CPCTC, $\angle RMS \cong \angle TMS$. Thus, $\overline{SM} \cong \overline{RT}$ (segments intersect to form congruent adjacent angles).

19. $\triangle RBS \cong \triangle ASR$ (HL $\cong$ HL) so $\overline{SB} \cong \overline{RA}$ and $\angle BSR \cong \angle ARS$ (CPCTC). Using the converse of the Base Angles Theorem, $\overline{TS} \cong \overline{TR}$ and, by the subtraction property, $\overline{TB} \cong \overline{TA}$.

20. Given: Isosceles $\triangle ABC$, $\overline{AB} \cong \overline{AC}$, altitudes $\overline{CX}$ and $\overline{AY}$. Prove: $\overline{CX} \cong \overline{AY}$. Show $\triangle AXC \cong \triangle CYA$ by AAS $\cong$ AAS. Then $\overline{CX} \cong \overline{AY}$ by CPCTC.

Lesson 3.4

1. **a.** $\triangle RLS \cong \triangle RLT$ (Hy-Leg).

b. $SL \cong TL$ and $\angle SLR \cong \angle TLR$ (CPCTC); $\angle SLW \cong \angle TLW$ (supplements of congruent angles are congruent); $\overline{LW} \cong \overline{LW}$ (reflexive property); $\triangle SLW \cong \triangle TLW$ (SAS); $\angle SWL \cong \angle TWL$ (CPCTC); $\overline{WL}$ bisects $\angle SWT$ (definition of bisector).

2. $\triangle KQM \cong \triangle LQN$ (SAS $\cong$ SAS). By CPCTC, $\angle K \cong \angle L$ and $\overline{KM} \cong \overline{LN}$. Then $\triangle KPM \cong \triangle LRN$ (ASA $\cong$ ASA) so $\overline{PM} \cong \overline{NR}$ by CPCTC.

3. **a.** $\triangle ABE \cong \triangle ADE$ (HL $\cong$ HL).

b. $\angle EAB \cong \angle EAD$ by CPCTC; $\overline{AEC} \cong \overline{AEC}$. $\triangle ABC \cong \triangle ADC$ (SAS $\cong$ SAS) so $\angle 3 \cong \angle 4$ by CPCTC.

4. $\triangle ABC \cong \triangle ADB$ (ASA $\cong$ ASA) so $\overline{AB} \cong \overline{AD}$ by CPCTC. Then $\triangle ABE \cong \triangle ADE$ (SAS $\cong$ SAS) so $\angle 1 \cong \angle 2$ by CPCTC.

5. $\triangle AFD \cong \triangle DFE$ (AAS); $\overline{FD} \cong \overline{FE}$ (CPCTC); $\overline{BF} \cong \overline{BF}$ (reflexive property); right triangle $FDB \cong$ right triangle FEB (Hy-Leg); $\angle FBD \cong \angle FEB$ (CPCTC); $\overline{BF}$ bisects $\angle DBE$ (definition of bisector).
6. $\triangle BDF \cong \triangle BEF$ by SSS $\cong$ SSS. By CPCTC, $\angle BDF \cong \angle BEF$, so $\angle ADF \cong \angle CEF$ (supplements of congruent angles are congruent). Also, $\angle AFD \cong \angle CFE$ (vertical angles are congruent). $\triangle AFD \cong \triangle CFE$ by ASA $\cong$ ASA. Then $\overline{AF} \cong \overline{CF}$ by CPCTC, so $\triangle AFC$ is isosceles.
7. **a.** $\triangle ADF \cong \triangle CEF$ (HL $\cong$ HL).
   b. By CPCTC, $\overline{DF} \cong \overline{EF}$ and $\angle A \cong \angle C$; $\overline{AB} \cong \overline{BC}$ and, using the subtraction property, $\overline{DB} \cong \overline{BE}$. Since $\overline{BF} \cong \overline{BF}$, $\triangle BDF \cong \triangle BEF$ (SSS $\cong$ SSS). Then $\angle BFD \cong \angle BFE$ (CPCTC) so $\overline{BF}$ bisects $\angle DFE$.
8. Because $\overline{AB} \cong \overline{AC}$, $\angle 1 \cong \angle 2$ so $\angle ABD \cong \angle ACE$ (supplements of congruent angles are congruent). Then $\triangle ABD \cong \triangle ACE$ (SAS $\cong$ SAS) and, by CPCTC, $\angle D \cong \angle E$. Thus, $\triangle DFB \cong \triangle EGC$ (AAS $\cong$ AAS) so $\overline{DF} \cong \overline{EG}$.
9. First prove $\triangle BCF \cong \triangle DCF$: $\angle FBC \cong \angle FDC$ (given), $\angle BFC \cong \angle DFC$ (definition of angle bisector), and $\overline{CGF} \cong \overline{CGF}$, so $\triangle BCF \cong \triangle DCF$ by AAS $\cong$ AAS. Prove $\triangle ACG \cong \triangle ECG$: $\overline{BC} \cong \overline{CD}$ (by CPCTC) and $\overline{AB} \cong \overline{ED}$ (given), so, using the addition property, $\overline{AC} \cong \overline{CE}$ (side); by CPCTC, $\angle ACG \cong \angle ECG$ (angle); $\overline{CG} \cong \overline{CG}$ (side). Therefore, $\triangle ACG \cong \triangle ECG$ by SAS $\cong$ SAS.

Lesson 3.5

1. (4)	**4.** (3)	**7.** (1)	**10.** (3)
2. (2)	**5.** (2)	**8.** (2)	**11.** (2)
3. (1)	**6.** (3)	**9.** (1)	**12.** (1)

13. $16 \nless 7 + 8$
14. $m\angle S = 51$ so $\overline{AT}$ is the shortest side.
15. $m\angle A = m\angle ACB > m\angle F$ so $m\angle A > m\angle F$. In $\triangle ADF$, $DF > AD$.
16. $m\angle CAB = m\angle B$ and $m\angle ASF > m\angle B$, $m\angle ASF > m\angle CAB$ and $m\angle CAB > m\angle FAS$ so $m\angle ASF > m\angle FAS$. In $\triangle ASF$, $AF > FS$.
17. $m\angle 1 > m\angle A$ and $m\angle 1 = m\angle 2$ so $m\angle 2 > m\angle A$. In $\triangle ADE$, $AD > ED$.
18. By the Base Angles Theorem, $m\angle CAB = m\angle CBA$ and $m\angle 1 = m\angle 2$ so, using the subtraction property, $m\angle 3 = m\angle 4$. Since $m\angle AED > m\angle 4$, $m\angle AED > m\angle 3$, making $AD > DE$.
19. Since $AD > BD$, $m\angle 3 > m\angle 2$, $m\angle 1 = m\angle 2$, and, by substition, $m\angle 3 > m\angle 1$. But $m\angle 4 > m\angle 3$, so $m\angle 4 > m\angle 1$, making $AC > DC$.

Lesson 3.6

1. Assume $\overline{AT} \perp \overline{CD}$. Then $\triangle ABT$ contains two right angles, which is impossible. Since the assumption that $\overline{AT} \perp \overline{CD}$ is false, $\overline{AT}$ is *not* perpendicular to $\overline{CD}$ as this is the only other possibility.
2. Assume $\overline{AB} \cong \overline{BC}$. Since $\angle 1 \cong \angle 2$, $\overline{AD} \cong \overline{CD}$ so $\triangle ABD \cong \triangle CBD$ by SSS $\cong$ SSS. By CPCTC, $\angle ABD \cong \angle CBD$. Since this contradicts the Given that $\overline{BD}$ does not bisect $\angle ABC$, the assumption that $\overline{AB} \cong \overline{BC}$ is false, $\overline{AB} \ncong \overline{BC}$ as this is the only other possibility.

3. Assume $\overline{JK} \cong \overline{ML}$. Then $\triangle JLM \cong \triangle MKJ$ by Hy-Leg $\cong$ Hy-Leg, so $\angle AJM \cong \angle AMJ$ by CPCTC. Since $\overline{AJ} \cong \overline{AM}$ contradicts the Given that $\triangle JAM$ is scalene, $\overline{JK} \not\cong \overline{ML}$.
4. Given: Scalene triangle ABC with M the midpoint of $\overline{AB}$, $\overline{MX} \perp \overline{AC}$, and $\overline{MY} \perp \overline{BC}$. Prove: $\overline{MX} \not\cong \overline{MY}$. Assume $\overline{MX} \cong \overline{MY}$. Right triangles AXM and BYM are congruent by Hy-Leg $\cong$ Hy-Leg, so $\angle A \cong \angle B$. Since $\overline{AC} \cong \overline{BC}$ contradicts the Given that $\triangle ABC$ is scalene, $\overline{MX} \not\cong \overline{MY}$.
5. Assume $\overline{AC}$ bisects $\angle BAD$. Then $\angle BAC \cong \angle DAC$. Because $\overline{BC} \parallel \overline{AD}$, $\angle DAC \cong \angle C$. By the transitive property, $\angle BAC \cong \angle C$ so $\overline{AB} \cong \overline{BC}$ (converse of the Base Angles Theorem). But this contradicts the Given that $\triangle ABC$ is not isosceles. Hence, the assumption that $\overline{AC}$ bisects $\angle BAD$ is false, so $\overline{AC}$ does *not* bisect $\angle BAD$.
6. Assume $\overline{BE} \cong \overline{EC}$. Since $\overline{AB} \cong \overline{AC}$ and $\overline{AE} \cong \overline{AE}$, $\triangle AEB \cong \triangle AEC$ by SSS $\cong$ SSS. By CPCTC, $\angle AEB \cong \angle AEC$, so $\angle BED \cong \angle CED$ (supplements of congruent angles are congruent). Because $\overline{ED} \cong \overline{ED}$, $\triangle BED \cong \triangle CED$, so $\overline{BD} \cong \overline{CD}$ by CPCTC. But this contradicts the Given that $\overline{BD} \not\cong \overline{CD}$. Hence the assumption that $\overline{BE} \cong \overline{EC}$ is false, so $\overline{BE} \not\cong \overline{EC}$.
7. Assume $\overline{PO} \cong \overline{OQ}$. Then $\triangle PON \cong \triangle QOM$ (SAS $\cong$ SAS) and $\angle P \cong \angle Q$ by CPCTC. Since alternate interior angles are congruent, $\overline{MQ} \parallel \overline{PN}$. But this contradicts the Given that $\overline{MQ}$ is *not* parallel to $\overline{PN}$. Hence, the assumption that $\overline{PO} \cong \overline{OQ}$ is false, so $\overline{PO} \not\cong \overline{OQ}$.
8. Assume $\overline{AB} \parallel \overline{DE}$. Then $\angle B \cong \angle D$. Since $\overline{DE} \cong \overline{CE}$, $\angle D \cong \angle DCE$. By transitivity, $\angle B \cong \angle DCE$. Since $\angle ABC \cong \angle DCE$, $\angle B \cong \angle ACB$, implying $\overline{AC} \cong \overline{AB}$. But this contradicts the Given ($AC > AB$). Hence, $\overline{AB}$ is *not* parallel to $\overline{DE}$.

CHAPTER 4

Lesson 4.1

1. (1) **2.** (4) **3.** (1) **4.** (4) **5.** (3) **6.** (2) **7.** (4) **8.** (3) **9.** (1) **10.** (3) **11.** (3) **12.** 120 **13.** 54

14. $\overline{EB} \cong \overline{AB}$. Since opposite sides of a parallelogram are congruent, $\overline{AB} \cong \overline{CD}$. Thus, $\overline{EB} \cong \overline{CD}$. Vertical angles EFB and DFC are congruent. Because $\overline{ABE} \parallel \overline{CD}$, $\angle EBF \cong \angle DCF$. Then $\triangle EFB \cong \triangle DFC$ (AAS $\cong$ AAS) and $\overline{EF} \cong \overline{FD}$ by CPCTC.
15. $\angle E \cong \angle H$ and $\angle EAF \cong \angle HCG$ (supplements of congruent angles). Because $\overline{AB} \cong \overline{CD}$ and $\overline{BF} \cong \overline{DG}$, $\overline{AF} \cong \overline{CG}$ (subtraction property). Thus, $\triangle EAF \cong \triangle HCG$ (AAS $\cong$ AAS) and $\overline{EF} \cong \overline{HG}$ by CPCTC.
16. **a.** $\angle CDK \cong \angle ADK$ and $\angle ADK \cong \angle CKD$ (congruent alternate interior angles) so $\angle CDK \cong \angle CKD$ (transitive property) and $\overline{CK} \cong \overline{CD}$ (converse of Base Angles Theorem).
    b. $\overline{BK} \cong \overline{AB}$ and $\overline{AB} \cong \overline{CD}$ so $\overline{BK} \cong \overline{CD}$. Because $\overline{CK} \cong \overline{CD}$ (from part a), $\overline{BK} \cong \overline{CK}$, so K is the midpoint of $\overline{BC}$.

17. Since $AD > DC$, $m\angle 2 > m\angle 1$. Because $\overline{AB} \parallel \overline{CD}$, $m\angle 3 = m\angle 2$ and by substitution, $m\angle 3 > m\angle 1$.

18. $m\angle 1 > m\angle ACB$ and $m\angle ACB = m\angle 2$ so $m\angle 1 > m\angle 2$.

19. $\triangle AED \cong \triangle CFB$ by SAS $\cong$ SAS. By CPCTC, $\overline{AD} \cong \overline{BC}$ and $\angle DAE \cong \angle BCF$ so $\overline{AD} \parallel \overline{BC}$. Since quadrilateral *ABCD* has a pair of sides that are both congruent and parallel, *ABCD* is a parallelogram.

20. $\overline{LD} \cong \overline{BM}$ and $\angle DLM \cong \angle BML$ so $\angle ALF \cong \angle CMB$ (supplements of congruent angles are congruent). $\triangle ALD \cong \triangle CMB$ by SAS $\cong$ SAS. By CPCTC, $\overline{AD} \cong \overline{BC}$ and $\angle DAL \cong \angle BCM$ so $\overline{AD} \parallel \overline{BC}$. Since quadrilateral *ABCD* has a pair of sides that are both congruent and parallel, *ABCD* is a parallelogram.

21. $\triangle BAL \cong \triangle DCM$ by ASA $\cong$ ASA. By CPCTC, $\overline{BL} \cong \overline{DM}$ and $\angle ALB \cong \angle CMD$ so $\overline{BL} \parallel \overline{DM}$. Since quadrilateral *BMDL* has a pair of sides that are both congruent and parallel, *BMDL* is a parallelogram.

22.

Statement		Reason
1. $\overline{NJ}$ and $\overline{DU}$ bisect each other at *K*.		1. Given.
2. $\overline{DK} \cong \overline{UK}$.	*Side*	2. Definition of segment bisector.
3. $\angle DKN \cong \angle UKJ$.	*Angle*	3. Vertical angles are congruent.
4. $\overline{NK} \cong \overline{JK}$.	*Side*	4. Same as 2.
5. $\triangle DKN \cong \triangle UKJ$.		5. SAS $\cong$ SAS.
6. $\angle 1 \cong \angle 2$ and $\overline{DN} \cong \overline{UJ}$.		6. CPCTC.
7. $\overline{AD} \parallel \overline{UQ}$.		7. If two lines form congruent alternate interior angles, the lines are parallel.
8. *N* is the midpoint of $\overline{DA}$; *J* is the midpoint of $\overline{QU}$.		8. Given.
9. $\frac{1}{2}(AD) = DN$ and $\frac{1}{2}(UQ) = UJ$.		9. Definition of the midpoint of a line segment.
10. $AD = 2(DN)$ and $UQ = 2(UJ)$.		10. Multiplication property.
11. $\overline{AD} \cong \overline{UQ}$.		11. Doubles of congruent segments are congruent.
12. *QUAD* is a parallelogram.		12. If one pair of sides of a quadrilateral are both parallel and congruent, then the quadrilateral is a parallelogram.

23. **a.** $\triangle FGC \cong \triangle EGA$ (SAS $\cong$ SAS). By CPCTC, $\angle 3 \cong \angle 4$. $\triangle ABC \cong \triangle CDA$ (AAS $\cong$ AAS) so $\overline{BC} \cong \overline{DA}$ (CPCTC).

b. Because $\angle 3 \cong \angle 4$, $\overline{BC} \parallel \overline{DA}$. Since quadrilateral *ABCD* has a pair of sides that are both congruent and parallel, *ABCD* is a parallelogram.

Lesson 4.2

1. (2) **2.** (4) **3.** (1) **4.** (3) **5.** (2) **6.** (3) **7.** (2) **8.** (2) **9.** (8, –1) **10.** $k = -3, x = 7$

11. Midpoint $\overline{MT}$ = midpoint $\overline{AH} = \left(\frac{5}{2}, 3\right)$. Since the diagonals bisect each other, *MATH* is a parallelogram.

12. **a.** $CA = CT = \sqrt{68}$.

b. Midpoint $\overline{AT} = \left(\frac{6+0}{2}, \frac{4+(-2)}{2}\right)$ = (3, 1) so point *S*(3, 1) is the midpoint of $\overline{AT}$. Slope of $\overline{AT}$ = 1 and slope of $\overline{CS}$ = –1 so $\overline{CS} \perp \overline{AT}$.

13. Midpoint $\overline{AC}$ = (–1, 2) and midpoint $\overline{BD}$ = (1, 10). Since diagonals do not bisect each other, *ABCD* is not a parallelogram.

14. **a.** Slope $\overline{NS}$ = slope $\overline{TS} = \frac{1}{2}$ so points *N*, *T*, and *S* are collinear.

b. Midpoint of $\overline{NS} = \left(\frac{-2+10}{2}, \frac{-1+5}{2}\right)$ = *T*(4, 2) so $\overline{YT}$ is a median to side $\overline{NS}$. Slope $\overline{NS} = \frac{1}{2}$ and slope of $\overline{YT}$ = –2 so $\overline{YT} \perp \overline{NS}$ and, as a result, $\overline{YT}$ is an altitude to side $\overline{NS}$.

15. **a.** Slope $\overline{BC} = -\frac{1}{2}$ and slope $\overline{AC}$ = 2 so $\overline{BC} \perp \overline{AC}$. Thus, $\angle C$ is a right angle so $\triangle ABC$ is a right triangle.

b. Hypotenuse *AB* = 10.The midpoint of $\overline{AB}$ is *M*(8, 3) and *CM* = 5 so $CM = \frac{1}{2}AB$.

16. **a.** Slope $\overline{AB} = \frac{3}{2}$ and slope $\overline{AC} = -\frac{2}{3}$ so $\overline{AB} \perp \overline{AC}$. Thus, $\angle A$ is a right angle so $\triangle ABC$ is a right triangle.

b. The midpoint of hypotenuse $\overline{BC}$ is *M*(4, 3). Since *AM* = *BM* = $CM = \sqrt{26}$, all three vertices of the triangle are equidistant from point *M*.

Lesson 4.3

1. (4) **2.** (4) **3.** (2) **4.** (3) **5.** (4) **6.** (1) **7.** (4) **8.** (2) **9.** (2) **10.** (3) **11.** (2) **12.** 17

13. Show slope $\overline{MA}$ = slope $\overline{TH} = \frac{7}{3}$ and slope $\overline{AT}$ = slope $\overline{MH} = -\frac{3}{7}$, so *MATH* is a parallelogram. Since the slopes of adjacent sides are negative reciprocals, *MATH* contains four right angles and is, therefore, a rectangle. Use the distance formula to show that a pair of adjacent sides are congruent.

14. Because midpoint $\overline{AC}$ = midpoint $\overline{BD} = \left(3, -\frac{1}{2}\right)$, *ABCD* is a parallelogram. Since $AB = \sqrt{117}$ and $BC = \sqrt{90}$, $AB \neq BC$, so *ABCD* is *not* a rhombus.

15. Slope $\overline{AB}$ = slope $\overline{CD} = \frac{3}{8}$, so $\overline{AB} \parallel \overline{CD}$. Slope $\overline{BC}$ = slope $\overline{AD} = \frac{5}{2}$ so $\overline{BC} \parallel \overline{AD}$ and *ABCD* is a parallelogram. Since the slopes of adjacent sides are not negative reciprocals, adjacent sides are not perpendicular so *ABCD* is *not* a rectangle.

16. Since midpoint $\overline{RC}$ = midpoint $\overline{ET} = \left(\frac{9}{2}, 6\right)$, *RECT* is a parallelogram.

As diagonal RC = diagonal $ET = \sqrt{125}$, parallelogram *RECT* is a rectangle. Because $RE = 10$ and $RT = 5$, $RE \neq RT$ so rectangle *RECT* is *not* a square.

17. $DE = EF = FG = DG = \sqrt{20}$ so *DEFG* is a rhombus.

18. $TE = EA = AM = TM = \sqrt{58}$ so *TEAM* is a rhombus. Since the slope of $\overline{TE} = \frac{7}{3}$ and the slope of $\overline{TM} = \frac{3}{7}$, $\overline{TE} \not\perp \overline{TM}$ because their slopes are not negative reciprocals. Thus, rhombus *TEAM* is *not* a square.

19. **a.** Show midpoint $\overline{AC} \neq$ midpoint $\overline{BD}$.
b. If *P*, *Q*, *R*, and *S* are the midpoints of $\overline{AB}$, $\overline{BC}$, $\overline{CD}$, and $\overline{AD}$, respecitvely, show midpoint $\overline{PR}$ = midpoint $\overline{QS}$.

20. $\triangle RQU \cong \triangle AMD$ (AAS $\cong$ AAS) so $\angle 1 \cong \angle 2$ by CPCTC. Using the addition property, $\angle QUA \cong \angle ADQ$. If the opposite angles of a quadrilateral are congruent, the quadrilateral is a parallelogram. Hence, *QUAD* is a parallelogram.

21. $\triangle ABL \cong \triangle BCM$ by SSS $\cong$ SSS, so $\angle ABL \cong \angle BCM$. Since $\overline{AB} \parallel \overline{CD}$, angles *ABL* and *BCM* are supplementary and congruent, making each a right angle, so *ABCD* is a square.

22. $\triangle VWT \cong \triangle SXT$ by ASA $\cong$ ASA since $\angle WTV = \angle XTS$ (vertical angles), $\overline{VT} \cong \overline{ST}$ (rhombus is equilateral, $\angle WVT \cong \angle XST$ (subtract corresponding sides of the two equations $m\angle RSX \cong m\angle RVW$ and $m\angle RST = m\angle TVR$). By CPCTC, $\overline{TX} \cong \overline{TW}$.

23. Use an indirect proof. Assume $ABCD$ is a rectangle. Then $\triangle BAD \cong \triangle CDA$ (SAS). $\angle 1 \cong \angle 2$ (PCPTC), contradicting the Given.

24. $ABCD$ is a square (given); $\angle B$ and $\angle D$ are right angles (a square contains four right angles); $\overline{AB} \cong \overline{AD}$ (adjacent sides of a square are congruent); $\angle 1 \cong \angle 2$ (given); $\overline{AE} \cong \overline{AF}$ (converse of the Base Angles Theorem); $\triangle ABE \cong \triangle ADF$ (Hy-Leg); $\overline{BE} \cong \overline{DF}$ (CPCTC).

25. $\triangle ABE \cong \triangle ADF$ by AAS $\cong$ AAS since $\angle B \cong \angle D$, $\angle BEA \cong \angle DFA$ (by subtracting corresponding sides of the two equations $m\angle BEF = m\angle DFE$ and $m\angle 1 = m\angle 2$), and $\overline{EA} \cong \overline{FA}$ (since $\angle 1 \cong \angle 2$). By CPCTC, $\overline{AB} \cong \overline{AD}$. Since an adjacent pair of sides of rectangle $ABCD$ are congruent, $ABCD$ is a square.

26. **a.** $\triangle ABP \cong \triangle DCN$ (HL $\cong$ HL)

b. So $\angle APB \cong \angle DNC$ by CPCTC. In $\triangle PEN$, $\overline{PE} \cong \overline{NE}$ (converse of Base Angles Theorem). Subtracting corresponding sides of $\overline{AP} \cong \overline{DN}$ and $\overline{PE} \cong \overline{NE}$ gives $\overline{AE} \cong \overline{DE}$.

27. Since a rhombus is equilateral, $\overline{BC} \cong \overline{CD}$. A diagonal of a rhombus bisects its angles so $\angle ACB \cong \angle ACD$. Because supplements of congruent angles are congruent, $\angle BCE \cong \angle DCE$. $\triangle BCE \cong \triangle DCE$ (SAS $\cong$ SAS) so $\overline{BE} \cong \overline{DE}$ by CPCTC.

28. Since $\overline{AD} \cong \overline{DB}$ and $\overline{CD} \cong \overline{ED}$, the diagonals of quadrilateral $AEBC$ bisect each other so $AEBC$ is a parallelogram. Because $\angle C$ is a right angle, $AEBC$ is a rectangle.

Lesson 4.4

1. (2) **2.** (2) **3.** (4) **4.** (2) **5.** (3)

6. Since slope $\overline{JK} = \frac{1}{2}$, slope $\overline{KL} = -\frac{4}{7}$, slope $\overline{LM} = \frac{1}{2}$, and slope $\overline{JM} = -2$, $\overline{JK} \parallel \overline{LM}$ and $\overline{KL} \nparallel \overline{JM}$ so $JKLM$ is a trapezoid. Since $KL = \sqrt{65}$ and $JM = \sqrt{45}$, $KL \neq JM$ so $JKLM$ is *not* an isosceles trapezoid.

7. Show slope $\overline{JA}$ = slope $\overline{KE} = 0$, slope $\overline{AK} \neq$ slope $\overline{JE}$, and $JK = AE = 5a$, so $JAKE$ is an isosceles trapezoid.

8. **a.** Show slope $\overline{BC}$ = slope $\overline{AD} = 1$, and slope $\overline{AB} \neq$ slope $\overline{CD}$.

b. $h = 3$, $k = 2$.

c. Since slope $\overline{AB} \times$ slope $\overline{BE} = -1$, $\angle ABE$ is a right angle, so $ABED$ is a rectangle.

9. Since parallel lines are everywhere equidistant, $\overline{BE} \cong \overline{CF}$. $\triangle AEB \cong \triangle DCF$ by SAS $\cong$ SAS. By CPCTC, $\overline{AB} \cong \overline{CD}$, so trapezoid $ABCD$ is isosceles.

10. $\triangle RSW \cong \triangle WTR$ by SAS. By CPCTC, $\angle TRW \cong \angle SWR$, so $\overline{RP} \cong \overline{WP}$ (converse of the Base Angles Theorem), and $\triangle RPW$ is isosceles.

11. $\triangle AGB \cong \triangle DCF$ by SAS $\cong$ SAS since $\overline{BG} \cong \overline{CF}$, $\angle BGA \cong \angle CFD$ (Base Angles Theorem), and $\overline{AG} \cong \overline{DF}$ (addition property). By CPCTC, $\overline{AB} \cong \overline{CD}$, so trapezoid $ABCD$ is isosceles.

12. Since $\angle 1 \cong \angle 2$, $\overline{BK} \cong \overline{BA}$. $\overline{BA} \cong \overline{CD}$ so $\overline{BK} \cong \overline{CD}$ by the transitive property. Similarly, $\angle 1 \cong \angle CDA$ so $\angle 2 \cong \angle CDA$, making $\overline{BK} \parallel \overline{CD}$. Since the same pair of sides of *BKDC* are both parallel and congruent, *BKDC* is a parallelogram.

13. Use an indirect proof.

Statement		Reason
1. Trapezoid *ROSE* with $\overline{OS} \parallel \overline{RE}$, and diagonals $\overline{RS}$ and $\overline{EO}$ intersection at point *M*.		1. Given.
2. Diagonals $\overline{RS}$ and $\overline{EO}$ bisect each other.		2. Assume this is true.
3. $\overline{OM} \cong \overline{EM}$	Side	3. A bisector divides a segment into two congruent segments.
4. $\angle OMR \cong \angle SMR$.	Angle	4. Vertical angles are congruent.
5. $\overline{RM} \cong \overline{SM}$.	Side	5. A bisector divides a segment into two congruent segments.
6. $\triangle ORM \cong \triangle SEM$.		6. SAS $\cong$ SAS.
7. $\angle ORM \cong \angle ESM$.		7. CPCTC
8. $\overline{OR} \parallel \overline{SE}$.		8. If two lines are cut by a transversal and alternate interior angles are congruent, the lines are parallel.
9. Satement 8 contradicts the given.		9. A trapezoid has exactly one pair of parallel sides.
10. Statement 2 is false.		10. A statement that leads to a contradiction is false.
11. $\overline{RS}$ and $\overline{EO}$ do *not* bisect each other.		11. The opposite of a false statement is true.

Lesson 4.5

1. (3) **2.** (4) **3.** 23 **4.** 45

5. $EF = 19$ and $RT = 38$

6. a. $\triangle DEB \cong \triangle CEF$ by SAS $\cong$ SAS so $\overline{BD} \cong \overline{CF}$. Since $\overline{AD} \cong \overline{BD}$, $\overline{AD} \cong \overline{CF}$. Angles *BDE* and *CFE* are congruent by CPCTC, which makes $\overline{AD} \parallel \overline{CF}$ so *ADFC* is a parallelogram. Since opposite sides of a parallelogram are parallel, $\overline{DE} \parallel \overline{AC}$.

b. By definition of midpoint, $DE = \frac{1}{2}DF$. Since opposite sides of a parallelogram have the same length, $DF = AC$. Using substitution property, $DE = \frac{1}{2}AC$.

7. $DE = \frac{1}{2}BC$, $DF = \frac{1}{2}AB$ so $\frac{1}{2}AB = \frac{1}{2}BC$. Using the multiplication property, $\overline{AB} \cong \overline{BC}$ so $\triangle ABC$ is isosceles.
8. By the Base Angles Theorem, $\angle PLM \cong \angle PML$. Since $\overline{LM}$ is a median, $\overline{LM} \cong \overline{AD}$, so $\angle PLM \cong \angle APL$ and $\angle PML \cong \angle DPM$. By the transitive property, $\angle APL \cong \angle DPM$. Show $\triangle LAP \cong \triangle MDP$ by SAS. By CPCTC, $\angle A \cong \angle D$, so trapezoid $ABCD$ is isosceles.
9. In $\triangle WST$, since B and C are midpoints, $\overline{BC} \parallel \overline{WT}$ and $BC = \frac{1}{2}WT$.

   Since $RSTW$ is a parallelogram, $\angle AWB \cong \angle CSB$ so $\triangle AWB \cong \triangle CSB$

   by ASA $\cong$ ASA. As $\overline{AB} \cong \overline{BC}$, $AB + BC = \frac{1}{2}WT + \frac{1}{2}WT$ so

   $\overline{AC} \cong \overline{WT}$. Since extensions of parallel lines are parallel, $\overline{AC} \parallel \overline{WT}$. Because $WACT$ has one pair of sides that are both congruent and parallel, it is a parallelogram.
10. The diagonals of a parallelogram bisect each other so $\overline{AX} \cong \overline{CX}$ and $\overline{BX} \cong \overline{DX}$. Because it is given that E, F, G, and H are midpoints of $\overline{AX}$, $\overline{BX}$, $\overline{CX}$, and $\overline{DX}$, respectively, $\overline{EX} \cong \overline{GX}$ and $\overline{FX} \cong \overline{HX}$ so $EFGH$ is a parallelogram. Since $EF = \frac{1}{2}AB$, $FG = \frac{1}{2}BC$, and $\overline{AB} \cong \overline{BC}$ (a rhombus is equilateral), $\overline{EF} \cong \overline{FG}$ so $EFGH$ is a rhombus.
11. Given rectangle $ABCD$, points P, Q, R, and S are the midpoints of $\overline{AB}$, $\overline{BC}$, $\overline{CD}$, and $\overline{AD}$, respectively. Draw diagonal $\overline{BD}$. $\overline{PS} \parallel \overline{QR}$ and $\overline{PS} \cong \overline{QR}$ since each segment is parallel to, and one-half the length of, $\overline{BD}$. Thus, $PQRS$ is a parallelogram. Since $\triangle PAS \cong \triangle PBQ$ by SAS $\cong$ SAS, $\overline{PS} \cong \overline{QR}$, which makes $PQRS$ a rhombus.

Lesson 4.6

1. Slope $\overline{AB}$ = slope $\overline{CD} = 0$ so $\overline{AB} \parallel \overline{CD}$. Slope $\overline{AD} \neq$ slope $\overline{BC}$ so $\overline{AD} \nparallel \overline{BC}$. Therefore, $ABCD$ is a trapezoid. $AD = BC = \sqrt{a^2 - 2ab + b^2 + c^2}$ so $ABCD$ is an isosceles trapezoid.
2. Using $Y(a, b)$, $S(3a, b)$, $N(2a, 0)$, $YN = SN = \sqrt{a^2 + b^2}$.
3. Midpoint $\overline{QS}$ = midpoint $\overline{RT} = \left(\frac{a+c}{2}, \frac{b}{2}\right)$.
4. For rectangle $A(0, 0)$, $B(0, b)$, $C(d, b)$, and $D(d, 0)$, $AC = BC = \sqrt{b^2 + d^2}$.
5. **a.** $AB = BC = \sqrt{4r^2 + 4s^2}$.

   b. If $L(-r, s)$ is the midpoint of $\overline{AB}$ and $M(r, s)$ is the midpoint of $\overline{BC}$, then $AM = CL = \sqrt{9r^2 + s^2}$.

6. a. $C(t, t)$

b. $AC = BC = t\sqrt{2}$. Slope $\overline{BD} \times$ slope $\overline{AC} = -1 \times 1 = -1$.

7. Midpoint $\overline{MT}$ = midpoint $\overline{AH} = \left(\frac{s}{2}, \frac{t}{2}\right)$

Show $MT \neq AH$.

8. a. $h = a - b, k = c$.

b. Show $AC = BD = \sqrt{(a-b)^2 + c^2}$.

9. a. $y = \sqrt{a^2 - b^2}$.

b. Slope $\overline{AC} = \frac{y}{a+b}$ and slope $\overline{BD} = \frac{y}{b-a}$. Lines are perpendicular if the product of their slopes is –1:

$$\text{Slope } \overline{AC} \times \text{slope } \overline{BD} = \frac{y}{a+b} \times \frac{y}{b-a} = \frac{\left(\sqrt{a^2 - b^2}\right)^2}{-(a^2 - b^2)} = -1$$

10. $AB = CD = s$ and $BC = AD = \sqrt{t^2 + s^2}$ so $ABCD$ is a parallelogram but *not* a rhombus because it is not equilateral.

Chapter 5

Lesson 5.1

1. (3)	**4.** (4)	**7.** (1)	**10.** –22	**13.** 75	**16.** 21
2. (3)	**5.** (3)	**8.** –6	**11.** 5	**14.** 36	**17.** 84
3. (1)	**6.** (4)	**9.** 20	**12.** 25	**15.** 48	**18.** 45

Lesson 5.2

1. (2)	**4.** (3)	**7.** (2)	**10.** (3)	**13.** 65	**16.** 10
2. (3)	**5.** (4)	**8.** (2)	**11.** (3)	**14.** 319	**17.** 40
3. (4)	**6.** (3)	**9.** (1)	**12.** 24	**15.** 12 ft	**18.** 15

19. a. 2

b. $\frac{1}{2}h$

c. 1

20. a. $\frac{80}{x} = \frac{15}{171 - x}$

b. 144

21. Because $\angle S \cong \angle A$ (opposite angles of a parallelogram are congruent) and $\angle 1 \cong \angle 2$ (all right angles are congruent), $\triangle SKT \sim \triangle ALT$.

22. Since $\angle A \cong \angle 1$ (all right angles are congruent) and $\angle ABD \cong \angle CDE$ (congruent alternate interior angles), $\triangle BAD \sim \triangle DEC$ so $\frac{EC}{AD} = \frac{CD}{BD}$. Since $\overline{AB} \cong \overline{CD}$, $\frac{EC}{AD} = \frac{AB}{BD}$.

23. Since $\overline{AG} \cong \overline{AE}$, $\angle AGH \cong \angle AEH$. $\overline{AC}$ bisects $\angle FAB$, so $\angle GAH \cong \angle EAH$. Because, $\overline{AB} \parallel \overline{CD}$, $\angle AEH \cong \angle CDH$ and $\angle EAH \cong \angle DCH$. By the transitive property, $\angle AGH \cong \angle CDH$ and $\angle GAH \cong \angle DCH$, so $\triangle AHG \sim \triangle CHD$.

24. Show $\triangle RMN \sim \triangle RAT$. $\angle RMN \cong \angle A$, and $\angle RNM \cong \angle T$. Write $\frac{MN}{AT} = \frac{RN}{RT}$. By the converse of the Base Angles Theorem, $NT = MN$. Substitute NT for MN in the proportion.

25. Show $\triangle SQP \sim \triangle WRP$. $\angle SPQ \cong \angle WPR$ (angle). Since $\overline{SR} \cong \overline{SQ}$, $\angle SRQ \cong \angle SQR$. $\angle SRQ \cong \angle WRP$ (definition of angle bisector). By the transitive property, $\angle SQR \cong \angle WRP$ (angle).

26. Show $\triangle PMQ \sim \triangle MKC$. Right triangle $MCK \cong$ right triangle PMQ. Since $\overline{TP} \cong \overline{TM}$, $\angle TPM \cong \angle TMP$.

27. Show $\triangle PMT \sim \triangle JKT$. Since $\overline{MP} \cong \overline{MQ}$, $\angle MPQ \cong \angle MQP$. Since $\overline{JK} \parallel \overline{MQ}$, $\angle J \cong \angle MQP$ so, by the transitive property, $\angle J \cong \angle MPQ$ (angle). $\angle K \cong \angle QMT$ (congruent alternate interior angles). $\angle QMT \cong \angle PMT$ (since $\triangle MTP \cong \triangle MTQ$ by SSS). By the transitive property, $\angle K \cong \angle PMT$ (angle). Write the proportion $\frac{PM}{JK} = \frac{PT}{JT}$. Substitute TQ for PT (see the Given) in the proportion.

28. **a.** $\angle C \cong \angle A$ and $\angle 2 \cong \angle 6$ so $\triangle FEC \sim \triangle DFA$.

b. $\angle 4$ is complementary to $\angle 1$ and $\angle C$ is complementary to $\angle 1$ so $\angle 4 \cong \angle C$; $\angle 2 \cong \angle 6$ so $\triangle EDF \sim \triangle DFA$.

c. $\triangle FEC \sim \triangle EDF$. If two triangles are similar to the same triangle, then they are similar to each other.

Lesson 5.3

1. (4) **2.** (2) **3.** (2) **4.** (3) **5.** (1) **6.** (3) **7.** 160 **8.** 45 in^2 **9.** 117 in^2

Lesson 5.4

1. (3) **2.** (2) **3.** (1) **4.** (4) **5.** (2) **6.** 6 **7.** 3 **8.** 25 **9.** 6 **10.** 28 **11.** 9 **12.** 13.7

13. Since $\angle ACB \cong \angle DCE$ (vertical angles are congruent) and $\frac{AC}{EC} = \frac{BC}{DC}$ (given), $\triangle ABC \sim \triangle DEC$ (SAS Similarity Theorem) so $\angle B \cong \angle D$ since corresponding angles of similar triangles are congruent.

14. Since $\overline{BCE} \cong \overline{FD}$, $\frac{BF}{AF} = \frac{CD}{AD}$ (if a line parallel to one side of a triangle intersects the other two sides, then it divides those sides proportionally). Since $\angle 1 \cong \angle 2$ and $\angle 1 \cong \angle ECD$, $\angle 2 \cong \angle ECD$ so $\overline{CD} \cong \overline{ED}$ and, by substitution, $\frac{BF}{AF} = \frac{ED}{AD}$.

15. Points S and T are midpoints so $\overline{TS} \parallel \overline{JK}$ and $TS = \frac{1}{2}KJ$. Because $\overline{RS} \cong \overline{RS}$, $\angle KRS \cong \angle TSR$ (congruent alternate interior angles), and $\overline{KR} \cong \overline{TS}$ (since $KR = \frac{1}{2}KJ$ and $TS = \frac{1}{2}KJ$), $\triangle KRS \cong \triangle TSR$ (SAS $\cong$ SAS) so $\angle 1 \cong \angle 2$ by CPCTC.

16. In right triangle KHT, hypotenuse $TK = 10$. Since $\frac{TK}{SK} = \frac{10}{5} = 2$ and $\frac{KH}{RH} = \frac{6}{3} = 2$, $\frac{TK}{SK} = \frac{KH}{RH}$. Because the sides that include congruent right angles RHK and SKT are in proportion, $\triangle RHK \sim \triangle SKT$ by the SAS Similarity Theorem.

Lesson 5.5

1. Show $\triangle AEH \sim \triangle BEF$. $\angle BEF \cong \angle HEA$ and $\angle EAH \cong \angle EBF$ (halves of equals are equal).

2. First prove the proportion $\frac{JY}{XZ} = \frac{YX}{ZL}$ by proving $\triangle JYX \sim \triangle XZL$. To prove these triangles are similar, show $\angle JYX \cong \angle XZL$ and $\angle J \cong \angle ZXL$. Since $KYXZ$ is a parallelogram, $KZ = YX$, so $\frac{JY}{XZ} = \frac{YX}{ZL} = \frac{KZ}{ZL}$. In the proportion $\frac{JY}{XZ} = \frac{KZ}{ZL}$, cross-multiplying gives the desired product.

3. Show $\triangle EIF \sim \triangle HIG$. Since $\overline{EF}$ is a median, $\overline{EF} \parallel \overline{AD}$, so $\measuredangle FEI \cong \measuredangle GHI$ and $\measuredangle EFI \cong \measuredangle HGI$.

4. Rewrite the product as $\frac{RW}{RV} = \frac{TW}{SV}$. Prove $\triangle RSV \sim \triangle RTW$. Since RVW bisects $\angle SRT$ (given), $\angle SRV \cong \angle TRW$ (angle). Since $\overline{TW} \cong \overline{TV}$ (given), $\angle TVW \cong \angle W$. Also, $\angle RVS \cong \angle TVW$ (vertical angles are congruent). By the transitive property, $\angle RVS \cong \angle W$ (angle). Hence, $\triangle RSV \sim \triangle RTW$ by the AA similarity postulate.

5. a. $\overline{EF} \parallel \overline{AC}$, $\measuredangle ACF \cong \measuredangle GFE$ since parallel lines form congruent alternate interior angles. Similarly, since $\overline{DE} \parallel \overline{AB}$, $\measuredangle AFC \cong \measuredangle EGF$. Hence, $\triangle CAF \sim \triangle FEG$.

b. $\triangle CAF \sim \triangle CDG$. By the transitive property, $\triangle CDG \sim \triangle FEG$, so $\frac{DG}{EG} = \frac{GC}{GF}$, implying $DG \times GF = EG \times GC$.

6. Draw right triangle ABC with altitude $\overline{CH}$ drawn to hypotenuse $\overline{AB}$. Prove $AC \times BC = AB \times CH$ by showing $\triangle ABC \sim \triangle ACH$, using $\angle A \cong \angle A$ right $\angle ACB \cong$ right $\angle AHC$.

Chapter 6

Lesson 6.1

1. (1)	**4.** (4)	**7.** (1)	**10.** 13	**13.** $9\sqrt{5} + 15$
2. (4)	**5.** (1)	**8.** $8\sqrt{5}$	**11.** 25	**14a.** 4
3. (2)	**6.** (1)	**9.** 11.8	**12.** $3\sqrt{3}$	**b.** $4\sqrt{5}$

Lesson 6.2

1. (4)	**6.** (2)	**11.** (3)	**16.** 168 cm^2	**21a.** 3.8
2. (3)	**7.** (1)	**12.** (2)	**17.** 36, 48	**b.** 13.2
3. (4)	**8.** (3)	**13.** 127.3	**18.** 5	**22.** 41.7
4. (3)	**9.** (3)	**14.** 308	**19.** 14.3	**23.** 9.4
5. (1)	**10.** (3)	**15.** 14.9	**20.** 2.8	**24.** 8.4

Lesson 6.3

1. (3)	**5.** (4)	**9.** (1)	**13.** 42 cm^2
2. (3)	**6.** (1)	**10.** (4)	**14.** 6.9
3. (2)	**7.** (3)	**11.** (4)	**15.** $G(-2, 11)$ or $G(-2,-3)$
4. (1)	**8.** (3)	**12.** $50\sqrt{3}$ in^2	**16.** 22

Lesson 6.4

1. (2)	**6.** 68.7°	**11.** 19.0	**15.** 76.8
2. (1)	**7.** 28°	**12.** 57	**16a.** 129.7
3. (2)	**8.** 15.5	**13a.** 2	**b.** 57.5
4. (2)	**9.** 60°	**b.** 70.5°	**17.** 210
5. (2)	**10.** 136	**14.** 86.6	

CHAPTER 7

Lesson 7.1

1. (1) **2.** (2) **3.** (2) **4.** (3) **5.** 120 **6.** 80 **7.** 142 **8.** 80

9. Since *OXEY* is a square, $\overline{OX} \perp \overline{OT}$, $\overline{OY} \perp \overline{PJ}$, $OX = OY$, so $\overline{QT} \cong \overline{JP}$. Subtracting corresponding sides of $\text{m}\,\widehat{QT} = \text{m}\,\widehat{JP}$ and $\text{m}\,\widehat{PT} = \text{m}\,\widehat{PT}$ gives $\widehat{QP} \cong \widehat{JT}$.

10. Since $\overline{DE} \cong \overline{FG}$, $\overline{PB} \cong \overline{PC}$, right triangle $APB \cong$ right triangle ACP by Hy-Leg $\cong$ Hy-Leg. By CPCTC, $\angle BAP \cong \angle CAP$, so $\overline{PA}$ bisects $\angle FAD$.

11. a. Since $\widehat{AD} \cong \widehat{BC}$, $\angle AOD \cong \angle BOC$ so $\triangle AOE \cong \triangle BOF$ by SAS $\cong$ SAS.

b. Because points *E* and *F* are midpoints of radii $\overline{OD}$ and $\overline{OC}$, $\overline{DE} \cong \overline{CF}$ (halves of congruent segments are congruent); $\angle ADE \cong \angle BCF$ (CPCTC from part a); $\overline{AD} \cong \overline{BC}$ (congruent arcs have congruent chords). Thus, $\triangle ADE \cong \triangle BCF$ by SAS $\cong$ SAS, so $\angle DAE \cong \angle CBF$ by CPCTC.

Lesson 7.2

1. (1) **2.** (4) **3.** (2) **4.** (3) **5.** (4) **6.** (1) **7.** (2) **8.** (4) **9.** 25.5 **10.** 11 **11.** $x = 10$, $y = 13$ **12.** 17 in **13.** 6 **14.** 13

15. Draw $\overline{OB}$ and radius $\overline{OA}$. Assume $\overline{OB}$ is perpendicular to $\overline{PA}$. Since $\overline{OA}$ is also perpendicular to $\overline{PA}$, $\triangle OAB$ contains two right angles, which is impossible. Hence, the assumption that $\overline{OB}$ is perpendicular to $\overline{PA}$ is false. Thus, $\overline{OB}$ is *not* perpendicular to $\overline{PA}$ as this is the only other possibility.

16. Since congruent circles have congruent radii, $\overline{OA} \cong \overline{O'B}$; $\angle OAC \cong \angle O'BC$ (right angles are congruent); $\angle OCA \cong \angle O'CB$ (vertical angles are congruent). Thus, $\triangle OAC \cong \triangle O'BC$ by AAS $\cong$ AAS. By CPCTC, $\overline{OC} \cong \overline{O'C}$.

Lesson 7.3

1. (1) **2.** (2) **3.** (2) **4.** (2) **5.** (1) **6.** (4) **7.** (1) **8.** (1) **9.** (3) **10.** $\frac{4}{3}\pi$ **11.** 9.43 **12.** 94 **13.** 12.6 **14a.** 54.6 **b.** 2,752

Lesson 7.4

1. (1)	**7.** (3)	**13.** 50	**19.** 35	**c.** 15
2. (2)	**8.** (1)	**14.** 22	**20.** 90	**d.** 135
3. (3)	**9.** (4)	**15.** 60	**21.** 6.8	**24a.** 30
4. (2)	**10.** (3)	**16.** 20	**22.** 2:1	**b.** 75
5. (3)	**11.** 64	**17.** 70	**23a.** 90	**25.** 25
6. (2)	**12.** 66	**18.** 80	**b.** 120	

26. $m\overset{\frown}{GF} = 30$, $m\angle BHD = 65$, $m\angle BDG = 75$, $m\angle GDE = 55$, $m\angle C = 35$, $m\angle BOD = 100$

27. In $\triangle APB$, $\overline{PDA} \cong \overline{PCB}$, so $\angle PAB \cong \angle PBA$. Inscribed angles DAC and CBD intercept the same arc, so they are congruent. Using the subtraction property, $\angle FAB \cong \angle FBA$. It follows that $\overline{AF} \cong \overline{BF}$ (converse of the Base Angles Theorem).

28. In $\triangle PBD$, $\overline{PAB} \cong \overline{PCD}$ so $\angle B \cong \angle D$. Since congruent inscribed angles intercept congruent arcs, $\overset{\frown}{ACD} \cong \overset{\frown}{CAB}$. Subtracting $\overset{\frown}{AC}$ from both $\overset{\frown}{ACD}$ and $\overset{\frown}{CAB}$ makes $\overset{\frown}{AB} \cong \overset{\frown}{CD}$, so $\overline{AB} \cong \overline{CD}$. Thus, $\triangle AEB \cong \triangle CFD$ by AAS $\cong$ AAS. By CPCTC, $\overline{AE} \cong \overline{CF}$.

Lesson 7.5

1. (2)	**4.** (1)	**7.** (1)	**10.** 10	**13.** 6
2. (4)	**5.** (2)	**8.** (1)	**11.** 5.5	**14.** 12
3. (3)	**6.** (1)	**9.** 12	**12.** 16	**15.** 10

16. Inscribed angles A and C intercept the same arc, $\overset{\frown}{BD}$, so they are congruent. Inscribed angles B and D intercept the same arc, $\overset{\frown}{AC}$, so they too are congruent. Hence, $\triangle AEB \sim \triangle CED$ by the AA Similarity Postulate.

17. $\triangle KLP \sim \triangle KJM$ since right triangle $KLP \cong$ right triangle KJM and $\angle LKP \cong \angle MKJ$ (they intercept congruent arcs).

18. Show $\triangle KLP \sim \triangle KJM$. Right triangle $KLP \cong$ right triangle KJM. Vertical angles are congruent so $\angle KPL \cong \angle JPM$. Since $\overline{JP} \cong \overline{JM}$, $\angle JPM \cong \angle JMK$ so, by the transitive property $\angle KPL \cong \angle JMK$.

19. Show $\triangle HBW \sim \triangle MBL$. Because $m\angle ABW$ and $m\angle H$ are each equal to one-half the measure of the same arc, $\angle H \cong \angle ABW$. Since $ABLM$ is a parallelogram, $\overline{AB} \parallel \overline{ML}$, so $\angle ABW \cong \angle BML$. By the transitive property of congruence, $\angle H \cong \angle BML$ (angle). Angle HBW is contained in both triangles. Hence, $\angle HBW \cong \angle MBL$ (angle). Therefore, $\triangle HBW \sim \triangle MBL$, so $\dfrac{BL}{BW} = \dfrac{BM}{BH}$.

20. Show $\triangle BCD \sim \triangle ABE$. Because $m\angle DBC$ and $m\angle A$ are each equal to one-half the measure of the same arc, $\angle DBC \cong \angle A$. Angle ABE is a right angle, since a diameter is perpendicular to a chord at the point of tangency. Since $\overline{AB} \cong \overline{CD}$ and interior angles on the same side of the transversal are supplementary, $\angle DCB$ is a right angle, which means that $\angle DCB \cong \angle ABE$. Therefore, $\triangle BCD \sim \triangle ABE$, so $\frac{BD}{AE} = \frac{CD}{BE}$.

21. Show $\triangle CDA \sim \triangle BAF$. Since $\overline{BGF} \parallel \overline{CDE}$, $\angle C \cong \angle ABF$. Since B is the midpoint of $\overset{\frown}{AD}$, $\overset{\frown}{AB} \cong \overset{\frown}{BD}$ so $\angle CAD \cong \angle F$ (inscribed angles that intercept congruent arcs are congruent) and the required pair of triangles are similar. Because the lengths of corresponding sides of similar triangles are in proportion, $\frac{CD}{BA} = \frac{CA}{BF}$ and, as a result, $CD \times BF = CA \times BA$.

22. Show $\triangle MPN \sim \triangle QPR$. Angles MPN and QPR are congruent; $\angle M \cong \angle NPT$ since they are each measured by one-half of the measure of the same arc; for the same reason, $\angle Q \cong \angle RPS$. Since $\angle NPT \cong \angle RPS$, $\angle M \cong \angle Q$, which makes $\triangle MPN \sim \triangle QPR$. Because the lengths of corresponding sides of similar triangles are in proportion, $\frac{MN}{QR} = \frac{PN}{PR}$, and, as a result, $MN \times PR = PN \times QR$.

23. Show $\triangle WTK \sim \triangle JTW$. Since $\angle T$ is contained in both triangles, $\angle WTK \cong \angle JTW$ (angle). Because $m\angle NTJ$ and $m\angle JWT$ are equal to one-half of the measure of the same arc, $\angle JWT \cong \angle NTJ$. It is given that $\overline{TK}$ bisects $\angle NTW$, which means that $\angle NTJ \cong \angle JTW$. Hence, $\angle JTW \cong \angle JWT$. It is also given that $\overline{WK} \cong \overline{WT}$, so $\angle K \cong \angle JTW$. Since $\angle JTW \cong \angle JWT$ and $\angle K \cong \angle JTW$, by the transitive property of congruence, $\angle K \cong \angle JWT$ (angle). Then $\triangle WTK \sim \triangle JTW$, so $\frac{JT}{TW} = \frac{TW}{TK}$. Setting the cross-products equal gives $(TW) \times (TW) = JT \times TK$ or, equivalently, $(TW)^2 = JT \times TK$.

24. Show $\triangle OAP \sim \triangle OEB$. Right angles OAP and OEB are congruent. Because a diameter perpendicular to a chord bisects the arcs of the chord, $\overset{\frown}{AD} \cong \overset{\frown}{BD}$ so angles AOP and BOE are congruent, and the required pair of triangles are similar.

25. **a.** 36
b. 39

26. **a.** $m\angle CAD = \frac{1}{2} m\overset{\frown}{CBD}$ and $m\angle CEA = \frac{1}{2}(m\overset{\frown}{AC} + m\overset{\frown}{BD})$. Because $\overline{AC} \cong \overline{BC}$, $\overset{\frown}{AC} \cong \overset{\frown}{BC}$, so by substition, $m\angle CEA = \frac{1}{2}(m\overset{\frown}{BC} + m\overset{\frown}{BD}) = \frac{1}{2} m\overset{\frown}{CBD}$. Because $m\angle CAD = \frac{1}{2} m\overset{\frown}{CBD}$ and $m\angle CEA = \frac{1}{2} m\overset{\frown}{CBD}$, $\angle CAD \cong \angle CEA$.

b. $\angle CAD \cong \angle CEA$ (from part a) and $\angle ACD \cong \angle ACE$ (reflexive property of congruence), so $\triangle CDA \sim \triangle ACE$ by AA $\cong$ AA. Thus, $\frac{CD}{AC} = \frac{AC}{CE}$ and, as a result, $(AC)^2 = CE \times CD$.

CHAPTER 8

Lesson 8.1

1. (1) **4.** (3) **7.** (2) **10.** (2)
2. (1) **5.** (3) **8.** (3) **11.** (4)
3. (3) **6.** (4) **9.** (3)

12. If a transformation is an isometry, then it is a line reflection. Counterexample: A *translation* is an isometry, but it is not a reflection.

13. **a.** $\triangle HOC$
b. Direct

Lesson 8.2

1. (4) **3.** (3) **5.** (3) **7.** (1) **9.** (3)
2. (4) **4.** (4) **6.** (2) **8.** (4) **10.** (1)

Lesson 8.3

1. (4) **3.** (1) **5.** (4) **7.** (4) **9.** (1)
2. (3) **4.** (4) **6.** (4) **8.** (2) **10.** (4)

11. Reflection in $\overline{EMB}$.
12. (x, y) $(x - 7, y + 9)$
13. Glide reflection (a reflection in the x-axis followed by a translation)
14. Reflection in point M
15. $(-7, 6)$
16. $(4, -4)$
17. $P'(-1, 5)$
18. **a.** Graph $ABCD$ and $A'B'C'D'$, where $A'(-2, 6)$, $B'(-7, 8)$, $C'(-9, 3)$, and $D'(-4, 1)$.
b. 29
19. $A'(-1, 1)$, $B'(4, -2)$, $C'(3, -5)$, $D'(-2, -2)$. Show midpoint $\overline{A'C'}$ = midpoint $\overline{B'D'} = (1, -2)$.
20. **a.** $h = 5$, $k = 3$
b. 22.5
c. 90
21. **a.** Graph $\triangle P'E'N'$ where $P'(-1, 2)$, $E'(-3, 0)$, and $N'(-6, 4)$.
b. Graph $\triangle P''E''N''$ where $P''(5, -1)$, $E''(7, -3)$, and $N''(10, 1)$.
22. **a.** Graph $\triangle S'A'M'$ where $S'(4, 3)$, $A'(-5, 3)$, and $M'(-2, -4)$.
b. Graph $\triangle S''A''M''$ where $S''(6, 8)$, $A''(6, -10)$, and $M''(-8, -4)$; 1:4.

Lesson 8.4

1. (1)
2. (3)
3. (1)
4. (2)
5. (4)
6. (3)
7. (2)
8. (3)
9. (4)
10. (1)
11. (1)
12. (2)
13. (4)
14. (4)
15. (3)
16. (3)
17. (1)
18. (1)
19. (–5, 1)
20. $P'(5, -4)$
21. $T_{3,2} \circ r_{y\text{-axis}}(\triangle RUN) \rightarrow \triangle GEM$
22. (3, 4)
23. Graph $A''(0, -5)$ and $B''(-2, 0)$; reflection in the origin.
24. **a.** Graph $A'(2, -1)$, $B'(2, 6)$, $C'(4, 3)$.
b. $r_{y=x}$
25. **a.** Graph $A'(0, 0)$, $B'(-1, 8)$, $C'(-4, 8)$. Choice (1).
b. Graph $A''(0, 0)$, $B''(8, 5)$, $C''(8, 8)$. Choice (4).
26. **a.** E
b. $\overline{DE}$
c. C

CHAPTER 9

Lesson 9.1

1. (4)
2. (4)
3. (2)
4. (3)
5. (1)
6. (2)
7. (1)
8. $x = 2$
9. $y = 2$
10. $y = 3$
11. $x = -4$
12. $x = \pm 2$
13. $x = -2, x = 8$
14. $y = -5, y = 3$

Lesson 9.2

1. (2)
2. (4)
3. (2)
4. (1)
5. (1)
6. (1)
7. (3)
8. (2)
9. (3)
10. (4)
11. (2)
12. (3)
13. (1)
14. (1)
15. (3)
16a. 0
b. 2
17. 2
18. 4
19. 4
20. 1
21. **a.** $x = 1, x = 5$
b. Circle with center at (3, 2) and radius of n units
c. 2
22. **b.** 3
23. **b.** (1) 2 (2) 0 (3) 4

Lesson 9.3

1. (3)
2. (4)
3. (1)
4. (3)
5. (2)
6. (4)
7b. 1
c. (2)

Lesson 9.4

1. (2) **2.** (4) **3.** (4) **4.** (2) **5.** (3)

6. a. $y - 5 = -\frac{2}{3}(x + 3)$

b. (0, 3)

7. $y = 3x + 4$

8. a. $y - 1 = \frac{3}{2}(x - 1)$

b. 10

9. a. $y + 6 = -\frac{2}{3}(x - 1)$

b. $y = \frac{3}{2}x$

c. $y = x + \frac{4}{3}$

10. a. $y = x + 1$

b. Show (8, 9) satisfies the equation in part a.

11. a. $m = 2$

b. $y = 2x - 1$

c. (2)

12. a. $y = -x + 14$

b. (7, 3)

13. a. $A'(8, 8)$; $y = x$

b. $B'(3, 6)$

c. Midpoint $\overline{AA'} = (5, 5)$ and midpoint $\overline{BB'} = (4.5, 4.5)$. Since the diagonals do not have the same midpoint, $ABA'B'$ is *not* a parallelogram.

14. a. $y = \frac{1}{3}x + \frac{10}{3}$

b. Slope $\overline{PM} = \frac{1}{3}$ and slope $\overline{QR} = -3$. Since slopes are negative reciprocals, the line segments are perpendicular.

c. Because $\triangle PQR$ is isosceles.

15. a. $y = -x + 6$

b. Show (5, 1) satisfies the equation in part a.

16. a. $y = -x + 4$

b. $y = 4$

c. $x = 0$

17. a. $y = 5x - 4$

b. $y = 5x - 27$

c. Lines have the same slope so they are parallel.

Lesson 9.5

1. (4) **3.** (4) **5.** (1) **7.** (1) **9.** (3)
2. (2) **4.** (2) **6.** (2) **8.** (1) **10.** (3)
11. $(x-2)^2+(y+3)^2=25$
12. a. Show $AB = BC = CD = AD = 10$. Because $ABCD$ is equilateral, it is a rhombus.
b. $(x-8)^2+(y-1)^2=100$
13. a. $x^2+y^2=25$
b. $y=\frac{3}{4}x+\frac{25}{4}$, y-intercept is $\frac{25}{4}$.
14. a. $y=2$
b. $(x-2)^2+(y+2)^2=16$. One point.
15. b. $y=\frac{4}{3}x-9$
16. a. $(x-3)^2+(y-2)^2=n^2$
b. $x=1, x=5$
c. (1) 0, (2) 2
17. a. Graph circle B: $(x-5)^2+(y+3)^2=9$
b. Graph circle C: $(x-5)^2+(y-3)^2=9$. Area $\triangle ABC = 15$ square units.
18. b. $(x+1)^2+(y-1)^2=16$
c. $y=-2x-1$
19. (4, 3), (0, 5)
20. (3, 2), (0, –1)
21. (10, –3), (8, –5)
22. (1, 3), (–7, 7)
23. (–4, –1), (1, 4)
24. (–4, 0), (1, 5)
25. (2, –3), (6, 5)
26. $x=7$

CHAPTER 10

Lesson 10.1

1. (3) **2.** (1) **3.** (1) **4.** (1) **5.** (4) **6.** (2)
7. a. 84 cm^2
b. 42 cm^3
8. a. 504 in^2
b. $336\sqrt{3}$ in^3
9. a. 120π in^2
b. 1,131.0 in^3
10. a. 260π cm^2
b. 2,654.6 cm^3
11. 720 cm^2

12. 288 in^3
13. 612π
14. 1,584 in^2
15. 7.25
16. 76.9
17. 45
18. 4.5 ft
19. $V = 288\pi$ in^3, SA = 168π in^2
20. 47

Lesson 10.2

1. (2) **2.** (4) **3.** (1) **4.** (1) **5.** (3)
6. 720 cm^2, 1,728 cm^3
7. **a.** 868 cm^2
b. 1,499.8 cm^3
8. 128π in^3
9. 765.2 cm^3
10. 1,082.8 ft^3
11. $\frac{512\pi}{3}$ in^3
12. 9,600 ft^3
13. **a.** 720 cm^2
b. 1,260 cm^3
14. 2,160 cm^2
15. 10,936 cm^3
16. **a.** 20 in
b. $V = 1{,}176\pi$ in^3
17. 40 in^3, 95 in^3

Lesson 10.3

1. (4) **3.** (1) **5.** (2) **7.** (3)
2. (1) **4.** (2) **6.** (3) **8.** (3)
9. **a.** 5-inch box
b. No. The surface area of the box is approximately 1.04 ft^2.
10. 13.5
11. 103 in^3
12. 3.75
13. 2,752 in^3
14. 3,654 cm^2
15. 576 cm^3
16. 353 mm^3
17. 12.3 in^3

Glossary of Geometry Terms

A

Acute angle An angle whose degree measure is less than 90 and greater than 0.

Acute triangle A triangle with three acute angles.

Adjacent angles Two angles with the same vertex, a common side, and no interior points in common.

Altitude In a triangle, a segment that is perpendicular to the side to which it is drawn.

Angle The union of two rays that have the same endpoint.

Angle of rotational symmetry The smallest positive angle through which a figure with rotational symmetry can be rotated to coincide with itself. For a regular n–polygon, this angle is $\frac{360}{n}$.

Apothem The radius of the inscribed circle of a regular polygon.

Arc A part of a circle whose endpoints are two distinct points of the circle. If the degree measure of the arc is less than 180, the arc is a **minor arc**. If the degree measure of the arc is greater than 180, the arc is a **major arc**. A **semicircle** is an arc whose degree measure is 180.

B

Biconditional A statement of the form "p if and only if q" where statement p is the hypothesis of a conditional statement and statement q is the conclusion. A biconditional represents the conjunction of a conditional statement and its converse. It is true only when both parts of the biconditional have the same truth values.

Bisect To divide into two congruent parts.

C

Center of a regular polygon The point in the interior of the polygon that is equidistant from each of the vertices. It is also the common center of its inscribed and circumscribed circles.

Center–radius equation of circle The equation $(x - h)^2 + (y - k)^2 = r^2$ where (h, k) is the center of a circle with radius r.

Central angle of a circle An angle whose vertex is at the center of a circle and whose sides contain radii.

Central angle of a regular polygon An angle whose vertex is the center of the polygon and whose sides are drawn to consecutive vertices of the polygon.

Centroid of a triangle The point of intersection of its three medians.

Chord of a circle A line segment whose endpoints are points on a circle.

Circle The set of points (x, y) in the plane that are a fixed distance r from a given point (h, k) called the *center*. An equation of the circle is

$$(x - h)^2 + (y - k)^2 = r^2$$

Circumcenter of a triangle The center of the circle that can be circumscribed about a triangle. It can be located by finding the point of intersection of the perpendicular bisectors of two of its sides.

Circumscribed circle about a polygon A circle that has each vertex of a polygon on it.

Circumscribed polygon about a circle A polygon that has all of its sides tangent to the circle.

Collinear points Points that lie on the same line.

Common external tangent A line tangent to two circles that does *not* intersect the line joining their centers at a point between the two circles.

Common internal tangent A line tangent to two circles that intersects the line joining their centers at a point between the two circles.

Complementary angles Two angles whose degree measures add up to 90.

Composition of transformations A sequence of transformations in which one transformation is applied to the image of another transformation.

Concave polygon A polygon in which there are two points in the interior of the polygon such that the line through them, when extended, intersects the polygon in more than two points. A concave polygon contains at least one interior angle that measures more than 180.

Concentric circles Circles in the same plane with the same center but unequal radii.

Conclusion In a conditional statement of the form "If . . . then . . . ," the statement that follows "then."

Concurrent When three or more lines intersect at the same point.

Conditional statement A statement of the form "If p, then q." A conditional statement is always true except in the single instance in which statement p (hypothesis) is true and statement q (the conclusion) is false.

Cone A solid figure with a circular base and a curved lateral surface that joins the base to a point in a different plane called the *vertex*.

Congruent angles (or sides) Angles (or sides) that have the same measure. The symbol "is congruent to" is $\cong$.

Congruent circles Circles with congruent radii.

Congruent parts Pairs of angles or sides that are equal in measure and, as a result, are congruent.

Congruent polygons Two polygons with the same number of sides such that all corresponding angles are congruent and all corresponding sides are congruent.

Congruent triangles Two triangles that agree in all of their corresponding parts. Two triangles are congruent if any one of the following conditions is true: the three sides of one triangle are congruent to the corresponding sides of the other triangle (SSS $\cong$ SSS); two sides and the included angle of one triangle are congruent to the corresponding parts of the other triangle (SAS $\cong$ SAS); two angles and the included side of one triangle are congruent to the corresponding parts of the other triangle (ASA $\cong$ ASA); two angles and the side opposite one of these angles is congruent to the corresponding parts of the other triangle (AAS $\cong$ AAS).

Conjunction A statement that uses "and" to connect two other statements called *conjuncts*. A conjunction is true only when both conjuncts are true.

Contrapositive The statement formed by interchanging and then negating the "If" and "then" parts of a conditional statement. Symbolically, the contrapositive of $p \rightarrow q$ is $\sim q \rightarrow \sim p$.

Converse The statement formed by interchanging the "If" and "then" parts of a conditional statement. Symbolically, the converse of $p \rightarrow q$ is $q \rightarrow p$.

Convex polygon A polygon for which any line that passes through it, provided the line is not tangent to a side or a corner point, intersects the polygon in exactly two points. Each interior angle of a convex polygon measures less than 180. A nonconvex polygon is called a *concave* polygon.

Coplanar A term applied to figures that lie in the same plane.

Corollary A theorem that is a direct consequence of another theorem and can be easily proved from it.

Counterexample A specific example that disproves a statement.

Cylinder A solid figure with congruent circular bases that lie in parallel planes connected by a curved lateral surface.

D

Deductive reasoning A logical chain of reasoning that uses general principles and accepted facts to reach a specific conclusion.

Degree A unit of angle measurement defined as $\frac{1}{360}$th of one complete rotation of a ray about its vertex.

Diagonal of a polygon A line segment whose endpoints are two nonconsecutive vertices of the polygon.

Diameter A chord of a circle that contains the center of the circle.

Dilation A transformation in which a figure is enlarged or reduced in size according to a given scale factor.

Direct isometry An isometry that preserves orientation.

Disjunction A statement that uses "or" to connect two other statements called *disjuncts*. A disjunction is true when at least one of the disjuncts is true.

E

Edge of a solid A segment that is the intersection of two faces of the solid.

Equidistant A term that means "same distance."

Equilateral triangle A triangle in which the three sides have the same length.

Equivalence relation A relation in which the reflexive ($a = a$), symmetric ($a = b$ and $b = a$), and transitive properties (if $a = b$ and $b = c$, then $a = c$) hold. Congruence, similarity, and parallelism are equivalence relations.

Exterior angle of a polygon An angle formed by a side of the polygon and the extension of an adjacent side of the polygon.

Externally tangent circles Circles that lie on opposite sides of their common tangent.

Extremes The first and last terms of a proportion. In the proportion $\frac{a}{b} = \frac{c}{d}$, the terms a and d are the extremes.

F

Frustrum The part of a cone that remains after a plane parallel to the base of the cone slices off a part of the cone below its vertex.

G

Geometric mean See *mean proportional.*

Glide reflection The composite of a line reflection and a translation whose direction is parallel to the reflecting line.

Great circle The largest circle that can be drawn on a sphere.

H

Hypotenuse In a right triangle, the side opposite the right angle.

Hypothesis In a conditional statement of the form "If . . . then . . . ," the statement that follows "If."

I

Image The result of applying a geometric transformation to an object called the *preimage*.

Incenter of a triangle The center of the inscribed circle of a triangle that represents the common point at which its three angle bisectors intersect.

Indirect proof A deductive method of reasoning that eliminates all but one possibility for the conclusion.

Inductive reasoning The process of reasoning from a few specific cases to a broad generalization. Conclusions obtained through inductive reasoning may or may not be true.

Inscribed angle An angle of a circle whose vertex is a point on a circle and whose sides are chords.

Inscribed circle of a triangle A circle drawn so that the three sides of the triangle are tangent to the circle.

Internally tangent circles Tangent circles that lie on the same side of their common tangent.

Inverse of a statement The statement formed by negating the "If" and "then" parts of a conditional statement. Symbolically, the inverse of $p \to q$ is $\sim p \to \sim q$.

Isometry A transformation that preserves distance. Reflections, translations, and rotations are isometries. A dilation is *not* an isometry.

Isosceles triangle A triangle in which two sides have the same length.

K

Kite A quadrilateral in which two pairs of adjacent, not opposite, sides are congruent.

L

Lateral edge An edge of a prism or pyramid that is not a side of a base.

Lateral face A face of a prism or pyramid that is not a base.

Leg of a right triangle Either of the two sides of a right triangle that include the right angle.

Linear pair Two adjacent angles whose exterior sides are opposite rays.

Line of symmetry A line that divides a figure into two congruent reflected parts.

Line reflection A transformation in which each point P on one side of the reflecting line is paired with a point P' on the opposite side of it so that the reflecting line is the perpendicular bisector of $\overline{PP'}$. If P is on the reflecting line, then P' coincides with P.

Line symmetry When a figure can be reflected in a line so that the image coincides with the original figure.

Locus The set of points, and only those points, that satisfy a given condition. The plural of *locus* is *loci*.

Logically equivalent Statements that always have the same truth values. A statement and its contrapositive are logically equivalent.

M

Major arc An arc whose degree measure is greater than 180 and less than 360.

Mapping A pairing of the elements of one set with the elements of another set. A mapping between sets A and B is one–to–one if every member of A corresponds to exactly one member of B, and every member of B corresponds to exactly one member of A.

Mean proportional When the second and third terms of a proportion are equal, either of these terms is the mean proportional between the first and fourth terms of the proportion. In $\frac{a}{x} = \frac{x}{d}$, x is the mean proportional between a and d where $x = \sqrt{ad}$.

Midsegment The line segment whose endpoints are the midpoints of two sides of a triangle.

Minor arc An arc whose degree measure is between 0° and 180°.

N

n-gon A polygon with n sides.

Negation of a statement The statement with the opposite truth value. Symbolically, the negation of statement p is $\sim p$.

Noncollinear points Points that do not all lie on the same line.

Non–Euclidean geometries Geometry systems that do not accept the Parallel Postulate.

O

Orthocenter of a triangle The point of intersection of the three altitudes of a triangle.

P

Parallel lines Coplanar lines that do not intersect.

Parallelogram A quadrilateral that has two pairs of parallel sides. In a parallelogram, opposite sides are congruent, opposite angles are congruent, consecutive angles are supplementary, and diagonals bisect each other.

Parallel Postulate Euclid's controversial assumption that through a point not on a line exactly one line can be drawn parallel to the given line.

Perpendicular bisector A line that is perpendicular to a line segment at its midpoint.

Perpendicular lines Two lines that intersect to form right angles.

Point–slope equation of a line An equation of a line with the form $y - b = m(x - a)$, where m is the slope of the line and (a, b) is a point on the line.

Point symmetry A figure has point symmetry if after it is rotated 180° (a half-turn) the image coincides with the original figure.

Polygon A closed plane figure bounded by line segments that intersect only at their endpoints.

Polyhedron A closed solid figure in which each side is a polygon.

Postulate A statement that is accepted as true without proof.

Preimage The original figure in a transformation. If A' is the image of A under a transformation, then A is the preimage of A'.

Prism A polyhedron whose faces, called *bases*, are congruent polygons in parallel planes.

Proportion An equation that states that two ratios are equal. In the proportion $\frac{a}{b} = \frac{c}{d}$, a and d are called the *extremes* and b and c are called the *means*. In a proportion, the product of the means is equal to the product of the extremes. Thus, $a \times d = b \times c$.

Pyramid A solid figure formed by joining the vertices of a polygon base to a point in a different plane known as the *vertex*.

Pythagorean Theorem In a right triangle, the square of the length of the hypotenuse is equal to the sum of the squares of the lengths of the two legs.

Q

Quadrilateral A polygon with four sides.

R

Ray Part of a line consisting of an endpoint and the set of all points on one side of the endpoint.

Rectangle A parallelogram with four right angles.

Reflection A transformation that produces an image that is the mirror image of the original object.

Reflection rules Reflections of points in the coordinate axes are given by the following rules: $r_{x\text{-axis}}(x, y) = (x, -y)$, $r_{y\text{-axis}}(x, y) = (-x, y)$. To reflect a point in the origin, use the rule $r_{\text{origin}}(x, y) = (-x, -y)$. See also *line reflection.*

Reflexive property A quantity is equal (or congruent) to itself.

Regular polygon A polygon that is both equilateral and equiangular, such as a square.

Rhombus A parallelogram with four congruent sides.

Right angle An angle whose degree measure is 90.

Right triangle A triangle that contains a right angle.

Rotation A transformation in which a point or figure is turned a given number of degrees about a fixed point.

Rotation rules The images of points rotated about the origin through angles that are multiples of 90° are given by the following rules: $R_{90°}(x, y) = (-y, x)$, $R_{180°}(x, y) = (-x, -y)$, and $R_{270°}(x, y) = (y, -x)$.

Rotational symmetry When a figure can be rotated through a positive angle of less than 360° so that the image coincides with the original figure.

S

Scalene triangle A triangle in which no two sides are congruent.

Secant line A line that intersects a circle in two different points.

Sector of a circle The interior region of a circle bounded by two radii and their intercepted arc.

Segment of a circle The interior region of a circle bounded by a chord of a circle and its intercepted arc.

Semicircle An arc whose degree measure is 180.

Similar polygons Polygons with the same shape. Similar polygons have congruent corresponding angles and corresponding sides that are in proportion.

Skew lines Lines in different planes that do not intersect but are not parallel.

Slope A numerical measure of the steepness of a line. The slope of a vertical line is undefined.

Slope–intercept equation of a line An equation of a line with the form $y = mx + b$, where m is the slope of the line and b is the y-intercept.

Sphere The set of all points in space that are at a fixed distance from a given point called the *center* of the sphere.

Square A parallelogram with four right angles and four congruent sides.

Substitution property A quantity may be replaced by its equal in an equation.

Supplementary angles Two angles whose degree measures add up to 180.

Symmetric property If $a = b$, then $b = a$.

T

Tangent circles Circles in the same plane that are tangent to the same line at the same point. *Internally tangent circles* lie on the same side of the common tangent. *Externally tangent circles* lie on opposite sides of the common tangent.

Tangent of a circle A line in the same plane as the circle that intersects it in exactly one point.

Theorem A mathematical generalization that can be proved.

Transformation A change in the position, size, or shape of a figure according to some given rule.

Transitive property If $a = b$ and $b = c$, then $a = c$.

Translation A transformation in which each point of a figure is shifted the same distance and in the same direction. The transformation $T_{h,k}$ slides a point h units horizontally and k units vertically. Thus, $T_{h,k}(x, y) = (x + h, y + k)$.

Transversal A line that intersects two or more other lines in different points.

Trapezoid A quadrilateral in which exactly one pair of sides is parallel. The nonparallel sides are called *legs*.

Triangle inequality In any triangle, the length of each side must be less than the sum of the lengths of the other two sides, and greater than their difference.

Truth value For a statement, either true or false, but not both.

V

Vertex angle In an isosceles triangle, the angle formed by the two congruent sides.

Vertex of a polygon The point at which two sides of the polygon intersect. The plural of *vertex* is *vertices*.

Vertical angles Opposite pairs of congruent angles formed when two lines intersect.

Volume The amount of space a solid figure occupies as measured by the number of nonoverlapping $1 \times 1 \times 1$ unit cubes that can exactly fill its interior.

Some Geometric Relationships Worth Remembering

Pairs of Angles

- Vertical angles are congruent:

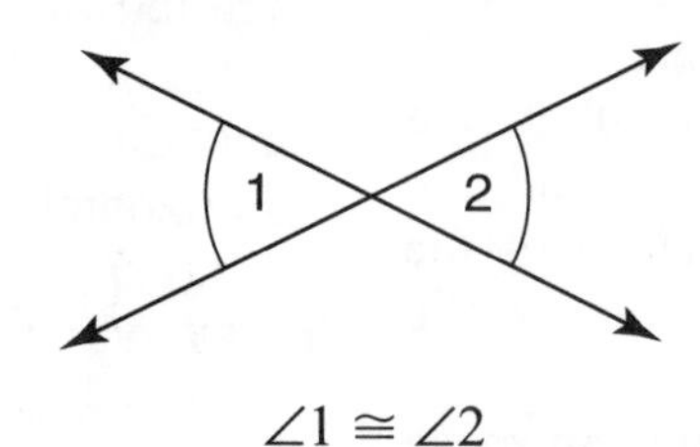

$\angle 1 \cong \angle 2$

- $p \parallel q$ If two parallel lines are cut by a transversal, any two of the eight angles that are formed are either congruent or supplementary.

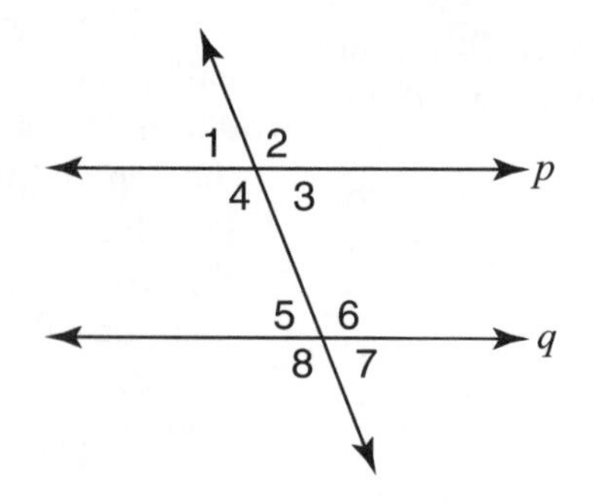

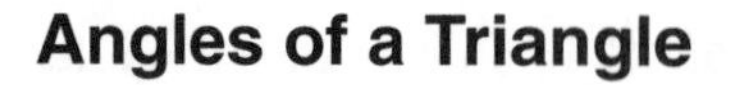

Angles of a Triangle

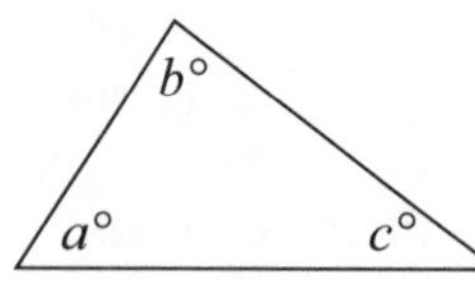

$a + b + c = 180$

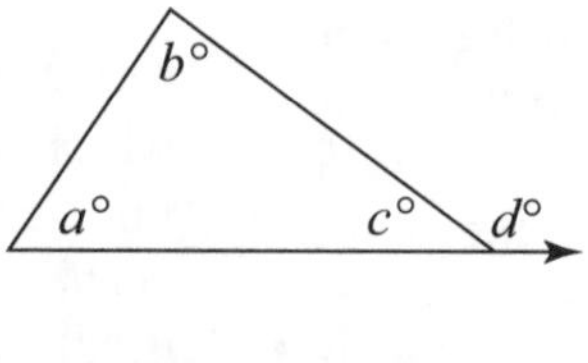

$d = a + b$

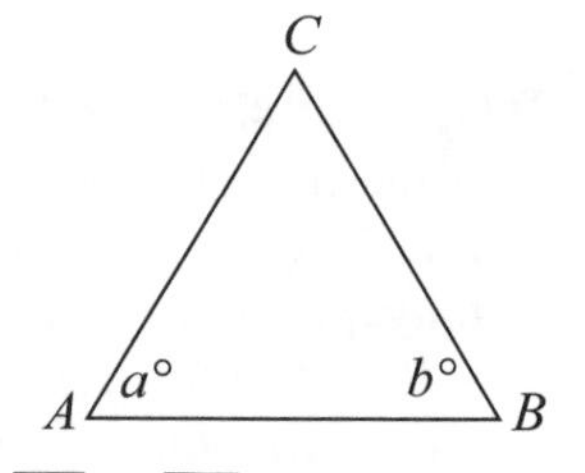

$\overline{AC} \cong \overline{BC} \Rightarrow a = b$
and
$a = b \Rightarrow \overline{AC} \cong \overline{BC}$

Angles of a Polygon

Polygon with n Sides	**Regular Polygon with n sides**
Sum of exterior angles = 360	Each exterior angle = $\frac{360}{n}$
Sum of interior angles = $(n-2) \times 180$	Each interior angle = $180 - \frac{360}{n}$

Inequalities in a Triangle

- If $x > y$, then $a > b$.
- If $a > b$, then $x > y$.
- $d > a$ and $d > c$.
- $z < x + y$, $x < y + z$, and $y < x + z$.

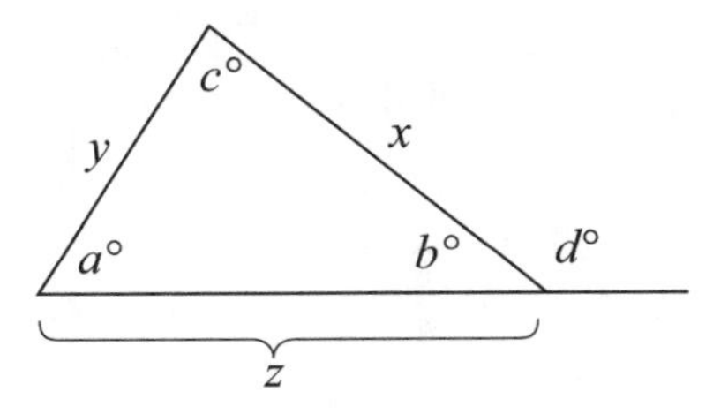

Proving Triangles Congruent

Two triangles are congruent if any one of the following is true:

- ASA ≅ ASA
- SAS ≅ SAS
- SSS ≅ SSS
- AAS ≅ AAS
- Hy–Leg ≅ Hy–Leg (only for right triangles)

Two triangles are *not* congruent when SSA ≅ SSA or AAA ≅ AAA.

Proving Triangles Similar

Two triangles are similar if any one of the following is true:

- AA ≅ AA
- Corresponding sides are in proportion.
- The lengths of two pairs of sides are in proportion and their included angles are congruent.

Properties of Parallelograms

In a parallelogram:

- Opposite sides are parallel.
- Opposite sides and opposite angles are congruent.
- Diagonals bisect each other.

Properties of Special Quadrilaterals

Property	Rectangle	Rhombus	Square	Isosceles Trapezoid
• All the properties of a parallelogram	✓	✓	✓	
• Equiangular (four right angles)	✓		✓	
• Equilateral (four congruent sides)		✓	✓	
• Congruent diagonals	✓		✓	✓
• Diagonals bisect opposite angles		✓	✓	
• Diagonals intersect at right angles		✓	✓	

Right Triangle Relationships

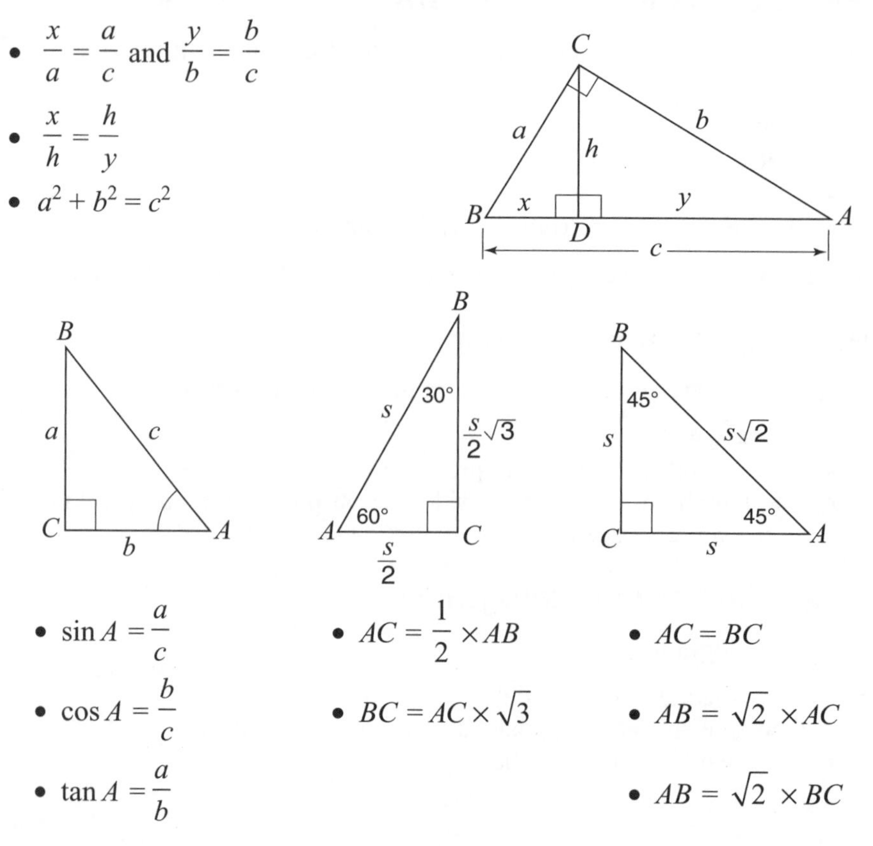

- $\frac{x}{a} = \frac{a}{c}$ and $\frac{y}{b} = \frac{b}{c}$
- $\frac{x}{h} = \frac{h}{y}$
- $a^2 + b^2 = c^2$

- $\sin A = \frac{a}{c}$
- $\cos A = \frac{b}{c}$
- $\tan A = \frac{a}{b}$

- $AC = \frac{1}{2} \times AB$
- $BC = AC \times \sqrt{3}$

- $AC = BC$
- $AB = \sqrt{2} \times AC$
- $AB = \sqrt{2} \times BC$

Midpoint and Centroid Relationships

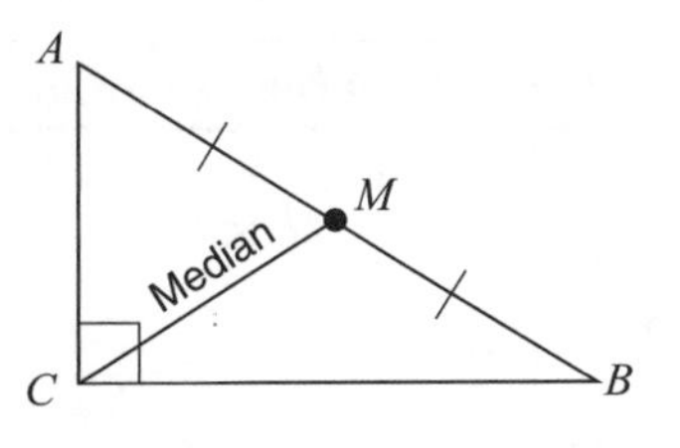

- In right $\triangle ABC$, $CM = \frac{1}{2}AB$.

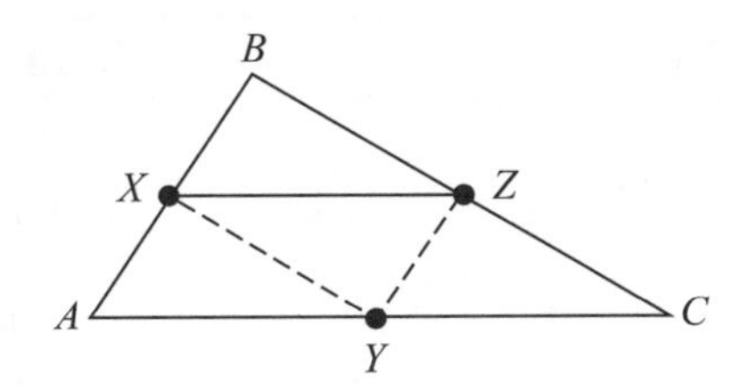

- If X and Z are midpoints, then $\overline{XZ} \parallel \overline{AC}$ and $YZ = \frac{1}{2}AC$.
- If X, Y, and Z are midpoints, then perimeter $\triangle XYZ = \frac{1}{2}$ perimeter of $\triangle ABC$.

The three medians of a triangle are concurrent at a point P, called the **centroid** of the triangle, such that each median is divided into segments whose lengths are in the ratio of 2:1. The distance from each vertex to the centroid is two-thirds of the length of the entire median drawn from that vertex.

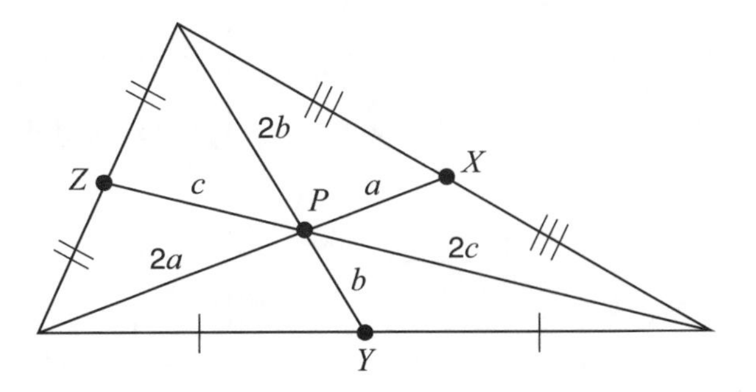

Area Formulas for Special Polygons

Polygon	Diagram	Area Formula
Triangle	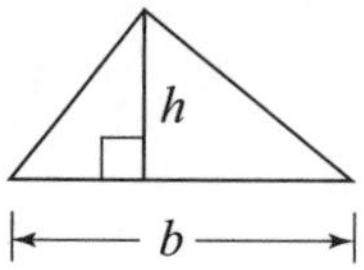 h, b	$A = \frac{1}{2} \times b \times h$
Equilateral triangle	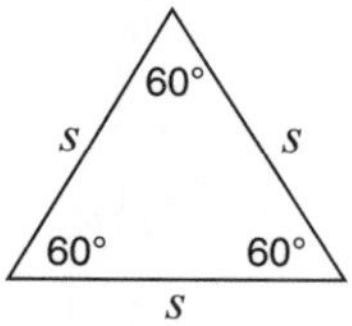 60°, 60°, 60°, s, s, s	$A = \frac{s^2}{4}\sqrt{3}$
Parallelogram and Rectangle	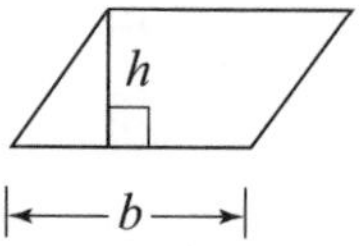 h, b	$A = b \times h$
Rhombus	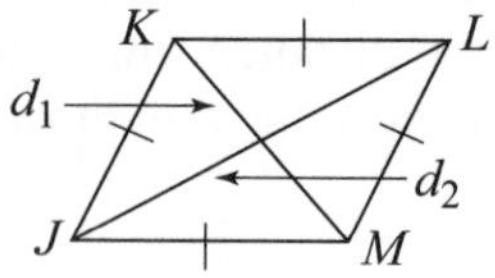 K, L, J, M, d_1, d_2	$A = \frac{1}{2} \times d_1 \times d_2$
Square	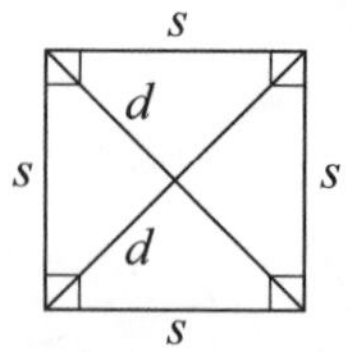 s, s, s, s, d, d	$A = s^2$ or $A = \frac{1}{2} \times d^2$
Trapezoid	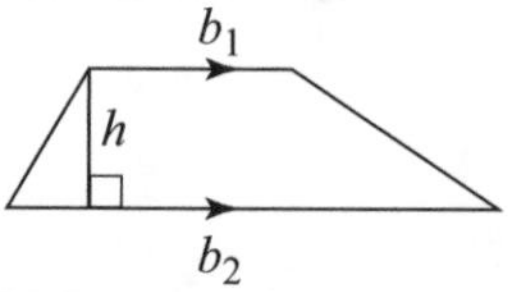 b_1, h, b_2	$A = \frac{1}{2} \times h(b_1 + b_2)$
Regular n-sided polygon	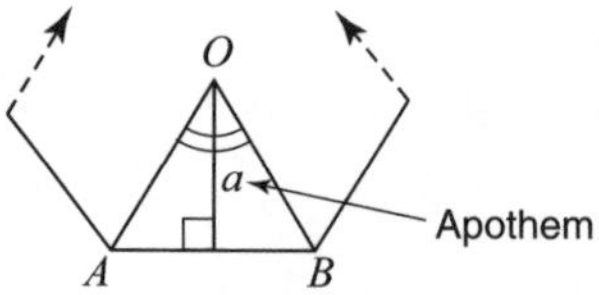 O, a, Apothem, A, B	Central $\angle AOB = \frac{360}{n}$ Area $= \frac{1}{2} \times a \times$ perimeter

Circle Formulas

- $L = \frac{n}{360} \times \overbrace{(2\pi r)}^{\text{Circumference}}$

- Area $AOB = \frac{n}{360} \times \overbrace{(\pi r^2)}^{\text{Area of circle}}$

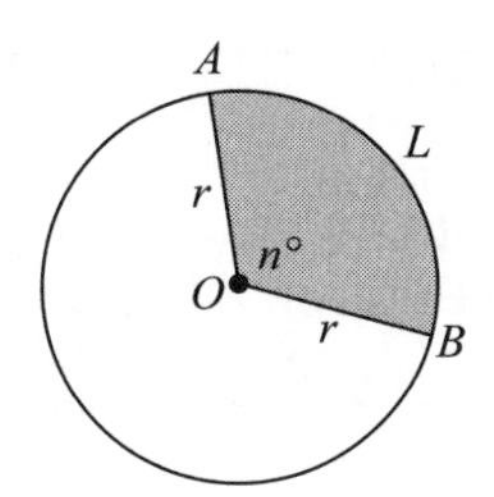

Chord, Tangent, and Secant Relationships

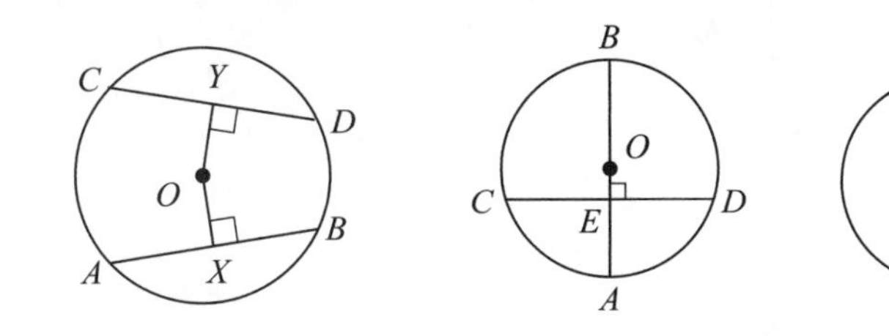

$\overline{AB} \cong \overline{CD} \Rightarrow OX = OY$
and
$OX = OY \Rightarrow \overline{AB} \cong \overline{CD}$

$\overline{AB} \perp \overline{CD} \Rightarrow \overline{CE} \cong \overline{DE}$, $\widehat{AC} \cong \widehat{AD}$, and $\widehat{BC} \cong \widehat{BD}$

$\overline{PA} \cong \overline{PB}$
and
$\overline{OP}$ bisects $\angle APB$

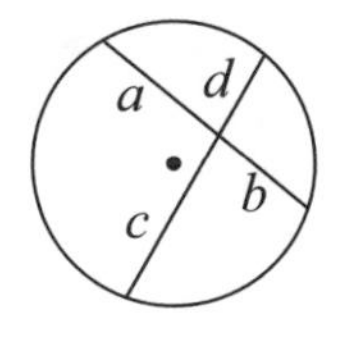

$a \times b = c \times d$

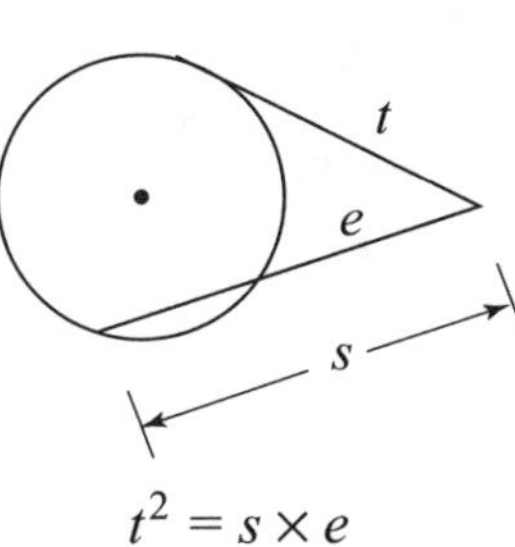

$t^2 = s \times e$

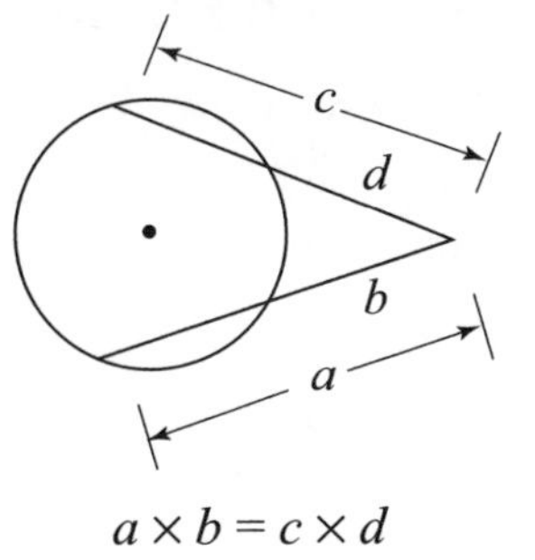

$a \times b = c \times d$

Circle Angle Measurement

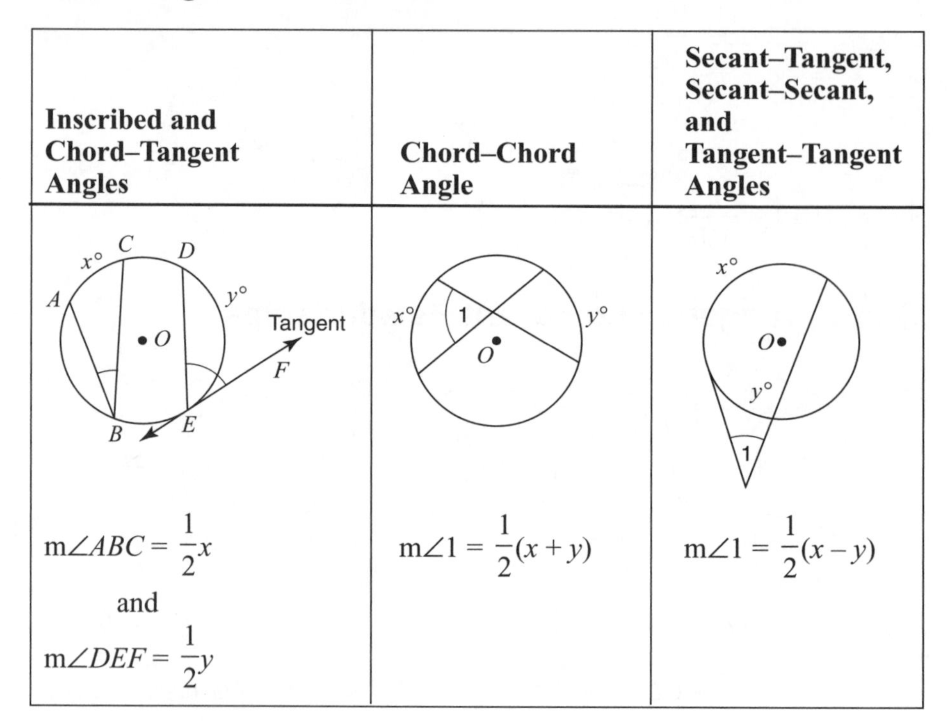

Inscribed and Chord–Tangent Angles	Chord–Chord Angle	Secant–Tangent, Secant–Secant, and Tangent–Tangent Angles
$m\angle ABC = \frac{1}{2}x$ and $m\angle DEF = \frac{1}{2}y$	$m\angle 1 = \frac{1}{2}(x + y)$	$m\angle 1 = \frac{1}{2}(x - y)$

Coordinate Formulas

Given $A(x_A, y_A)$ and $B(x_B, y_B)$:

- Midpoint of $\overline{AB} = \left(\frac{x_A + x_B}{2}, \frac{y_A + y_B}{2}\right)$
- Slope of $\overline{AB} = \frac{\Delta y}{\Delta x} = \frac{y_B - y_A}{x_B - x_A}$
- Length of $\overline{AB} = \sqrt{(x_B - x_A)^2 + (y_B - y_A)^2}$

General Equations:

- Slope–intercept equation of a line: $y = mx + b$
- Point–slope equation of a line: $y - y_A = m(x - x_A)$
- Circle with center (h, k) and radius r: $(x - h)^2 + (y - k)^2 = r^2$

Volume and Area Formulas for Solids

Let h = height, ℓ = slant height, p = perimeter, r = radius, and B = area of a base.

Solid Figure*	**Volume (V)**	**Area†**
Prism	$V = B \times h$	L.A. $= hp$
Pyramid	$V = \frac{1}{3} B \times h$	L.A. $= \frac{1}{2} p\ell$
Cylinder	$V = \pi r^2 h$	L.A. $= 2\pi rh$
Cone	$V = \frac{1}{3} \pi r^2 h$	L.A. $= \pi r\ell$
Sphere	$V = \frac{4}{3} \pi r^3$	S.A. $= 4\pi r^2$

*Volume formulas hold for both right and oblique solids (prisms, cylinders, and cones) and for both regular and nonregular pyramids.
†L.A. = lateral area and S.A. = surface area.

Properties of Transformations

Transformation	Properties Preserved	Isometry*	Coordinate Rules
Line Reflection	Collinearity Angle measure Distance	Opposite	$r_{x\text{-axis}}(x, y) = (x, -y)$ $r_{y\text{-axis}}(x, y) = (-x, y)$ $r_{\text{origin}}(x, y) = (-x, -y)$ $r_{y=x}(x, y) = (y, x)$
Translation	Collinearity Angle measure Distance Orientation	Direct	$T_{h,k}(x, y) = (x + h, y + k)$
Rotation	Collinearity Angle measure Distance Orientation	Direct	$R_{90^\circ}(x, y) = (-y, x)$ $R_{180^\circ}(x, y) = (-x, -y)$ $R_{270^\circ}(x, y) = (y, -x)$
Dilation	Collinearity Angle measure Orientation	Image is similar to the original figure	$D_k(x, y) = (kx, ky)$ where k is the scale factor
Glide reflection	Collinearity Angle measure Distance	Opposite	

*Isometry is a transformation that produces a congruent image. A *direct* isometry preserves orientation, and an *opposite* isometry reverses orientation.

Appendix I
More Facts About Planes

- If a line is perpendicular to each of two intersecting lines at their point of intersection, then the line is perpendicular to the plane determined by them.

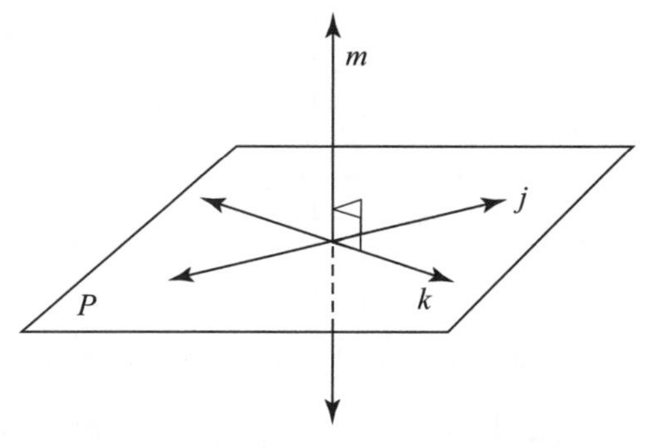

If $m \perp j$ and $m \perp k$, then $m \perp P$.

- Through a given point there passes one and only one plane perpendicular to a given line.

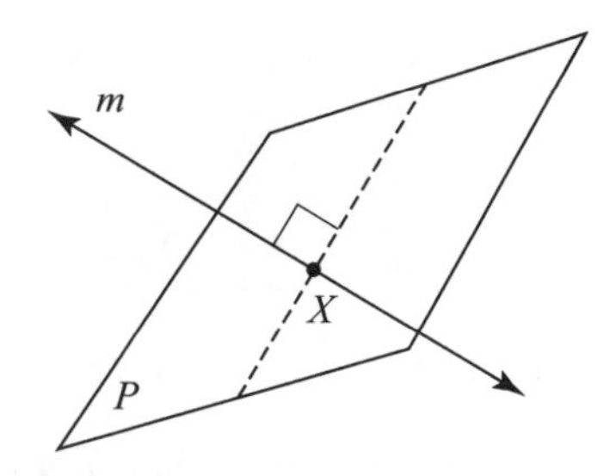

At point X on line m, only plane $P \perp m$.

- Through a given external point, there passes one and only one line perpendicular to a given plane.

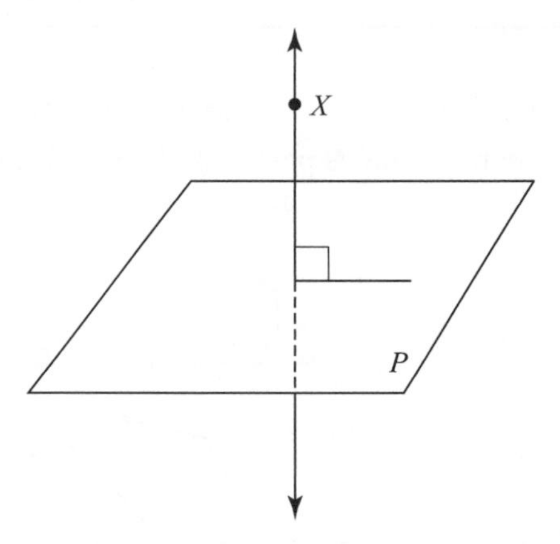

Only one line can be drawn through X and perpendicular to plane P.

- Two lines perpendicular to the same plane are coplanar.

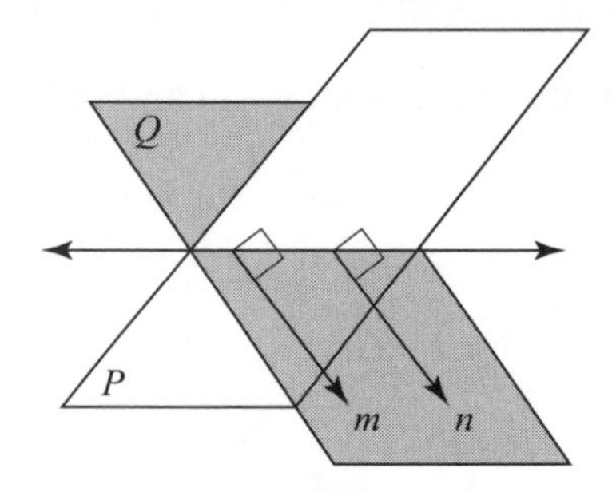

If $m \perp P$ and $n \perp P$, then lines m and n both lie in plane Q.

- Two planes are perpendicular to each other if and only if one plane contains a line perpendicular to the other plane.

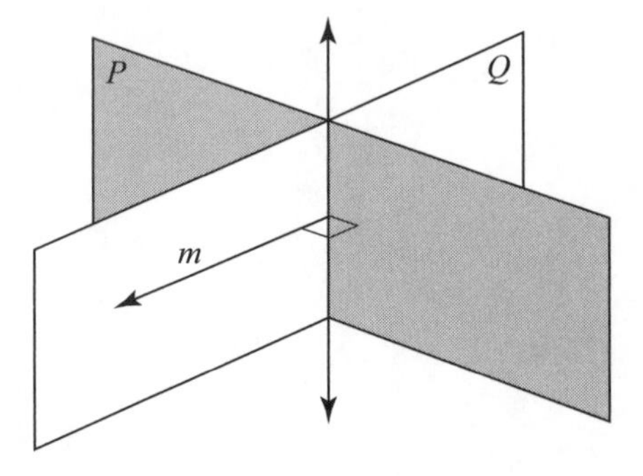

$P \perp Q$ if and only if Q contains m and $m \perp P$.

- If a line is perpendicular to a plane, then any line perpendicular to the given line at its point of intersection with the given plane is in the given plane.

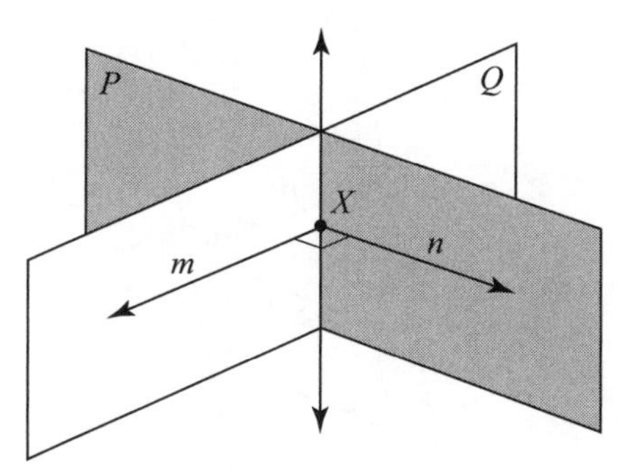

If $m \perp P$ at point X, then $n \perp m$ at X and n is in P.

- If a line is perpendicular to a plane, then every plane containing the line is perpendicular to the given plane.

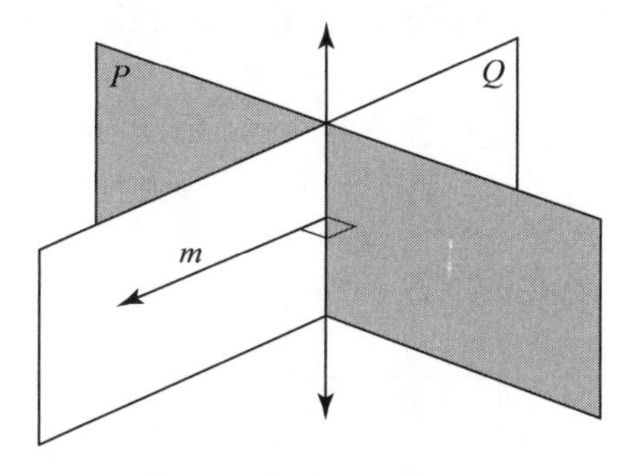

If $m \perp P$, then $Q \perp P$.

- If a plane intersects two parallel planes, then the intersection is two parallel lines.

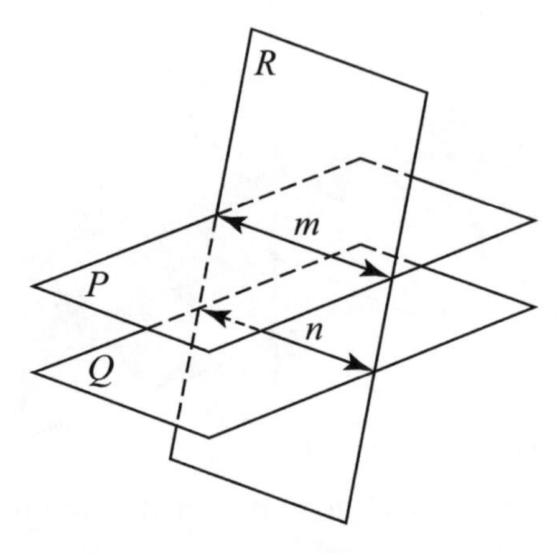

If plane R intersects parallel planes P and Q, then lines m and n are parallel.

- If two planes are perpendicular to the same line, they are parallel.

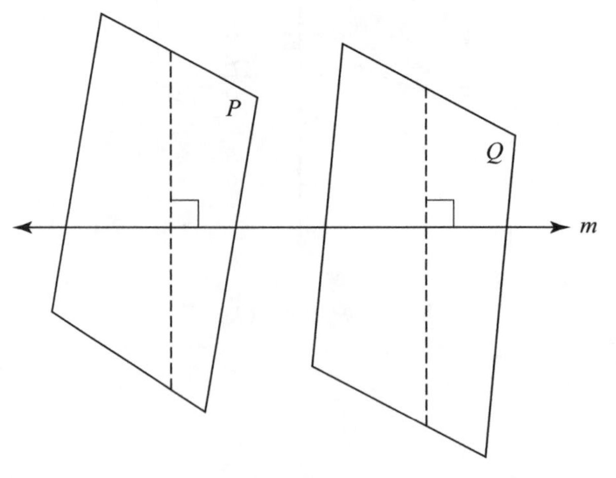

If $P \perp m$ and $Q \perp m$, then $P \parallel Q$.

Appendix II
Required Geometric Constructions

A geometric construction, unlike a drawing or sketch, is made only with a straightedge such as an unmarked ruler and a compass. The point at which the pivot point of the compass is placed is called the **center**, and the fixed compass setting that is used to draw an arc is the **radius length**.

Construction 1: To Construct an Equilateral Triangle

Objective: To construct an equilateral triangle using $\overline{AB}$ as one of its sides:

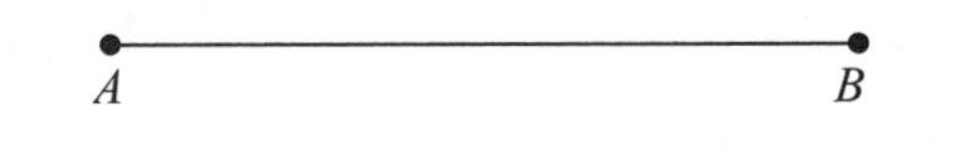

Step	Diagram
1. Set the radius length of your compass equal to the length of $\overline{AB}$ by placing the fixed point of the compass on A and the pencil point on B and draw an arc. 2. Using this compass setting and keeping the point of the compass on A, draw another arc. 3. Shift the compass point to B. Using the same radius length, draw an arc that intersects the first arc. Label the point of intersection, C. 4. Draw $\overline{AC}$ and $\overline{BC}$.	A B C A B

Conclusion: $\triangle ABC$ is equilateral.

Justification: The distance from A to any point on the first arc is AB *units*. Therefore, the third vertex of the triangle is the point C on this arc that is also a distance of AB units from B.

Construction 2: To Copy an Angle

Objective: To construct an angle congruent to *ABC* using *S* as a vertex.

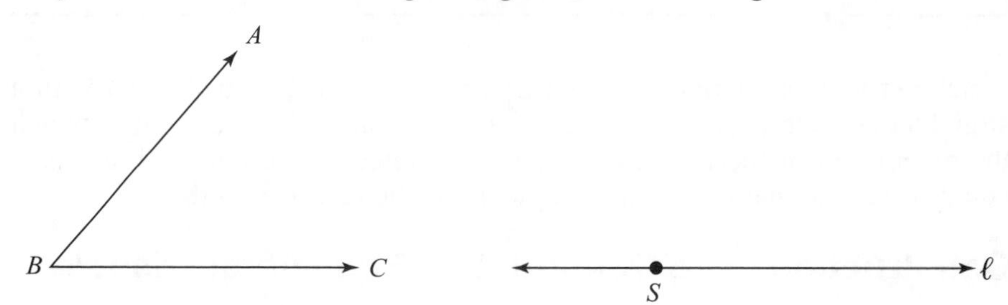

Step	Diagram
1. Place the compass point at vertex B and draw any arc that intersects the sides of $\angle ABC$. Label the points of intersection at points X and Y. 2. Using the same radius length, place the compass point on vertex S, and draw $\overset{\frown}{WT}$, intersecting line ℓ at T. 3. Place the compass point at X and adjust the radius length to the length of the line segment determined by points X and Y. 4. With this radius length, place the compass point at T and draw an arc intersecting $\overset{\frown}{WT}$ at point R. 5. Using a straightedge, draw $\overrightarrow{SR}$.	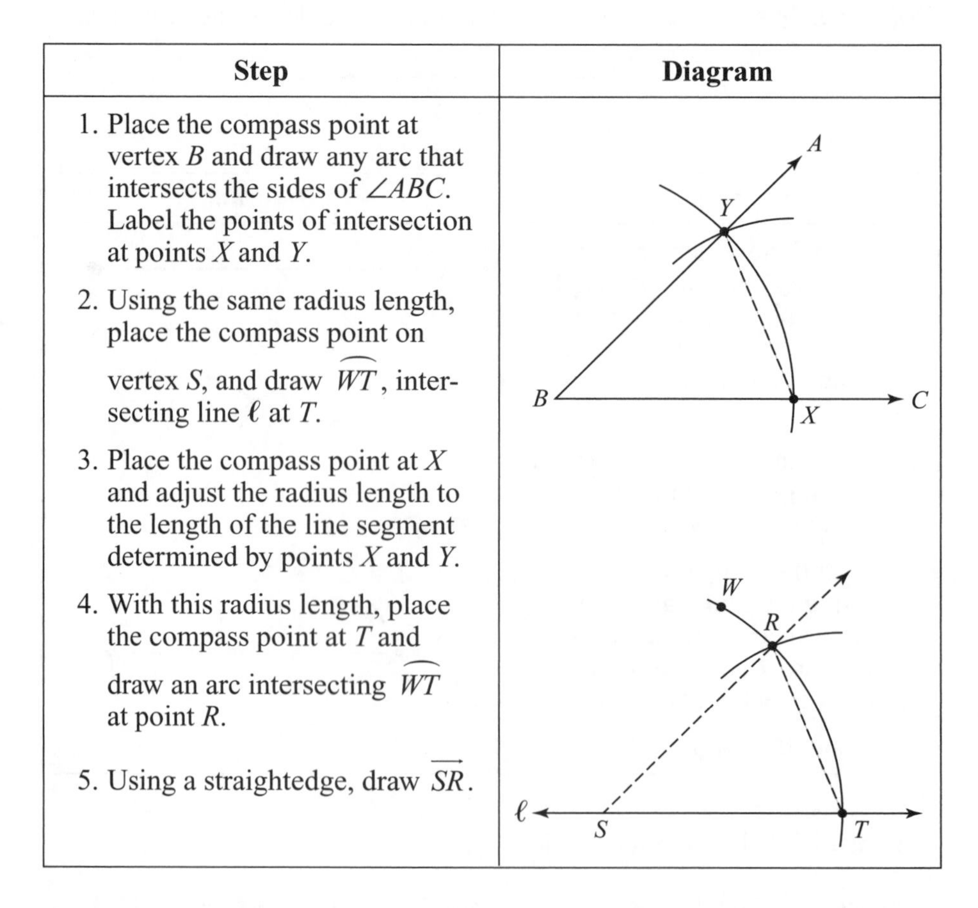

Conclusion: $\angle ABC \cong \angle RST$.

Construction 3: To Bisect an Angle

Objective: To construct the ray that bisects $\angle ABC$.

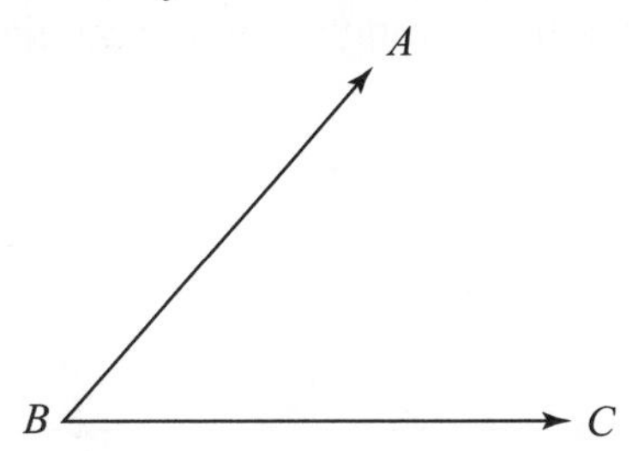

Step	Diagram
1. Using B as a center, construct an arc, using any convenient radius length, that intersects $\overrightarrow{BA}$ at point P and $\overrightarrow{BC}$ at point Q. 2. Using points P and Q as centers and the same radius length, draw a pair of arcs that intersect. Label the point at which the arcs intersect as D. 3. Draw $\overrightarrow{BD}$.	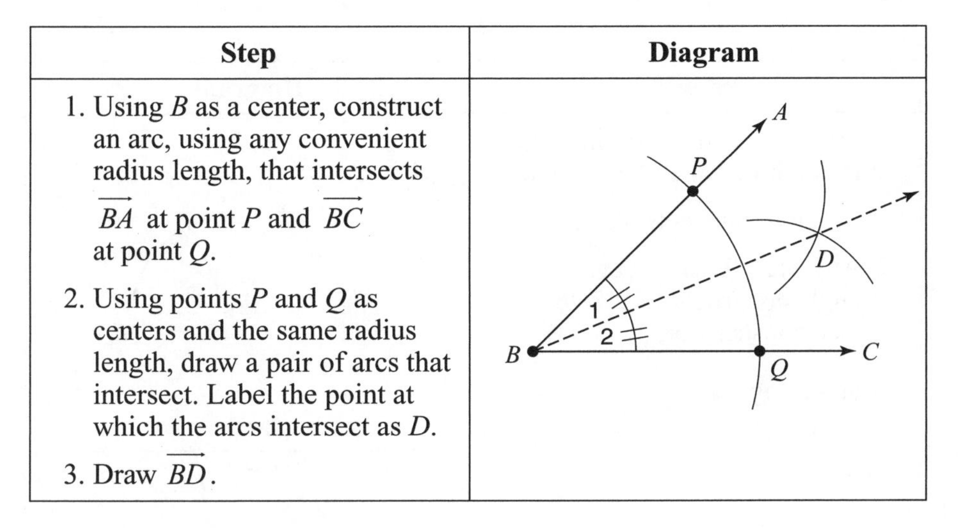

Conclusion: $\overrightarrow{BD}$ is the bisector of $\angle ABC$.

Justification: The arcs were constructed such that $\overline{BP} \cong \overline{BQ}$ and $\overline{PD} \cong \overline{QD}$. Since $\overline{BD} \cong \overline{BD}$, $\triangle BPD \cong \triangle BQD$. By CPCTC, $\angle 1 \cong \angle 2$, so $\overrightarrow{BD}$ is the bisector of $\angle ABC$.

Construction 4: Constructing a Line Parallel to a Given Line

Objective: To construct a line through P and parallel to $\overleftrightarrow{AB}$.

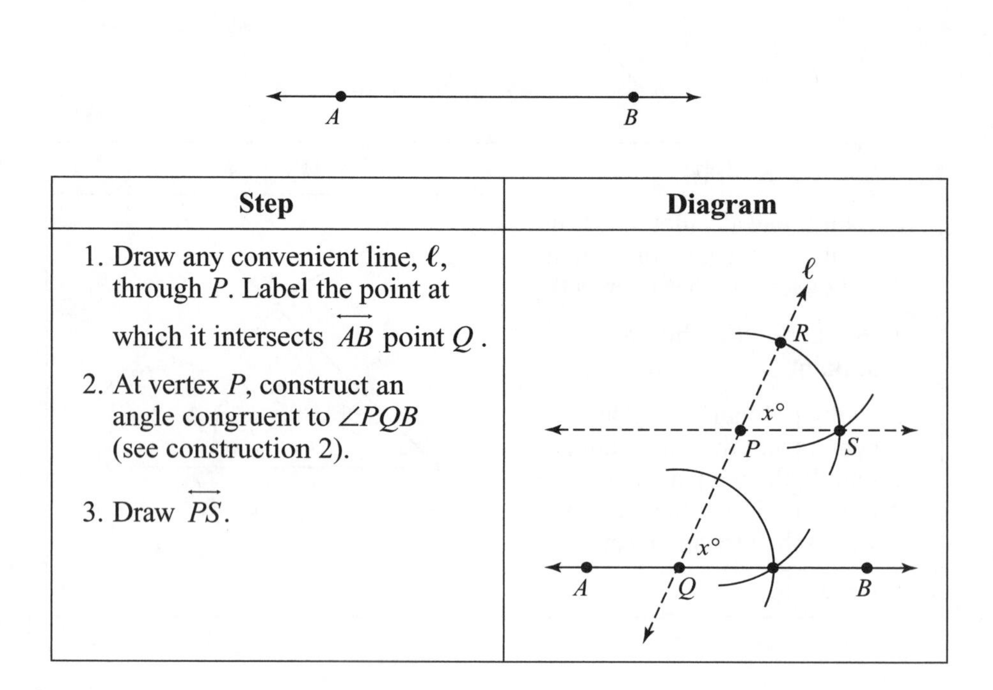

Step	Diagram
1. Draw any convenient line, ℓ, through P. Label the point at which it intersects $\overleftrightarrow{AB}$ point Q. 2. At vertex P, construct an angle congruent to $\angle PQB$ (see construction 2). 3. Draw $\overleftrightarrow{PS}$.	

Conclusion: $\overleftrightarrow{PS} \parallel \overleftrightarrow{AB}$.

Justification: According to the Parallel Postulate, there exists exactly one line through P and parallel to $\overleftrightarrow{AB}$. Angles PQB and RPS are corresponding angles. Because $\angle PQB \cong \angle RPS$, $\overleftrightarrow{PS} \parallel \overleftrightarrow{AB}$.

Construction 5: To Construct a Perpendicular Line at a Given Point on a Line

Objective: To construct a line that is perpendicular to line ℓ at point P.

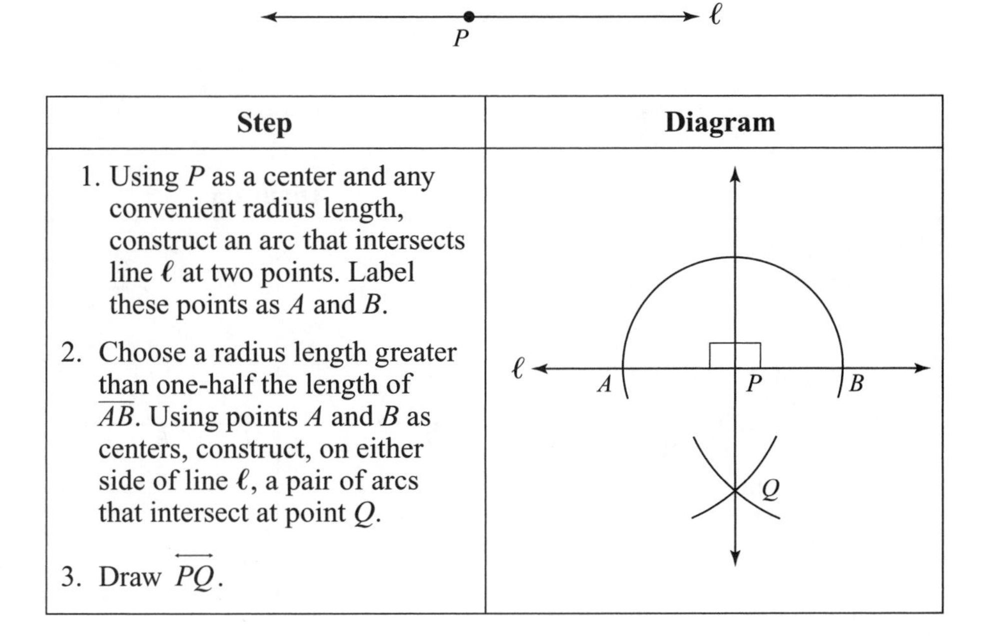

Step	Diagram
1. Using P as a center and any convenient radius length, construct an arc that intersects line ℓ at two points. Label these points as A and B. 2. Choose a radius length greater than one-half the length of $\overline{AB}$. Using points A and B as centers, construct, on either side of line ℓ, a pair of arcs that intersect at point Q. 3. Draw $\overleftrightarrow{PQ}$.	

Conclusion: $\overleftrightarrow{PQ} \perp \ell$.

Justification: The arcs were constructed such that $\overline{AP} \cong \overline{BP}$ and $\overline{AQ} \cong \overline{BQ}$.

- As $\overline{PQ} \cong \overline{PQ}$, $\triangle APQ \cong \triangle BPQ$ by SSS $\cong$ SSS.
- By CPCTC, $\angle APQ \cong \angle BPQ$.
- Because angles APQ and BPQ are congruent adjacent angles, $\overleftrightarrow{PQ} \perp \ell$.

Construction 6: To Construct a Perpendicular Line through a Point *Not* on a Line

Objective: To construct a line that is perpendicular to line ℓ and through point P.

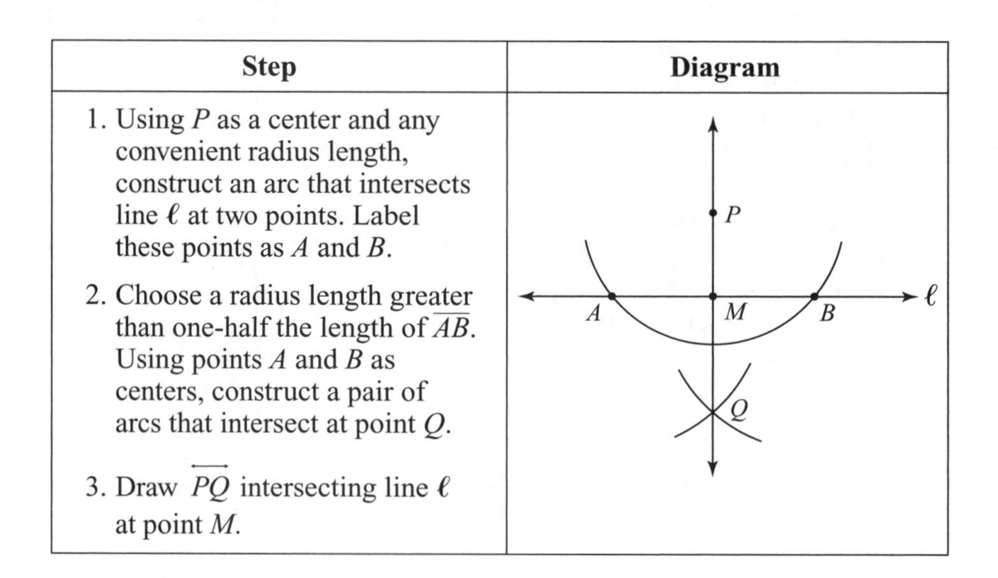

Step	Diagram
1. Using P as a center and any convenient radius length, construct an arc that intersects line ℓ at two points. Label these points as A and B. 2. Choose a radius length greater than one-half the length of $\overline{AB}$. Using points A and B as centers, construct a pair of arcs that intersect at point Q. 3. Draw $\overleftrightarrow{PQ}$ intersecting line ℓ at point M.	

Conclusion: $\overleftrightarrow{PQ} \perp \ell$ at point M.

Justification: The arcs were constructed such that $\overline{AB} \cong \overline{BP}$ and $\overline{AQ} \cong \overline{BQ}$.

- As $\overline{PQ} \cong \overline{PQ}$, $\triangle PAQ \cong \triangle PBQ$ by SSS $\cong$ SSS.
- By CPCTC, $\angle APM \cong \angle BPM$.
- $\triangle AMP \cong \triangle BMP$ by SAS $\cong$ SAS. By CPCTC, $\angle AMP \cong \angle BMP$.
- Because angles AMP and BMP are congruent adjacent angles, $\overleftrightarrow{PQ} \perp \ell$.

Construction 7: To Construct the Perpendicular Bisector of a Given Line Segment

Objective: To construct the perpendicular bisector of $\overline{AB}$.

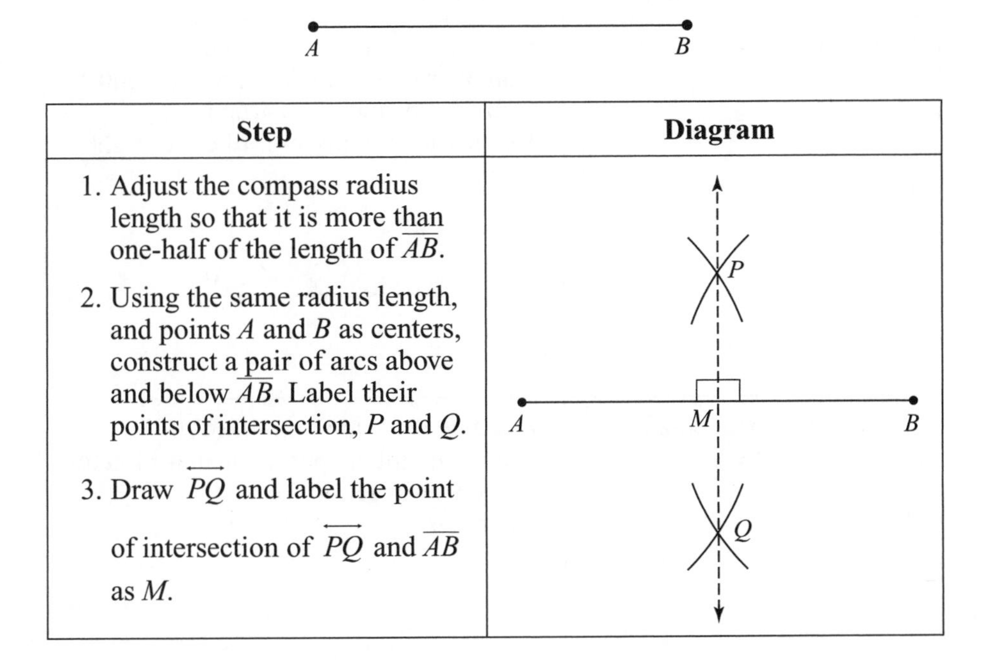

Step	Diagram
1. Adjust the compass radius length so that it is more than one-half of the length of $\overline{AB}$. 2. Using the same radius length, and points A and B as centers, construct a pair of arcs above and below $\overline{AB}$. Label their points of intersection, P and Q. 3. Draw $\overleftrightarrow{PQ}$ and label the point of intersection of $\overleftrightarrow{PQ}$ and $\overline{AB}$ as M.	

Conclusion: $\overleftrightarrow{PQ}$ is the perpendicular bisector of $\overline{AB}$.

Justification: Because the same compass radius length was used to draw the arcs, $\overline{AP} \cong \overline{AQ} \cong \overline{BP} \cong \overline{BQ}$.

- As $\overline{PQ} \cong \overline{PQ}$, $\triangle PAQ \cong \triangle PBQ$ by SSS $\cong$ SSS.
- By CPCTC, $\angle APM \cong \angle BPM$.
- $\triangle AMP \cong \triangle BMP$ by SAS $\cong$ SAS. By CPCTC, $\overline{AM} \cong \overline{BM}$ and $\angle AMP \cong \angle BMP$ (congruent adjacent angles), which makes $\overleftrightarrow{PQ}$ the perpendicular bisector of $\overline{AB}$.

Applying Constructions

You may be asked to construct a special angle, line segment, or triangle.

Required Construction	What to Do
To construct a 45° angle	• Construct a line perpendicular to another line at any convenient point on the line (see Construction 5). • Construct the bisector of either right angle. 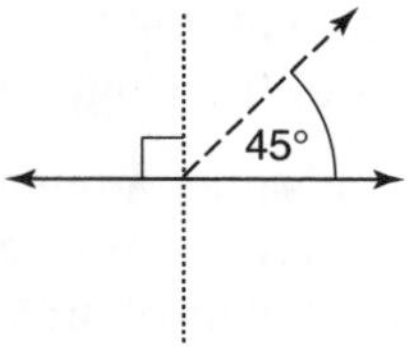 45°
To construct the median to side $\overline{AC}$ of $\triangle ABC$	• Locate the midpoint, M, of $\overline{AC}$ by constructing its perpendicular bisector (see Construction 7). • Draw $\overline{BM}$. 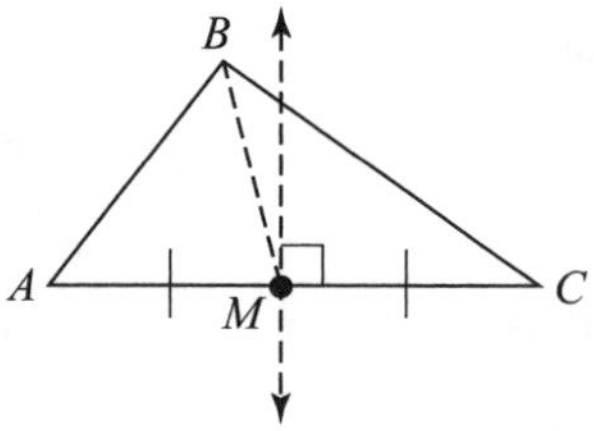 B A M C
To construct the altitude to side $\overline{AC}$ of $\triangle ABC$	• Construct the line through B and perpendicular to $\overline{AC}$ (see Construction 6). Label the point at which the perpendicular line intersects $\overline{AC}$, H. • Draw $\overline{BH}$. 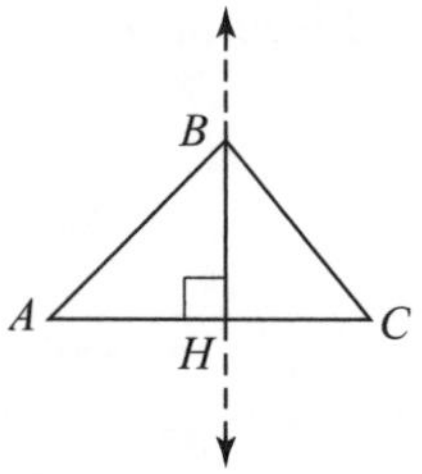 B A H C

Required Construction	What to Do
To construct a triangle congruent to $\triangle ABC$ 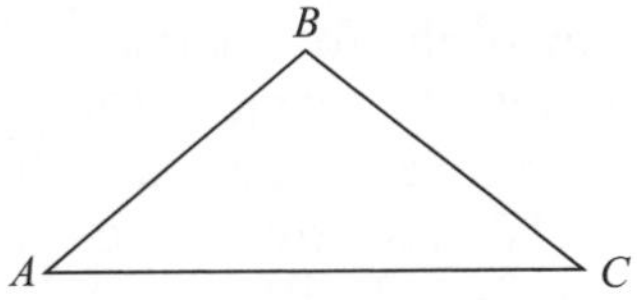	• At a point D on line ℓ, construct $\angle XDY \cong \angle A$ (see Construction 2). 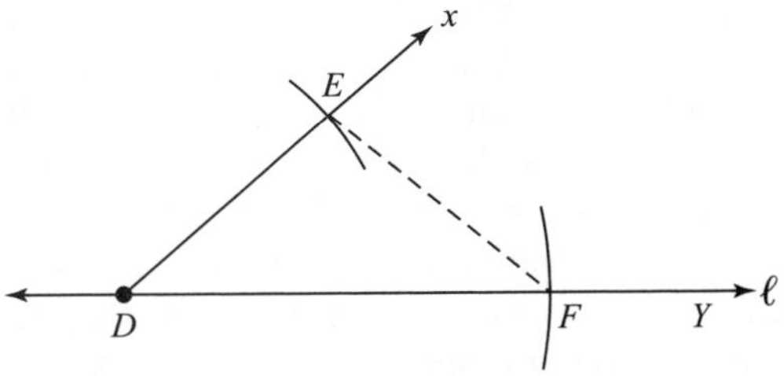 • On $\overrightarrow{DX}$, construct $\overline{DE} \cong \overline{AB}$ and on $\overrightarrow{DY}$, construct $\overline{DF} \cong \overline{AC}$. • Complete the triangle by drawing $\overline{EF}$. • $\triangle ABC \cong \triangle DEF$ by SAS $\cong$ SAS.
To circumscribe a circle about $\triangle ABC$	• Locate the center of the circumscribed circle by constructing the perpendicular bisectors of two sides (see Construction 7). Label their point of intersection O. 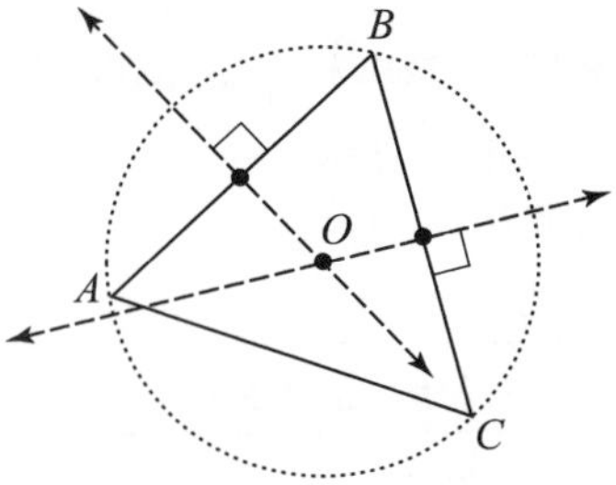 • Using a compass, draw a circle with radius OA.

Learning about the Geometry Regents Examination

The first administration of the new Regents Examination in Geometry will be June 2009. It is a three-hour exam that is divided into four parts with a total of 38 questions. No choice is permitted in any of the four parts. All 38 questions must be answered. Part I consists entirely of regular multiple-choice questions. Parts II, III, and IV each contain a set of questions that must be answered directly in the question booklet. You are required to show how you arrived at the answers for the questions in Parts II, III, and IV. The accompanying table shows how the exam breaks down.

Question Type	Number of Questions	Credit Value
Part I: Multiple choice	28	$28 \times 2 = 56$
Part II: 2-credit open ended	6	$6 \times 2 = 12$
Part III: 4-credit open ended	3	$4 \times 3 = 12$
Part IV: 6-credit open ended	1	$1 \times 6 = 6$
	Total = 38 questions	Total = 86 points

How Are Credits Distributed by Topic Area?

The table below shows the percentage of total credits that will be aligned to each major topic area.

Content Band	% of Total Credits
Geometric relationships	8–12
Constuctions	3–7
Locus	4–8
Informal and formal proofs	41–47
Transformational geometry	8–13
Coordinate geometry	23–28

How Is the Exam Scored?

- Each of your answers to the 28 multiple-choice questions in Part I will be scored as either right or wrong.
- Solutions to questions in Parts II, III, and IV that are not completely correct may receive partial credit according to a special rating guide that is provided by the New York State Education Department. In order to receive full credit for a correct answer to a question in Parts II, III, or IV, you must show or explain how you arrived at your answer by indicating the key steps taken, including appropriate formula substitutions, diagrams, graphs, and charts. A correct numerical answer with no work shown will receive only 1 credit.
- The raw scores for the four parts of the test are added together. The maximum total raw score for the Geometry Regents is 86 points. Using a special conversion chart that is provided by the New York State Education Department, your total raw score will be equated to a final test score that falls within the usual 0 to 100 scale.

What Type of Calculator Is Required?

Graphing calculators are *required* for the Geometry Regents examination. During the administration of the Regents exam, schools are required to make a graphing calculator available for the exclusive use of each student. You will need to use your calculator to work with trigonometric functions of angles, find roots of numbers, and perform routine calculations.

Knowing how to use a graphing calculator gives you more options when deciding how to solve a problem. Rather than solving a problem algebraically with pen and paper, it may be easier to solve the same problem using a graph or table created by a graphing calculator. A graphical or numerical solution using a calculator can also be used to help confirm an answer obtained by solving the problem algebraically.

Are Any Formulas Provided?

The Regents Examination in Geometry will include a reference sheet containing the formulas specified below.

Volume	Cylinder	$V = Bh$, where B is the area of the base
	Pyramid	$V = \frac{1}{3}Bh$, where B is the area of the base
	Right circular cone	$V = \frac{1}{3}Bh$, where B is the area of the base
	Sphere	$V = \frac{4}{3}\pi r^3$

Lateral Area (L.A.)	Right circular cylinder	L.A. = $2\pi rh$
	Right circular cone	L.A. = $\pi r\ell$, where ℓ is the slant height

Surface Area (S.A.)	Sphere	S.A. = $4\pi r^2$

What Else Should I Know?

- Do not omit any questions from Part I. Since there is no penalty for guessing, make certain that you record an answer for each of the 28 multiple-choice questions.
- If the method of solution is not stated in the problem, choose an appropriate method (numerical, graphical, or algebraic) with which you are most comfortable.
- If you solve a problem in Parts II, III, or IV using a trial-and-error approach, show the work for at least *three* guesses with appropriate checks. Should the correct answer be reached on the first trial, you must further illustrate your method by showing that guesses below and above the correct guess do *not* work.
- Avoid rounding errors when using a calculator. Unless otherwise directed, the (pi) key on a calculator should be used in computations involving the constant π rather than the common rational approximation of 3.14 or $\frac{22}{7}$. When performing a sequence of calculations in which the result of one calculation is used in a second calculation, do not round off. Instead, use the full power/display of the calculator by performing a "chain" calculation, saving intermediate results in the calculator's memory. Unless otherwise specified, rounding, if required, should be done only when the *final* answer is reached.
- Check that each answer is in the requested form. If a specific form is not required, answers may be left in any equivalent form, such as $\sqrt{75}$, $5\sqrt{3}$, or 8.660254038 (the full power/display of the calculator).
- If a problem involves finding the volume or lateral area of a solid, look for the appropriate formula in the reference sheet that is included in the test booklet.
- Clearly write any formula you use before making any appropriate substitutions. Then evaluate the formula in step-by-step fashion.
- For any problem solved in Parts II, III, and IV using a graphing calculator, you must indicate how the calculator was used to obtain the answer such as by copying graphs or tables created by your calculator together with the equations used to produce them. When copying graphs, label each graph with its equation, state the dimensions of the viewing window, and identify the intercepts and any points of intersection with their coordinates. Whenever appropriate, indicate the rationale of your approach.

Examination

Geometry

Important Note: As of the publication date of this book, an actual Regents Examination in Geometry was not yet available as the first administration of this examination is scheduled for June 2009. The test that follows is *not* an actual Regents Examination. It is a practice test that has the same format, credit distribution, and topic coverage as an actual Regents Examination in Geometry.

PART I

Answer all questions in this part. Each correct answer will receive 2 credits. No partial credit will be allowed. For each question, write in the space provided the numeral preceding the word or expression that best completes the statement or answers the question. [56]

1 One piece of the birdhouse that Natalie is building is shaped like a regular pentagon, as shown in the accompanying diagram. If side $\overline{AE}$ is extended to point F, what is the measure of exterior angle DEF?

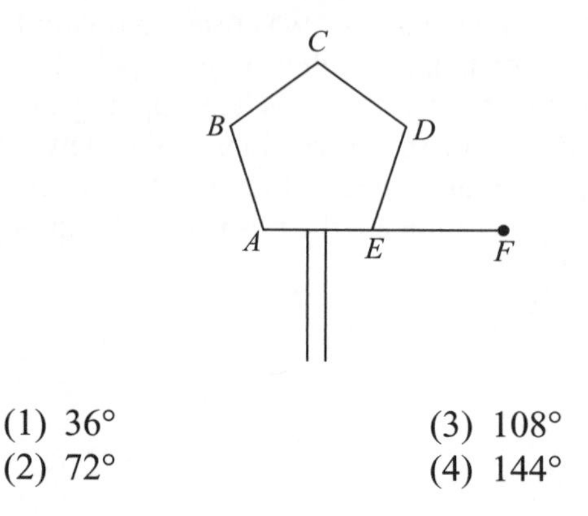

(1) 36°　　(3) 108°
(2) 72°　　(4) 144°　　1 _____

2 What are the coordinates of point (2, –3) after it is reflected over the x-axis?

(1) (2, 3)
(2) (–2, 3)
(3) (–2, –3)
(4) (–3, 2)

2 _____

3 In the accompanying diagram, line ℓ is perpendicular to line m at A, line k is perpendicular to line m at B, and lines ℓ, m, and k are in the same plane.

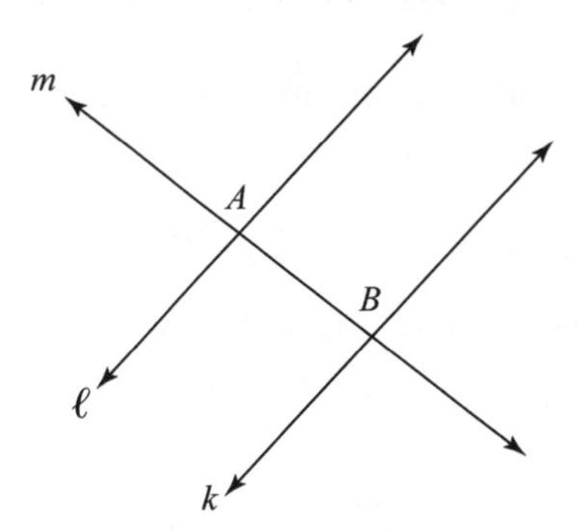

Which statement is the first step in an indirect proof to prove that ℓ is parallel to k?

(1) Assume that ℓ, m, and k are not in the same plane.
(2) Assume that ℓ is perpendicular to k.
(3) Assume that ℓ is not perpendicular to m.
(4) Assume that ℓ is not parallel to k.

3 _____

4 A circular sunflower with a diameter of 4 centimeters is placed on a coordinate grid such that its center is at (–2, 1). Which equation represents the sunflower?

(1) $(x - 2)^2 + (y + 1)^2 = 2$
(2) $(x + 2)^2 + (y - 1)^2 = 4$
(3) $(x - 2)^2 + (y - 1)^2 = 4$
(4) $(x + 2)^2 + (y - 1)^2 = 2$

4 _____

5 Which statement is logically equivalent to "If a triangle is an isosceles triangle, then it has two congruent sides"?

(1) If a triangle does not have two congruent sides, then it is an isosceles triangle.
(2) If a triangle does not have two congruent sides, then it is not an isosceles triangle.
(3) If a triangle is not an isosceles triangle, then it has two congruent sides.
(4) If a triangle is an isosceles triangle, then it does not have two congruent sides.

5 ______

6 In the accompanying diagram, $\overleftrightarrow{AB} \parallel \overleftrightarrow{CD}$. From point E on $\overleftrightarrow{AB}$, transversals $\overrightarrow{EF}$ and $\overrightarrow{EG}$ are drawn, intersecting $\overleftrightarrow{CD}$ at H and I, resectively.

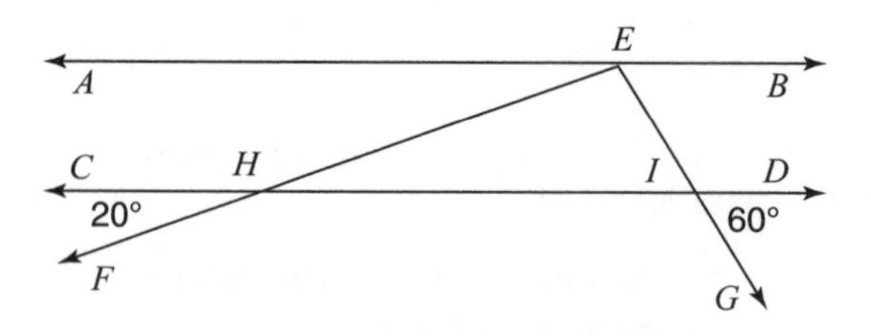

If m$\angle CHF$ = 20 and m$\angle DIG$ = 60, What is m$\angle HEI$?

(1) 60
(2) 80
(3) 100
(4) 120

6 ______

7 $\overleftrightarrow{AB}$ and $\overleftrightarrow{CD}$ intersect at point E, m$\angle AEC = 6x + 20$, and m$\angle DEB = 10x$. What is the value of x?

(1) $4\frac{3}{8}$
(2) 5
(3) 10
(4) $21\frac{1}{4}$

7 ______

8 The amount of light produced by a cylindrical-shaped fluorescent light bulb depends on its lateral area. A certain cylindrical-shaped fluorescent light bulb is 48 inches in length with a 1 inch diameter. If this fluorescent light bulb is manufactured to produce 0.226 watts of light per square inch, what is the best estimate for the total amount of light it is able to produce?

(1) 32 watts
(2) 34 watts
(3) 35 watts
(4) 70 watts

8 _____

9 In the accompanying diagram, $\overrightarrow{PA}$ is tangent to circle O at A, $\overline{PBC}$ is a secant, $PB = 4$, and $BC = 8$.

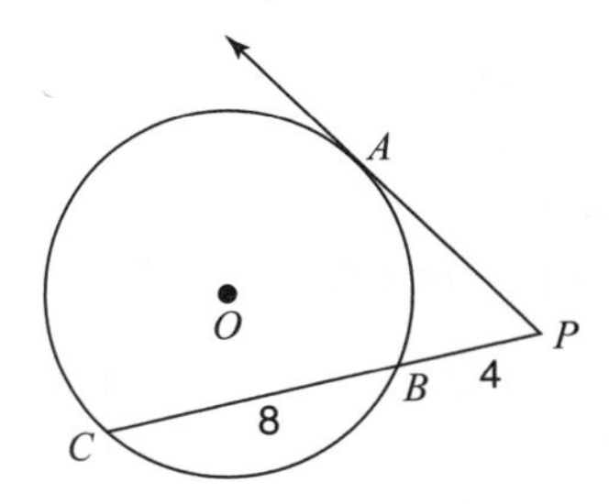

What is the length of $\overline{PA}$?

(1) $4\sqrt{6}$
(2) $4\sqrt{2}$
(3) $4\sqrt{3}$
(4) 4

9 _____

10 Which letter has point symmetry but *not* line symmetry?

(1) **H**
(2) **S**
(3) **T**
(4) **X**

10 _____

11 In the accompanying diagram of a construction, what does $\overline{PC}$ represent?

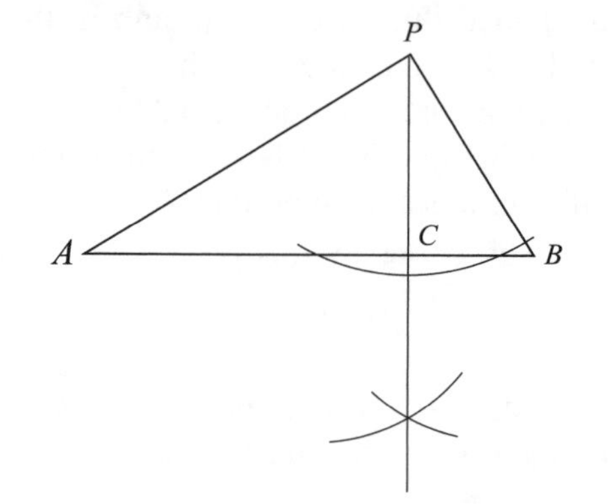

(1) an altitude drawn to $\overline{AB}$
(2) a median drawn to $\overline{AB}$
(3) the bisector of $\angle APB$
(4) the perpendicular bisector of $\overline{AB}$

11 _____

12 The image of the origin under a certain translation is the point (2, –6). The image of point (–3, –2) under the same translation is the point

(1) (–6, 12)
(2) (–5, 4)
(3) $\left(-\frac{3}{2}, \frac{1}{3}\right)$
(4) (–1, –8)

12 _____

13 In the accompanying diagram, $\overline{AB}$ and $\overline{CD}$ intersect at point E such that $\overline{AC}$ is parallel to $\overline{DB}$. If $AC = 3$, $DB = 4$, and $AB = 14$, what is AE?

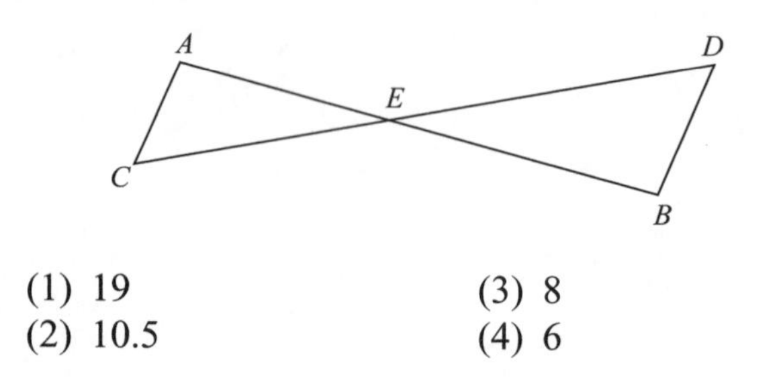

(1) 19
(2) 10.5
(3) 8
(4) 6

13 _____

14 If in $\triangle RST$, $m\angle R = 71$ and $m\angle S = 37$, then

(1) $ST > RS$
(2) $RS > RT$
(3) $RS = ST$
(4) $RT > ST$

14 _____

15 Tangents $\overline{PA}$ and $\overline{PB}$ are drawn from external point P to circle O at points A and B, respectively. Angle PAB is always

(1) a right angle
(2) an acute angle
(3) the supplement of angle OAB
(4) congruent to angle APB

15 _____

16 Which type of transformation is illustrated in the accompanying diagram?

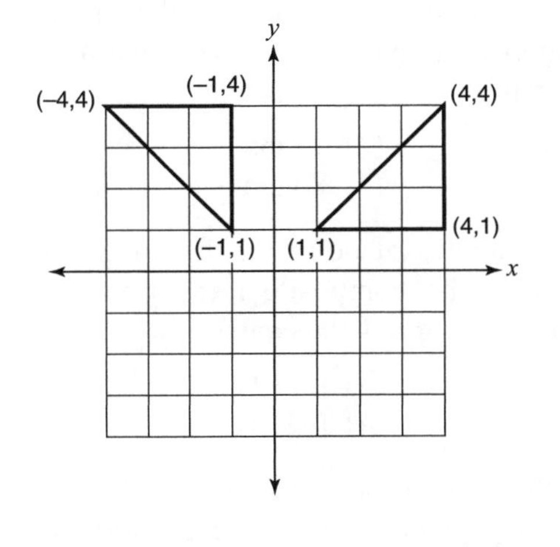

(1) dilation
(2) reflection
(3) translation
(4) rotation

16 _____

17 Two chords intersect in a circle whose radius is 8. The product of the lengths of the segments of one of the chords can *not* be

(1) 80
(2) 48
(3) 32
(4) 16

17 _____

18 In the accompanying diagram of parallelogram $FLSH$, $\overline{FGAS}$ is a diagonal, $\overline{LG} \perp \overline{FS}$, and $\overline{HA} \perp \overline{FS}$.

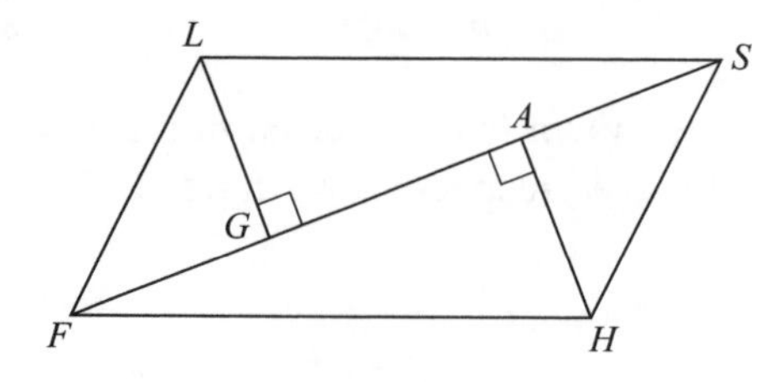

Which expression represents the most direct method that could be used to prove $\triangle LGS \cong \triangle HAF$?

(1) SAS $\cong$ SAS
(2) ASA $\cong$ ASA
(3) AAS $\cong$ AAS
(4) HL $\cong$ HL

18 _____

19 The perimeter of a rhombus is 60 cm, and the length of its longer diagonal measures 24 cm. The length of the shorter diagonal is

(1) 9 cm
(2) 15 cm
(3) 18 cm
(4) 20 cm

19 _____

20 What are the coordinates of point A', the image of point $A(-4, 1)$ after the composite transformation $R_{90°} \circ r_{y\text{-axis}}$ where the origin is the center of rotation?

(1) (1, 4)
(2) (–4, –1)
(3) (4, –1)
(4) (–1, 4)

20 _____

21 If the altitude to the hypotenuse of a right triangle is 8, the lengths of the segments of the hypotenuse formed by the altitude may be

(1) 8 and 12
(2) 2 and 32
(3) 3 and 24
(4) 6 and 8

21 _____

22 In the accompanying figure, $\overline{AB} \cong \overline{AC}$. It can be proved that $\overline{CD} \cong \overline{BE}$ if it is also known that

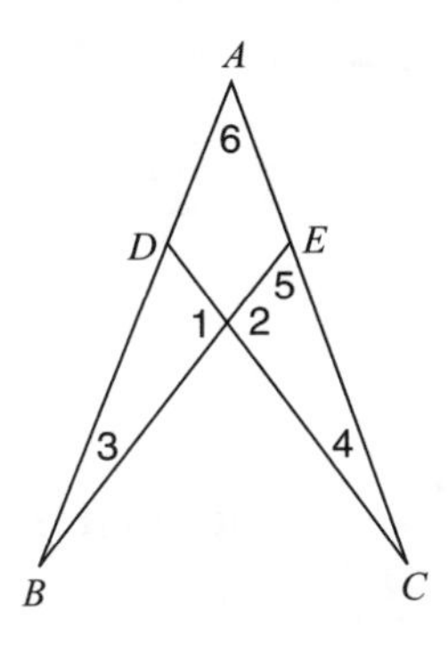

(1) $\angle 1 \cong \angle 2$
(2) $\angle 3 \cong \angle 4$
(3) $\angle 3 \cong \angle 5$
(4) $\angle 4 \cong \angle 6$

22 _____

23 If the midpoints of the sides of a triangle are connected, the area of the triangle formed is what part of the area of the original triangle?

(1) $\frac{1}{4}$
(2) $\frac{1}{3}$
(3) $\frac{3}{8}$
(4) $\frac{1}{2}$

23 _____

24 The center of the circle that can be circumscribed about a scalene triangle is located by constructing the

(1) medians of the triangle
(2) altitudes of the triangle
(3) perpendicular bisectors of the sides of the triangle
(4) bisectors of the angles of the triangle

24 _____

25 If two circles are externally tangent, what is the total number of common tangents that can be drawn to the circles?

(1) 1
(2) 2
(3) 3
(4) 0

25 _____

26 In the accompanying diagram, point P lies 3 centimeters from line ℓ.

How many points are both 2 centimeters from line ℓ and 1 centimeter from point P?

(1) 1
(2) 2
(3) 0
(4) 4

26 _____

27 Which statement best describes the lines whose equations are $4y + x = 12$ and $3y = 12x + 1$?

(1) They are segments.
(2) They are perpendicular to each other.
(3) They intersect each other.
(4) They are parallel to each other.

27 _____

28 If the intersection of the perpendicular bisectors of the sides of a triangle lies on one of the sides of the triangle, the triangle must be

(1) acute
(2) obtuse
(3) scalene
(4) right

28 _____

PART II

Answer all questions in this part. Each correct answer will receive 2 credits. Clearly indicate the necessary steps, including appropriate formula substitutions, diagrams, graphs, charts, etc. For all questions in this part, a correct numerical answer with no work shown will receive only 1 credit. [12]

29 Matthew is a fan of the Air Force's Thunderbirds flying team and is designing a jacket patch for the team, as shown in the accompanying diagram.

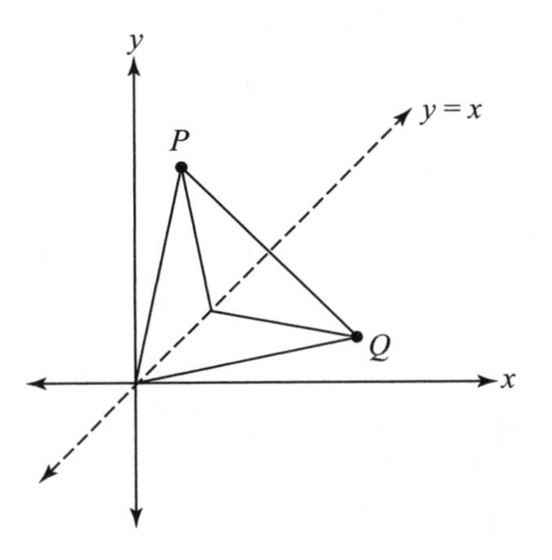

If the coordinates of point P are (a, b), express in terms of a and b:

(1) the coordinates of Q, the reflection of P in the line $y = x$

(2) an equation of the line that contains points P and Q

30 The accompanying diagram shows the path of a cart traveling on a circular track of radius 2.40 meters. The cart starts at point A and stops at point B, moving in a counterclockwise direction. What is the length of minor arc AB, over which the cart traveled, to the *nearest tenth of a meter*?

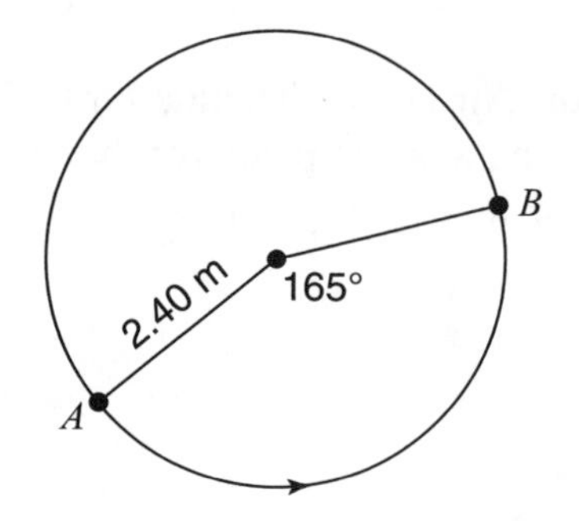

31 In the accompanying diagram, $\triangle LRS$ is inscribed in circle O. If $\overline{RP} \perp \overline{LOS}$, $OS = 7$, and $LP = 3$, find the length of $\overline{RP}$ to the *nearest tenth*.

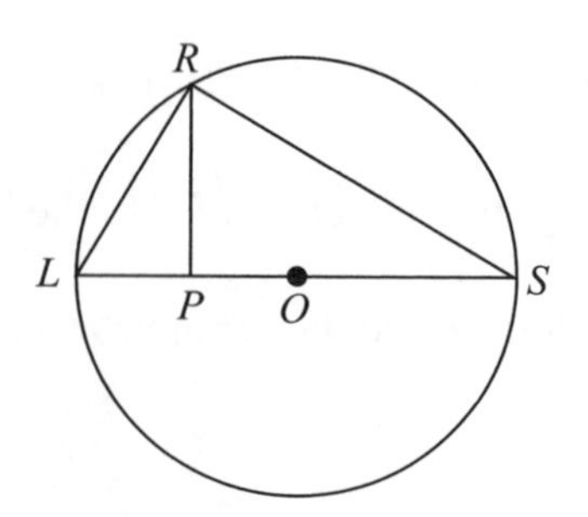

32 In the accompanying diagram of circle O, diameter $\overline{AOB}$ is extended through B to external point P, tangent $\overline{PC}$ is drawn to point C on the circle, and

$m\overset{\frown}{AC} : m\overset{\frown}{BC} = 7 : 2$. Find $m\angle CPA$.

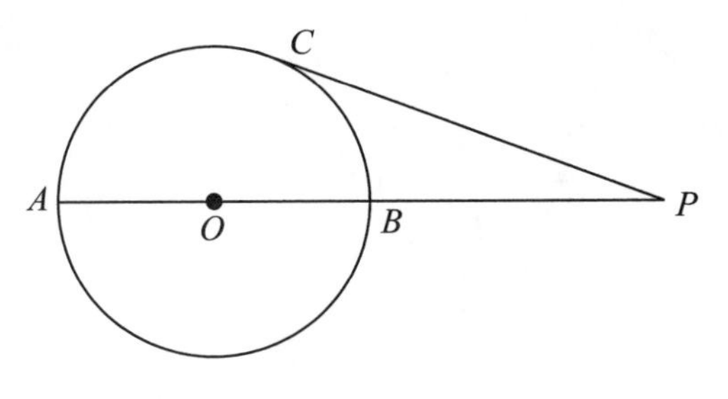

(Not drawn to scale)

33 In the accompanying diagram, $\overline{ABE} \parallel \overline{CD}$. Write an explanation or an informal proof that shows $m\angle 1 > m\angle 2$.

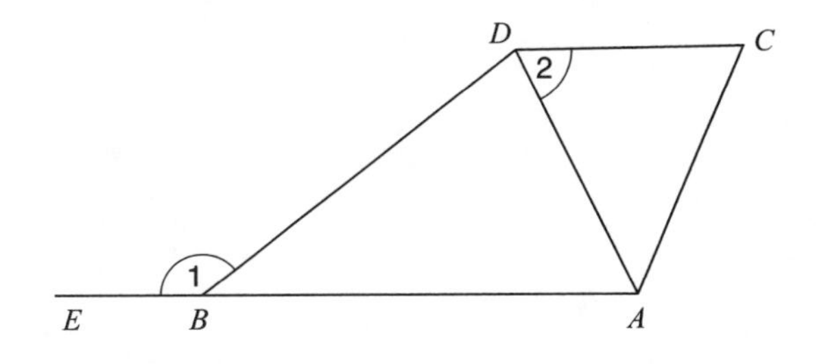

34 In the accompanying diagram, $m\overset{\frown}{BR} = 70$, $m\overset{\frown}{YD} = 70$, and $\overline{BOD}$ is a diameter of circle O. Write an explanation or an informal proof that shows $\triangle RBD$ and $\triangle YDB$ are congruent.

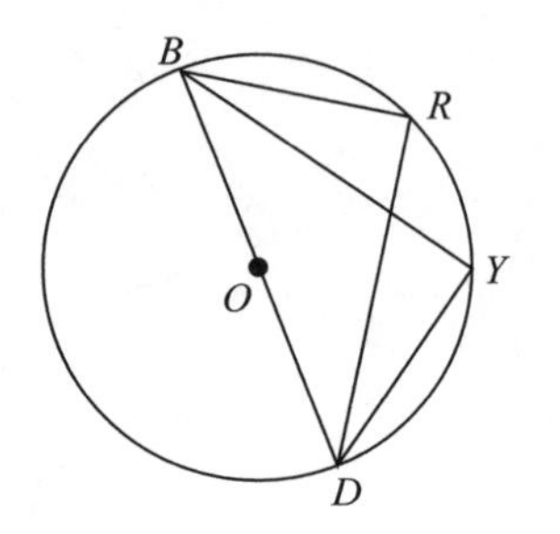

PART III

Answer all questions in this part. Each correct answer will receive 4 credits. Clearly indicate the necessary steps, including appropriate formula substitutions, diagrams, graphs, charts, etc. For all questions in this part, a correct numerical answer with no work shown will receive only 1 credit. [12]

35 In the accompanying diagram of circle O, diameter $\overline{AOB}$ is drawn, tangent $\overline{CB}$ is drawn to the circle at B, E is a point on the circle, and $\overline{BE} \parallel \overline{ADC}$.

Prove: $\frac{AC}{AB} = \frac{AB}{BE}$.

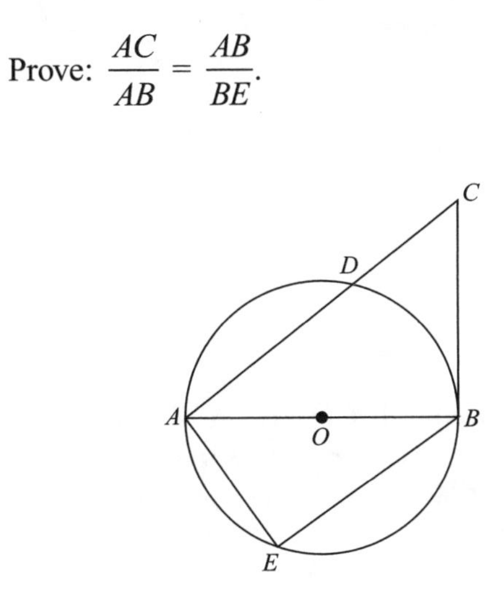

36 In the accompanying diagram, $\triangle ABC$ is *not* isosceles. Give a mathematical argument or proof that shows if altitude $\overline{BD}$ were drawn to side $\overline{AC}$, it would *not* bisect $\overline{AC}$.

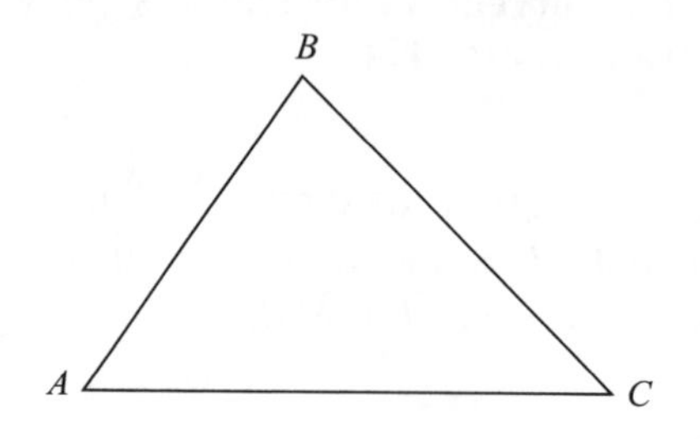

37 The vertices of quadrilateral $ABCD$ are $A(-2, 2)$, $B(6, 5)$, $C(4, 0)$, $D(-4, -3)$. Prove by means of coordinate geometry that quadrilateral $ABCD$ is a parallelogram but *not* a rectangle.

PART IV

A correct answer will receive 6 credits. Clearly indicate the necessary steps, including appropriate formula substitutions, diagrams, graphs, charts, etc. A correct numerical answer with no work shown will receive only 1 credit. [6]

38 Given: $\triangle ABC$, $\overline{AB} \cong \overline{BC}$,
$\overline{ED} \parallel \overline{AC}$,
$\overline{DG} \perp \overline{AB}$,
$\overline{EF} \perp \overline{BC}$,
$\overline{BGEA}$ and $\overline{BFDC}$.

Prove: $\overline{GE} \cong \overline{FD}$.

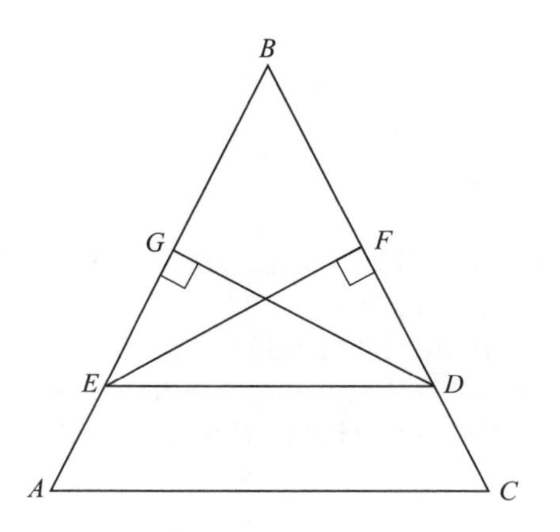

Answers to Sample Examination

PART I

1. (2)	**8.** (1)	**15.** (2)	**22.** (2)
2. (1)	**9.** (3)	**16.** (4)	**23.** (1)
3. (4)	**10.** (2)	**17.** (1)	**24.** (3)
4. (2)	**11.** (1)	**18.** (3)	**25.** (3)
5. (2)	**12.** (4)	**19.** (3)	**26.** (1)
6. (3)	**13.** (4)	**20.** (4)	**27.** (2)
7. (2)	**14.** (4)	**21.** (2)	**28.** (4)

PART II

29. (1) $(-b, a)$ (2) $y = -x + (a + b)$

30. 6.9 m

31. 5.7

32. 50°

33. $m\angle 1 > m\angle DAE$ and $m\angle DAE = m\angle 2$ since $\overline{ABE} \parallel \overline{CD}$. By substitution, $m\angle 1 > m\angle 2$.

34. Angles BRD and DYB are right angles since they are inscribed in a semicircle. $\overline{BR} \cong \overline{YD}$ (Leg) since congruent arcs have congruent chords and $\overline{BD} \cong \overline{BD}$ (Hyp). Hence, $\triangle RBD \cong \triangle YDB$ by HL $\cong$ HL.

PART III

35. Since $\overline{BE} \parallel \overline{ADC}$, alternate interior angles CAB and ABE are congruent. Angle E is a right angle since it is inscribed in a semicircle. Angle ABC is a right angle since it is formed by a tangent and a radius. Hence, $\angle E \cong \angle ABC$ since all right angles are congruent. Thus, $\triangle ABC \sim \triangle BEA$ and $\frac{AC}{AB} = \frac{AB}{BE}$ since the lengths of corresponding sides of similar triangles are in proportion.

36. Use an indirect proof by assuming altitude $\overline{BD}$ bisects $\overline{AC}$. Then $\triangle BDA \cong \triangle BDC$ because SAS ≅ SAS and $\overline{AB} \cong \overline{BC}$ by CPCTC. Because this contradicts the Given that $\triangle ABC$ is not isosceles, the assumption that altitude $\overline{BD}$ bisects $\overline{AC}$ is not correct. Thus, altitude $\overline{BD}$ does *not* bisect $\overline{AC}$ as this is the only other possibility.

37. Slope $\overline{AB}$ = slope $\overline{CD} = \frac{3}{8}$ and slope $\overline{BC}$ = slope $\overline{AD} = \frac{5}{2}$.

Since lines that have the same slope are parallel, $\overline{AB} \parallel \overline{CD}$ and $\overline{BC} \parallel \overline{AD}$. Hence, quadrilateral $ABCD$ is a parallelogram. Because the slopes of a pair of adjacent sides are *not* negative reciprocals, parallelogram $ABCD$ does not contain a right angle so it is *not* a rectangle.

PART IV

38. Write a two-column proof.

Statement	Reason
1. $\overline{AB} \cong \overline{BC}$.	1. Given.
2. $\angle A \cong \angle C$.	2. If two sides of a triangle are congruent, the angles opposite them are congruent.
3. $\overline{ED} \parallel \overline{AC}$.	3. Given.
4. $\angle GED \cong \angle A$ and $\angle FDE \cong \angle C$.	4. If two lines are parallel, then corresponding angles are congruent.
5. $\angle GED \cong \angle FDE$. Angle	5. Transitive property of congruence.
6. $\overline{DG} \perp \overline{AB}$ and $\overline{EF} \perp \overline{BC}$.	6. Given.
7. Angles EGD and DFE are right angles.	7. Perpendicular lines intersect to form right angles.
8. $EGD \cong DFE$. Angle	8. All right angles are congruent.
9. $\overline{ED} \cong \overline{ED}$. Side	9. Reflexive property of congruence.
10. $\triangle EGD \cong \triangle DFE$.	10. AAS ≅ AAS.
11. $\overline{GE} \cong \overline{FD}$.	11. CPCTC.

Index